AF772835

Physics of Condensed Matter

Physics of Condensed Matter

Edited by
Jaron Finley

Willford Press
www.willfordpress.com

Published by Willford Press,
118-35 Queens Blvd., Suite 400,
Forest Hills, NY 11375, USA

Copyright © 2018 Willford Press

This book contains information obtained from authentic and highly regarded sources. Copyright for all individual chapters remain with the respective authors as indicated. All chapters are published with permission under the Creative Commons Attribution License or equivalent. A wide variety of references are listed. Permission and sources are indicated; for detailed attributions, please refer to the permissions page and list of contributors. Reasonable efforts have been made to publish reliable data and information, but the authors, editors and publisher cannot assume any responsibility for the validity of all materials or the consequences of their use.

Trademark Notice: Registered trademark of products or corporate names are used only for explanation and identification without intent to infringe.

ISBN: 978-1-68285-431-0

Cataloging-in-Publication Data

Physics of condensed matter / edited by Jaron Finley.
p. cm.
Includes bibliographical references and index.
ISBN 978-1-68285-431-0
1. Condensed matter. 2. Physics. I. Finley, Jaron.
QC173.454 .P49 2018
530.41--dc23

For information on all Willford Press publications
visit our website at www.willfordpress.com

Willford Press

Printed in China.

Contents

Preface

This book has been a concerted effort by a group of academicians, researchers and scientists, who have contributed their research works for the realization of the book. This book has materialized in the wake of emerging advancements and innovations in this field. Therefore, the need of the hour was to compile all the required researches and disseminate the knowledge to a broad spectrum of people comprising of students, researchers and specialists of the field.

The physical properties of condensed matter can be studied by applying the laws of electromagnetism, quantum mechanics and statistical mechanics. It can be divided into hard condensed matter physics and soft matter physics. Condensed matter physics can be applied in the formulation of laser technology, semiconductor transistors, etc. This book unravels the recent studies in this field. Coherent flow of topics, student-friendly language and extensive use of examples make this book an invaluable source of knowledge.

At the end of the preface, I would like to thank the authors for their brilliant chapters and the publisher for guiding us all-through the making of the book till its final stage. Also, I would like to thank my family for providing the support and encouragement throughout my academic career and research projects.

Editor

1

Detection of nanoscale electron spin resonance spectra demonstrated using nitrogen-vacancy centre probes in diamond

L.T. Hall[1], P. Kehayias[2], D.A. Simpson[1], A. Jarmola[2], A. Stacey[3], D. Budker[2] & L.C.L. Hollenberg[1,3]

Electron spin resonance (ESR) describes a suite of techniques for characterizing electronic systems with applications in physics, chemistry, and biology. However, the requirement for large electron spin ensembles in conventional ESR techniques limits their spatial resolution. Here we present a method for measuring ESR spectra of nanoscale electronic environments by measuring the longitudinal relaxation time of a single-spin probe as it is systematically tuned into resonance with the target electronic system. As a proof of concept, we extracted the spectral distribution for the P1 electronic spin bath in diamond by using an ensemble of nitrogen-vacancy centres, and demonstrated excellent agreement with theoretical expectations. As the response of each nitrogen-vacancy spin in this experiment is dominated by a single P1 spin at a mean distance of 2.7 nm, the application of this technique to the single nitrogen-vacancy case will enable nanoscale ESR spectroscopy of atomic and molecular spin systems.

[1] School of Physics, University of Melbourne, Parkville, Victoria 3010, Australia. [2] Department of Physics, University of California, Berkeley, California 94720, USA. [3] Centre for Quantum Computation and Communication Technology, School of Physics, University of Melbourne, Parkville, Victoria 3010, Australia. Correspondence and requests for materials should be addressed to L.T.H. (email: lthall@physics.unimelb.edu.au) or to L.C.L.H. (email: lloydch@unimelb.edu.au).

Techniques to detect electron spin resonance (ESR) have long been used to study materials and systems containing unpaired electron spins, such as metal complexes and organic radicals. From an operational viewpoint, the low spin density and high decay rates of such radicals *in-situ*, together with the limited sensitivity of ESR detection systems, often makes the task of obtaining an ESR signal above the detection limit difficult. At present, the best spatial resolution achieved using conventional induction-based ESR detection of solid samples is at the micron level[1]. This is further complicated by the fact that a major component of any biological tissue is water, which has an absorption band in the same region of the electromagnetic spectrum (microwave) as the signals emitted in ESR experiments[2]. As such, there is a great need for a highly sensitive, highly localized technique, which may be used to obtain ESR spectra from paramagnetic systems without the need for microwave control of the sample.

The nitrogen-vacancy (NV) centre point defect in diamond (see refs 3,4 for extensive reviews), comprised of a substitutional atomic nitrogen impurity and an adjacent vacancy (Fig. 1a), is a promising candidate for such an ideal nanoscale-ESR probe, and is the focus of this work; although we note that the technique outlined in this work may potentially be realized by using other solid-state spin systems such as phosphorous donors in silicon, or gallium-arsenide and indium-arsenide quantum dots. The energy level scheme of the C_{3v}-symmetric NV system (Fig. 1b) consists of ground ($^3A_2$), excited (^3E) and singlet electronic states. The ground-state spin-1 manifold has three spin sublevels ($|0\rangle, |\pm 1\rangle$), which at zero-field are split by 2.87 GHz. Since these sublevels have different fluorescence intensities upon illumination, we can achieve spin-state readout optically[5,6]. The degeneracy between the $|\pm 1\rangle$ states may be lifted with the application of an external magnetic field, B_0, with a corresponding separation of $2\gamma B_0$ (where $\gamma = 17.6 \times 10^6$ rad s^{-1} G^{-1} or 2.80 MHz G^{-1} is the NV gyromagnetic ratio), permitting all three states to be accessible via microwave control. By isolating either the $|0\rangle \leftrightarrow |+1\rangle$ or $|0\rangle \leftrightarrow |-1\rangle$ transitions, the NV spin system constitutes a controllable, addressable spin qubit.

Much attention has been focused on using measurements of the NV quantum phase interference between its spin sublevels[7–10] or dephasing rates[11–16] to characterize dynamic processes occurring in external magnetic environments. These protocols have been shown to have remarkable sensitivity to frequencies in the kHz–MHz range, and are thus well suited to characterizing nuclear spin environments[17–24]. However, to achieve the desired sensitivity to the more rapidly fluctuating (GHz) fields associated with electron spin environments, and more importantly, the ability to be frequency-selective, complex and technologically challenging pulse sequences would be necessary[13,25,26].

Alternatively, if a transition frequency of an environmental spin approaches that of the NV spin, they will exchange magnetization and their longitudinal spin relaxation rates will increase. This enables the detection of environmental spins by monitoring changes in NV T_1 times[27–32]. This approach can be more sensitive because NV T_1 times can be up to three orders of magnitude longer than spin-echo-based T_2 times. For example, Steinert, *et al.*[27] ($T_1 = 1.2$ ms, $T_2 = 1.9$ μs) demonstrated that relaxation-based sensing gave an improvement in the signal-to-noise ratio (SNR) by a factor of 388 over spin-echo methods for the detection of external Gd^{3+} electron spins.

The large NV zero-field splitting is fortuitous in the present context in that transitions in the ground-state spin triplet manifold are far off-resonance from the ~MHz transition frequencies of the chief magnetic defects in diamond (nitrogen electron-donor defects, referred to hereafter as P1 centres, and ^{13}C nuclei), leaving NV transitions unable to be excited unless

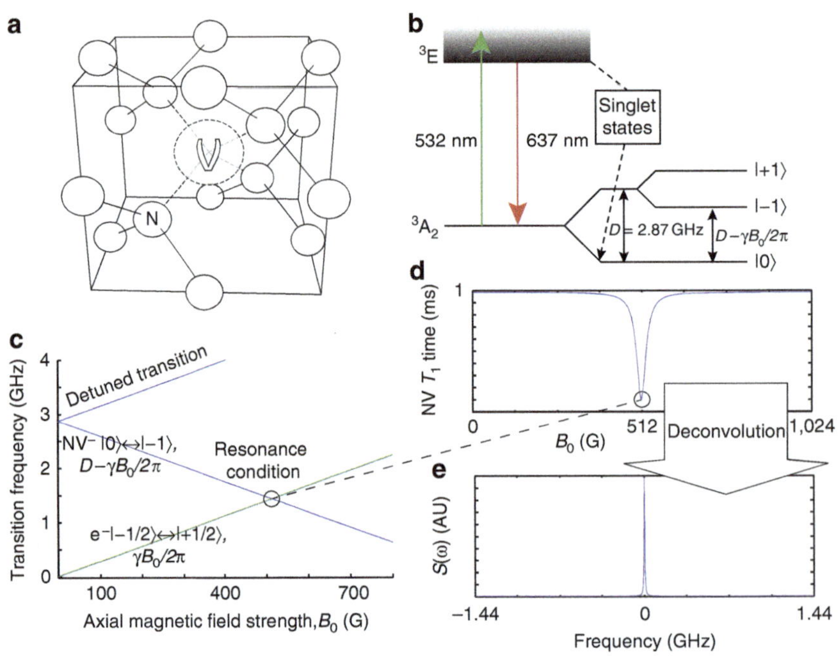

Figure 1 | Overview of nitrogen-vacancy and environmental spin energies. (**a**) The nitrogen-vacancy centre point defect in a diamond lattice, comprised of a substitutional nitrogen atom (N) and an adjacent crystallographic vacancy (V). Blank spheres represent carbon atoms. (**b**) The NV ground-state spin sublevels are separated by $D = 2.87$ GHz. Upon optical excitation at 532 nm, the population of the $|0\rangle$ state may be read out by monitoring the intensity of the emitted red light. (**c**) Transition frequencies of the NV spin, and that of a single electron as a function of external magnetic field strength, B_0. At $B_0 = 512$ G, the $|0\rangle \leftrightarrow |-1\rangle$ transition of the NV is resonant with the electron transition, allowing them to exchange energy. (**d**) Measurements of the NV T_1 time versus axial magnetic field strength, B_0, show a dramatic decrease near the $B_0 = 512$ G resonance point. (**e**) The inherent broadening of the T_1 resonance in **d** due to the NV-P1 interaction may be removed by deconvolving it from the NV filter function to give the environmental spectral density.

brought into resonance using an axial magnetic field ($B_0 \approx 512$ G, for an electron spin environment). The weak spin–orbit coupling to crystal phonons (and low phonon occupancy, owing to the large Debye temperature of diamond) leads to longitudinal relaxation of the spin state on timescales of roughly $T_1 \sim 1$–10 ms at room temperature, and hundreds of seconds when limited by NV–NV cross-relaxation at temperatures below ~ 20 K (ref. 33). On the other hand, the transverse relaxation time (T_2) of the NV spin is the result of dipole–dipole coupling to other spin impurities in the diamond crystal, which occur on timescales of 0.1–1 μs in type 1b (1 of 4) diamond. Sensing based on the much longer longitudinal spin relaxation times can thus be significantly more sensitive than those based on T_2. Projection-noise statistics dictate that the minimum magnetic field, b_{min}, detectable over a total interrogation time $T = N\tau$, comprised of N cycles of dark time τ, is proportional to $b_{\mathrm{min}} \propto 1/\sqrt{T\tau}$. As the dark time, τ, is limited by T_1 in relaxation-based protocols, the latter offer an improvement in sensitivity by a factor of $\sqrt{T_1/T_2}$ over the former for a fixed interrogation time, T. Hence, the relatively long relaxation time of the NV ground state and inherent sensitivity to GHz frequencies, together with room temperature operation and optical readout, make it an ideal system for the ESR spectral mapping protocol discussed in this work.

Here we focus on a new, and potentially simpler, technique for extracting ESR spectra by measuring the longitudinal spin relaxation time, T_1, of a nearby NV centre. Using a controlled external magnetic field, B_0, the transition energies of the NV probe, $\omega_{\mathrm{NV}} = 2\pi D \pm \gamma B_0$, can be brought into resonance with environmental spin energies, ω_{E}, via the Zeeman effect, and the corresponding dependence of T_1 on B_0 may be measured. We establish a general and robust method to determine the environmental spectral distribution $S(\omega_{\mathrm{E}})$ by measuring the relaxation time of the NV probe as a function of the external field strength, $T_1(B_0)$, and discuss how this may be used to perform nanoscale ESR spectroscopy on arbitrary electron spin environments. Our method is demonstrated experimentally by using an ensemble of NV spins to measure the ESR spectrum of the substitutional nitrogen (P1 centres) donor electron spins in type 1b (2 of 4) diamond over the range ± 150 MHz. Comparison with theoretical expectations for this known system confirms the validity of this general approach for measuring the ESR spectrum using the NV centre spin probe in a range of nanoscale applications.

Results

Relaxation of an NV spin in an arbitrary magnetic environment. At zero-field, both NV transitions, $|0\rangle \leftrightarrow |+1\rangle$ and $|0\rangle \leftrightarrow |-1\rangle$, occur at ~ 2.87 GHz. These transitions are then split symmetrically by the presence of an axial magnetic field and any axial NV–environment couplings, and move to higher and lower frequencies, respectively, with the resulting shift being directly proportional to the strength of the field (Fig. 1b). When the $^3A_2$ transition frequencies of the spin-1 NV system are brought into resonance with those of a particular environmental frequency (Fig. 1c), the two will be able to exchange polarization via their magnetic dipole interaction. As the axial shifts induced by the environment on the NV spin (typically of order MHz) are constantly fluctuating, they act to broaden the NV's transition frequencies, as characterized by their resulting inhomogeneous linewidth, or transverse spin relaxation rate, Γ_{NV} (which may be either $\Gamma_{\mathrm{NV}} = 1/T_2^*$ or $1/T_2$ depending on the whether or not additional microwave π pulses are used to refocus the environment). This results in a resonant enhancement of the NV spin relaxation rates to environmental fields fluctuating within Γ_{NV} of the NV frequency (Fig. 1d), from which the spectral distribution of the environment may be extracted (Fig. 1e). A theoretical example of the case of an NV spin coupled to a spin $-\frac{5}{2}$ system (such as Mn(II) or Fe(III)) is shown in Fig. 2a,b; and that of a random distribution of five electrons in a $1 \times 1 \times 1\ \mathrm{nm}^3$ cube external to the diamond crystal is shown in Fig. 2c,d.

To analytically model the response of the NV spin to such an environment, we must account for two dominant processes: energy exchange between the NV and the environment (effective coupling rate b), which changes the population of the magnetic sublevels of the NV ground state; and the destruction of the phase coherence between these sublevels (dephasing, occurring at a rate of Γ_{NV}). Depending on the relative strengths of these processes (that is, how many energy exchanges may occur before the NV spin is dephased), the NV spin can exhibit diverse behaviour (Fig. 3). However, as there are typically more sources of dephasing than energy exchange (with the latter effect further decreasing when the NV spin and environment are away from resonance), we have $\Gamma_{\mathrm{NV}} \gg b$ for the cases considered in this work.

Under this regime, if the NV is initially polarized in its $|0\rangle$ state and only the $|0\rangle \leftrightarrow |-1\rangle$ transition is being excited, the subsequent population at time t when coupled to a spin system of transition energy ω_{E} is given by (see Supplementary Note 1 for details)

$$P_0(\omega_{\mathrm{E}}, B_0, b, t) = \frac{1}{2} + \frac{1}{2}\exp\left(-\frac{b^2 t}{2}\frac{\Gamma_{\mathrm{NV}}}{\delta^2 + \Gamma_{\mathrm{NV}}^2}\right), \qquad (1)$$

where $\delta = \omega_{\mathrm{NV}} - \omega_{\mathrm{E}}$ is the difference in transition frequencies between the NV (ω_{NV}) and the environment (ω_{E}). We note also that this expression does not include the effect of diamond lattice phonons; however, these are included in the analysis of the experimental demonstration below.

Typical electron spin environments will exhibit a distribution of coupling strengths (b) and frequencies (ω_{E}), denoted $P_b(b)$ and $S(\omega_{\mathrm{E}}, B_0)$ of roughly the MHz–GHz regime[15,28,34], which must be taken into account to analytically determine the full response of an NV ensemble. This response is given by

$$\langle P_0(B_0, t)\rangle = \int G(\omega_{\mathrm{E}}, B_0, t) S(\omega_{\mathrm{E}}, B_0)\mathrm{d}\omega_{\mathrm{E}},$$

where we identify the function

$$G(\omega_{\mathrm{E}}, B_0, t) = \int P_0(\omega_{\mathrm{E}}, B_0, b, t) P_b(b)\mathrm{d}b \qquad (2)$$

as an environmental spectral filter with a Lorentzian point-spread function centred on the NV transition frequency (equation 1). The NV-relaxation filter, G, is tunable via the Zeeman interaction, meaning that it may be directly tuned to specific parts of the environment's spectral distribution, $S(\omega_{\mathrm{E}}, B_0)$, by choosing the strength of the external magnetic field, B_0.

In the following, we discuss how we may take advantage of the relaxation filter to reconstruct the spectral density, $S(\omega_{\mathrm{E}}, B_0)$ of an arbitrary environment. If the environment is comprised of some distribution of single electrons, their transition frequencies, ω_{E} (including spectral features such as hyperfine interactions, but centred about the γB_0 Zeeman shift), may be brought into resonance with those of the $|0\rangle \leftrightarrow |-1\rangle$ transition of the NV (that is, $\omega_- \sim 2\pi D - \gamma B_0$) by choosing a magnetic field strength B_0 such that $B_0 = \pi D/\gamma \approx 512$ G. In other cases, the environment may contain spin-1 or greater systems possessing their own zero-field splitting (see Fig. 2 for the case of a spin $-\frac{5}{2}$ system or (ref. 28) for the case of spin $-\frac{7}{2}$ Gd spins coupled to individual NV centres). If the environmental zero-field splittings are greater than that of the NV, the $|0\rangle \leftrightarrow |+1\rangle$ transition of the NV will need to be utilized to ensure that the respective energies are brought into resonance.

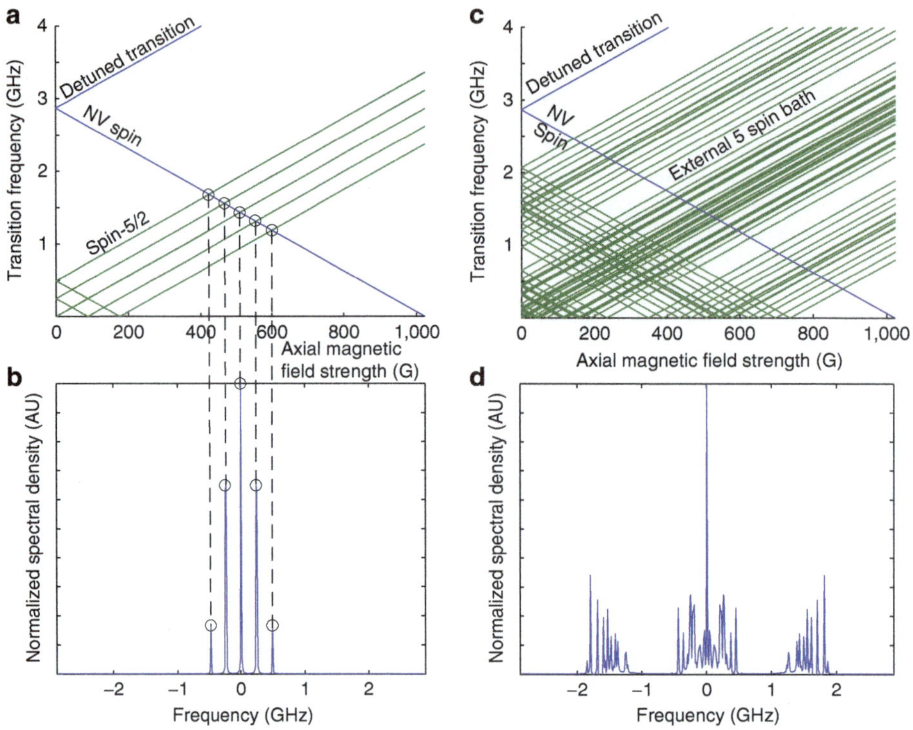

Figure 2 | Numerical examples of T_1-based ESR spectroscopy of practically relevant nanoscale environments. (**a**) Transition frequencies of the NV spin and an external spin $-\frac{5}{2}$ system such as Fe(III) or Mn(II). (**b**) Resulting ESR spectrum of the system in **a**. (**c**) Transition frequencies of the NV spin and an example external system of five randomly placed interacting electrons in a $1\times1\times1\,\mathrm{nm}^3$ cube. (**d**) As in **b** but for the system of five electrons outlined in **c**.

Reconstruction of the environmental spectral density. As noted above, the region of the spectral density sampled by the NV filter functions may be tuned by controlling the strength of the static external field. This suggests that, by sweeping the filter function across the entire spectrum, we can reconstruct it by measuring the relaxation rate of the NV spin for an appropriate range of external field strengths (see Fig. 4a–e).

We denote an arbitrary given spectrum at zero-field by $S_0(\omega_E)\equiv S(\omega_E, 0)$. The distribution at some finite external field B_0 is then $S_0(\omega_E-\gamma B_0)$. For most cases of practical interest, we assume the shape of the distribution does not change with γB_0, although this case can be handled by extension. Furthermore, we also assume that one of the NV transitions is sufficiently off-resonance that it is not sensitive to the environment, making the overlap with the spectrum insignificant. Even if this is not true, the non-resonant filter function will translate with γB_0 at the same rate as $S_0(\omega_E-\gamma B_0)$ (Fig. 1d), and thus produce a constant shift in the overall measurement that does not change with B_0, which may be later subtracted.

The measured response of the NV-relaxation rate, $M(\omega_0, t)$, to S_0 for some external field strength, B_0 ($\omega_0\equiv\gamma B_0$), is then given by (see Fig. 4f)

$$M(\omega_0, t) = \int_{-\infty}^{\infty} S_0(\omega_E-\omega_0)G(\omega_E-2\pi D+\omega_0, t)\mathrm{d}\omega_E, \quad (3)$$

where, for brevity, we have put $M(\omega_0, t)\equiv\langle P_0(t, B_0)\rangle_{\mathrm{meas}}$. By introducing the frequency-space variable $\Omega=\omega_E-\omega_0$, and the parameter, $\Omega_0=2\pi D-2\omega_0$, and making use of the symmetry properties of the Lorentzian function, we may write this integral as a Fourier-space convolution,

$$\begin{aligned} M(\Omega_0, t) &= \int_{-\infty}^{\infty} S_0(\Omega)G(\Omega_0-\Omega, t)\mathrm{d}\Omega \\ &= (S_0 * G)_t(\Omega_0). \end{aligned} \quad (4)$$

Given that the filter function is known, the spectral density may thus be reconstructed using an appropriate deconvolution algorithm (see Fig. 4g). Owing to the inherent noise that exists within the acquired data set, the Wiener deconvolution method[35] was found to be ideal for our purposes. An alternative to this would be to fit the data set with an assumed (smooth) functional form, although the accuracy of this approach will depend on having some *a priori* knowledge of the environmental spectrum.

To perform this deconvolution, it is necessary for the function $G(\omega_E)$ to be known accurately. As this function is the frequency-dependent response of the NV spin to a lateral oscillating magnetic field, $G(\omega_E)$ may be mapped to arbitrary precision by measuring the response of the NV longitudinal relaxation rate to an applied oscillatory magnetic field of frequency ω_E, provided the microwave field strength, ω_x, is such that $\omega_x \ll 1/T_2^*$ (to avoid microwave broadening of G).

In the following section, we provide an experimental example of this protocol applied to the P1 electron spin bath of type 1b (3 of 4) diamond.

Experimental demonstration. To demonstrate the reconstruction of an environmental spectrum using this technique, we measured the relaxation rate of an ensemble of NV centres subject to a bath of substitutional nitrogen donor (or P1 centre) electron spins. The sample chosen was a synthetic high-pressure high-temperature (HPHT) grown diamond, with ~50 p.p.m. substitutional nitrogen and 1–10 p.p.m. NV^- concentrations.

An NV spin ensemble, as opposed to a single-NV spin, was chosen for the T_1 probe to obtain the average nanoscale P1 spectrum, thereby affording a much more straightforward comparison with the average spectrum discussed in the theoretical component of this work. While the motivation for developing this technique is to use a single-NV spin to perform ESR spectroscopy on proximate nanoscale environments, realization of the P1 distribution surrounding a single-NV centre would exhibit too much variation (owing to the randomized positions occupied by P1 centres relative to a given NV, as well as the different possible orientations and switching timescales of the P1 axes) to perform such a comparison. Despite this, we emphasize that the response of each individual NV centre is the result of its individual nanoscale environment, and that the ensemble-averaged spectrum we obtain is, therefore, the average P1 ESR spectrum due to a nanoscale region surrounding an NV centre.

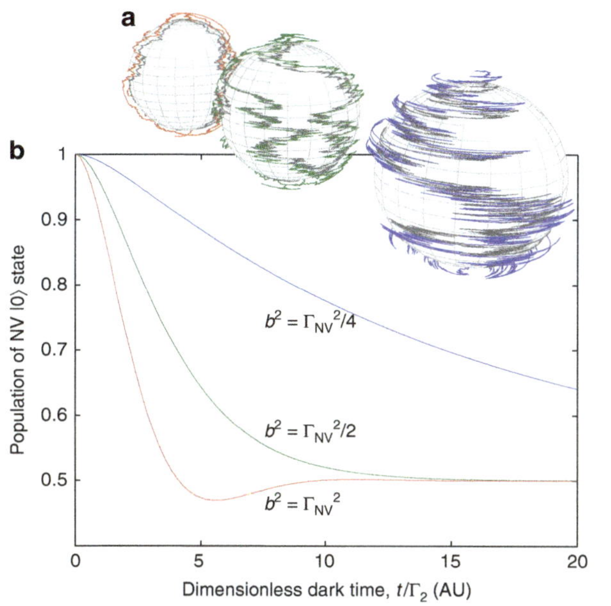

Figure 3 | Bloch vector behaviour at constant dephasing rate for three different cases of relative NV-environment couplings. (a) Example realizations of Bloch sphere trajectories for cases of under-damped ($b>\Gamma_{NV}/\sqrt{2}$, red), critically damped ($b=\Gamma_{NV}/\sqrt{2}$, green) and over-damped ($b<\Gamma_{NV}/\sqrt{2}$, blue) NV-environment couplings, with the $+z$ axis representing the $|0\rangle$ state of the NV spin, and the $-z$ axis representing the $|-1\rangle$ state. **(b)** Population of the NV $|0\rangle$ state of the NV spin ($P_0(t)$). In the under-damped case the Bloch vector is able to rotate appreciably before its transverse projection decays. In the over-damped case, the transverse projection of the Bloch vector is always pulled back to the z axis before it can appreciably rotate. Most practical cases reside in the latter regime because of the intrinsic dephasing arising from strong coupling of the NV to paramagnetic defects in the diamond crystal.

To estimate the size of this region, we note that spins within it are coupled to the NV spin via the variance of their effective magnetic field strength, b^2 (see equation 1), which is the sum of field variances due to each of the individual P1 spins, $b^2=\sum_k b_k^2$. If we order these spins according to their distance from a given NV centre, we may obtain the ensemble-averaged signal by averaging over the probability distributions associated with each successive NV–P1 separation (see Methods section). By examining the average relative contribution of the first to the Nth P1 spin to the total coupling, $\sum_k^N b_k^2/\sum_k^\infty b_k^2$ (Fig. 5a, blue curve), we see that the coupling is dominated by the nearest P1 centre, which resides at an average distance of 2.7 nm from the NV centre, and that 91% of the signal comes from the first four P1 spins surrounding the NV centre with the fourth spin residing at an average distance of 4.6 nm from the NV centre.

The localization of the NV–P1 interaction is seen to be even more pronounced if we consider the entire time-dependent population of the NV $|0\rangle$ state rather than the effective field strength alone. This measure is more accurate, since the measured changes in the NV fluorescence are proportional to the population of this state. By decomposing the field variance in the same manner as above, we can see that the exponential term in equation 1 factorizes into individual contributions from each P1 centre. Applying the same ensemble averaging approach as above, we can determine the cumulative contribution to the overall decay of the NV $|0\rangle$ state population as each successive P1 centre is incorporated, as shown in Fig. 5b. Finally, by examining this contribution at the relaxation ($1/e$) time of the NV spin,

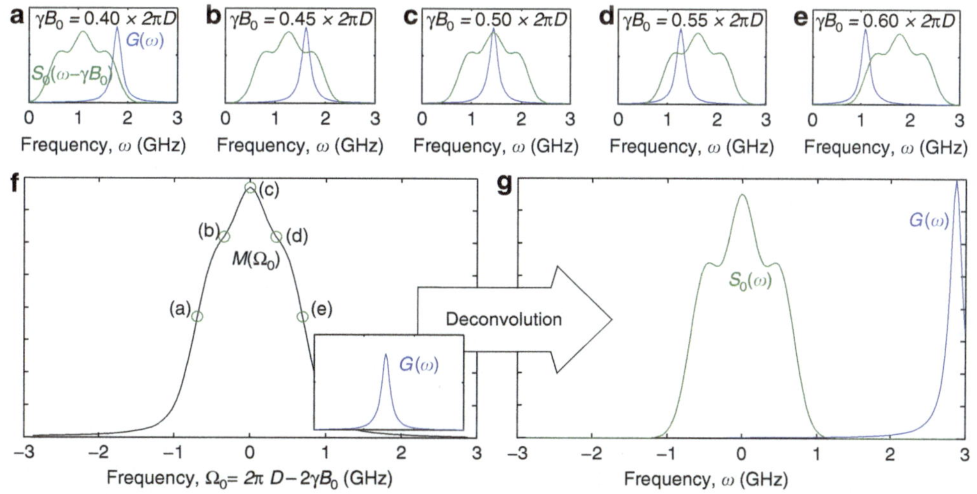

Figure 4 | Relationship between the NV filter function, environmental spectral density and the resulting NV-relaxation rate. (a–e) By controlling the external field strength, B_0, the NV filter function, G may be tuned to filter specific regions of the spectral density, S. The overlap integral of G and S is proportional to the relaxation rate at that field strength. **(f)** The resulting measured relaxation rate, $M(\Omega_0)$, is the convolution of the NV filter function, G, and the spectral density of the environment, S. **(g)** Given that G is known, the spectral density, S, may be reconstructed by deconvolving S and G from M.

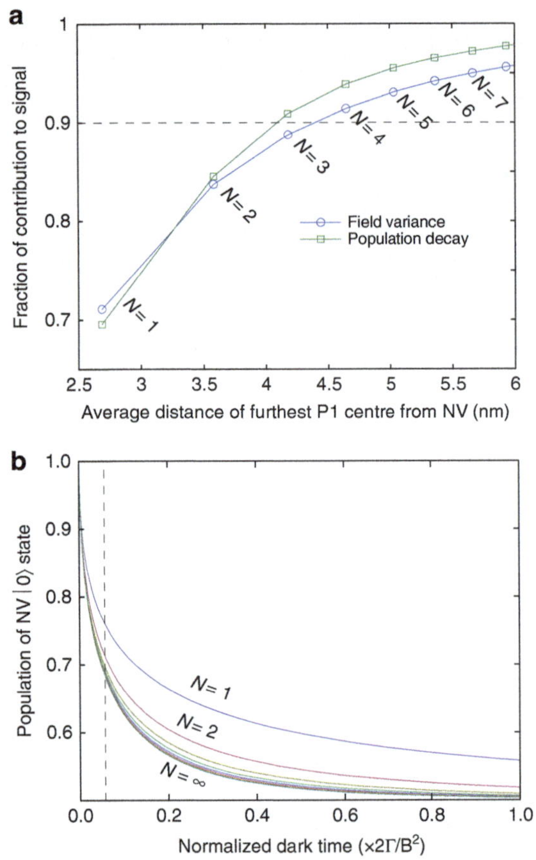

Figure 5 | Localization of the effective environmental magnetic field causing relaxation of the NV spin. (**a**) Plot showing the ensemble-averaged fraction of environmental coupling (blue), and decay of the NV $|0\rangle$ state population (green), accounted for by considering increasing numbers (N) of P1 centres surrounding the NV centre. The horizontal axis shows the averaged distance to the furthest (Nth) P1 spin. (**b**) Plots showing the ensemble-averaged decay of the NV $|0\rangle$ state population due to increasing numbers of P1 centres surrounding the NV centre. The relaxation time of $T_1 \approx 0.1 \times \Gamma/b^2$ is shown by the dashed line, along which the data for the green curve in (**a**) is obtained.

$T_1 \approx 0.1 \times \Gamma/b^2$, we can see that 91% of the signal is attributable to the nearest three P1 spins, with the third spin residing an average distance of 4.2 nm from the NV centre (Fig. 5a, green curve). These results demonstrate the nanoscale nature of this detection technique. A discussion of the spatial resolution of this technique when applied to ESR spectroscopy of spins external to the diamond crystal may be found in the Discussion section below.

NV centres are oriented along all four symmetry axes, and a solenoid was used to apply an adjustable magnetic field along one of these orientations. After initializing NV spins in the $|0\rangle$ state via optical pumping (or the $|-1\rangle$ state with an additional microwave π pulse), we measured the fluorescence following a variable amount of 'dark time' (for which the pump laser was off). NVs not aligned with the external field and other defects contribute to the diamond fluorescence, which may decay at rates different from that of the aligned NVs. To remove this effect, we employed a form of common-mode rejection (described in ref. 33). After initializing to either the $|0\rangle$ or $|-1\rangle$ state, we measured the fluorescence contrast after each dark time, and subtracted the results for each initial state. The resulting decay of the fluorescence difference indicates the loss of NV population from the initial state, and the regions of the environmental spectrum these populations are sensitive to. We note here that for cases in which the application of microwaves may be impractical, this technique may be performed in an all-optical manner, although this approach will also require the characterization of other sources of fluorescence within the sample.

The P1 centre, the target of our demonstration, is a substitutional defect of a nitrogen atom in place of a carbon atom in a diamond lattice. The single unbonded electron can reside along any one of the four crystallographic bond axes, giving rise to four possible orientations of the hyperfine coupling to the nitrogen nucleus. For field strengths above $\sim$100 G, the quantization axis of the P1 electron spin is set by its Zeeman interaction, effectively reducing the number of possible P1 species from four to two. Hence, there exists a 25% chance that the delocalization axis of the P1 centre is aligned with the NV axis, giving an axial hyperfine coupling of 114 MHz, and a 75% chance that the delocalization axis is $\arccos\left(-\frac{1}{3}\right) \approx 109°$ to the NV axis, producing an axial hyperfine coupling of 86 MHz (the spin properties of this defect are discussed in detail in Supplementary Note 2). Figure 6a,b show the overlap of the transition frequencies associated with both the NV spin and the P1 centre.

Measurements of the longitudinal spin relaxation of the NV ensemble, $\langle P_0(t, B_0)\rangle_{\text{meas}}$, were taken at 500 different external magnetic field strengths between 480 and 540 G (Fig. 6c–e). Before we deconvolve the data set to determine the spectral density, we extract and inspect the magnetic field-dependent component of the NV-relaxation rate due to spin–spin relaxation with the environment. Accordingly, the data was fitted using the function given by

$$P_0^{\text{fit}}(B_0, t) = \exp\left(-\sqrt{\Gamma_1(B_0)t} - Rt\right), \quad (5)$$

where $R \approx 360$ Hz is the 'baseline' component of the relaxation because of diamond lattice phonons (see Methods for the derivation of this fitting form). The resulting spin–spin relaxation rates are plotted in Fig. 6f. As expected, the measurements show that the NV-relaxation rate increases when $D - 2\omega_0/2\pi = 0$, $\pm$86 MHz, $\pm$114 MHz or $B_0 = 490$, 495, 510, 525 and 530 G, corresponding to the conditions under which the NV spin may directly exchange magnetization with P1 electron spins, leaving the projection of the P1 nuclear spin unchanged. Other features (as observed previously in (ref. 36)) are evident at $B_0 = 499$, 502, 517 and 520 G, and correspond to a two-step process. Firstly, the P1 electron undergoes a mutual flip-flop with the P1 nuclear spin due to their hyperfine interaction. This step is not energy conserving and is heavily suppressed according to the ratio of the hyperfine strength to the strength of the P1 electron's Zeeman interaction. The P1 energy deficit is then paid for by flipping the NV electron spin (as mediated by the $x - z$ and $y - z$ components of the NV–P1 electron dipole–dipole interaction—see Supplementary Note 2). Because these transitions are comparatively weak, they are partially obscured by the large spin–phonon relaxation effect of the NV spin, resulting from two-phonon Orbach[37] and two-phonon[38] Raman processes.

To obtain the spectral distribution, we carry out the deconvolution over the entire set of data, $\langle P_0(t, B_0)\rangle_{\text{meas}}$, and plot the spectral distribution in frequency space, $S(f)$, in Fig. 6g, demonstrating good agreement with the corresponding theoretical expectations (see Supplementary Note 2 for details). The deconvolution of the spectral density is shown to remove much of the broadening seen in the raw relaxation data in Fig. 6f, and thus provides a better measure of the environmental dynamics. Some small discrepancies between the resulting spectrum and the theoretical result are evident; however, we note that while the theory incorporates effects such as hyperfine couplings and dephasing rates, more complicated effects, such as g-factor anisotropy and enhancement, interactions with other

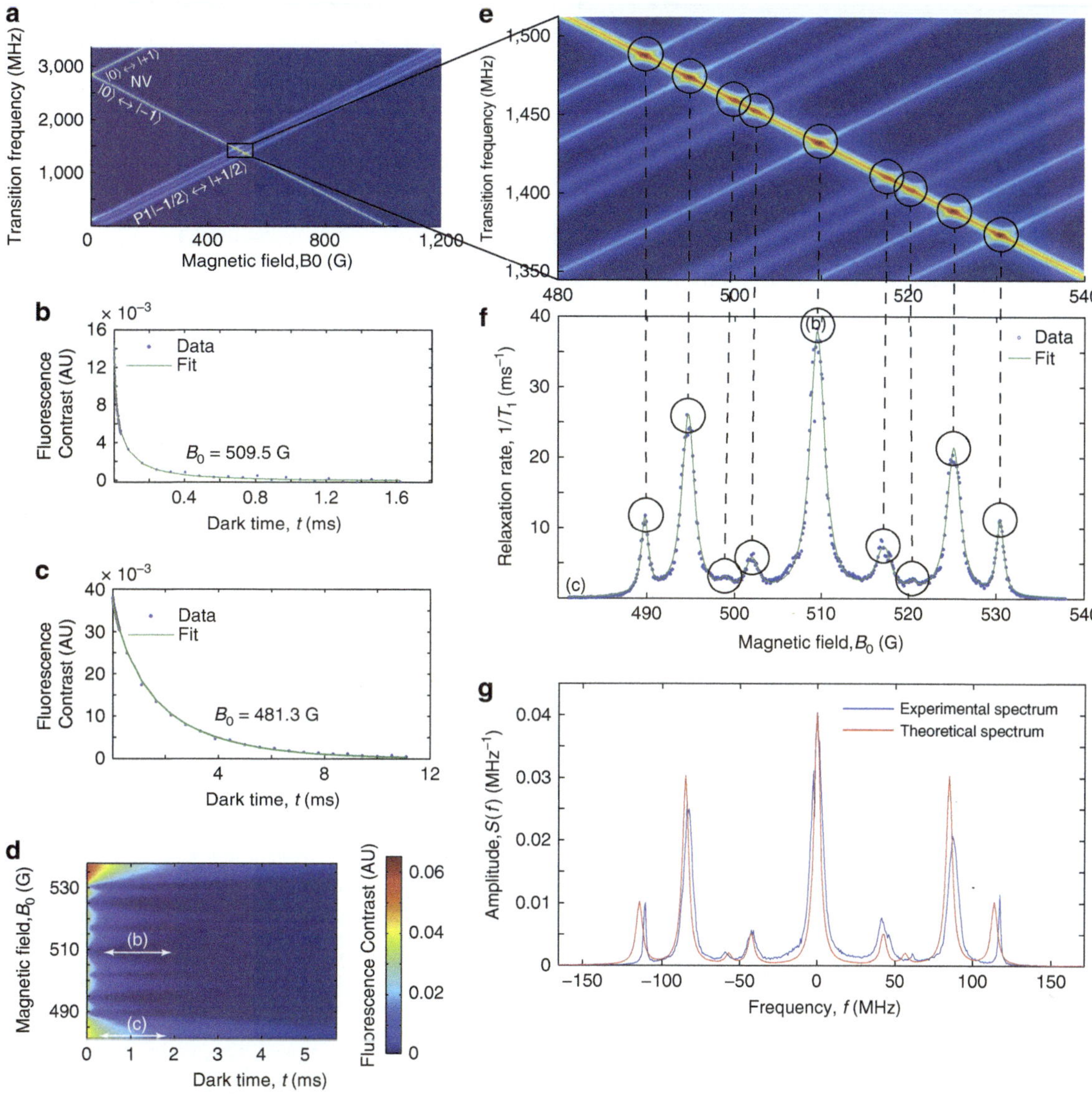

Figure 6 | 'Experimental demonstration of the reconstruction of an environmental spectrum using an ensemble of NV spins. (**a**) Theoretical plot showing the transition frequencies of the NV centre and P1 electron spins versus the strength of an external field aligned along the ⟨111⟩ axis. Energy exchange between an NV spin and a nearby P1 spin is achieved when two transition frequencies approach resonance. Broadening of these lines is caused by interactions with other P1 spins and spin impurities within the diamond crystal. (**b,c**) Plots of NV centre fluorescence contrast versus dark time at 509.5 and 481.3 G, respectively. In the case of the former, NV and P1 electron energies are on resonance, resulting in a comparatively rapid decay. In the case of the latter, NV and P1 transition energies are too far from resonance to facilitate a dipole-mediated resonant energy exchange, meaning that the depolarization of the NV spin ensemble is dominated by interactions with crystal phonons. (**d**) Measurements of curves such as those shown in **b,c** for 500 magnetic field strengths between 480 and 540 G. Data contours used for plots **b,c** are highlighted. (**e**) Zoomed plot of that in **a** highlighting the hyperfine structure and the corresponding spin conserving and non-spin conserving transitions of the P1 centre. (**f**) Plot of the external field-dependent relaxation rate, $\Gamma_1(B_0)$, extracted from the data shown in **e**. Correspondences of the observed features with the P1 spin transitions in **e** are indicated explicitly by dashed lines, and relaxation rates corresponding to examples in **b,c** are labelled as such. (**g**) Application of the deconvolution procedure to the data in **f** yields the spectral density of the spin bath environment surrounding the NV centres in the diamond sample (blue), showing good agreement with theoretical predictions (red).

paramagnetic impurities and other strain-related phenomena, have not been included. Such effects are highly sample-dependent and thus difficult to predict in general terms, although the technique developed in this work provides an ideal means to facilitate their investigation. Finally, we note that although this demonstration involves ensembles of NV and P1 centres, the detection is highly local as the response of each NV is dominated by its nearest P1 centre: for this sample (50 p.p.m. P1) the mean distance to the nearest P1 centre is about 2.7 nm (ref. 34).

Discussion

In this work, we have presented a general method for extracting the spectral distribution of an arbitrary electronic environment based on tuning a spin-1 NV probe system, via controlled application of an external field, into resonance with the transitions of the target electronic system. The method was tested using an ensemble of NV centres in a type 1b diamond sample to determine the spectral distribution of the P1 spin bath, showing excellent agreement with the theoretical expectations.

The P1 electron spin bath was chosen as our example environment because it is well understood both theoretically and experimentally, and thus provides an ideal system with which to compare our results; however, practical application of this technique is ultimately intended for spectroscopy of spin systems external to the diamond surface. We note that the ability to couple to such systems is ultimately limited by the intrinsic T_1 of the NV spin: a few milliseconds when phonon limited at room temperature and up to hundreds of seconds when determined by NV–NV interactions at temperatures below 77 K[33]. Such relaxation times would permit room temperature resonance-based spectroscopy of single electron spins up to 100 nm from the NV centre, with distances as far as a few μm possible below 77 K. As NV centres having T_1 times of 1.4 ms have been observed within 3–4 nm of the diamond surface[39] (with one centre residing stably at 1.9 nm below the surface, although its T_1 time was not reported).

Our relaxation-based ESR method has a number of advantages over the existing techniques. Measurements of the NV relaxation in general do not require microwave control, and thus require no manipulation of the sample. With relaxation times much longer than dephasing times, T_1-based protocols can be significantly more sensitive to ESR detection. Finally, even in the ensemble case demonstrated here, the NV spin relaxation is dominated by local interactions with the environment affords an effective spatial resolution of a few nanometres. By extending this technique to the single-NV probe case as well as other solid-state single-spin systems such as phosphorous donors in silicon and gallium-arsenide quantum dots, determination and characterization of nanoscale ESR spectra of single electronic systems will be possible.

Methods

Spatial resolution. The total magnetic variance felt by the NV centre spin is the sum of variances from individual sources,

$$\langle b^2 \rangle = \sum_k \langle b_k^2 \rangle, \quad (6)$$

where $b_k = \frac{\beta}{r_k^3}\frac{3}{2}\sin^2(\theta_k)$ is the coupling rate to the components of environmental spin k that induce NV spin transitions (see Supplementary Note 2 for a derivation of how this term arises from the magnetic dipole–dipole coupling between the NV spin and spins in the environment), r_k is their spatial separation and $\beta = \frac{\mu_0}{4\pi}\hbar\gamma^2$.

The probability distribution for the distance from the NV centre to the kth nearest P1 centre, r_k, is given by[34]

$$\mathrm{p}_k(r_k) = \frac{4\pi n r_k^2}{(k-1)!}\left(\frac{4\pi n r_k^3}{3}\right)^{k-1} \exp\left(-\frac{4\pi n r_k^3}{3}\right). \quad (7)$$

Using this distribution, we find the average distance from a given NV centre to its kth nearest P1 centre is given by

$$\langle r_k \rangle = \left(\frac{3}{4\pi n}\right)^{\frac{1}{3}} \frac{\Gamma\left(k+\frac{1}{3}\right)}{\Gamma(k)}, \quad (8)$$

where Γ is the Gamma function, which, for a 50 p.p.m. P1 sample ($n = 8.8 \times 10^{24}\,\mathrm{m}^{-3}$), gives an average distance to the nearest P1 of

$$\bar{r} \equiv \langle r_1 \rangle \pm \sqrt{\langle r_1^2 \rangle - \langle r_1 \rangle^2} = (2.7 \pm 0.9)\ \mathrm{nm}. \quad (9)$$

If we substitute equation 6 into equation 1, we see that the time-dependent population of the NV $|0\rangle$ state factorizes into individual contributions from each individual field source, k. By again ordering these sources in terms of their separation from the NV centre, and averaging over the respective distributions of these separations, we obtain the ensemble-averaged NV $|0\rangle$ state population due to field sources 1–N,

$$\langle \mathrm{P}_0(t) \rangle_N = \prod_{k=1}^{k=N} \int \exp\left(-\frac{b_k^2(r_k)t}{2}\frac{\Gamma_{\mathrm{NV}}}{\delta^2+\Gamma_{\mathrm{NV}}^2}\right)\mathrm{p}_k(r_k)\mathrm{d}r_k. \quad (10)$$

The relative contribution of sources 1–N to the overall population decay is then given by $\langle \mathrm{P}_0(t)\rangle_N / \langle \mathrm{P}_0(t)\rangle_\infty$.

Fitting form. To determine the expected fitting form for the ensemble population of the NV $|0\rangle$ state (equation 5), we analytically evaluate the integral in equation 10 for $N \rightarrow \infty$, giving

$$\langle \mathrm{P}_0(t) \rangle = \prod_{k=1}^{k=\infty} \frac{\pi^{3/2}2^{k-1}}{\Gamma(k)}\left(2\frac{{}_0\tilde{F}_2\left(;\frac{1}{2}-\frac{k}{2},1-\frac{k}{2};-\tau\right)}{\sin(\pi k)} + \tau^{k/2}\left[\sqrt{\tau}\frac{{}_0\tilde{F}_2\left(;\frac{3}{2},\frac{k+3}{2};-\tau\right)}{\cos\left(\frac{\pi k}{2}\right)} - \frac{{}_0\tilde{F}_2\left(;\frac{1}{2},\frac{k}{2}+1;-\tau\right)}{\sin\left(\frac{\pi k}{2}\right)}\right]\right),$$

where ${}_0\tilde{F}_2$ is the regularized, generalized hypergeometric function[40], and $\tau = \frac{4\pi^2}{9}\beta^2 n^2 t \frac{\Gamma_{\mathrm{NV}}}{(\Gamma_{\mathrm{NV}}^2+\delta^2)}$. Expanding about $\tau = 0$, we find

$$\langle \mathrm{P}_0(t) \rangle \sim \exp\left(-2\sqrt{\pi\tau}\right), \quad (11)$$

where the dependence of the decay rate on the axial magnetic field, B_0 occurs via the detuning, δ (see main text). To account for the contribution to the NV spin relaxation from environmental phonons, we multiply this result by an additional decay term, $\exp(-Rt)$, where R is independent of the axial field, to give the form used to fit the relaxation data (equation 5).

References

1. Blank, A., Suhovoy, E., Halevy, R., Shtirberg, L. & Harneit, W. ESR imaging in solid phase down to sub-micron resolution: methodology and applications. *Phys. Chem. Chem. Phys.* **11**, 6689–6699 (2009).
2. Bertie, J. E. & Lan, Z. Infrared intensities of liquids XX: The intensity of the oh stretching band of liquid water revisited, and the best current values of the optical constants of H_2O(l) at 25° between 15,000 and 1 cm^{-1}. *Appl. Spectrosc.* **50**, 1047–1057 (1996).
3. Doherty, M. W. *et al.* The nitrogen-vacancy colour centre in diamond. *Phys. Rep.* **528**, 1–45 (2013).
4. Schirhagl, R., Chang, K., Loretz, M. & Degen, C. L. Nitrogen-vacancy centers in diamond: nanoscale sensors for physics and biology. *Annu. Rev. Phys. Chem.* **65**, 83–105 (2014).
5. Jelezko, F. *et al.* Single spin states in a defect center resolved by optical spectroscopy. *Appl. Phys. Lett.* **81**, 2160–2162 (2002).
6. Jelezko, F. & Wrachtrup, J. Single defect centres in diamond: a review. *Phys. Status Solidi Appl. Mater. Sci.* **203**, 3207–3225 (2006).
7. Degen, C. L. Scanning magnetic field microscope with a diamond single-spin sensor. *Appl. Phys. Lett.* **92**, 243111 (2008).
8. Taylor, J. M. *et al.* High-sensitivity diamond magnetometer with nanoscale resolution. *Nat. Phys.* **4**, 810–816 (2008).
9. Balasubramanian, G. *et al.* Nanoscale imaging magnetometry with diamond spins under ambient conditions. *Nature* **455**, 648–651 (2008).
10. Maze, J. R. *et al.* Nanoscale magnetic sensing with an individual electronic spin in diamond. *Nature* **455**, 644–647 (2008).
11. Cole, J. H. & Hollenberg, L. C. Scanning quantum decoherence microscopy. *Nanotechnology* **20**, 495401 (2009).
12. Hall, L. T., Cole, J. H., Hill, C. D. & Hollenberg, L. C. L. Sensing of fluctuating nanoscale magnetic fields using nitrogen-vacancy centers in diamond. *Phys. Rev. Lett.* **103**, 220802 (2009).
13. Hall, L. T., Hill, C. D., Cole, J. H. & Hollenberg, L. C. L. Ultrasensitive diamond magnetometry using optimal dynamic decoupling. *Phys. Rev. B* **82**, 045208 (2010).
14. Hall, L. T. *et al.* Monitoring ion-channel function in real time through quantum decoherence. *Proc. Natl Acad. Sci. USA* **107**, 18777–18782 (2010).
15. McGuinness, L. P. *et al.* Ambient nanoscale sensing with single spins using quantum decoherence. *New. J. Phys.* **15**, 073042 (2013).
16. Luan, L. *et al.* Decoherence imaging of spin ensembles using a scanning single-electron spin in diamond. *Sci. Rep.* **5**, 8119 (2015).
17. Zhao, N., Hu, J.-L., Ho, S.-W., Wan, J. T. K. & Liu, R. B. Atomic-scale magnetometry of distant nuclear spin clusters via nitrogen-vacancy spin in diamond. *Nat. Nano* **6**, 242–246 (2011).
18. Zhao, N. *et al.* Sensing single remote nuclear spins. *Nat. Nano* **7**, 657–662 (2012).
19. Taminiau, T. H. *et al.* Detection and control of individual nuclear spins using a weakly coupled electron spin. *Phys. Rev. Lett.* **109**, 137602 (2012).
20. Kolkowitz, S., Unterreithmeier, Q. P., Bennett, S. D. & Lukin, M. D. Sensing distant nuclear spins with a single electron spin. *Phys. Rev. Lett.* **109**, 137601 (2012).
21. Staudacher, T. *et al.* Nuclear magnetic resonance spectroscopy on a (5-nanometer)^3 sample volume. *Science* **339**, 561–563 (2013).
22. Mamin, H. J. *et al.* Nanoscale nuclear magnetic resonance with a nitrogen-vacancy spin sensor. *Science* **339**, 557–560 (2013).
23. Shi, F. *et al.* Sensing and atomic-scale structure analysis of single nuclear-spin clusters in diamond. *Nat. Phys.* **10**, 21–25 (2014).
24. DeVience, S. J. *et al.* Nanoscale NMR spectroscopy and imaging of multiple nuclear species. *Nat. Nano* **10**, 129–134 (2015).
25. Cywinski, L., Lutchyn, R. M., Nave, C. P. & Das Sarma, S. How to enhance dephasing time in superconducting qubits. *Phys. Rev. B* **77**, 174509 (2008).
26. Bar-Gill, N. *et al.* Suppression of spin-bath dynamics for improved coherence of multi-spin-qubit systems. *Nat. Commun.* **3**, 858 (2012).

27. Steinert, S. *et al.* Magnetic spin imaging under ambient conditions with sub-cellular resolution. *Nat. Commun.* **4,** 1607 (2013).
28. Kaufmann, S. *et al.* Detection of atomic spin labels in a lipid bilayer using a single-spin nanodiamond probe. *Proc. Natl Acad. Sci. USA* **110,** 10894–10898 (2013).
29. Tetienne, J. P. *et al.* Spin relaxometry of single nitrogen-vacancy defects in diamond nanocrystals for magnetic noise sensing. *Phys. Rev. B* **87,** 235436 (2013).
30. Sushkov, A. O. *et al.* All-optical sensing of a single-molecule electron spin. *Nano Lett.* **14,** 6443–6448 (2014).
31. Pelliccione, M., Myers, B. A., Pascal, L. M. A., Das, A. & Bleszynski Jayich, A. C. Two-dimensional nanoscale imaging of gadolinium spins via scanning probe relaxometry with a single spin in diamond. *Phys. Rev. Appl.* **2,** 054014 (2014).
32. Kolkowitz, S. *et al.* Quantum electronics. Probing Johnson noise and ballistic transport in normal metals with a single-spin qubit. *Science* **347,** 1129–1132 (2015).
33. Jarmola, A., Acosta, V. M., Jensen, K., Chemerisov, S. & Budker, D. Temperature- and magnetic-field-dependent longitudinal spin relaxation in nitrogen-vacancy ensembles in diamond. *Phys. Rev. Lett.* **108,** 197601 (2012).
34. Hall, L. T., Cole, J. H. & Hollenberg, L. C. L. Analytic solutions to the central-spin problem for nitrogen-vacancy centers in diamond. *Phys. Rev. B* **90,** 075201 (2014).
35. Papoulis, A. *Signal Analysis* (McGraw-Hill, 1977).
36. Armstrong, S., Rogers, L. J., McMurtrie, R. L. & Manson, N. B. NV-NV electronelectron spin and NV-N_S electron - electron and electron-nuclear spin interaction in diamond. *Phys. Procedia* **3,** 1569–1575 (2010).
37. Redman, D. A., Brown, S., Sands, R. H. & Rand, S. C. Spin dynamics and electronic states of N-V centers in diamond by EPR and four-wave-mixing spectroscopy. *Phys. Rev. Lett.* **67,** 3420–3423 (1991).
38. Walker, M. B. A. T^5 spin-lattice relaxation rate for non-Kramers ions. *Can. J. Phys.* **46,** 1347–1353 (1968).
39. Loretz, M., Pezzagna, S., Meijer, J. & Degen, C. L. Nanoscale nuclear magnetic resonance with a 1.9-nm-deep nitrogen-vacancy sensor. *Appl. Phys. Lett.* **104,** 033102 (2014).
40. Abramowitz, M. & Stegun, I. A. *Handbook of Mathematical Functions* 9th edn, book section 15, 555–566 (Dover, 1972).

Acknowledgements

We gratefully acknowledge discussions with L. McGuinness and J. Wood. This work was supported in part by the Australian Research Council (ARC) under the Centre of Excellence scheme (project No. CE110001027). L.C.L.H. acknowledges the support of an ARC Laureate Fellowship (project no. FL130100119). D.B. acknowledges support from the AFOSR/DARPA QuASAR program, DFG through the DIP program (FO 703/2-1) and NSF grant No. ECCS-1202258.

Author contributions

L.T.H. conceived the idea of T1 based spectroscopy. L.T.H. and L.C.L.H. developed the general theoretical NV-ESR framework. L.T.H., D.A.S., A.S. and L.C.L.H. developed the theoretical model of the P1 centre. P.K., A.J. and D.B. carried out the experiments. L.T.H., D.A.S., P.K. and D.B. analysed the experimental results. All authors discussed the results and participated in writing the manuscript.

Additional information

Competing financial interests: The authors declare no competing financial interests.

2

Nanoelectronic primary thermometry below 4 mK

D.I. Bradley[1], R.E. George[1], D. Gunnarsson[2], R.P. Haley[1], H. Heikkinen[2], Yu. A. Pashkin[1,3], J. Penttilä[4], J.R. Prance[1], M. Prunnila[2], L. Roschier[3] & M. Sarsby[1]

Cooling nanoelectronic structures to millikelvin temperatures presents extreme challenges in maintaining thermal contact between the electrons in the device and an external cold bath. It is typically found that when nanoscale devices are cooled to ~10 mK the electrons are significantly overheated. Here we report the cooling of electrons in nanoelectronic Coulomb blockade thermometers below 4 mK. The low operating temperature is attributed to an optimized design that incorporates cooling fins with a high electron-phonon coupling and on-chip electronic filters, combined with low-noise electronic measurements. By immersing a Coulomb blockade thermometer in the $^3He/^4He$ refrigerant of a dilution refrigerator, we measure a lowest electron temperature of 3.7 mK and a trend to a saturated electron temperature approaching 3 mK. This work demonstrates how nanoelectronic samples can be cooled further into the low-millikelvin range.

[1]Department of Physics, Lancaster University, Bailrigg, Lancaster LA1 4YB, UK. [2]VTT Technical Research Centre of Finland, P.O. Box 1000, 02044 VTT Espoo, Finland. [3]Lebedev Physical Institute, Moscow 119991, Russia. [4]Aivon Oy, Valimotie 13A, 00380 Helsinki, Finland. Correspondence and requests for materials should be addressed to J.R.P. (email: j.prance@lancaster.ac.uk) or to M.P. (email: mika.prunnila@vtt.fi) or to L.R. (email: leif.roschier@iki.fi).

Understanding how to obtain and measure electron temperatures approaching 1 mK has the potential to open a new regime for studying nanoelectronics and pave the way towards pioneering sub-millikelvin techniques[1]. This would benefit numerous areas of activity; for example, investigations of the fractional quantum Hall effect in two-dimensional electron gases[2,3] and solid-state quantum technologies including superconducting and semiconducting qubits[4,5]. To access these temperatures one must minimize parasitic heating and internal Joule heating, and maximize the coupling to cold contact wires and phonons in the host lattice, all the while overcoming the decrease in electron–phonon coupling and electrical heat conduction as temperatures drop[6].

Here we study Coulomb blockade thermometers (CBTs) that have been designed to operate significantly below 10 mK and demonstrate cooling of electrons in a nanoelectronic device to below 4 mK. A CBT consists of an array of Coulomb-blockaded metallic islands connected by tunnel junctions. The conductance of the array is temperature dependent, due to the balance between thermal excitations and an electrostatic barrier to single electron tunnelling across the islands[7,8]. CBTs typically function over a decade of temperature and have previously been demonstrated to work at temperatures as low as 7.5 mK (refs 9,10). Perhaps most importantly, they can be viewed as a primary thermometer of their internal electron temperature. We have used CBTs as a diagnostic tool to quantify and optimize the thermal environment of a nanoelectronic device and to unambiguously determine the local electron temperature. In the future, similar CBTs could be used to probe the electron temperature of nearby samples with thermal contact provided by direct electrical connections. Furthermore, the techniques that are used to cool the CBTs can be used to cool other nanoelectronic samples.

Results

Structure of the thermometer devices. The structure of the CBTs studied here is shown in Fig. 1. Devices are fabricated using an *ex situ* tunnel junction process[11], which provides excellent tunnel junction uniformity[12], and has also been used to fabricate superconducting qubits[13] (see Methods for details). Efficient thermal coupling between electrons and phonons in the metallic islands of the CBT is critical for reaching low electron temperatures[14,15]. The electron–phonon heat flow P_{ep} is described by the material-dependent electron–phonon coupling constant Σ and the volume of the metallic island Ω,

$$P_{ep} = \Sigma\Omega\left(T_e^5 - T_p^5\right), \qquad (1)$$

where T_e is the electron temperature and T_p is the phonon temperature[16]. To minimize T_e, the island volume should be large and the material chosen to maximize Σ. We use electroplated Au on top of the CBT islands to increase their volume to nominal $5 \times 205 \times 38.5\,\mu m^3$ (see Methods for details). The effective electron–phonon coupling in these islands, with a relatively large volume and a high coupling constant[17] in Au $\Sigma = 2.4 \times 10^9\,WK^{-5}m^{-3}$, is estimated to be more than two orders of magnitude larger than in previous CBTs fabricated using the *ex situ* junction process[18].

In addition to efficient thermalization of the CBT itself, it is important to cool the incoming leads through robust thermal anchoring and heavy electromagnetic filtering[19]. We improve the chain of thermalization and filtering by including on-chip resistive meander structures in line with all electrical contacts. These form a distributed resistive-capacitive chain with a cutoff frequency ≈ 40 MHz. Similar filters based on a large area capacitor and tunnel junctions in series have previously been incorporated in a CBT[18].

CBT characteristics above 7 mK. Figure 2 shows the behaviour of a CBT fabricated using the process described above, focusing on temperatures between 7 and 80 mK. The sensor is measured in both a commercial cryogen-free dilution refrigerator (Bluefors Cryogenics LD250) with a base temperature ≈ 7 mK and in a custom dilution refrigerator manufactured at Lancaster University[20] with a base temperature ≈ 2.5 mK. The conductance of the CBT is measured in a current-driven four-wire configuration, with the drive current and voltage amplification provided by an Aivon PA10 amplifier. A small AC excitation (typically $5\,pA \leq I_{AC} \leq 100\,pA$) is added to the DC bias I_{DC}, allowing the differential conductance G to be measured with a lock-in amplifier.

In both refrigerators, the CBT is in a vacuum and housed in a gold-plated copper package (Aivon SH-1) that is attached to the mixing chamber plate. The package includes RC filters with a cutoff frequency ≈ 300 kHz. Electrical contacts to the CBT are thermalized in additional cold *RC* filters potted in Eccosorb CR

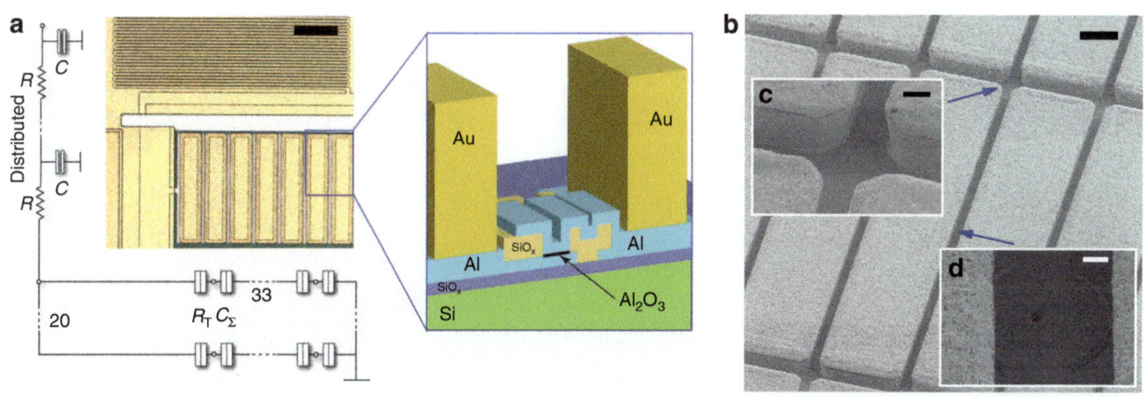

Figure 1 | Details of the CBT device structure. (**a**) Optical micrograph of the CBT with equivalent circuit diagram and schematic cross-section of the structure. Scale bar on the optical micrograph, 10 μm. The CBT is formed of 32 × 20 metallic islands of capacitance C_Σ connected in an array by tunnel junctions of resistance R_T, as shown in the circuit diagram. Connection to the array is made via on-chip *RC* filters comprising a meandering electrode sandwiched between large-area grounded metal films, separated by 250 nm SiO_x. Each filter has a distributed resistance $R \approx 500\,\Omega$ and capacitance $C \approx 10$ pF. The schematic cross-section shows one Al_2O_3 tunnel junction connecting two Al islands, with Au thermalization blocks on top of each. (**b**) Scanning electron microscopic (SEM) image showing Au thermalization blocks. Scale bar, 20 μm. (**c**) The sidewalls of four adjacent Au blocks. Scale bar, 4 μm. (**d**) One tunnel junction connecting two adjacent blocks. Scale bar, 1 μm.

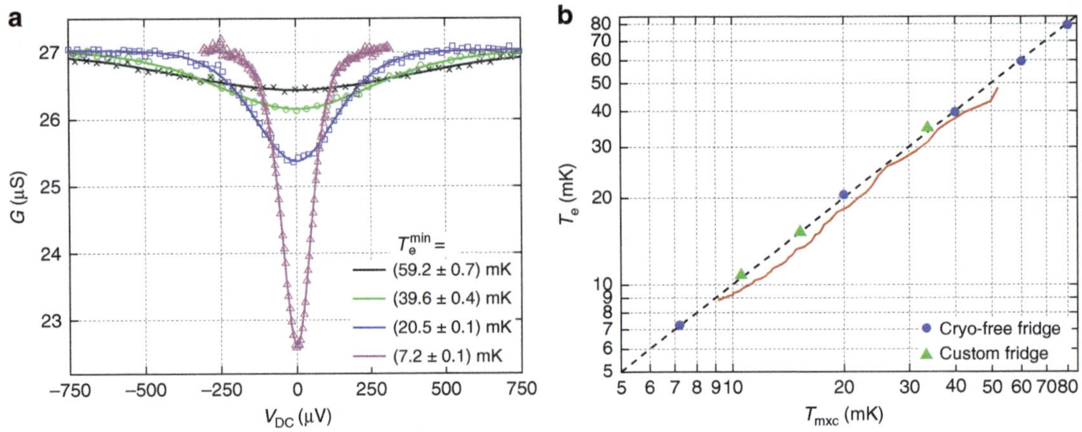

Figure 2 | CBT behaviour between 80 and 7 mK in two dilution refrigerators. (**a**) CBT conductance G versus measured bias voltage V_{DC} at four temperatures. Symbols show measured values and lines show best fits to the calculated ideal conductance. The warmest measurements (crosses, circles and squares) are fitted simultaneously to calibrate the sensor, giving $C_\Sigma = 236.6$ fF and $R_T = 22.42$ kΩ. The coldest measurement (triangles) is fitted using this calibration. The minimum electron temperatures T_e^{min} are in close agreement with the refrigerator temperature measured by a RuO_2 thermometer: 59.9, 40.1, 20.0 and 7.2 mK respectively. The uncertainties on T_e^{min} are calculated from uncertainties in the fitted parameters. (**b**) CBT electron temperature T_e versus refrigerator temperature T_{mxc}. Symbols show T_e^{min} from fits to conductance dips measured in the cryogen-free refrigerator (circles) and the custom refrigerator (triangles). Error bars are within the symbols. The solid curve shows T_e determined by monitoring the conductance G_0 (at $V_{DC} = 0$) as the cryo-free fridge cools over 35 min from 52 to 9 mK, showing that the CBT has a stronger thermal link to the refrigerator than the RuO_2 thermometer, leading to the thermal lag ($T_{mxc} \geq T_e$) during this time. The dashed line shows $T_e = T_{mxc}$.

124 (Aivon Therma), which are also attached to the mixing chamber plate. The filters have a cutoff frequency ≈15 kHz.

As shown in Fig. 2a, the CBT conductance dips around zero bias and the dip becomes deeper and narrower at lower temperatures. Its full-width at half maximum is related to temperature by $V_{1/2} \approx 5.439 N k_B T/e$, where N is the number of tunnel junctions in series[7]. This result does not account for self-heating in the sensor and so is only applicable when $T_e = T_p \equiv T$. A more practical parameter to determine temperature is the zero-bias conductance G_0, which has an approximate analytic relation to temperature[21]

$$G_0 \approx G_T\left(1 - 1/6u_N - 1/60u_N^2 + 1/630u_N^3\right), \qquad (2)$$

where $u_N \equiv E_C/k_B T$ is the dimensionless inverse temperature, $E_C \equiv [(N-1)/N]e^2/C_\Sigma$ is the charging energy of the system, C_Σ is the total capacitance of each island and G_T is the asymptotic conductance. When $u_N < 2.5$, the temperature measurement error is <2.5% (ref. 15). Thus, if C_Σ and G_T are known, it is possible to determine T by measuring only G_0. The most complete method to determine temperature is using a full tunnelling model to calculate $G(V_{DC})$ numerically[7]. We use this last approach to find C_Σ and G_T for the device, and these parameters are then used to find the temperature from subsequent measurements.

Numerical calculations of conductance are made using an algorithm derived from the free, open-source library pyCBT (see Methods for further details). In addition, we account for overheating in the sensor by predicting the electron temperature T_e in the islands using a model for the heat flow P into each island,

$$P = \frac{V_{DC}^2}{R_T} - \Sigma\Omega\left(T_e^5 - T_p^5\right) + P_0, \qquad (3)$$

where the first term is Joule heating at tunnel junctions of resistance R_T, the second term is heat flow to phonons and P_0 accounts for parasitic heating. For a given phonon temperature, the minimum electron temperature T_e^{min} is found at $V_{DC} = 0$. If P_0 is small, then $T_e^{min} \approx T_p$.

Figure 2a shows the result of fitting the calculated $G(V_{DC})$ simultaneously to three measurements made between 20 and 60 mK. The fit parameters are R_T, C_Σ and T_p for each measurement. The fit is found to be insensitive to the value of P_0 and hence parasitic heating is assumed to be negligible at these temperatures. Having calibrated the CBT, the fitted C_Σ is used to fit further measurements with T_p and R_T as the free parameters. An example is given in Fig. 2a, where the electron temperature is found to be 7.2 ± 0.1 mK at a refrigerator temperature of 7.2 mK. Figure 2b shows that the electron temperature measured by the CBT is in close agreement with the refrigerator temperature T_{mxc} in both refrigerators between 7 and 80 mK. Details of how T_{mxc} was measured in each fridge can be found in the Methods section.

CBT characteristics below 7 mK. We investigate the behaviour of the CBT below 7 mK in the custom dilution refrigerator. Figure 3a shows the electron temperature of the sensor when in vacuum gradually cooling below 4 mK, while the refrigerator was held at $T_{mxc} \lesssim 2.8$ mK. Here, T_e is determined from the value of G_0, which is found by measuring G over a small range of V_{DC} (≈30 μV) close to $V_{DC} = 0$. We observe an extremely long equilibration time (over 3 days) but a rapid cooling of the CBT following a heating event (a refill of the liquid helium bath that briefly heats the CBT above 5.5 mK and the fridge above 3.5 mK). This suggests that the thermal contact between the CBT and the refrigerator is relatively strong, and that its slow cooling is probably due to heat leak from an external warm object.

A second CBT was immersed in the $^3He/^4He$ refrigerant of the custom dilution refrigerator to improve thermal coupling and better isolate external sources of heating. A schematic of the immersion cell is shown in the Fig. 3c. Sintered silver blocks increase the thermal contact between the refrigerant and the incoming contact wires[2]. Several sinters are attached to the sensor package, to each of the four measurement wires and to a grounding wire for the package and the RC filters. The immersed CBT is found to equilibrate much faster, as shown in Fig. 3a,

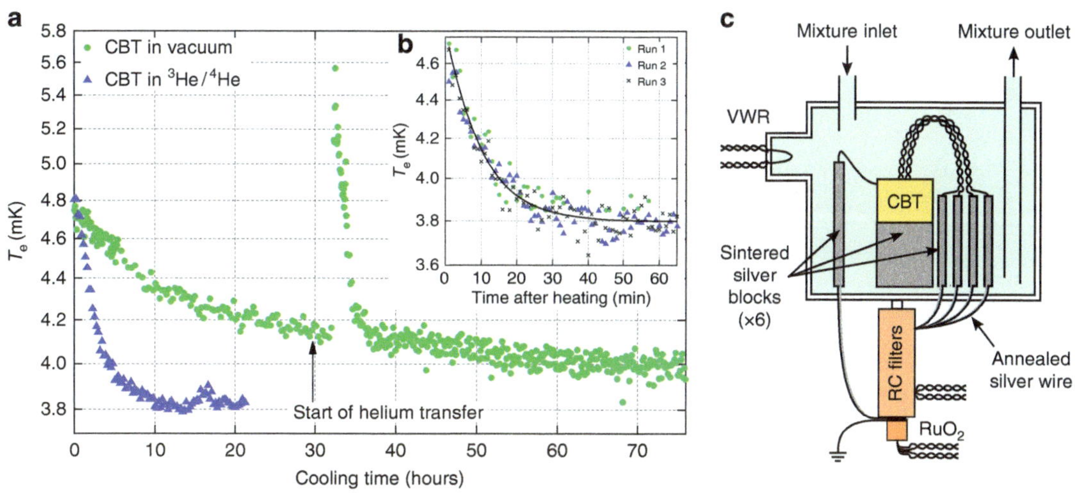

Figure 3 | Thermalization of two CBTs at a refrigerator temperature ≤ 2.8 mK. (**a**) Cooling of one CBT in vacuum (circles) and one immersed in the ^{3}He/^{4}He refrigerant of the dilution refrigerator (triangles). In both cases, the CBTs are cooling after being warmed above 10 mK by temporarily increasing the refrigerant temperature. The CBT in vacuum is extremely slow to thermalize. By comparison, the CBT immersed in ^{3}He/^{4}He thermalizes significantly faster. (**b**) Cooling of the immersed CBT after it has been heated by a large DC drive current (50, 40 and 30 nA for run 1, 2 and 3, respectively). Fitting to an exponential decay (solid line) yields a time constant of 570 s and a saturation temperature of 3.8 mK. (**c**) Schematic of the immersion cell used to cool a CBT in the ^{3}He/^{4}He mixture of a dilution refrigerator.

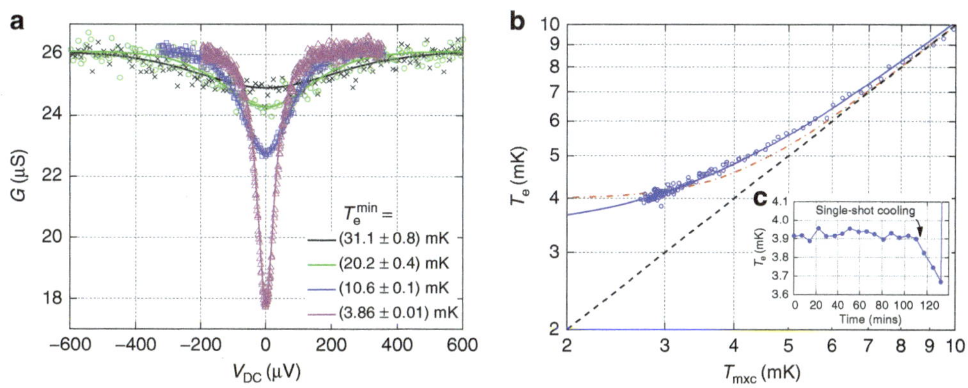

Figure 4 | Characteristics of a CBT immersed in ^{3}He/^{4}He refrigerant. (**a**) Fitting to the warmest three measurements gives $C_\Sigma = 209.5$ fF and $R_T = 23.21$ kΩ. The fitted minimum electron temperatures T_e^{min} for the warmest three curves are in reasonable agreement with the refrigerator temperatures as measured by the vibrating wire resonator (VWR) thermometer: 29.4, 19.0 and 10.5 mK, respectively. (**b**) Measured electron temperature in the CBT as the refrigerator cooled steadily from 10 to 2.7 mK over a period of 12 h. The solid curve shows a fit of the form $T_e^x = T_{mxc}^x + c$, yielding an exponent $x = 2.7$. The dot-dashed curve shows a best fit of $T_e^5 = T_{mxc}^5 + c$. The dashed line shows $T_e = T_{mxc}$. (**c**) The measured T_e as the refrigerator was temporarily cooled in single-shot mode to 2.2 mK, reaching a lowest T_e below 3.7 mK.

reaching $T_e \approx 3.8$ mK at $T_{mxc} \approx 2.7$ mK. It is important to note that even when T_e is elevated above T_p and T_{mxc}, the CBT remains a primary thermometer of its internal electron temperature.

To study the CBT at $|V_{DC}| > 0$, the time needed for the sensor to reach thermal equilibrium after a change of Joule heating needs to be known. Figure 3b shows the relaxation of T_e after the CBT has been heated by a large drive current for long enough to reach thermal equilibrium (>30 min). The subsequent value of T_e is measured by scanning close to $V_{DC} = 0$, where Joule heating should be negligible. The relaxation of T_e is found to have a time constant of 570 s.

Figure 4a shows the calibration of the immersed sensor. The three warmest measurements are fitted simultaneously to determine $C_\Sigma = 209.5$ fF and $R_T = 23.21$ kΩ. The fitted temperatures agree with T_{mxc} to within 6%. Given the agreement between the fitted T_e^{min} and T_{mxc}, we can assume that parasitic heating is still negligible down to 10 mK.

The coldest measurement in Fig. 4a is fitted using the above values, yielding a minimum electron temperature of 3.86 ± 0.01 mK. This measurement was made over a period of 7 h, to ensure that the CBT was in thermal equilibrium at each value of V_{DC}. At these temperatures, the parasitic heating of the CBT is now significant and T_e^{min} does not match the refrigerator temperature of $T_{mxc} = 2.7$ mK. To fit this conductance dip, the thermal model, equation 3, is used with $T_p = T_{mxc}$, and with the parasitic heating P_0 and the electron–phonon coupling $\Sigma\Omega$ as free parameters.

Figure 4b shows how the CBT electron temperature diverges from the refrigerator temperature below ≈7 mK. Here the value of T_e is found by measuring G_0 close to $V_{DC} = 0$ and so Joule heating can be neglected. The lowest temperature reported by the

CBT is below 3.7 mK when operating the fridge in single-shot mode (see Fig. 4c).

Discussion

The overheating of the sensor at $V_{DC}=0$ constrains the value of $P_0(\Sigma\Omega)^{-1}$ in the fit to the coldest measurement in Fig. 4a. However, the parasitic heating is not large enough to reliably separate the values of P_0 and $\Sigma\Omega$ in the fitting. Qualitatively, the fits suggest that $P_0 \geq 300$ aW per island and $\Sigma\Omega$ is at least four times larger than expected from the nominal size of the thermalization blocks and the literature value of Σ for Au[17]. It is not possible to determine an upper bound on P_0 without constraining $\Sigma\Omega$. It is worth noting that the power required to measure the CBT conductance ($\sim$1 aW per island due to Joule heating from I_{AC}) is much lower than our estimate of P_0. As such, we believe that CBTs of this type can be operated at still lower temperatures by further reducing the parasitic heating.

The functional form of T_e versus T_{mxc}, as shown in Fig. 4b, should have the same temperature dependence as the dominant thermalization mechanism, that is, T^5 for electron–phonon coupling. However, other power laws have been observed[9]. Here we find that the best fit of $T_e^x = T_{mxc}^x + c$ gives $x=2.7$ and a saturated T_e of $c^{1/x}=3.4$ mK. The fitted exponent x cannot be confirmed by fits to the conductance dips in Fig. 4a, because the overheating is still relatively weak, even at the lowest temperatures, and there is little effect on the shape of the dip. We find that a thermal model with a T^3 thermalization term fits the conductance dips equally well as a model using T^5.

The saturation of the measured CBT temperature below 7 mK could be caused by parasitic heating of the islands or excess voltage noise across the tunnel junctions. It is possible that the operating temperature could be lowered by reducing parasitic heating through better shielding and by lowering the voltage noise in the measurement circuit. To understand the cause of saturation in more detail, or to test an improved measurement environment, this sensor would need to be cooled closer to 1 mK.

In conclusion, the CBTs described here have been shown to operate as reliable primary thermometers of electron temperature down to 3.7 mK. The large thermalization blocks incorporated in the device and a relatively low level of parasitic heating ensure that the electron subsystem in the sensor is well coupled to the phonon subsystem down to $\approx$7 mK. An immersion cell is shown to improve thermal coupling between a CBT and a dilution refrigerator. This allows the onset of overheating to be observed below 7 mK, and although the presence of overheating can be seen, the effect is sufficiently weak that the sensor will need to be cooled further to fully characterize the thermalization mechanisms.

Methods

Device fabrication. The CBT devices are fabricated using an *ex situ* tunnel junction process[11]. The Al films that define the CBT circuit have a thickness of 250 nm. Tunnel junctions between sections of Al are formed by an insulating layer of 250 nm SiO_x deposited by plasma-enhanced chemical vapour deposition (PECVD). The junctions have a nominal diameter 0.8 μm and a resistivity $\sim$10 kΩ μm^2. The substrate is undoped Si with 300 nm thermal oxide on the surface.

The island thickness achievable with the *ex situ* tunnel junction process or other deposition techniques used for tunnel junction devices is typically up to 1 μm. Thicker films suffer from stress build-up, causing poor adhesion between the film and the substrate. This is a severe problem at mK temperatures where poor adhesion can lead to poor thermalization and even mechanical failure due to thermal motion during cool down. We avoid these problems by using a combination of the *ex situ* process followed by masked electroplating of Au on top of the CBT islands[22], which we refer to as thermalization blocks. Electroplating can produce $\sim$10 μm-thick, low-stress films and here we choose a nominal thickness of 5 μm for the thermalization blocks.

Refrigerator thermometry. Two different dilution refrigerators are used in the experiments described above. In the commercial refrigerator, the mixing chamber temperature T_{mxc} is measured by a calibrated RuO_2 resistor (Sensor model RU-1000-BF0.007 supplied and calibrated by Bluefors Cryogenics) in contact with the mixing chamber plate. In the custom refrigerator, T_{mxc} is measured using a conventional vibrating wire resonator viscometer immersed in the saturated dilute phase of the ^{3}He/^{4}He refrigerant in the mixing chamber[23,24]. The vibrating wire resonator is validated by comparison with a calibrated RuO_2 resistor (calibrated to 20 mK and supplied by Lake Shore Cryogenics). This resistor is thermally connected to the refrigerant via an immersed pad of sintered silver.

References

1. European Microkelvin Collaboration. EU-funded Integrating Activity project http://www.microkelvin.eu/ (2015).
2. Pan, W. *et al.* Exact quantization of the even-denominator fractional quantum Hall state at v = 5/2 Landau level filling factor. *Phys. Rev. Lett.* **83,** 3530–3533 (1999).
3. Samkharadze, N. *et al.* Integrated electronic transport and thermometry at milliKelvin temperatures and in strong magnetic fields. *Rev. Sci. Instrum.* **82,** 053902 (2011).
4. Hanson, R., Kouwenhoven, L. P., Petta, J. R., Tarucha, S. & Vandersypen, L. M. K. Spins in few-electron quantum dots. *Rev. Mod. Phys.* **79,** 1217–1265 (2007).
5. Devoret, M. H. & Schoelkopf, R. J. Superconducting circuits for quantum information: an outlook. *Science* **339,** 1169–1174 (2013).
6. Giazotto, F., Heikkilä, T. T., Luukanen, A., Savin, A. M. & Pekola, J. P. Opportunities for mesoscopics in thermometry and refrigeration: physics and applications. *Rev. Mod. Phys.* **78,** 217–274 (2006).
7. Pekola, J. P., Hirvi, K. P., Kauppinen, J. P. & Paalanen, M. A. Thermometry by arrays of tunnel junctions. *Phys. Rev. Lett.* **73,** 2903–2906 (1994).
8. Meschke, M., Engert, J., Heyer, D. & Pekola, J. Comparison of Coulomb blockade thermometers with the international temperature scale PLTS-2000. *Int. J. Thermophys.* **32,** 1378–1386 (2011).
9. Casparis, L. *et al.* Metallic Coulomb blockade thermometry down to 10mK and below. *Rev. Sci. Instrum.* **83,** 083903 (2012).
10. Scheller, C. P. *et al.* Silver-epoxy microwave filters and thermalizers for millikelvin experiments. *Appl. Phys. Lett.* **104,** 211106 (2014).
11. Prunnila, M. *et al.* Ex situ tunnel junction process technique characterized by Coulomb blockade thermometry. *J. Vac. Sci. Technol. B* **28,** 1026–1029 (2010).
12. Hahtela, O. M. *et al.* Investigation of uncertainty components in Coulomb blockade thermometry. *AIP Conf. Proc.* **1552,** 142–147 (2013).
13. Gunnarsson, D. *et al.* Dielectric losses in multi-layer Josephson junction qubits. *Supercond. Sci. Technol.* **26,** 085010 (2013).
14. Meschke, M., Pekola, J., Gay, F., Rapp, R. & Godfrin, H. Electron thermalization in metallic islands probed by Coulomb blockade thermometry. *J. Low Temp. Phys.* **134,** 1119–1143 (2004).
15. Feshchenko, A. *et al.* Primary thermometry in the intermediate Coulomb blockade regime. *J. Low Temp. Phys.* **173,** 36–44 (2013).
16. Wellstood, F. C., Urbina, C. & Clarke, J. Hot-electron effects in metals. *Phys. Rev. B* **49,** 5942–5955 (1994).
17. Echternach, P. M., Thoman, M. R., Gould, C. M. & Bozler, H. M. Electron-phonon scattering rates in disordered metallic films below 1 K. *Phys. Rev. B* **46,** 10339–10344 (1992).
18. Roschier, L. *et al.* Effect of on-chip filter on Coulomb blockade thermometer. *J. Phys. Conf. Ser.* **400,** 052029 (2012).
19. Bladh, K. *et al.* Comparison of cryogenic filters for use in single electronics experiments. *Rev. Sci. Instrum.* **74,** 1323–1327 (2003).
20. Bradley, D. I., Follows, M., Miller, I., Oswald, R. & Ward, M. Simple design for dilution refrigerator with base temperature of 2.3mK. *Cryogenics* **34,** 549–550 (1994).
21. Farhangfar, S. *et al.* One dimensional arrays and solitary tunnel junctions in the weak Coulomb blockade regime: CBT thermometry. *J. Low Temp. Phys.* **108,** 191–215 (1997).
22. Xu, H. *et al.* Wafer-level SLID bonding for MEMS encapsulation. *Adv. Manuf.* **1,** 226–235 (2013).
23. Zeegers, J., de Waele, A. & Gijsman, H. Viscosity of saturated ^{3}He-^{4}He mixture below 200 mK. *J. Low Temp. Phys.* **84,** 37–47 (1991).
24. Pentti, E., Rysti, J., Salmela, A., Sebedash, A. & Tuoriniemi, J. Studies on helium liquids by vibrating wires and quartz tuning forks. *J. Low Temp. Phys.* **165,** 132–165 (2011).

Acknowledgements

We thank Jukka Pekola and Matthias Meschke for useful discussions, and Andrey Shchepetov for producing the scanning electron microscopic images in Fig. 1. This research is supported by the U.K. EPSRC (EP/K01675X/1 and EP/L000016/1), the European FP7 Programme MICROKELVIN (project number 228464), Tekes project FinCryo (grant number 220/31/2010), Academy of Finland (project number 252598) and the Royal Society. J.R.P. acknowledges support of the People Programme (Marie Curie Actions) of the European FP7 Programme under REA grant agreement 618450. Yu.A.P. acknowledges support by the Royal Society and the Wolfson Foundation.

Author contributions

D.G., H.H., J.P., M.P. and L.R. designed, fabricated and packaged the devices. J.P. and L.R. developed custom measurement instrumentation and methods. D.I.B., R.E.G., R.P.H., Yu.A.P., J.R.P., L.R. and M.S. performed measurements and calculations. J.R.P. drafted the manuscript. All authors discussed the results and implications, and commented on the manuscript at all stages.

Additional information

Competing financial interests: The authors declare no competing financial interests.

3

Enhanced spin-orbit torques by oxygen incorporation in tungsten films

Kai-Uwe Demasius[1,2,*], Timothy Phung[1,*], Weifeng Zhang[1,*], Brian P. Hughes[1], See-Hun Yang[1], Andrew Kellock[1], Wei Han[1], Aakash Pushp[1] & Stuart S.P. Parkin[1,3]

The origin of spin-orbit torques, which are generated by the conversion of charge-to-spin currents in non-magnetic materials, is of considerable debate. One of the most interesting materials is tungsten, for which large spin-orbit torques have been found in thin films that are stabilized in the A15 (β-phase) structure. Here we report large spin Hall angles of up to approximately -0.5 by incorporating oxygen into tungsten. While the incorporation of oxygen into the tungsten films leads to significant changes in their microstructure and electrical resistivity, the large spin Hall angles measured are found to be remarkably insensitive to the oxygen-doping level (12-44%). The invariance of the spin Hall angle for higher oxygen concentrations with the bulk properties of the films suggests that the spin-orbit torques in this system may originate dominantly from the interface rather than from the interior of the films.

[1] IBM Almaden Research Center, San Jose, California 95120, USA. [2] Dresden University of Technology, 01062 Dresden, Germany. [3] Max Planck Institute for Microstructure Physics, 06120 Halle (Saale), Germany. * These authors contributed equally to this work. Correspondence and requests for materials should be addressed to K.-U.D. (email: Kai-Uwe.Demasius@mailbox.tu-dresden.de) or to T.P. (email: phungt@us.ibm.com) or to S.S.P.P. (email: stuart.parkin@mpi-halle.mpg.de).

There has been considerable recent interest in the controlled manipulation of magnetic moments using spin–orbit torques for the purposes of building novel spintronic devices[1–4]. Spin–orbit torques are derived from pure spin currents that are generated from charge currents flowing through non-magnetic metals via spin–orbit coupling[5–8]. The conversion efficiency of the charge-to-spin current can be described by the spin Hall angle (SHA). Although small effects were first observed in semiconductors a decade ago ($\sim 2\times10^{-4}$ in GaAs[9]), recently much larger SHAs have been surprisingly observed in simple metals, making them potentially useful for spintronic applications. Indeed, reliable methods to measure the spin currents involve their diffusion into adjacent magnetic layers on which they can exert significant spin torques[4,10,11]. These spin–orbit torques have been shown to be sufficiently large to induce the motion of magnetic domain walls[12–16], excite precessional magnetization dynamics[10,17] or switch a uniform magnetic layer[3,4,18].

Although the exact origins responsible for these spin–orbit torques in conventional metals are not understood in detail, existing theories ascribe them to combinations of intrinsic mechanisms and extrinsic spin-dependent scattering from within the bulk of the materials studied[19]. Most typically, extrinsic mechanisms will likely dominate because of innate bulk disorder in the thin films that have been studied to date. Largely neglected has been the role of interface scattering (which is known to dominate the origin of giant magnetoresistance[20]), although electronic discontinuities at interfaces between two distinct materials are known to give rise to Rashba effects that can result in charge-to-spin conversion[21–23].

One of the most efficient materials exhibiting spin–orbit torques is the highly resistive β-phase of tungsten, where SHAs of up to approximately −0.35 have been reported[24–26]. In contrast, the α-phase of tungsten exhibits a much smaller SHA. Consequently, one would expect that the SHA should scale strongly with significant changes in the microstructure of the materials under consideration. A detailed understanding of the role of materials' microstructure on the spin–orbit torques is thus crucial for enhancing the efficiency of these effects, as well as giving insight towards the underlying mechanisms responsible for the spin–orbit torques.

Here we demonstrate that by doping oxygen into tungsten thin films, large spin–orbit torques are attained that are rather insensitive to considerable changes in the resistivity and film microstructure. The spin–orbit torques that we observe, when quantified as a SHA, are the largest to be reported thus far for a conventional metal-based system (approximately −0.5), and could be of considerable technological interest for spintronic applications. Furthermore, our findings suggest that these very-efficient spin–orbit torques are largely interfacial in origin.

Results

Spin torque ferromagnetic resonance of W(O) | CoFeB structures. The films in this study were deposited at room temperature using d.c. magnetron sputtering onto oxidized silicon substrates consisting of the layer-structure Si substrate | SiO_x (25) | W(O) (6) | $Co_{40}Fe_{40}B_{20}$ (6) | TaN (2) (Fig. 1a,b; with film thicknesses in nm in parentheses). Hereafter, we shall refer to oxygen-doped tungsten as W(O) and $Co_{40}Fe_{40}B_{20}$ as CoFeB. The amount of oxygen gas flow in the reactive mixture during the sputtering, Q, was varied between 0 and 3%. A 20-Å-thick highly resistive TaN layer (5.08 Ωcm for 50 nm thickness) was used for capping the magnetic CoFeB film (109 μΩ cm for 6 nm thickness; Fig. 1a,b). The TaN layer was $\sim 10^4$ times more resistive than the most resistive W(O) film considered in this study, ensuring that current

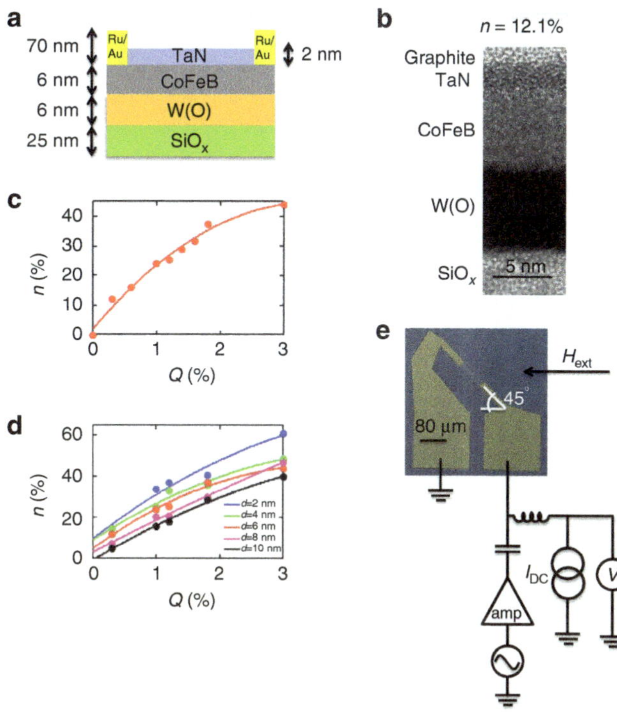

Figure 1 | STFMR measurement. (**a**) Device structure consisting of substrate | W(O) (6) | $Co_{40}Fe_{40}B_{20}$ (6) | TaN (2) (with film thickness in parantheses in nm). (**b**) Cross-section transmission electron microscope (TEM) image of the device structure for $n = 12.1\%$ ($Q = 0.3\%$). (**c**) Oxygen concentration n is plotted against gas flow Q as determined by Rutherford backscattering spectrometry (RBS). (**d**) Oxygen concentration versus gas flow Q for W(O) films of different thicknesses. (**e**) Microscope image and schematic of the electrical circuit used for STFMR measurements.

primarily flows in the W(O) and CoFeB layers. The atomic oxygen concentration n was determined in films specially deposited on graphite substrates using Rutherford backscattering spectrometry (see Fig. 1c). n increases monotonically with increasing gas flow Q reaching $\sim$44 atomic % at the maximum oxygen gas flow used, for 6-nm-thick W(O) films. Furthermore, we found that the amount of oxygen incorporation was increased for thinner films (Fig. 1d). This finding is consistent with previous studies of sputtered tungsten thin films, and one explanation for it may arise from the compressive stress, which increases with film thickness for W films deposited at comparable pressures to ours[27]. The increased compressive stress leads to a smaller lattice constant and hence reduced oxygen incorporation.

We use the spin transfer torque (STT) ferromagnetic resonance technique[10] to determine an effective SHA, θ_{SH}^{eff}, which we use to assess the value of the damping-like spin–orbit torque. Figure 1e shows a microscope image of a representative microstrip device (10 μm × 80 μm) with a 45° tilt (for more information on the fabrication, see Methods) and a schematic illustration of the experimental set-up used for this measurement. A radiofrequency (RF) current is passed through the device, which generates spin–orbit torques as well as an Oersted field, in the presence of an externally applied magnetic field H_{ext}. These torques cause a sustained precession of the magnetization of the magnetic layer, which is measured through the mixing d.c. voltage V_{mix} that is generated from the oscillating anisotropic magnetoresistance and spin Hall magnetoresistance[28] signals and the applied RF current. The RF source has been set to an output power of 5 dBm and then amplified to 22 dBm. We extract the magnitude of θ_{SH}^{eff} by performing two different types of analyses based on fitting V_{mix} to a Lorentzian function consisting of symmetric (F_S) and asymmetric (F_A) components:

$$V_{mix} = V_0(SF_S(H_{ext}) + AF_A(H_{ext})), \quad (1)$$

where

$$V_0 = -\frac{1}{4}\frac{dR}{d\varphi}\frac{\gamma\mu_0 I_{rf,tot}\cos\varphi}{\Delta 2\pi(df/dH_{ext})|_{H_{ext}=H_0}}, \quad (2)$$

$$F_S(H_{ext}) = \frac{\Delta^2}{\Delta^2 + (H_{ext} - H_0)^2}, \quad (3)$$

and

$$F_A(H_{ext}) = F_S(H_{ext})\frac{(H_{ext} - H_0)}{\Delta} \quad (4)$$

Here Δ is the linewidth and H_0 is the resonance field. The prefactors, S and A, are the contributions of the symmetric and asymmetric parts, respectively, and V_0 is a constant prefactor. In equation (2), R represents the resistance of the device, which, because of a combination of anisotropic magnetoresistance and spin Hall magnetoresistance, depends on the angle φ between the current and the magnetization (which for our measurements is either at 45° or −135°, depending on the applied field direction), γ is the gyromagnetic ratio, μ_0 is the magnetic permeability of free space, $I_{rf,tot}$ is the RF current through the device and f is the frequency of the applied RF current. Figure 2a shows representative data for V_{mix} ($n = 12.1\%$, $Q = 0.3\%$) for an RF current applied at 9 GHz, along with the symmetric and asymmetric components of the data extracted from the fitting.

The first type of analysis, which we call here the line-shape analysis (θ_{LS}^{eff}), associates the symmetric part of the Lorentzian function to the antidamping-like STT and the asymmetric part to the Oersted field in the system. Meanwhile, the line-width analysis is based on measuring changes in the linewidth (θ_{LW}^{eff}),

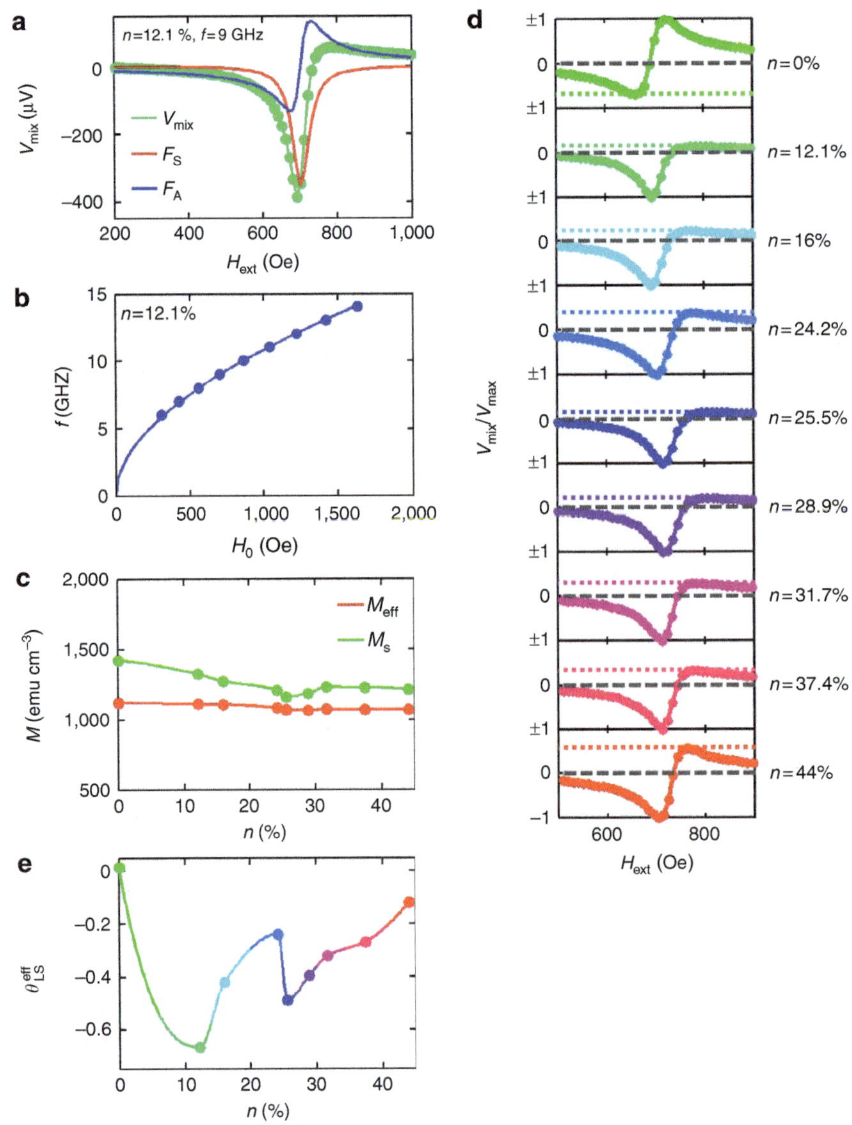

Figure 2 | Line-shape analysis. (**a**) V_{mix} along with the fitted (green), symmetric (F_S, red) and asymmetric (F_A, blue) Lorentzian functions used for the fitting for $n = 12.1\%$ ($Q = 0.3\%$). (**b**) Frequency as a function of the resonant field for $n = 12.1\%$ ($Q = 0.3\%$) used in the Kittel formula fitting. (**c**) The magnetization M_S determined by VSM and effective demagnetization field M_{eff} from the Kittel formula fitting versus oxygen concentration n. (**d**) V_{mix} normalized to either its minimum or maximum value for different oxygen concentrations at 9 GHz. The coloured dotted line denotes the corresponding maximum/minimum value of V_{mix}. The dotted line (dashed grey line) indicates the zero level. (**e**) SHA calculated from the line-shape analysis θ_{LS}^{eff} as a function of the oxygen concentration n.

that is, damping, as a function of a superimposed d.c. current. We note that the SHAs we report here are effective values because we do not take into account the transparency[29], whose role is minor in the materials under consideration here because of their relatively high resistivities.

Line-shape analysis. The determination of the SHA based on the line-shape analysis is described by the following formula[10]:

$$\theta_{\mathrm{LS}}^{\mathrm{eff}} = \frac{J_{\mathrm{S,rf}}}{J_{\mathrm{C,rf}}} = \frac{S}{A}\frac{2e\mu_0 M_{\mathrm{S}}t}{\hbar}\beta_{\mathrm{Oe}}\sqrt{1+\frac{M_{\mathrm{eff}}}{H_0}} \quad (5)$$

$$\beta_{\mathrm{Oe}} = \frac{d}{2}, \quad (6)$$

where e represents the electron charge and $\hbar$ represents Planck's constant above. We deduce the thicknesses of the magnetic (t) and non-magnetic (d) layers by determining the film deposition rates from measurements of the thicknesses of nominally 50-nm-thick calibration films with a profilometer. The saturation magnetization M_{S} was measured by vibrating sample magnetometry (VSM) and the effective demagnetization field M_{eff} was calculated by fitting the frequency versus the resonance field, according to the Kittel formula (Fig. 2b; see Supplementary Table 1 for a summary of the parameters above for the films under study):

$$f = \frac{\gamma\mu_0}{2\pi}\sqrt{H_0(H_0+M_{\mathrm{eff}})} \quad (7)$$

Figure 2c illustrates a weak decrease in M_{S} with increasing oxygen concentration.

Figure 2d shows the fitting of normalized V_{mix} (H_{ext}) at 9 GHz for different amounts of oxygen incorporated into the W(O) film. The data are normalized either to the minimum or maximum value of V_{mix}. The corresponding maximum/minimum of V_{mix} is indicated by the coloured dotted line, as a guide to the eye. Using the data analysis given above, large SHAs imply a large symmetric component relative to the asymmetric component. Consequently, the closer the dotted line is to zero (grey dashed line in figure), the higher the SHA is. The highest value we reach with this analysis is $\theta_{\mathrm{LS}}^{\mathrm{eff}} = -0.67$. Moreover, the variation of the SHA as a function of the oxygen concentration is rather significant (Fig. 2e).

We further note that this commonly used analysis technique requires careful attention from possible artefact voltages. In particular, the line-shape analysis assumes that the symmetric component arises completely from the antidamping STT. However, a contribution to the symmetric component from spin pumping[30] that induces an inverse spin Hall effect[31] voltage can occur that can diminish and even change the sign of the observed SHA, which is what we find for $n = 0\%$ (see Supplementary Note 1 and Supplementary Figs 1 and 2). Another caveat of this analysis is that the asymmetric component is assumed to arise completely from the Oersted field-generated torque; however, for the $n = 0\%$ sample, we find this torque to be smaller than the field-like STTs (see Supplementary Note 1 and Supplementary Fig. 2). Thus, a complete quantitative analysis based on this technique requires accounting for the field-like STTs as well as a consideration of the size of the spin-pumping contribution to the symmetric signal. Nevertheless, we perform the line-shape analysis to gain a qualitative sense of the dependence of the SHA with the oxygen concentration in our films, but use a line-width analysis to more reliably assess the value of the SHA that is generated solely from the antidamping STT.

Line-width analysis. The Gilbert damping α can be deduced from the dependence of Δ on frequency that is given by

$$\Delta = \Delta_0 + \frac{2\pi f}{\gamma}\alpha \quad (8)$$

where Δ_0 represents the inhomogeneous broading. We observe that the Gilbert damping remains relatively constant with respect to the oxygen concentration, with an average value of $\alpha \sim 0.0085$ (Fig. 3). We have also compared the change in the Gilbert damping with the presence of the W(O) layer to deduce the effective spin-mixing conductance as a function of n, and find it to be relatively constant (see Supplementary Note 2 and Supplementary Fig. 3).

We further measured the dependence of Δ on an applied d.c. current through the device (Fig. 3b)

$$\Delta = \frac{2\pi f}{\gamma}\left(\alpha + \frac{\sin\varphi}{(H_{\mathrm{ext}}+0.5M_{\mathrm{eff}})\mu_0 M_{\mathrm{S}}t}\frac{\hbar}{2e^2}J_{\mathrm{S}}\right), \quad (9)$$

To determine $\theta_{\mathrm{LW}}^{\mathrm{eff}}$ originating from the antidamping-like STT[10,32] one has to use the slopes $\delta\Delta/\delta I_{\mathrm{d.c.}}$ from Fig. 3b:

$$\theta_{\mathrm{LW}}^{\mathrm{eff}} = \frac{\delta\Delta/\delta I_{\mathrm{DC}}}{\frac{2\pi f}{\gamma}\frac{\sin\varphi}{(H_{\mathrm{ext}}+0.5M_{\mathrm{eff}})\mu_0 M_{\mathrm{S}}t}\frac{\hbar}{2e^2}}\frac{R_{\mathrm{FM}}+R_{\mathrm{W(O)}}}{R_{\mathrm{FM}}}A_{\mathrm{C}} \quad (10)$$

R_{FM} and $R_{\mathrm{W(O)}}$ are the resistances of the ferromagnetic layer and W(O) layer, respectively, and A_{C} is the cross-sectional area of the device. From this line-width analysis we find that $\theta_{\mathrm{LW}}^{\mathrm{eff}}$ increases abruptly on the introduction of oxygen, but varies slightly with the addition of more oxygen into the tungsten film, as shown in Fig. 3c. Although the dependence of $\theta_{\mathrm{LW}}^{\mathrm{eff}}$ and $\theta_{\mathrm{LS}}^{\mathrm{eff}}$ on oxygen concentration is qualitatively similar, the variation in $\theta_{\mathrm{SH}}^{\mathrm{eff}}$ with higher oxygen concentrations is much more pronounced when using the line-shape analysis (Fig. 2e). The SHA values reached here are at most $\theta_{\mathrm{LW}}^{\mathrm{eff}} = -0.49$ where $n = 12.1\%$. The SHA in the pure tungsten film, by comparison, is $\theta_{\mathrm{LW}}^{\mathrm{eff}} = -0.14$, which is consistent with a previous report for a 6-nm-thick film where both the α- and β-phase tungstens were found to co-exist[24]. The difference between the line-shape and line-width analyses stems mostly from the contributions of the spin-pumping and field-like torques. To quantitatively reconcile the values based on the two analysis techniques, an accurate determination of the RF current through the device is required (see Supplementary Note 1 and Supplementary Figs 1 and 2)[33–35]. Figure 3d shows the comparison between the line-shape and line-width analyses after accounting for the spin-pumping and field-like torques. Here we shall focus on the line-width analysis when drawing conclusions regarding the SHA.

The role of thickness of the W(O) on the SHA was examined for a gas flow $Q = 1.2\%$ (Fig. 3e). However, one important point to note is that the amount of oxygen that is actually incorporated in the film varies significantly with the thickness of the grown film, as mentioned above (Fig. 1d). Thus, the SHA of these films can be plotted as a function of oxygen content, irrespective of their thickness, and compared with data from above where the thickness remains constant and the oxygen content is intentionally varied (Fig. 3f). On the basis of this analysis, we find that the SHA does not change significantly despite large changes in thickness. Furthermore, based on a volume origin of the SHE, the SHA as a function of thickness varies as

$$\frac{J_{\mathrm{S}}(d)}{J_{\mathrm{S}}(\infty)} = 1 - \mathrm{sech}\frac{d}{\lambda_{\mathrm{s}}} \quad (11)$$

Our data thus imply that either the spin diffusion length λ_{S} would have to be significantly smaller compared with 4.4 nm or alternatively the spin–orbit torque has an interfacial origin.

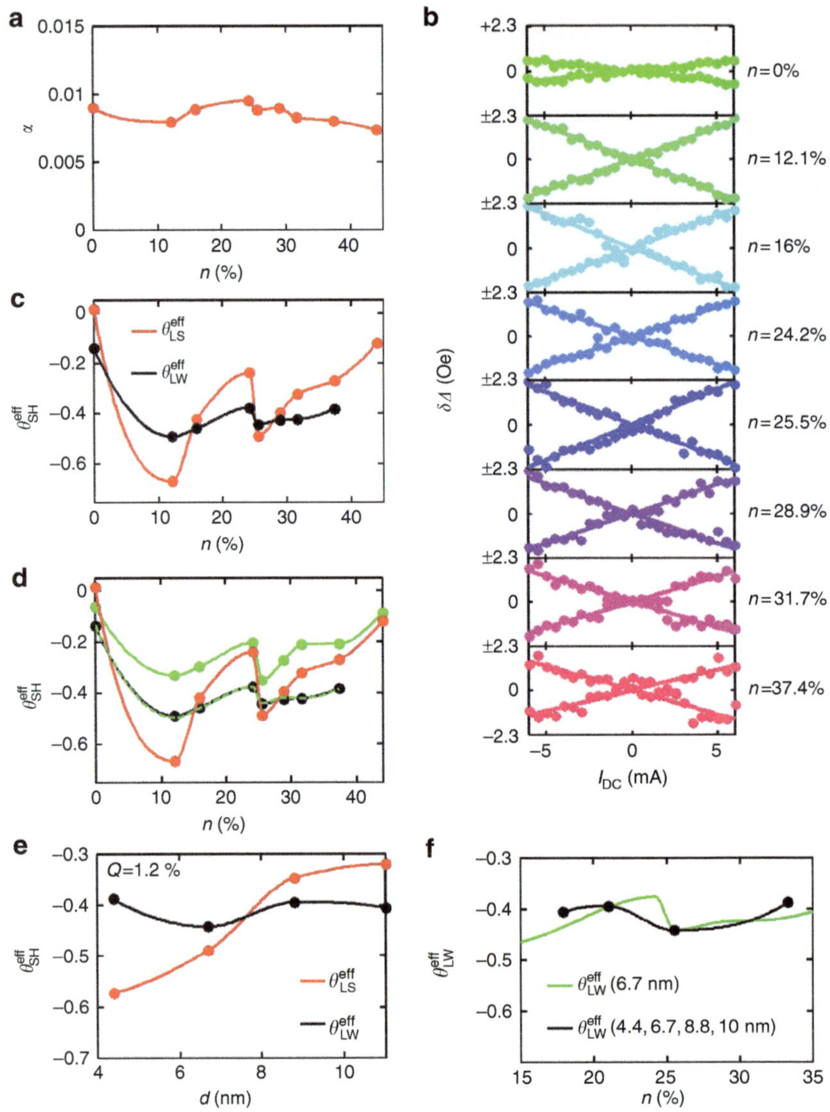

Figure 3 | Line-width analysis. (**a**) Dependence of the Gilbert damping on *n*. (**b**) Change of linewidth ($\delta\Delta$) versus d.c. current $I_{d.c.}$ for different *n*. (**c**) Damping-like spin Hall angle (θ^{eff}_{LW}) versus *n* (black curve) calculated from d.c.-current-dependent linewidth. θ^{eff}_{LS} (red curve) is shown for comparison purposes. (**d**) Correction of θ^{eff}_{LS} (red) with the RF current measured with a network analyser (green curve, $\theta^{eff}_{corr}(I_{rf,tot} = \text{measured})$) and correction with a slightly adjusted RF current (dashed green curve, $\theta^{eff}_{corr}(I_{rf,tot} = \text{adjusted})$) to match the black curve (θ^{eff}_{LW}). (**e**) Thickness dependence of θ^{eff}_{LS} and θ^{eff}_{LW} for $Q = 1.2\%$. (**f**) θ^{eff}_{LW} for ~6.7-nm-thick films as a function of oxygen concentration *n*, and θ^{eff}_{LW} of films with $Q = 1.2\%$ of various thicknesses.

Material characterization. To understand the origin of the observed trends in the SHA, we performed resistivity and X-ray diffraction measurements. To increase the signal-to-noise ratio, X-ray diffraction was performed on multilayers with the stack sequence $4 \cdot [SiO_x\,(2.5)\,|\,W(O)\,(6)\,|\,CoFeB\,(6)\,|\,TaN\,(2.5)]$ (with film thicknesses in parenthesis in nm). The X-ray diffraction measurements (Fig. 4a,b) reveal that the tungsten film grown without any oxygen exhibits a predominantly $\alpha - W$ phase with a (100) orientation (strong peak at $2\theta = 40.3°$) with some contributions from a $\beta - W$ structure with (200) and (211) crystal orientations ($2\theta = 35.4°$ and 43.7°, respectively). The sample with the highest SHA, corresponding to $n = 12.1\%$, exhibits a peak at $2\theta = 39.8°$ that is distinct from the $\alpha - W$ (100) peak and corresponds to the (210) peak for $\beta - W$. Thus, the X-ray data show that the amount of $\beta - W$ is increased significantly for $n = 12.1\%$ compared with $n = 0\%$ and there is no evidence for any remaining $\alpha - W$. The X-ray diffraction measurements also show that increasing the oxygen concentration in the films leads to eventually significantly broader and weaker X-ray diffraction peaks, indicating an increasing nanocrystallization of the W(O) films. The grain size can be calculated using the Scherrer equation

$$L_{Grain} = \frac{K\lambda}{\beta \cos\theta}, \qquad (12)$$

with $K = 1$ (shape factor), wavelength $\lambda = 0.154$ nm, β being the half intensity width and θ being diffraction angle. The grain size versus oxygen concentration is shown in Fig. 4c. Interestingly, we observe a sudden decrease in the grain size at $n = 25.5\%$ to an oxygen concentration-independent grain size for higher *n*. The transition indicates that the W(O) material becomes nanocrystalline at a critical *n*. The X-ray diffraction measurements of the actual films used in the STT ferromagnetic resonance technique study as well as much thicker films, nominally ~50 nm thick, are shown in the Supplementary Section, and show similar trends to the multilayer films (see Supplementary Note 3 and Supplementary Fig. 4).

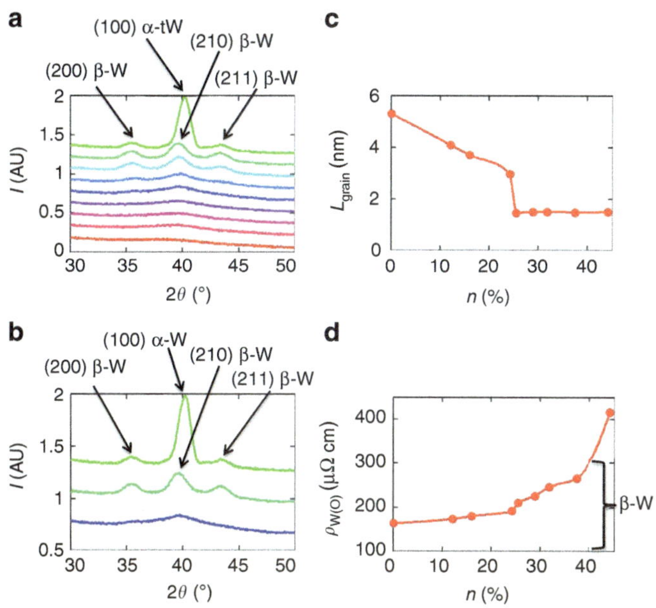

Figure 4 | Materials' characterization. The colour schemes used here for n, indicating the oxygen concentration, are the same as in Figs 2 and 3. (**a**) X-ray diffraction for W(O) films with different n. The crystallographic orientation and the corresponding phase (α or β) are indicated by arrows. (**b**) Detailed X-ray diffraction on films with three noteworthy oxygen concentrations: pure W, $n=12.1\%$ and $n=25.5\%$. There is a small shift in the peak at $2\theta=40°$ from $n=0\%$ to $n=12.1\%$, indicating a larger amount of $\beta-W$ for $n=12.1\%$. (**c**) Grain size L_{Grain} versus n. (**d**) Resistivity versus n. $\beta-W$ has a resistivity between 100 and 300 μΩ cm as indicated by the bracket.

We further investigated the dependence of the resistivity on the oxygen concentration in our W(O) films (Fig. 4d). The resistivity of Si substrate | SiO_x (25) | W(O) (6) | TaN (2) (with film thicknesses in parenthesis in nm) films were measured using a four-point measurement. The resistivity shows a continuous increase with an increasing oxygen content. However, we note that the resistivity does not increase significantly until the films become nanocrystalline. Moreover, there is a sharp increase in resistivity (at $n\sim37\%$). We speculate that this could arise from the formation of an oxygen network, which encloses metallic tungsten grains (causing the resistance to suddenly increase). The resistivity of $\beta-W$ is in the range of 100–300 μΩ cm (ref. 24), whereas $\alpha-W$ is 11.2 μΩ cm, which corroborates our X-ray diffraction measurements that indicate that the 6-nm-thick pure tungsten film contains $\beta-W$. We have also performed a resistance versus temperature study to investigate transport mechanisms underlying the observed effects (see Supplementary Note 4 and Supplementary Fig. 5)[26,36]. With an increasing oxygen content, 50-nm-thick films become less metallic, until for when n is greater than $\sim25.5\%$, the film resistance increases with a decreasing temperature. For the thinner films, the resistance increases with decreasing temperature for all films. For $n<31.7\%$, indications of superconductivity are observed below ~2 K that give strong evidence of the A15 phase that is known to exhibit a superconducting transition near this temperature.

Discussion

Our materials' characterization shows that the incorporation of oxygen stabilizes the $\beta-W$. This is consistent with previous studies, which have shown that the formation of $\beta-W$ is extremely sensitive to the oxygen incorporation during the growth process[37–41], and an oxygen concentration of only 10% is sufficient to stabilize $\beta-W$. In our experiments, at $n=12.1\%$, we obtain a SHA of $\theta_{LW}^{eff}=-0.49$, which is higher than any previous measurements in conventional metal-based systems.

With the addition of further oxygen-doping, we observe the formation of a nanocrystalline structure. Despite the large changes we have observed in the bulk microstructure (as evidenced by a grain size change by more than a factor of 2), and a resistivity change by a factor of 2, we see that the SHA still remains very large. Our observation of the independence of the SHA despite significant changes in the bulk properties, namely the oxygen concentration, the resistivity, the microstructure as well as thickness, suggests that the mechanism responsible for the observed highly efficient spin–orbit torque may originate at the W(O)/ferromagnet interface. Our observations are consistent with theoretical proposals of antidamping-like torques from interfacial spin–orbit coupling such as the Rashba effect[42–46]. Such interfacial spin–orbit torques have also been reported in the Pt | oxidized CoFeB system[47]. One important difference in our experiment is that we do not intentionally oxidize the CoFeB. Furthermore, W is a heavy metal with large spin–orbit coupling, which gives rise to a large Rashba effect in combination with a 3*d* ferromagnet, such as CoFeB (ref. 48). For the Gd(0001) surface, it has been shown that on oxidation the Rashba parameter was enhanced[49]. Such a phenomenon may also play a role in our experiments. It has been postulated that spin memory loss (SML) at the interface could reduce the magnitude of the spin current[50]. It is difficult to estimate the magnitude of any SML but from the weak dependence of G_{eff} on oxygen content (Supplementary Fig. 3), it therefore follows that the SML would also be weakly dependent on the oxygen content, and, therefore cannot account for the variation of effective SHA with oxygen content that we observe. In any case, any contribution from SML would simply mean that the values of the SHA that we report would be even larger. We have also examined the role of the inhomogeneous broadening on the SHA, and observe that it is uncorrelated to any changes we observe in the SHA (see Supplementary Note 5 and Supplementary Fig. 6).

We thus postulate that the first SHA peak at $n=12.1\%$ is caused by β-phase stabilization and the large SHA observed at the higher oxygen concentrations to arise from interfacial spin–orbit torques. It is still possible, however, that for $n=12.1\%$ there is already a considerably large interfacial spin–orbit torque that explains why we obtain larger values for $\beta-W$ compared with previous studies. Furthermore, we note that one alternative explanation for the large SHA at high oxygen content may arise from amorphization of the W(O) material (Fig. 4c) and the SHA could be enhanced because of extrinsic effects in analogy with the large anomalous Hall effects found in amorphous magnetic materials[51]. However, in this case, one would expect the SHA to increase with higher oxygen concentration and scale with the resistivity. We instead observe a nearly constant SHA towards very high oxygen concentrations and thus believe that it is more likely that interfacial spin–orbit torques account for our observations.

Our results illustrate an intriguing path towards enhancing the magnitude of spin–orbit torques, and also serve to bridge the link between oxide electronics with spintronics.

Methods

Sample growth and preparation. The films were sputtered in the presence of Argon gas on undoped oxidized silicon with (100) orientation. Two-inch-diameter targets were used with the target to substrate distance kept fixed at 12.5 cm. The base pressure before deposition was less than 10^{-8} torr and the pressure during deposition was 3 mTorr. The W(O) layers were formed by introducing oxygen gas into the Ar-sputtering gas atmosphere during the film deposition. The gas flow of oxygen Q was varied between 0 and 3%. The highly resistive TaN-capping layer was fabricated by introducing 50% Nitrogen into the Argon gas flow.

Device-patterning was performed with optical lithography and ion milling, followed by a 14-nm-thick AlO_x refill. Subsequently, 5 nm ruthenium and 65 nm gold were deposited and patterned for electrical contacts to the device.

References

1. Brataas, A. & Hals, K. M. D. Spin-orbit torques in action. *Nat. Nanotechnol.* **9**, 86–88 (2014).
2. Gambardella, P. & Miron, I. M. Current-induced spin-orbit torques. *Philos. Trans. R Soc. Lond. A Math. Phys. Eng. Sci* **369**, 3175–3197 (2011).
3. Miron, I. M. *et al.* Perpendicular switching of a single ferromagnetic layer induced by in-plane current injection. *Nature* **476**, 189–193 (2011).
4. Liu, L. *et al.* Spin-torque switching with the giant spin Hall effect of Tantalum. *Science* **336**, 555–558 (2012).
5. Hirsch, J. E. Spin Hall effect. *Phys. Rev. Lett.* **83**, 1834–1837 (1999).
6. Zhang, S. Spin hall effect in the presence of spin diffusion. *Phys. Rev. Lett.* **85**, 393–396 (2000).
7. Manchon, A. & Zhang, S. Theory of spin torque due to spin-orbit coupling. *Phys. Rev. B* **79**, 94422 (2009).
8. Manchon, A. & Zhang, S. Theory of nonequilibrium intrinsic spin torque in a single nanomagnet. *Phys. Rev. B* **78**, 212405 (2008).
9. Kato, Y. K., Myers, R. C., Gossard, A. C. & Awschalom, D. D. Observation of the spin Hall. *Science* **306**, 1910–1913 (2004).
10. Liu, L., Moriyama, T., Ralph, D. C. & Buhrman, R. a. Spin-torque ferromagnetic resonance induced by the spin Hall effect. *Phys. Rev. Lett.* **106**, 036601 (2011).
11. Pi, U. H. *et al.* Tilting of the spin orientation induced by Rashba effect in ferromagnetic metal layer. *Appl. Phys. Lett.* **97**, 2–5 (2010).
12. Ryu, K.-S., Thomas, L., Yang, S.-H. & Parkin, S. Chiral spin torque at magnetic domain walls. *Nat. Nanotechnol.* **8**, 527–533 (2013).
13. Khvalkovskiy, A. V. *et al.* Matching domain-wall configuration and spin-orbit torques for efficient domain-wall motion. *Phys. Rev. B* **87**, 020402 (2013).
14. Emori, S., Bauer, U., Ahn, S.-M., Martinez, E. & Beach, G. S. D. Current-driven dynamics of chiral ferromagnetic domain walls. *Nat. Mater.* **12**, 611–616 (2013).
15. Yang, S.-H., Ryu, K.-S. & Parkin, S. Domain-wall velocities of up to 750 m s − 1 driven by exchange-coupling torque in synthetic antiferromagnets. *Nat. Nanotechnol.* **10**, 221–226 (2015).
16. Miron, I. M. *et al.* Fast current-induced domain-wall motion controlled by the Rashba effect. *Nat. Mater.* **10**, 419–423 (2011).
17. Liu, L., Pai, C.-F., Ralph, D. C. & Buhrman, R. A. Magnetic oscillations driven by the spin Hall effect in 3-terminal magnetic tunnel junction devices. *Phys. Rev. Lett.* **109**, 186602 (2012).
18. Liu, L., Lee, O. J., Gudmundsen, T. J., Ralph, D. C. & Buhrman, R. a. Current-induced switching of perpendicularly magnetized magnetic layers using spin torque from the spin Hall effect. *Phys. Rev. Lett.* **109**, 096602 (2012).
19. Hoffmann, A. Spin Hall effects in metals. *IEEE Trans. Magn.* **49**, 5172–5193 (2013).
20. Parkin, S. S. P. Origin of enhanced magnetoresistance of magnetic multilayers: spin-dependent scattering from magnetic interface states. *Phys. Rev. Lett.* **71**, 1641–1644 (1993).
21. Dresselhaus, G. Spin-orbit coupling effects in zinc blende structures. *Phys. Rev.* **100**, 580–586 (1955).
22. Bychkov, Y. A. & Rashba, E. I. Properties of a 2D electron gas with lifted spectral degeneracy. *JETP Lett.* **39**, 78–81 (1984).
23. Edelstein, V. M. Spin polarization of conduction electrons induced by electric current in two-dimensional asymmetric electron systems. *Solid State Commun.* **73**, 233–235 (1990).
24. Pai, C.-F. *et al.* Spin transfer torque devices utilizing the giant spin Hall effect of tungsten. *Appl. Phys. Lett.* **101**, 122404 (2012).
25. Hao, Q. & Xiao, G. Giant spin Hall effect and switching induced by spin-transfer torque in a $W/Co_{40}Fe_{40}B_{20}/MgO$ structure with perpendicular magnetic anisotropy. *Phys. Rev. Appl.* **3**, 034009 (2015).
26. Hao, Q., Chen, W. & Xiao, G. Beta (β) tungsten thin films: structure, electron transport, and giant spin Hall effect. *Appl. Phys. Lett.* **106**, 182403 (2015).
27. Shen, Y. G. *et al.* Residual stress, microstructure, and structure of tungsten thin films deposited by magnetron sputtering. *J. Appl. Phys.* **87**, 177 (2000).
28. Nakayama, H. *et al.* Spin Hall magnetoresistance induced by a nonequilibrium proximity effect. *Phys. Rev. Lett.* **110**, 1–5 (2013).
29. Zhang, W., Han, W., Jiang, X., Yang, S.-H. & Parkin, S. S. P. Role of transparency of platinum–ferromagnet interfaces in determining the intrinsic magnitude of the spin Hall effect. *Nat. Phys.* **11**, 496–502 (2015).
30. Tserkovnyak, Y., Brataas, A. & Bauer, G. E. W. Enhanced gilbert damping in thin ferromagnetic films. *Phys. Rev. Lett.* **88**, 117601 (2002).
31. Saitoh, E., Ueda, M., Miyajima, H. & Tatara, G. Conversion of spin current into charge current at room temperature: inverse spin-Hall effect. *Appl. Phys. Lett.* **88**, 182509 (2006).
32. Ando, K. *et al.* electric manipulation of spin relaxation using the spin Hall effect. *Phys. Rev. Lett.* **101**, 36601 (2008).
33. Nan, T. *et al.* Comparison of spin-orbit torques and spin pumping across NiFe/Pt and NiFe/Cu/Pt interfaces. *Phys. Rev. B* **91**, 214416 (2015).
34. Mellnik, A. R. *et al.* Spin-transfer torque generated by a topological insulator. *Nature* **511**, 449–451 (2014).
35. Wang, Y., Deorani, P., Qiu, X., Kwon, J. H. & Yang, H. Determination of intrinsic spin Hall angle in Pt. *Appl. Phys. Lett.* **105**, 152412 (2014).
36. Bond, W. L. *et al.* Superconductivity in films of tungsten and other transition metals. *Phys. Rev. Lett.* **15**, 260–261 (1965).
37. O'Keefe, M. J., Grant, J. T. & Solomon, J. S. Magnetron sputter deposition of A-15 and bcc crystal structure tungsten thin films. *J. Electron. Mater.* **24**, 961–967 (1995).
38. Weerasekera, I. a., Shah, S. I., Baxter, D. V. & Unruh, K. M. Structure and stability of sputter deposited beta-tungsten thin films. *Appl. Phys. Lett.* **64**, 3231–3233 (1994).
39. Narasimham, A. J. *et al.* Fabrication of 5-20 nm thick β-W films. *AIP Adv.* **4**, 117139 (2014).
40. Petroff, P. Microstructure, growth, resistivity, and stresses in thin tungsten films deposited by rf sputtering. *J. Appl. Phys.* **44**, 2545 (1973).
41. Shen, Y. & Mai, Y. Influences of oxygen on the formation and stability of A15 β-W thin films. *Mater. Sci. Eng. A* **284**, 176–183 (2000).
42. Wang, X. & Manchon, A. Diffusive spin dynamics in ferromagnetic thin films with a Rashba interaction. *Phys. Rev. Lett.* **108**, 117201 (2012).
43. Kim, K.-W., Seo, S.-M., Ryu, J., Lee, K.-J. & Lee, H.-W. Magnetization dynamics induced by in-plane currents in ultrathin magnetic nanostructures with Rashba spin-orbit coupling. *Phys. Rev. B* **85**, 180404 (2012).
44. Pesin, D. A. & MacDonald, A. H. Quantum kinetic theory of current-induced torques in Rashba ferromagnets. *Phys. Rev. B* **86**, 014416 (2012).
45. van der Bijl, E. & Duine, R. A. Current-induced torques in textured Rashba ferromagnets. *Phys. Rev. B* **86**, 094406 (2012).
46. Haney, P. M., Lee, H. W., Lee, K. J., Manchon, A. & Stiles, M. D. Current induced torques and interfacial spin-orbit coupling: semiclassical modeling. *Phys. Rev. B.* **87**, 1–13 (2013).
47. Qiu, X. *et al.* Spin–orbit-torque engineering via oxygen manipulation. *Nat. Nanotechnol.* **10**, 333–338 (2015).
48. Park, J.-H., Kim, C. H., Lee, H.-W. & Han, J. H. Orbital chirality and Rashba interaction in magnetic bands. *Phys. Rev. B* **87**, 041301 (2013).
49. Krupin, O. *et al.* Rashba effect at magnetic metal surfaces. *Phys. Rev. B* **71**, 201403 (2005).
50. Rojas-Sánchez, J.-C. *et al.* Spin pumping and inverse spin Hall effect in platinum: the essential role of spin-memory loss at metallic interfaces. *Phys. Rev. Lett.* **112**, 106602 (2014).
51. Coey, J. M. D. *Magnetism and Magnetic Materials* (Cambridge University Press, 2010).

Acknowledgements

We thank T. Topuria and E.A. Delenia for transmission electron microscope data and analysis. We also acknowledge fruitful discussions with M. Ali. The internship of K.-U.D. was financed by the PROMOS programme of the German Academic Exchange Service (DAAD).

Author contributions

K.-U.D., T.P. and S.S.P.P. conceived the experiments. K.-U.D., T.P. and W.Z. performed the experiments and analysed the data. B.P.H. fabricated the samples. S.-H.Y. grew the films. A.K. performed the Rutherford backscattering spectrometry studies. K.-U.D., T.P. and S.S.P.P. wrote the manuscript. S.S.P.P. supervised the project. All authors discussed the results and implications.

Additional information

Competing financial interests: The authors declare no competing financial interests.

4

Manipulating the interfacial structure of nanomaterials to achieve a unique combination of strength and ductility

Amirhossein Khalajhedayati[1], Zhiliang Pan[2] & Timothy J. Rupert[1,2]

The control of interfaces in engineered nanostructured materials has met limited success compared with that which has evolved in natural materials, where hierarchical structures with distinct interfacial states are often found. Such interface control could mitigate common limitations of engineering nanomaterials. For example, nanostructured metals exhibit extremely high strength, but this benefit comes at the expense of other important properties like ductility. Here, we report a technique for combining nanostructuring with recent advances capable of tuning interface structure, a complementary materials design strategy that allows for unprecedented property combinations. Copper-based alloys with both grain sizes in the nanometre range and distinct grain boundary structural features are created, using segregating dopants and a processing route that favours the formation of amorphous intergranular films. The mechanical behaviour of these alloys shows that the trade-off between strength and ductility typically observed for metallic materials is successfully avoided here.

[1] Department of Chemical Engineering and Materials Science, University of California, Irvine, California 92697, USA. [2] Department of Mechanical and Aerospace Engineering, University of California, 4200 Engineering Gateway, Irvine, California 92697, USA. Correspondence and requests for materials should be addressed to T.J.R. (email: trupert@uci.edu).

The vast majority of engineering materials are polycrystalline and tailoring the interfaces, or grain boundaries, between the many crystals within a given sample volume offers a promising approach for the discovery of new materials with improved properties. Two recent strategies for realizing such gains have been (i) increasing the number of interfaces or effectively the volume fraction of material located in the interfacial regions and (ii) tailoring the structure of grain boundaries by changing interfacial chemistry. The first strategy is characterized by a dramatic reduction in the size of the crystalline grains that comprise the material and can be termed 'nanostructuring' when this reduction brings the average grain size below ~100 nm. The second strategy uses segregating dopant atoms to tailor the local equilibrium structure of interfaces and can be termed 'complexion engineering'[1]. Although nanostructuring and complexion engineering each offer opportunities on their own, these two material design approaches have not yet been combined in a complementary manner. Natural materials, such as dental enamel[2] and nacre[3], often have evolved to contain such hierarchical nanostructures with distinct interfacial features that serve to dramatically influence their performance. Such a strategy applied to engineering materials should allow for the realization of unique property sets, where a material has the benefits of a nanostructured material with regards to one property yet avoids the usual degradation of other properties through complexion engineering.

Nanostructuring has been used extensively in recent years to create a wide range of nanocrystalline metals and ceramics[4,5]. These materials take advantage of the fact that atoms in the grain boundaries behave differently than atoms in the grain interior, because of lower packing fractions and the lack of long-range crystalline order. For example, grain boundaries resist the transmission of plastic strain in the form of moving dislocations, as these linear defects in crystals cannot glide through the interfacial region easily. Therefore, a large reduction in grain size is equivalent to adding more obstacles to dislocation motion, leading to a dramatic increase in the material's strength following a well-known scaling law known as Hall–Petch scaling[6,7]. Wang *et al.*[8] reported on nanocrystalline Cu with an average grain size (d) of 30 nm and a yield strength of 760 MPa, or more than ten times higher than that of Cu with an average grain size in the micrometre range (55 MPa)[9]. Unfortunately, the addition of a large volume fraction of interfaces can have negative consequences as well. For instance, these same nanocrystalline metals are almost always brittle[9,10], with very little global plastic strain before failure, negating the value of their high strength in many cases.

Separately, it has been recognized that a grain boundary can be described as a type of phase-like structure in many cases, although these interfacial features differ from traditional bulk phases because their structure is dependent on the two abutting crystals and are therefore known as complexions[1,11]. However, similar to bulk phases, interface complexions can be analysed with equilibrium thermodynamic theories and may transform between different structures as temperature or chemical composition varies[1,12]. Complexions are difficult to observe in materials made from a single element[13], but recent research has shown that distinct interfacial structures become more accessible in alloys[14,15]. Dillon *et al.* created a categorization scheme based on transmission electron microscopy (TEM) observations in doped Al_2O_3 that relies on complexion thickness[11]. Six discrete complexion types were found, with increasing levels of structural disorder, which are categorized as (i) single dopant layers, (ii) nominally 'clean' grain boundaries, (iii) bilayers, (iv) multilayers, (v) intergranular films with equilibrium thicknesses in the nanometre range and (vi) wetting films with arbitrary thicknesses. Complexion types V and VI are often disordered films without any long-range crystalline order, in which case they can both be classified as amorphous intergranular films (AIFs). An AIF is an example of a complexion that is both distinct in its local structure as well as its effect on material properties, being suggested as the cause for the previously unexplained phenomenon of activated sintering in refractory metals[14].

To date, complexions have been studied and utilized predominantly in either coarse-grained polycrystalline materials or simple bicrystal samples. Here, we report on a strategy for combining the concepts of nanostructuring and complexion engineering, to enable a new class of materials: nanomaterials with tunable grain boundary structure. We begin by developing a processing route for making such materials, with an eye on techniques that are flexible enough to create a wide variety of chemistries while also scalable so that they can eventually be used to produce bulk quantities of material. Since AIFs are a complexion type that will have atomic structure and properties which are unique compared with normal, clean grain boundaries, we focus on inducing and characterizing these features in a nanoscale grain structure. Then, the mechanical properties of the alloy are probed, to demonstrate the clear advantage of this new materials design strategy. Specifically, we find that the ductility and toughness of nanostructured metals can be greatly improved using complexion engineering without sacrificing any strength, breaking the paradigm of a direct strength-ductility trade-off that has dominated prior observations.

Results

Materials selection, processing and characterization. Cu-based alloys were selected for this investigation, as Cu is a common face-centred cubic material and our findings should be broadly applicable to many other structural metals. In addition, the deformation physics and properties of pure nanocrystalline Cu have been well-studied, giving reference points to compare against[16]. As doping of the interfacial regions is required to induce complexions, alloying elements that segregate to the grain boundaries are sought. Empirical considerations such as large atomic size mismatch and limited solubility in the bulk can give rough guidelines for dopant selection, whereas grain boundary segregation enthalpy calculations give a more direct prediction of segregation tendency[17]. As this study is concerned with mechanical behaviour, embrittlement is to be avoided and there is a need to preserve metallic bonding at the interfaces. Non-metals or poor metals such as Bi and Pb are eliminated from consideration. Finally, thermodynamic modelling tools developed by Luo and co-workers[12,18] suggest that AIFs will form if the total energy of an amorphous layer with finite thickness and two new amorphous-crystalline interfaces (that is, the entire complexion) is lower than the energy associated with the interface between two crystalline phases (that is, the original grain boundary). This means that alloy combinations that have a low volumetric free energy penalty for a liquid-like structure and can easily form amorphous solids are promising for the formation of AIFs. With all of these considerations in mind, Zr was chosen as our alloying element, as it will segregate to the grain boundaries[17], is a transition metal and binary Cu-Zr alloys have a high glass-forming ability[19].

Mechanical alloying with a high-energy ball mill produces powders with particle sizes of micrometre-scale diameter, with each particle containing many individual nanometre-scale grains. This technique can make nanostructured powders with a wide variety of chemical compositions and scale-up to bulk parts through powder metallurgy techniques is a possibility[20]. As a control, unalloyed nanocrystalline Cu with $d = 30$ nm was first produced, as shown in Fig. 1a,d. The selected area electron diffraction pattern in Fig. 1d shows that the material is a random polycrystalline sample. A Cu-3 atomic % Zr alloy was then created, as full segregation of this amount of Zr to the interfaces will put the grain boundary composition into the glass-forming range for Cu-Zr metallic glasses[21]. In the as-milled state, the alloy sample also has $d = 30$ nm (Fig. 1e). The bottom frame of Fig. 1e presents a compositional line scan through the as-milled Cu-Zr alloy, showing that Zr is found throughout the sample, mixed into both crystal interior and grain boundary regions (shown schematically in Fig. 1b). Three grain boundaries are labelled in this figure with black arrows. A small amount of grain boundary segregation is occasionally found even in the as-milled sample, as shown by the slightly elevated Zr concentration at the first boundary noted, but significant amounts of Zr were inside of the grains as well. As Zr has negligible solubility (~0.12 atomic % (ref. 22)) in the Cu lattice according to the bulk phase diagram, this structure is a supersaturated solid solution. To induce segregation, the powders were annealed at 950 °C for 1 h. This temperature is extremely high for Cu and Cu-Zr alloys, being ~90% of the melting temperature of pure Cu and ~98% of the solidus temperature (where the material begins to melt) of Cu-3 atomic % Zr[22]. During annealing, Zr diffuses to the grain boundaries, as shown by the compositional line scan in Fig. 1f, where two obvious examples of grain boundaries are labelled. Grain interiors are depleted of Zr after annealing, whereas the interfacial regions show intense Zr enrichment (shown schematically in Fig. 1c). Additional evidence of Zr grain boundary segregation was found by using high-angle annular dark-field TEM imaging (see Supplementary Fig. 1 for details). The Cu-Zr alloy experiences very little grain growth, only coarsening to an average grain size of 45 nm (Fig. 1f), because of a reduction in grain boundary energy with Zr segregation[23], whereas the grain size of a pure Cu sample annealed under the same conditions coarsens to the micrometre range. Although not the explicit focus of this study, the observed thermal stability at such a high temperature suggests that this material can be consolidated into bulk pieces[20].

The high-temperature annealing treatment used to induce Zr segregation is also useful for promoting AIF formation. Shi and Luo[12] recently developed interfacial thermodynamic models and grain boundary diagrams, showing that higher temperatures usually promote thicker AIFs. A set of annealed Cu-Zr powders were quickly quenched by dropping into a large water bath in less than 1 s, freezing in any structures which are in equilibrium at 950 °C. Fresnel fringe imaging[24] was used to identify interfacial films, followed by high-resolution TEM for detailed characterization of grain boundary structure and measurement of AIF thickness. A representative example of an AIF is presented in Fig. 2a. The areas in the bottom left and top right of Fig. 2a are crystalline, as shown by the presence of lattice fringes in the image as well as sharp spots in the fast Fourier transform patterns, which denote periodic order associated with the lattice. In contrast, the region at the interface, between the two dashed lines, is amorphous and disordered with a thickness of 5.7 nm. The fast Fourier transform pattern shows no sign of long-range crystalline order in this case and is completely featureless. An estimation of the diffusion length for this system shows that these

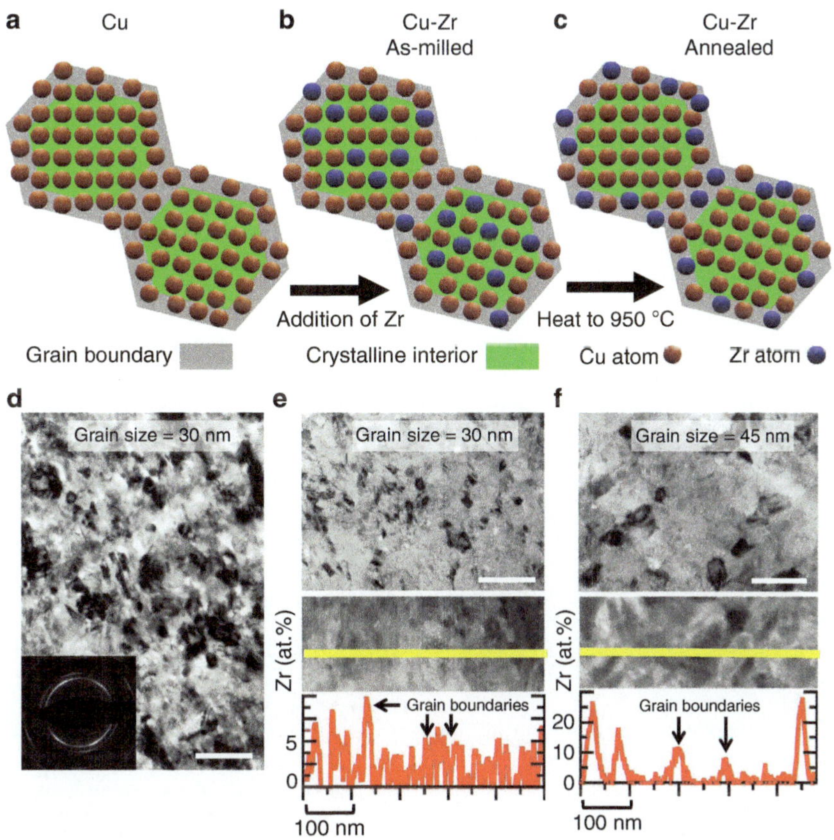

Figure 1 | Alloy design strategy for adding segregating dopants. Pure Cu (**a**) can be converted to an alloy by adding Zr during ball milling (**b**). Although the Zr is mixed throughout the grain structure, annealing treatments can then be used to induce preferential Zr segregation to the grain boundaries (**c**). Both the (**d**) pure Cu and (**e**) Cu-Zr as-milled samples have an average grain size of 30 nm (scale bars, 100 nm). (**f**) Annealing the Cu-Zr sample at 950 °C for 1 h allows for segregation and only causes coarsening to a grain size of 45 nm (scale bar, 100 nm). The same level of Zr segregation was found in both the quenched and air-cooled samples.

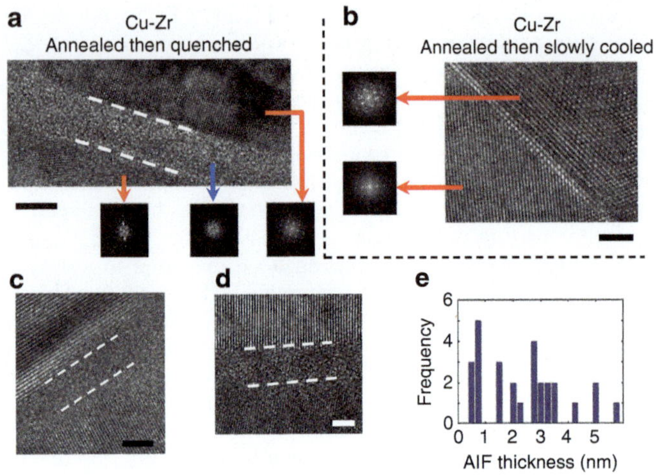

Figure 2 | High-resolution TEM images of grain boundary structure in nanocrystalline Cu-Zr alloys. (**a**) An amorphous intergranular film with thickness of 5.7 nm was observed at a grain boundary after quickly quenching from 950 °C (scale bar, 5 nm). (**b**) In contrast, grain boundaries in a slowly cooled sample, with structures that are in equilibrium near ambient temperatures, are all ordered interfaces (scale bar, 2 nm). Insets are fast Fourier transform patterns, highlighting the disordered nature of the interface in **a**. (**c**,**d**) Additional examples of amorphous complexions in the quenched sample (scale bars, 2 nm), whereas **e** summarizes the measurements from the 28 interfacial films found here.

AIFs cannot be a metastable phase, as the high temperature used for annealing provides more than enough kinetic driving force for equilibrium chemical distributions to form within the nanocrystalline grain structure (see Supplementary Note 1 for detailed calculations).

To ensure that the observed AIFs were truly complexions, driven by alloy thermodynamics with a combination of grain boundary dopants and elevated temperatures required, and did not form through another process such as solid-state amorphization, which is purely driven by local composition, another set of powders was annealed at 950 °C and then slowly cooled back to room temperature over a period of ~5 min, or ~300 times longer than the quenched sample. Such a cooling schedule gives a sample with interfaces that are in equilibrium at ambient temperatures. No AIFs were observed in this specimen and all interfaces studied were ordered, with a representative example shown in Fig. 2b. Crystalline order is present all of the way up to the boundary. Zr dopants were still found to have segregated to the grain boundaries, with compositional line scans appearing nominally identical to the one shown in Fig. 1f. Although the boundaries were still heavily doped with Zr, the lack of AIFs proves that the amorphous structure shown in Fig. 2a is only in equilibrium at high temperature, in line with thermodynamic theories of disordered complexions.

Additional examples of AIFs from the quenched Cu-Zr sample are shown in Fig. 2c,d with thicknesses of 2.2 and 2.7 nm, respectively. Twenty-eight AIFs were studied with high-resolution TEM, with a distribution of measured AIF thicknesses shown in Fig. 2e. In addition, a number of ordered grain boundaries were observed in this sample as well. As a whole, this means that not all grain boundaries have the same structure in the quenched Cu-Zr alloy, but rather there is a distribution of complexion types. Possible explanations for these variations include differences in grain boundary energy, local segregation state or slight differences in local cooling rates during quenching. The existence of amorphous structures below the solidus temperature of the alloy suggests that these features are stable AIFs with thicknesses determined by a combination of local grain boundary chemical composition and high temperature (type V), although some of the thicker AIFs could possibly be wetting films (type VI). Even though the annealing temperature was below the solidus temperature for Cu-3 atomic % Zr, it was above the eutectic temperature, the lowest possible melting temperature over all mixing ratios, which occurs at higher Zr concentrations. It is possible that some boundaries have higher levels of Zr segregation and move into a different region of the phase diagram where wetting films can form at 950 °C. However, previous observations of complexions in Al_2O_3 showed that film thickness changes significantly along individual wetting films[25], whereas each individual boundary studied here showed no measureable variation in thickness. In addition, wetting films are often much thicker (at least 10 nm thick) than the films observed here. Even if there is some uncertainty in the complexion type of the thickest AIFs, the existence of many nanometre-thick amorphous interfacial structures is clear.

Mechanical behaviour and the effect of complexions. To demonstrate the potential utility of the material described in this report, a nanostructured metal with disordered amorphous grain boundaries, we turn our attention to mechanical behaviour. Nanocrystalline metals have very high strengths, but are generally also very brittle and exhibit little ductility, with elongations or strain-to-failure values of only a few percent typically reported[9,10]. This brittle nature often restricts the usage of nanocrystalline materials in technological applications. Prior attempts to increase the ductility of nanocrystalline metals, for example, by encouraging mechanically driven grain coarsening[26] or by inducing superplasticity due to sulfur contamination[27,28] or inclusion of a phase with low melting temperature[29], have come with dramatic reductions in strength. The high strength of a nanocrystalline material comes from the fact that traditional dislocation plasticity mechanisms are shut off at these extremely fine crystallite sizes, requiring dislocations to be nucleated at an interface, propagate across the nanograin and then be absorbed at the opposite grain boundary. The repeated dislocation absorption required for such a physical mechanism may be the cause of the brittle nature of nanocrystalline metals, as atomistic simulations show that a single absorption event leads to high local stresses at the grain boundary[30], whereas multiple absorption events result in crack nucleation[31]. Such a hypothesis is supported by *in situ* TEM deformation studies, which show that nanometre size cracks nucleate at grain boundary sites during plastic deformation[32]. As a result, damage tolerant grain boundaries should be the key to designing ductile nanocrystalline metals, making control of grain boundary structure a priority.

Small cylindrical specimens were created in the powder particles using a focused ion beam microscope, then tested in compression to quantify strength and in bending to quantify strain-to-failure. Representative examples from each test for pure nanocrystalline Cu are shown in Fig. 3a–c, demonstrating the typical response of a nanocrystalline metal to which we can then compare our Cu-Zr results. Figure 3a shows that pure Cu has a high-yield strength of 740 MPa but the pillar cracks and fails during compression, resembling the failure of a brittle ceramic. Compression experiments are not able to quantify the ductility of a material, which is defined as the ability to deform under tensile stress, so pillar bending experiments were used to measure plastic strain-to-failure (see Supplementary Methods for detailed experimental techniques). Videos of representative beam bending experiments for pure Cu and Cu-Zr with AIFs are shown in Supplementary Movies 1 and 2, respectively. Figure 3b,c shows

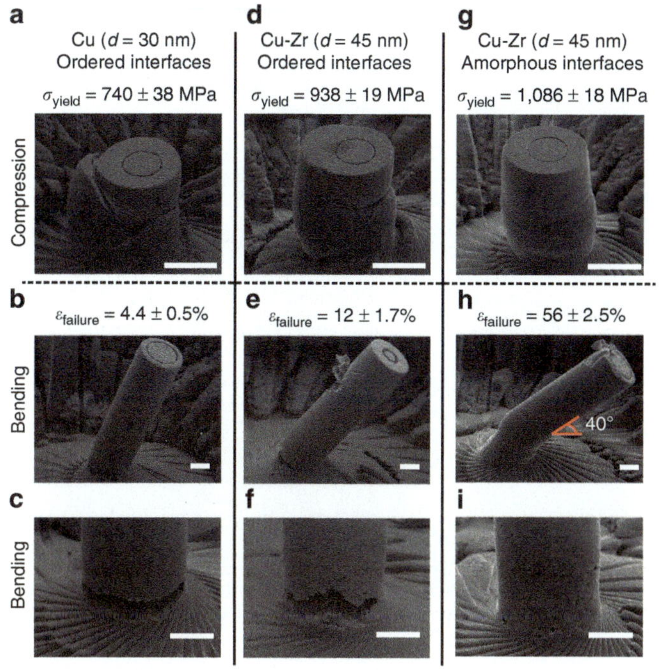

Figure 3 | Mechanical testing results from nanocrystalline Cu and Cu-Zr samples. Failed micropillars after compression, as well as yield strength measurements, are shown for **(a)** pure Cu with ordered interfaces, **(d)** Cu-Zr with ordered interfaces and **(g)** Cu-Zr with amorphous intergranular films (scale bars, 5 μm). Failed micropillars after bending, as well as strain-to-failure measurements, are shown for **(b,c)** pure Cu with ordered interfaces, **(e,f)** Cu-Zr with ordered interfaces and **(h,i)** Cu-Zr with amorphous intergranular films (scale bars, 2 μm). A nanocrystalline alloy with AIFs can be both stronger and more ductile than its traditional, pure metal counterpart.

this same brittle response during bending, where a crack forms early during the experiment and there is little plastic strain built up at the base of the pillar, from which strain-to-failure can be measured as 4.4%. This low value is nearly identical to a previous report for nanocrystalline Cu, produced by a different method and tested in uniaxial tension, that had the same average grain size (30 nm) and a similar yield strength (760 MPa)[8]. Representative examples for each test from our Cu-Zr alloy with AIFs are shown in Fig. 3g–i, providing evidence of a strong yet ductile response. Figure 3g shows that the Cu-Zr with AIFs is even stronger than the pure Cu, having a yield strength of 1,086 MPa, but more importantly the pillar compresses in a stable, homogeneous manner. Figure 3h shows that the pillar is ductile during the bending experiment as well and a strain-to-failure of 56% was measured for these alloys, more than ten times larger than the pure Cu value. A magnified view of the pillar base is presented in Fig. 3i, showing that extensive plasticity has occurred and that, even at this large plastic strain, there are only small, unconnected cracks instead of a crack completely across the surface, suggesting that this strain-to-failure value may even be slightly conservative for this alloy. To highlight the importance of AIFs for the observed behaviour, mechanical tests were also performed on the Cu-Zr sample that was slowly cooled in order to have ordered grain boundaries. These results, presented in Fig. 3d–f, demonstrate that this alloy is brittle as well, mimicking the pure Cu sample by crumbling in compression and failing at a small strain in bending. As the only difference between the Cu-Zr samples is the grain boundary structure, with grain size and Zr segregation state identical for both samples, the increased ductility of the Cu-Zr with AIFs can be attributed to the addition of amorphous interfacial films alone.

Discussion

To understand why AIFs act as damage tolerant grain boundaries, molecular dynamics simulations of repeated dislocation absorption at room temperature were carried out on ordered interfaces in both pure Cu and Cu-Zr, as well as Cu-Zr with AIFs. A hybrid Monte Carlo/molecular dynamics simulation technique[33] was used to create these structures and ensure that they were in equilibrium. For the ordered interfaces in pure Cu, shown in Fig. 4a, very few dislocations can be absorbed before a crack nucleates, leading to cracking at a low applied shear strain (γ). The crack then grows quickly after a small additional strain increment until it runs through the specimen. The ordered interfaces in Cu-Zr behaved the same way (see Supplementary Fig. 5 for details). On the other hand, the AIFs diffuse the strain concentration brought by dislocation absorption into a wider region within the boundary (Fig. 4b), delaying the crack nucleation event until a much larger applied shear strain is reached. Pan and Rupert studied the mechanics of this problem systematically in a recent paper, finding that crack nucleation as well as crack growth rate is suppressed as AIF thickness increases[34]. Brandl *et al.* even showed that an AIF can exert attractive forces on nearby dislocations as it is elastically softer than the crystalline lattice, pulling dislocations towards the interface where they can then be absorbed[35]. Experiments on crystalline-amorphous nanocomposite films support this concept as well, as the addition of a thin amorphous layer was found to improve strain-to-failure when compared with films with only a crystalline phase[36–38]. Although the nanocomposites outperform the crystalline films, the strain-to-failure values were only measured to be in the range of 4–14% (much lower than the values reported here), likely because the AIFs were only added in one direction of the film. Donohue *et al.*[39] reported on another nanolaminate comprised of 90-nm-thick layers of crystalline Cu separated by amorphous Pd-Si layers of 10 nm thickness. Although the monolithic amorphous films had negligible ductility by themselves, the nanolaminate structure had a strain-to-failure of 3%. In contrast, our materials have a fully three-dimensional grain boundary network decorated with AIFs. In our case, by adding amorphous interfacial features, it is possible to have the high strength characteristic of a nanocrystalline metal without the typical brittle failure (Fig. 4c). In fact, the strain-to-failure and plastic flow of a nanocrystalline metal with AIFs (Fig. 4d) resembles that of a coarse-grained or single crystalline metal (Fig. 4e).

The observation of high strength with large ductility fits a recent narrative that nanocrystalline materials are intrinsically ductile on the nanoscale[40], but suffer from premature failure due to strain localization or cracking[41]. If this premature failure can be suppressed, then the traditional trade-off between strength and ductility can be avoided. One recently developed path for suppression of this premature failure is to incorporate nanocrystalline metals as the outer layer in a gradient architecture, where grain size changes continuously from nanometres to micrometres to avoid early plastic necking[42,43]. The results presented here for Cu-Zr alloys created by a unique processing route can be interpreted as another example of this design strategy, with an explicit focus on resisting crack nucleation and growth through the introduction of grain boundary complexions. A strength-ductility synergy is achieved and our Cu-Zr alloys outperform traditional metals. A compilation of data from the literature for Cu and Cu alloys is presented in Fig. 5, where strain-to-failure is plotted as a function of yield strength. The properties

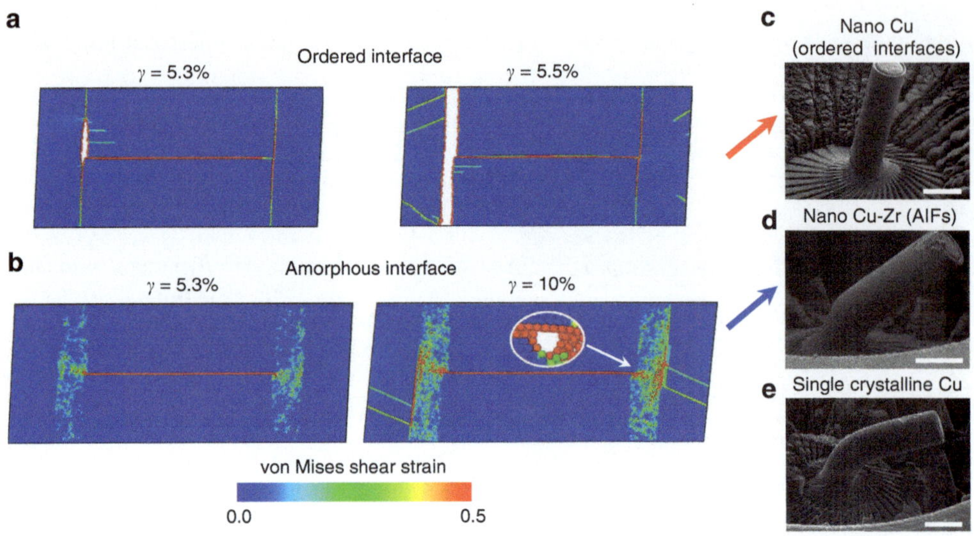

Figure 4 | Connection between interfacial structure and damage tolerance. Molecular dynamics simulations of dislocation absorption at (**a**) an ordered grain boundary and (**b**) an amorphous intergranular film show the formation and propagation of crack damage. The ordered interface quickly fractures, whereas the amorphous interface diffuses the strain concentration brought by dislocation absorption and fracture is delayed. The delay of failure explains why (**d**) nanocrystalline Cu-Zr with AIFs has ductility reminiscent of (**e**) coarse-grained Cu while retaining the high strength of (**c**) nanocrystalline Cu (scale bars, 5 μm).

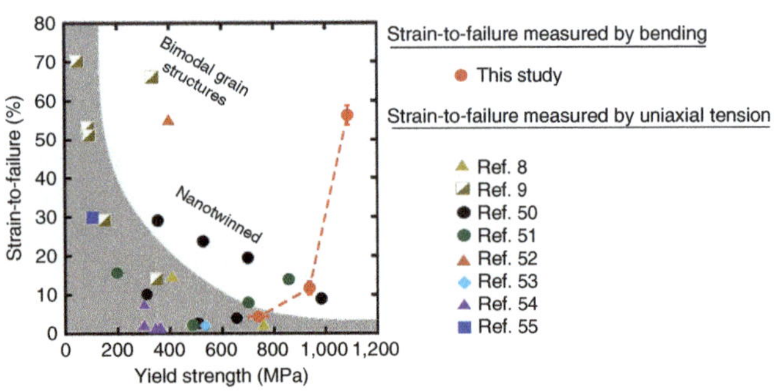

Figure 5 | Strain-to-failure and yield strength for Cu and Cu-based alloys. The vast majority of data fall within the grey envelope, with recent advanced alloys pushing slightly outside this limit. Our Cu-Zr alloy breaks the expected trend, with both higher strength and ductility than the pure Cu sample. All strain-to-failure data comes from material that failed under tensile stresses, either under uniaxial tension (literature data) or from the tensile side of a bending experiment (this study). Literature data are taken from refs 8,9,50–55.

of traditional Cu, both microcrystalline and nanocrystalline, fall within the grey envelope, demonstrating a clear trade-off between strength and ductility. Advanced alloys, such as Cu with a bimodal grain structure (micrometre-scale grains embedded in a matrix of nanocrystalline grains) and nanotwinned Cu, push beyond these limits and demonstrate improved combinations of these properties. The Cu and Cu-Zr alloys described in this paper are shown as red points on the plot. It is important to note that our strength measurements come from microcompression tests and strain-to-failure is taken from bending experiments, whereas the data from the literature comes from tensile experiments. The pure Cu sample exhibits classical behaviour and falls within the grey envelope. The strength and strain-to-failure measurements for our pure nanocrystalline Cu are remarkably similar to those of Wang *et al.*[8], taken from nanocrystalline Cu with an identical grain size created by surface mechanical attrition and appearing as a yellow triangle in Fig. 5. The close agreement between these values demonstrates that it is reasonable to compare our micropillar experiments to the uniaxial tension results available in the literature. Our Cu-Zr with AIFs breaks the traditional paradigm, being both stronger and more ductile than the pure Cu sample. Although the improved ductility has already been discussed, the increased strength comes from the segregation of Zr to the interfaces. Vo *et al.*[44] used atomistic simulation techniques to study how interfacial segregation altered the grain boundary energy and strength of nanocrystalline Cu. These authors found that boundary energy and strength were intimately related, with strength increasing monotonically as grain boundary energy was reduced by doping. Physically, this means reducing grain boundary energy makes it harder for dislocations to nucleate and then propagate through the grain (the deformation mechanisms, which have been tied to yield strength[45,46]). In our case, Zr segregation to grain boundary sites reduces grain boundary energy, which is the inherent driving force for segregation in general but is also supported by the increased thermal stability of Cu-Zr compared with the pure Cu sample. There is even a small increase in strength when going from Cu-Zr with ordered boundaries to Cu-Zr with AIFs. Complexion formation is fundamentally driven by the material seeking a reduction in total interface energy, so this should provide a strengthening contribution. Dislocation nucleation may also be affected by the shift from a crystalline–crystalline interface in the ordered grain boundary to an amorphous–crystalline interface in the case of the AIF. As a whole, the concurrent increase in both strength and ductility shows that control of grain boundary structure is a promising pathway for the design of optimized structural materials.

Our study shows that the previously separate techniques of nanostructuring and complexion engineering can be combined in a complementary manner, introducing a new feature into the design toolbox of materials scientists: grain boundary structure. Unique Cu-Zr alloys are reported, created through powder metallurgy techniques, with AIFs 'frozen' into the interfaces as a result of Zr segregation and rapid quenching from high-temperature ageing treatments. These new materials demonstrate a combination of strength and ductility that is not found in traditional Cu alloys, behaviour that can be attributed directly to the fracture resistance of AIFs. However, it is important to note that mechanical behaviour is one of many places where the materials described here may be useful. Even if only focusing on nanocrystalline metals with amorphous complexions, we envision materials that can be rapidly consolidated because of faster diffusion along AIFs or that are more tolerant to radiation damage with AIFs acting as sinks for residual point defects[47]. Amorphous interfaces could also act as fast transport paths for energy or electronic applications. We suggest that the introduction of nanocrystalline materials with designed grain boundary structures will dramatically broaden the suite of material properties that can be achieved.

Methods

Powder processing and characterization. Nanostructured metals were created by ball milling high-purity powders of the constituent elements with a hardened steel vial and milling media in an argon environment. Powders were then encapsulated in quartz tubes under vacuum and annealed to induce segregation of dopants, either followed by slow air cooling or fast quenching in a water bath. Milling in argon and annealing under vacuum was done to avoid oxidation of the powders. The phase content of the powders was measured using X-ray diffraction, whereas grain size and detailed grain boundary structure were characterized with TEM. The yield strength of each alloy was measured by making micrometre-sized pillars in each powder, followed by compression with an Agilent G200 nanoindenter using a flat triangular diamond tip and an engineering strain rate of $3.1 \times 10^{-4}\,s^{-1}$. Engineering stress–strain curves were calculated and are presented in Supplementary Fig. 2. Yield strengths were calculated based on a 0.7% plastic strain offset, following the work of Brandstetter *et al.*[48] Strain-to-failure was calculated from another set of micropillars that were bent with a micro-manipulator inside of a scanning electron microscope. The process for estimating strain-to-failure is illustrated in detail in Supplementary Fig. 3.

Atomistic modelling. Classical molecular dynamics simulations were performed using the Large-scale Atomic/Molecular Massively Parallel Simulator (LAMMPS) package[49]. The starting configurations for the atomistic simulations were created using a hybrid Monte Carlo/molecular dynamics method capable of finding equilibrium chemical composition and structural configuration. Further details of the processing, mechanical testing and simulation methods can be found in the Supplementary Methods section.

References

1. Cantwell, P. R. *et al.* Grain boundary complexions. *Acta Mater.* **62,** 1–48 (2014).
2. Gordon, L. M. *et al.* Amorphous intergranular phases control the properties of rodent tooth enamel. *Science* **347,** 746–750 (2015).
3. Sun, J. & Bhushan, B. Hierarchical structure and mechanical properties of nacre: a review. *RSC Adv* **2,** 7617–7632 (2012).
4. Gleiter, H. Nanocrystalline Materials. *Prog. Mater. Sci.* **33,** 223–315 (1989).
5. Gleiter, H. Nanostructured materials: basic concepts and microstructure. *Acta Mater.* **48,** 1–29 (2000).
6. Hall, E. O. The deformation and ageing of mild steel: III. Discussion of results. *Proc. Phys. Soc. Lond. B* **64,** 747–753 (1951).
7. Petch, N. J. The cleavage strength of polycrystals. *J. Iron Steel Inst* **174,** 25–28 (1953).
8. Wang, Y. M. *et al.* Microsample tensile testing of nanocrystalline copper. *Scr. Mater* **48,** 1581–1586 (2003).
9. Wang, Y., Chen, M., Zhou, F. & Ma, E. High tensile ductility in a nanostructured metal. *Nature* **419,** 912–915 (2002).
10. Meyers, M. A., Mishra, A. & Benson, D. J. Mechanical properties of nanocrystalline materials. *Prog. Mater. Sci.* **51,** 427–556 (2006).
11. Dillon, S. J., Tang, M., Carter, W. C. & Harmer, M. P. Complexion: a new concept for kinetic engineering in materials science. *Acta Mater.* **55,** 6208–6218 (2007).
12. Shi, X. M. & Luo, J. Developing grain boundary diagrams as a materials science tool: a case study of nickel-doped molybdenum. *Phys. Rev. B* **84,** 014105 (2011).
13. Hsieh, T. E. & Balluffi, R. W. Experimental study of grain boundary melting in aluminum. *Acta Metall.* **37,** 1637–1644 (1989).
14. Gupta, V. K., Yoon, D. H., Meyer, H. M. & Luo, J. Thin intergranular films and solid-state activated sintering in nickel-doped tungsten. *Acta Mater.* **55,** 3131–3142 (2007).
15. Shi, X. M. & Luo, J. Grain boundary wetting and prewetting in Ni-doped Mo. *Appl. Phys. Lett.* **94,** 251908 (2009).
16. Schiotz, J. & Jacobsen, K. W. A maximum in the strength of nanocrystalline copper. *Science* **301,** 1357–1359 (2003).
17. Murdoch, H. A. & Schuh, C. A. Estimation of grain boundary segregation enthalpy and its role in stable nanocrystalline alloy design. *J. Mater. Res.* **28,** 2154–2163 (2013).
18. Luo, J. Developing interfacial phase diagrams for applications in activated sintering and beyond: current status and future directions. *J. Am. Ceram. Soc.* **95,** 2358–2371 (2012).
19. Wang, D. *et al.* Bulk metallic glass formation in the binary Cu–Zr system. *Appl. Phys. Lett.* **84,** 4029–4031 (2004).
20. Tschopp, M. A., Murdoch, H. A., Kecskes, L. J. & Darling, K. A. 'Bulk' nanocrystalline metals: review of the current state of the art and future opportunities for copper and copper alloys. *JOM* **66,** 1000–1019 (2014).
21. Kang, D.-H. *et al.* Interfacial free energy controlling glass-forming ability of Cu-Zr alloys. *Sci. Rep* **4,** 5167 (2014).
22. Smithells, C. J., Gale, W. F. & Totemeier, T. C. *Smithells Metals Reference Book* 8th edn (Elsevier Butterworth-Heinemann, 2004).
23. Atwater, M. A., Scattergood, R. O. & Koch, C. C. The stabilization of nanocrystalline copper by zirconium. *Mater. Sci. Eng. A* **559,** 250–256 (2013).
24. Jin, Q., Wilkinson, D. S. & Weatherly, G. C. Determination of grain-boundary film thickness by the fresnel fringe imaging technique. *J. Eur. Ceram. Soc.* **18,** 2281–2286 (1998).
25. Dillon, S. J. & Harmer, M. P. Multiple grain boundary transitions in ceramics: a case study of alumina. *Acta Mater.* **55,** 5247–5254 (2007).
26. Gianola, D. S., Mendis, B. G., Cheng, X. M. & Hemker, K. J. Grain-size stabilization by impurities and effect on stress-coupled grain growth in nanocrystalline Al thin films. *Mater. Sci. Eng. A* **483,** 637–640 (2008).
27. McFadden, S. X. & Mukherjee, A. K. Sulfur and superplasticity in electrodeposited ultrafine-grained Ni. *Mater. Sci. Eng. A* **395,** 265–268 (2005).
28. McFadden, S. X., Mishra, R. S., Valiev, R. Z., Zhilyaev, A. P. & Mukherjee, A. K. Low-temperature superplasticity in nanostructured nickel and metal alloys. *Nature* **398,** 684–686 (1999).
29. Valiev, R. *et al.* Unusual super-ductility at room temperature in an ultrafine-grained aluminum alloy. *J. Mater. Sci.* **45,** 4718–4724 (2010).
30. Bitzek, E., Brandl, C., Weygand, D., Derlet, P. M. & Van Swygenhoven, H. Atomistic simulation of a dislocation shear loop interacting with grain boundaries in nanocrystalline aluminium. *Modell. Simul. Mater. Sci. Eng* **17,** 055008 (2009).
31. Pan, Z. & Rupert, T. J. Damage nucleation from repeated dislocation absorption at a grain boundary. *Comput. Mater. Sci.* **93,** 206–209 (2014).
32. Kumar, K. S., Suresh, S., Chisholm, M. F., Horton, J. A. & Wang, P. Deformation of electrodeposited nanocrystalline nickel. *Acta Mater.* **51,** 387–405 (2003).
33. Sadigh, B. *et al.* Scalable parallel Monte Carlo algorithm for atomistic simulations of precipitation in alloys. *Phys. Rev. B* **85,** 184203 (2012).
34. Pan, Z. & Rupert, T. J. Amorphous intergranular films as toughening structural features. *Acta Mater.* **89,** 205–214 (2015).
35. Brandl, C., Germann, T. C. & Misra, A. Structure and shear deformation of metallic crystalline–amorphous interfaces. *Acta Mater.* **61,** 3600–3611 (2013).
36. Wang, Y. M., Li, J., Hamza, A. V. & Barbee, T. W. Ductile crystalline-amorphous nanolaminates. *Proc. Natl Acad. Sci. USA* **104,** 11155–11160 (2007).
37. Nieh, T. G. & Wadsworth, J. Bypassing shear band nucleation and ductilization of an amorphous–crystalline nanolaminate in tension. *Intermetallics* **16,** 1156–1159 (2008).
38. Nieh, T. G., Barbee, T. W. & Wadsworth, J. Tensile properties of a free-standing Cu/Zr nanolaminate (or compositionally-modulated thin film). *Scr. Mater* **41,** 929–935 (1999).
39. Donohue, A., Spaepen, F., Hoagland, R. G. & Misra, A. Suppression of the shear band instability during plastic flow of nanometer-scale confined metallic glasses. *Appl. Phys. Lett.* **91,** 241905 (2007).
40. Hasnaoui, A., Van Swygenhoven, H. & Derlet, P. M. Dimples on nanocrystalline fracture surfaces as evidence for shear plane formation. *Science* **300,** 1550–1552 (2003).
41. Lu, K. Making strong nanomaterials ductile with gradients. *Science* **345,** 1455–1456 (2014).
42. Fang, T. H., Li, W. L., Tao, N. R. & Lu, K. Revealing extraordinary intrinsic tensile plasticity in gradient nano-grained copper. *Science* **331,** 1587–1590 (2011).
43. Wu, X., Jiang, P., Chen, L., Yuan, F. & Zhu, Y. T. Extraordinary strain hardening by gradient structure. *Proc. Natl Acad. Sci. USA* **111,** 7197–7201 (2014).
44. Vo, N. Q. *et al.* Reaching theoretical strengths in nanocrystalline Cu by grain boundary doping. *Scr. Mater* **65,** 660–663 (2011).
45. Van Swygenhoven, H., Derlet, P. M. & Froseth, A. G. Nucleation and propagation of dislocations in nanocrystalline fcc metals. *Acta Mater.* **54,** 1975–1983 (2006).
46. Rupert, T. J., Trenkle, J. C. & Schuh, C. A. Enhanced solid solution effects on the strength of nanocrystalline alloys. *Acta Mater.* **59,** 1619–1631 (2011).
47. Ludy, J. E. & Rupert, T. J. Amorphous intergranular films act as ultra-efficient point defect sinks during collision cascades. *Scr. Mater* **110,** 37–40 (2016).

48. Brandstetter, S. *et al.* From micro- to macroplasticity. *Adv. Mater.* **18,** 1545–1548 (2006).
49. Plimpton, S. Fast parallel algorithms for short-range molecular-dynamics *J. Comput. Phys.* **117,** 1–19 (1995).
50. Lu, L., Chen, X., Huang, X. & Lu, K. Revealing the maximum strength in nanotwinned copper. *Science* **323,** 607–610 (2009).
51. Dao, M., Lu, L., Shen, Y. F. & Suresh, S. Strength, strain-rate sensitivity and ductility of copper with nanoscale twins. *Acta Mater.* **54,** 5421–5432 (2006).
52. Valiev, R. Z., Alexandrov, I. V., Zhu, Y. T. & Lowe, T. C. Paradox of strength and ductility in metals processed by severe plastic deformation. *J. Mater. Res.* **17,** 5–8 (2002).
53. Legros, M., Elliott, B. R., Rittner, M. N., Weertman, J. R. & Hemker, K. J. Microsample tensile testing of nanocrystalline metals. *Philos. Mag. A* **80,** 1017–1026 (2000).
54. Sanders, P. G., Eastman, J. A. & Weertman, J. R. Elastic and tensile behavior of nanocrystalline copper and palladium. *Acta Mater.* **45,** 4019–4025 (1997).
55. Lu, L., Wang, L. B., Ding, B. Z. & Lu, K. High-tensile ductility in nanocrystalline copper. *J. Mater. Res.* **15,** 270–273 (2000).

Acknowledgements

This study was supported by the US Army Research Office under Grant W911NF-12-1-0511. Materials characterization was performed at the Laboratory for Electron and X-ray Instrumentation (LEXI) at UC Irvine, using instrumentation funded in part by the National Science Foundation Center for Chemistry at the Space-Time Limit (CHE-082913). We thank Professor Daniel S. Gianola (University of Pennsylvania), Professor Kevin J. Hemker (Johns Hopkins University) and Professor Christopher A. Schuh (Massachusetts Institute of Technology) for valuable discussions and feedback on this manuscript.

Author contributions

A.K. and T.J.R. designed the processing route and experiments. A.K. implemented materials synthesis and mechanical testing experiments. Z.P. performed the atomistic simulations. T.J.R. wrote the initial manuscript with input from all authors. All authors contributed to discussion of the results, provided input and approved the final manuscript.

Additional information

Competing financial interests: The authors declare no competing financial interests.

5

Ferroelasticity and domain physics in two-dimensional transition metal dichalcogenide monolayers

Wenbin Li[1] & Ju Li[2]

Monolayers of transition metal dichalcogenides can exist in several structural polymorphs, including 2H, 1T and 1T′. The low-symmetry 1T′ phase has three orientation variants, resulting from the three equivalent directions of Peierls distortion in the parental 1T phase. Using first-principles calculations, we predict that mechanical strain can switch the relative thermodynamic stability between the orientation variants of the 1T′ phase. We find that such strain-induced variant switching only requires a few percent elastic strain, which is eminently achievable experimentally with transition metal dichalcogenide monolayers. Calculations indicate that the transformation barrier associated with such variant switching is small (<0.2 eV per chemical formula unit), suggesting that strain-induced variant switching can happen under laboratory conditions. Monolayers of transition metal dichalcogenides with 1T′ structure therefore have the potential to be ferroelastic and shape memory materials with interesting domain physics.

[1] Research Laboratory of Electronics, Massachusetts Institute of Technology, Cambridge, Massachusetts 02139, USA. [2] Department of Nuclear Science and Engineering and Department of Materials Science and Engineering, Massachusetts Institute of Technology, Cambridge, Massachusetts 02139, USA. Correspondence and requests for materials should be addressed to J.L. (email: liju@mit.edu).

The discovery of two-dimensional (2D) atomic crystals[1] has fuelled intensive research efforts on this new class of materials, revealing fundamentally new physics and properties[2–6] that could be essential for next-generation nanoscale devices. Monolayers of group VI transition metal dichalcogenides (TMDs) with chemical formula MX_2, where M is Mo or W and X stands for S, Se or Te, have in particular attracted much recent attention due to their semiconducting, optical and valleytronic properties[4,7–9]. Owning to their atomic thickness, the TMD monolayers have extraordinary mechanical flexibility and strength, capable of sustaining up to 10% of elastic strain before failure[10,11], which enables significant dynamical tuning of their properties by strain engineering[12] and makes them attractive for application in ultrathin flexible electronics[13–15].

MX_2 monolayers can exist in several polytypic structures, including 2H, 1T and 1T′[16–18]. In the semiconducting 2H phase, the atomic stacking sequence within a single XMX monolayer is Bernal (ABA) and the M–X coordination is trigonal prismatic. In contrast, in the 1T phase, the XMX stacking sequence is rhombohedral (ABC), and the M and X atoms form octahedral coordination. The 1T phase is metallic, but was found to be unstable to Peierls distortion[19,20], where two adjacent lines of metal atoms along the highest symmetry directions can dimerize and form parallel chains of M atoms. This leads to the formation of 1T′ phase[17,18], in which the octahedral coordination between M and X atoms becomes distorted, and the symmetry of the crystal structure is reduced. While the thermodynamically stable phase of most group VI MX_2 monolayers under ambient conditions is 2H, the ground-state phase of WTe_2 has 1T′ structure[16,21]. For other MX_2 monolayers, the 1T′ phase is usually metastable, but large transition barriers of order 1 eV per formula unit exist between 1T′ and 2H (ref. 22), suggesting that the 1T′ phase can be stabilized under appropriate thermal or chemical conditions. In particular, the energetic difference between the 2H and 1T′ phase of $MoTe_2$ is rather small[19], suggesting that the 1T′ phase can be stabilized relatively easily. Indeed, single crystals and few-layer films of $MoTe_2$ in 1T′ phase have been synthesized on a large scale recently[20,23,24]. It has also been theoretically proposed that the 2H to 1T′ transition in $MoTe_2$ monolayers can be induced by experimentally accessible tensile strain[19].

The low-symmetry 1T′ phase of TMD monolayers harbours extraordinary properties that have only started to be revealed, which, for example, includes enhanced catalytic activities[25], large, non-saturating magnetoresistence[21] and quantum spin Hall effect[22].

A ferroelastic material is defined by the existence of two or more equally stable orientation variants, which can be switched from one variant to another without diffusion by the application of external stress[26,27]. A ferroelastic phase usually forms through a structural phase transition (or a hypothetical one) that reduces the symmetry of a prototype phase. The low-symmetry ferroelastic phase possesses several orientation states (domain variants) with different spontaneous strain[28], that is, the distortion of the unit cell relative to that in the prototype phase. The difference in spontaneous strain between different variants enables external stress to couple energetically with the strain state of the system and drive orientation switch, analogous to the switching of spontaneous polarization by external electric field in a ferroelectric material. In a ferroelastic crystal, domains of different orientations can coexist and form twin boundaries. On activation by appropriate external stress, those twin boundaries can move in a glissile fashion, resulting in the growth of one orientation state at the expense of another, as well as hysteretic stress–strain response[27].

In this article, we focus on the possibility of ferroelastic behaviours in 1T′–MX_2 monolayers. A notable feature associated with the 1T′ phase that has hitherto been overlooked is that it has three distinct orientation variants, resulting from the three equivalent directions of structural distortion in the parental 1T phase. Our density functional theory (DFT) calculations indicate that ferroelastic switching can occur between the different orientation variants of the 1T′ phase with a few percent of elastic strain, which is experimentally achievable for MX_2 monolayers.

Results

Crystal structures and transformation strains. We use WTe_2 monolayers as a representative of 1T′–MX_2 to illustrate the possibility of 2D ferroelasticity. Figure 1 shows the atomistic structures of 1T–WTe_2 and 1T′–WTe_2 monolayers. In the 1T phase, the W atoms arrange in 2D triangular lattice, which is sandwiched between two Te atomic layers. The 2D primitive cell of the 1T phase is a 120° rhombus with side length t_0. Due to Fermi surface nesting induced Peierls distortion[20], adjacent parallel lines of W atoms along the high-symmetry [100], [010] or [$\bar{1}$10] directions in the 1T structure can spontaneously dimerize and result in the formation of 1T′ phase, with distorted octahedral coordination. The 2D primitive cell of the 1T′ phase is a rectangle with dimensions $a \times b$, which corresponds to the $1\times\sqrt{3}$ supercell of the 1T phase. Because of the $P\bar{3}m2$ space group symmetry of the 1T phase, there are three symmetry-equivalent directions of structural distortion in the 1T phase. These directions are labelled on Fig. 2a as direction 1, 2 and 3 on the 2D triangular lattice formed by W atoms. The atomistic structures of the three orientation variants formed by structural distortion in the 1T phase along the three directions are shown in Fig. 2b–d. Hereafter, we refer to the three orientation states as the O1, O2 and O3 variant, respectively.

The spontaneous transformation strains associated with the 1T to 1T′ transformation can be compared between the three variants based on the $2\times2\sqrt{3}$ supercell of the prototype 1T phase. All the three orientation variants of the 1T′ phase, namely, O1, O2 and O3, can be derived through the Peierls distortion of this supercell and the atoms within the supercell along the corresponding orientation direction. Namely, the $2\times2\sqrt{3}$

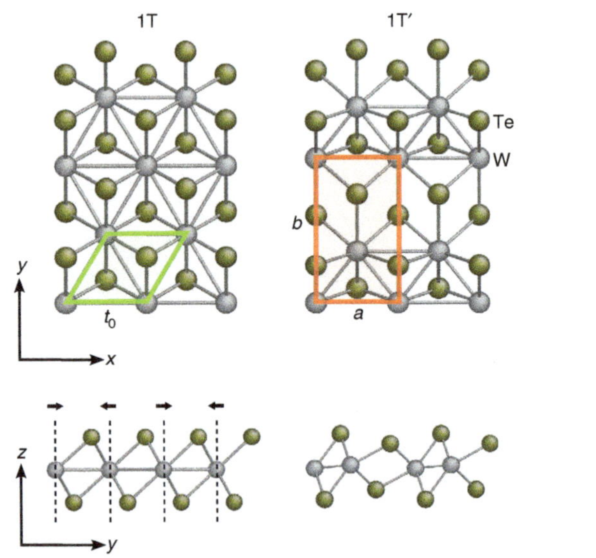

Figure 1 | Atomistic structure of of 1T-WTe_2 and 1T′-WTe_2 monolayers. The 2D primitive cells of 1T and 1T′ are highlighted in green and red, respectively. The primitive cell of 1T′ corresponds to the $1\times\sqrt{3}$ supercell of 1T. The 1T′ phase can be derived via the structural distortion of the 1T phase, which is schematically illustrated in the side views. These features are generic to all other group VI MX_2 monolayers.

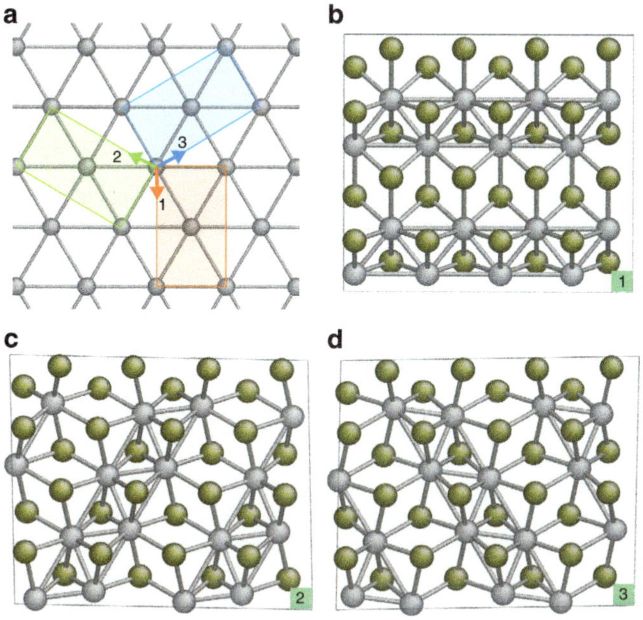

Figure 2 | Three orientation variants of 1T′-MX$_2$ monolayers. (**a**) The three symmetry-equivalent directions of structural distortion in the 1T structure are indicated by arrows and numerical labels on the triangular lattice formed by M atoms. The corresponding primitive cells of the 1T′ phase after structural distortion are represented as shaded rectangles. (**b**-**d**) Relaxed atomistic structure of the 1T′ phase after structural distortion along the three different directions, which are referred to as orientation variants O1, O2 and O3, respectively.

supercell of 1T can transform to become the supercells of all three variants of the 1T′ phase. In Cartesian coordinates, the 2D basis vectors $\mathbf{h}_1$ and $\mathbf{h}_2$ of the 1T supercell can be written as $\mathbf{h}_1 = 2t_0\hat{\mathbf{x}}$, $\mathbf{h}_2 = 2\sqrt{3}t_0\hat{\mathbf{y}}$, where $\hat{\mathbf{x}}$ and $\hat{\mathbf{y}}$ are the unit vectors along the x and y directions labelled on Fig. 1. A supercell matrix $\mathbf{H}_0 = \{\mathbf{h}_1, \mathbf{h}_2\}$ can be constructed, where $\mathbf{h}_1$ and $\mathbf{h}_2$ are treated as column vectors, that is,

$$\mathbf{H}_0 = \begin{pmatrix} 2t_0 & 0 \\ 0 & 2\sqrt{3}t_0 \end{pmatrix}. \quad (1)$$

After transforming to the 1T′ phase, the distorted supercell matrix corresponding to the O1, O2 and O3 variants will be denoted by $\mathbf{H}_1$, $\mathbf{H}_2$ and $\mathbf{H}_3$, respectively. These new supercell matrices can be related to the original supercell matrix by transformation matrices $\mathbf{J}_i$, which map the undistorted supercell to the distorted supercells. Namely, $\mathbf{H}_i = \mathbf{J}_i\mathbf{H}_0$, where the subscript i stands for the i-th orientation variant. The transformation strain matrices $\boldsymbol{\eta}_i$ associated with different variants can then be calculated from $\mathbf{J}_i$ based on the definition of Green-Lagrange strain tensor:

$$\eta_i = \frac{1}{2}\left(\mathbf{J}_i^{\mathrm{T}}\mathbf{J}_i - \mathbf{I}\right) = \frac{1}{2}\left[\left(\mathbf{H}_0^{-1}\right)^{\mathrm{T}}\mathbf{H}_i^{\mathrm{T}}\mathbf{H}_i\mathbf{H}_0^{-1} - \mathbf{I}\right]. \quad (2)$$

Here, the superscripts -1 and T denote matrix inversion and transposition, respectively. $\mathbf{I}$ is a 2×2 identity matrix. The 2D transformation strain tensor $\boldsymbol{\eta}_i$ has the following symmetric form:

$$\eta = \begin{pmatrix} \varepsilon_{xx} & \varepsilon_{xy} \\ \varepsilon_{xy} & \varepsilon_{yy} \end{pmatrix}, \quad (3)$$

where ε_{xx} and ε_{yy} are the tensile/compressive strain along x or y direction, and ε_{xy} is the shear strain component.

We have employed DFT calculations to obtain the equilibrium supercell vectors and the relaxed atomic coordinates of the O1, O2 and O3 variants, resulting from the distortion of $2\times2\sqrt{3}$ supercell in the 1T prototype phase. The supercell matrices $\mathbf{H}_i$ for different MX_2 monolayers are tabulated in Supplementary Table 1. From the supercell matrices, the spontaneous transformation strain tensors $\boldsymbol{\eta}_i$ can be evaluated, which are listed in Supplementary Table 2. For WTe_2 monolayers, the transformation strain matrices form 1T to 1T′ are

$$\eta_1 = \begin{pmatrix} -0.005 & 0.0 \\ 0.0 & 0.039 \end{pmatrix},$$
$$\eta_2 = \begin{pmatrix} 0.029 & -0.019 \\ -0.019 & 0.006 \end{pmatrix}, \quad (4)$$
$$\eta_3 = \begin{pmatrix} 0.029 & 0.019 \\ 0.019 & 0.007 \end{pmatrix}.$$

The difference in transformation strain between the three variants of 1T′ suggests that one may switch the relative thermodynamic stability between different variants by applying suitable external mechanical stress. Since the equilibrium structure of WTe_2 is 1T′, it is informative to directly compare the distorted supercell of the three variants by computing the relative supercell strain associated with the transformation from one variant to another. This can be carried out again using the supercells of the three variants derived from the common $2\times2\sqrt{3}$ supercell in the prototype 1T phase. The reference configuration for computing the supercell strain is now chosen to be the O1 variant of 1T′ phase, and we use ε_i^j to denote the transformation strain tensor from variant i to j. Calculations based on the same definition of strain tensor as in equation (2) give the transformation strain associated with O1–O2 and O1–O3 switching to be

$$\varepsilon_1^2 = \begin{pmatrix} 0.034 & -0.019 \\ -0.019 & -0.030 \end{pmatrix},$$
$$\varepsilon_1^3 = \begin{pmatrix} 0.033 & 0.019 \\ 0.019 & -0.030 \end{pmatrix}. \quad (5)$$

It then follows that, starting with the O1 variant of 1T′ in a strain-free state, after imposing an external strain of magnitude ε_1^2 on the monolayer, the system would be in a thermodynamically more favourable state by transforming to the O2 variant, since both O1 and O2 belong to the same 1T′ structure, but O1 will have higher strain energy than O2. The same argument applies to any other two variants. Hence, the relative energetic stability between the different orientation variants of 1T′ phase can be controlled by external stress or strain.

Variant energetics under biaxial and shear strain. To study in detail the relative thermodynamic stability of different variants when external mechanical deformation is imposed on a 1T′–MX_2 monolayer, we have used DFT to calculate the potential energy surfaces of the three variants of 1T′ as a function of 2D supercell dimensions. We first investigate the possibility of mechanically switching the O1 variant to O2 or O3 variant by applying biaxial strain to the system, again using WTe_2 monolayer as an example. The strain-free 2×2 supercell of the O1 variant, derived from the distortion of the aforementioned $2\times2\sqrt{3}$ supercell in the parental 1T phase, is chosen to be the reference system. The 2×2 supercell of the O1 variant has dimensions $2a\times2b$ within the x–y plane of 2D monolayer. We adjust the dimensions of the supercell along x and y directions independently, with the values of a and b range from -10 to 10% of engineering strain at an equal step of 2%. At each pair of (a,b), the atomic coordinates within the supercell are relaxed. We also compute the energies of O2 and O3 variants when their supercell dimensions are fixed to be the same as O1. The energies U of all three variants are computed on a

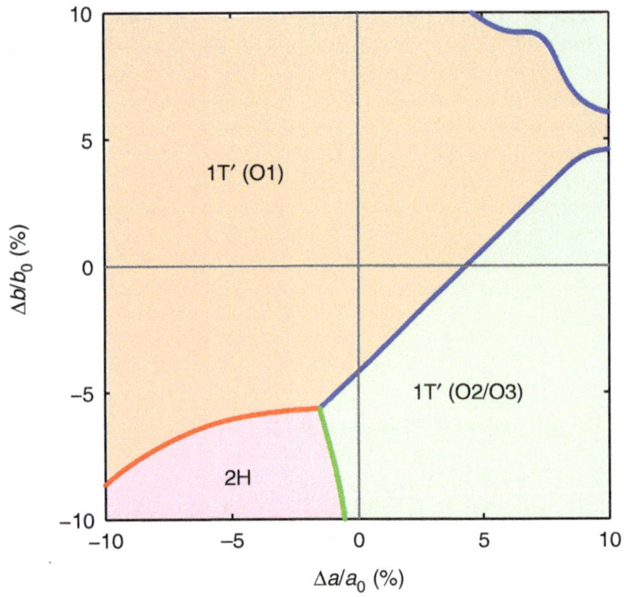

Figure 3 | Intersection contours of the energy surfaces between the different orientation variants of 1T′-$MoTe_2$ and between 2H and 1T′. The lattice constants *a* and *b*, corresponding to the dimensions of the rectangular primitive cell of the O1 variant in the 1T′ phase, are represented as percent engineering strain with respect to the equilibrium lattice constants a_0 and b_0. The regions of lower-energy phase/variant are labelled and shaded in different colours.

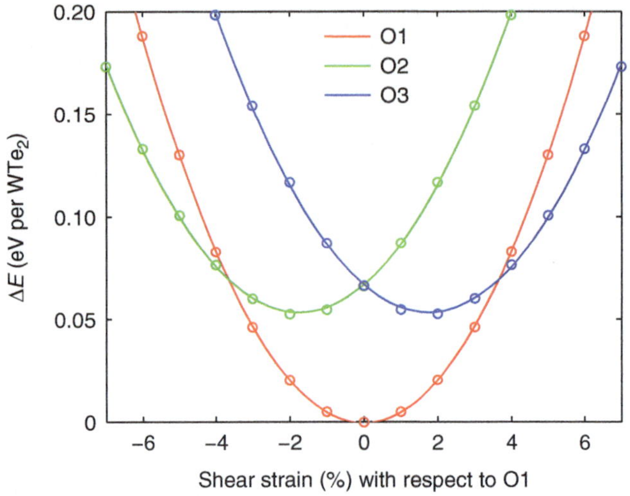

Figure 4 | Potential energies of the three variants of 1T′-WTe_2 monolayers as a function of shear strain with respect to the equilibrium supercell of the O1 variant. The O2 or O3 variant becomes energetically more favourable when negative- or positive-shear strain of a few percent is imposed on the O1 variant.

11×11 grid in the (a,b) space, giving a total number of 121 data points distributed evenly around the equilibrium lattice constants of the O1 variant. Smooth potential energy surfaces are then constructed by approximating the intermediate values of $U(a,b)$ using 2D spline interpolation, which allows us to directly compare the relative energetic stability of the O1, O2 and O3 variants in the full (a,b) space. In addition, the $U(a,b)$ for the 2H phase is computed for comparison, as a previous study indicates that strain-induced phase transformation between the 2H and 1T′ phases can happen in MX_2 monolayers[19].

After obtaining the potential energy surfaces for all the three variants of 1T′ as well as the 2H phase, the lowest-energy variants/phases in the (a,b) space are determined. The result is shown in Fig. 3, where we label the lowest-energy variant/phase in each region of phase space and plot the intersection boundaries between two neighbouring variants/phases. An important feature of Fig. 3 is that the potential energy surfaces of O1 and O2/O3 variants intersect at a few percent of biaxial supercell strain, which is experimentally achievable in MX_2 monolayers[10,11]. The O2 and O3 variants are grouped together in Fig. 3 because their potential energies in the (a,b) space are essentially the same. This can be rationalized by the fact that the supercells of both variants can be derived from the distortion of the $2\times2\sqrt{3}$ supercell in the 1T phase, and their distortion directions are related by mirror symmetry along the y direction in the 1T phase, as can be seen from Fig. 2. Since biaxial strain does not break the mirror symmetry of 1T phase along the y axis, the O2 and O3 variants are still mirror images of each other and have the same energy. We however expect that shear strain, which breaks the mirror symmetry, can distinguish the energies of all the three variants of 1T′. Indeed, Fig. 4 shows that, when shear strain ε_{xy} of magnitude $>3.5\%$ is imposed on the O1 variant, O3 becomes the lowest-energy variant within the strained supercell. If the sign of ε_{xy} is reversed, then the O2 variant has lower energy than both O1 and O3.

Figure 3 indicates that the 2H phase of WTe_2 monolayer only takes a small region in the (a,b) space as the lowest-energy phase. This result is different from the study by Duerloo *et al.*[19] of strain-induced phase transformation between the 2H and 1T′ phases of MX_2 monolayers, as the authors did not take into account the existence of orientation variant degrees of freedom in the 1T′ phase.

Figure 3 also shows that the fastest route to switching the energetic order between O1 and O2/O3 in the (a,b) space is by applying tensile strain along the a axis of the O1 variant, which is the direction of dimerized metal-atom chains, while simultaneously applying compressive strain along the b axis. It is however known that 2D MX_2 monolayers usually cannot sustain large compressive strain due to compression-induced buckling response and formation of ripplocations[29]. On the contrary, experiments have demonstrated that 2H–MX_2 monolayers can withstand tensile elastic strain as large as 10% before mechanical failure[10,11]. Hence, it may be experimentally more convenient to realize variant switching in 1T′–WTe_2 by uniaxially stretching it along the a axis, which is the direction of dimerized tungsten atoms. This axis can be identified by mechanical cleavage or by the anisotropic response to external fields that is expected for the low-symmetry 1T′ structure[21].

In Supplementary Fig. 1, we have also computed the intersection contours of the potential energy surfaces between the O1 and O2/O3 variants for other 1T′–MX_2 monolayers, including MoS_2, $MoSe_2$, $MoTe_2$, WS_2 and WSe_2. The results are very similar to WTe_2, indicating that strain-induced switching of thermodynamic stability between different orientation variants is generic to MX_2 monolayers with 1T′ structure.

Variant energetics under uniaxial tension. We emphasize that the strain at which the potential energy surfaces of different variants intersects is not the same as the strain at which variant switching becomes thermodynamically favourable. The system can minimize its free energy by choosing a state where different variants (or phases) coexist, akin to the two-phase region in chemical-composition phase diagrams. Under constant temperature and fixed external strain (supercell dimensions), the thermodynamic potential that determines the relative variant/

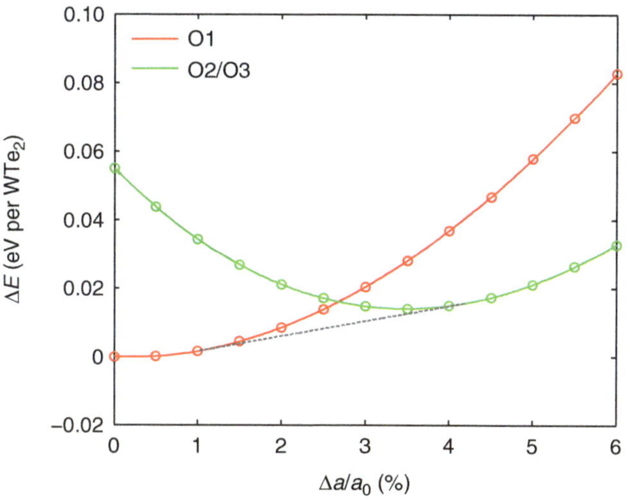

Figure 5 | Potential energy curves of the different variants of 1T′-WTe_2 uniaxially strained along the *a* axis. The dashed line is the common tangent between the two curves. The uniaxial strain is calculated with respect to the equilibrium supercell dimension of the O1 variant along the *x* direction.

phase stability is the Helmholtz free energy $F = E - TS$, where E is internal energy that includes both potential energy U and kinetic energy, T is temperature and S is entropy. Because all the three variants O1, O2 and O3 belong to the same 1T′ structure, and because the entropy of solids (mainly vibrational) is relatively insensitive to small deformation, we can use the potential energy U of different variants, computed by DFT at zero temperature, to compare the free energies of different variants at ambient conditions. In Fig. 5, we plot the potential energy curve of the O1 variant of 1T′-WTe_2 when it is uniaxially stretched along the *a* axis. Consistent with typical experimental set-ups for uniaxial deformation, the stress of the supercell along the *b* axis is relaxed to zero. This corresponds to free boundary, or zero stress ($\sigma_y = 0$) condition along the *b* axis. In Fig. 5, we also plot the potential energy of the O2/O3 variant in a rectangular supercell with the same dimension along the *a* axis and the same boundary condition along the *b* axis. A common tangent can be constructed between the energy curves of O1 and O2/O3, which intersects the two curves at uniaxial strains equal to 1% and 4%, respectively. Between these two values, the system can lower its energy by existing in a state where both O1 and O2/O3 variants coexist. This indicates that the formation of O2/O3 variants becomes thermodynamically favourable when the uniaxial strain along the *a* axis of O1 is as low as 1%.

Kinetic aspects of variant switching. Up to now, we have only considered the thermodynamic aspects of variant switching in the 1T′-MX_2 monolayers. Our results suggest that it becomes thermodynamically favourable for the O1 variant of 1T′-WTe_2 monolayers to switch to the other two variants when applying uniaxial strain around 1% along the direction of dimerized tungsten atom chains. However, if the kinetic barrier associated with variant switching is too high, such variant switching may not occur under normal experimental conditions and timescale, and the materials would still not be ferroelastic. We have therefore computed the transition barrier associated with the switching between the O1, O2 and O3 variants using climbing image nudged elastic band (NEB) method[30]. The result of our calculation for the variant switching between the O1 and O2 variants of WTe_2 monolayer is shown in Fig. 6. We find the transition barrier of variant switching is only 0.22 eV per formula unit. Very similar results are obtained for orientation switching between other variants, as presented in Supplementary Fig. 2. Note that to facilitate these NEB calculations, we impose supercell strains on the O2 or O3 variants such that they have the same supercell dimensions of the O1 variants. The strain energy results in the slightly higher energy of the O2 or O3 variant that would otherwise be energetically degenerate with the O1 variant. In Supplementary Fig. 3, we have also computed the transition barrier and the pathway between stress-free O1 and O2 variants using generalized solid-state NEB method[31], which allows both the atomic and supercell degrees of freedom to relax along the transition pathway. The results of the generalized solid-state NEB calculation are very close to those obtained using a fixed-supercell approach, with the calculated energy barrier of variant switching equals to 0.19 eV per formula unit. We note that, while the transition pathway illustrated in Fig. 6 may not be the only possible one, if other pathways exist, the transformation barrier of variant switching can only be smaller or equal than the values we have obtained.

In Supplementary Fig. 4, we have also computed the transition barriers for other MX_2 monolayers, and the barriers obtained are even lower than 1T′-WTe_2 monolayers. The transition barriers of variant switching are significantly lower than the barriers of phase transition between the 2H and 1T′ phase[19], which we computed to be 0.8 eV per formula unit for 1T′-WTe_2 monolayers at the equilibrium lattice constants of the 2H phase. The much smaller transition barriers associated with the variant switching within the 1T′ phase as compared with the 1T′ to 2H phase transition has an intuitive geometric explanation. Variant switching between the orientation variants of 1T′ phase only involves the distortion of M–X octahedral coordination, while the 1T’ to 2H phase transition requires the complete change of M–X coordination pattern from octahedral to trigonal prismatic.

According to transition-state theory, assuming a characteristic attempt frequency of 10 THz, which is the typical frequency of optical phonons in 1T′-MX_2 monolayers[22], a 0.2 eV barrier is associated with a timescale of around 0.2 ns. Although the actual barrier of forming a critical nucleus of new variant may involve multiple formula units, and other factors such as interfaces and pre-existing defects may also affect the transformation kinetics, the much smaller barrier associated with variant switching within the 1T′ phase as compared with 2H to 1T′ phase transition[19,20] suggests that ferroelastic variant switching in 1T′-MX_2 monolayers is very likely to happen under normal laboratory experimental conditions.

Ferroelastic domain boundaries. A direct consequence of strain-induced variant switching in 1T′-MX_2 monolayers is the formation of domain boundaries between different orientation variants. Strain-induced ferroelastic switching between the O1, O2 and O3 variants can lead to the formation of three possible types of coherent twin boundaries, between O1 and O2, O1 and O3, and between O2 and O3, which we refer to as O1–O2, O1–O3 and O2–O3, respectively. The DFT-relaxed atomistic structures of the three different types of twinning domain boundaries in 1T′-WTe_2 monolayers under zero external stress are shown in Fig. 7. The three domain boundaries are energetically degenerate, and they are related to each other by 120° rotational symmetry operation. Unlike their three-dimensional (3D) counterparts, where the domain boundaries are 2D, the boundaries formed between the domains of 2D MX_2 monolayers are quasi-one dimensional (1D) in nature, which may impart them unique properties. We have calculated the domain boundary energies associated with the three types of the 1D domain boundaries and found they have small formation energies. Our DFT calculations give the domain boundary energies of MoS_2,

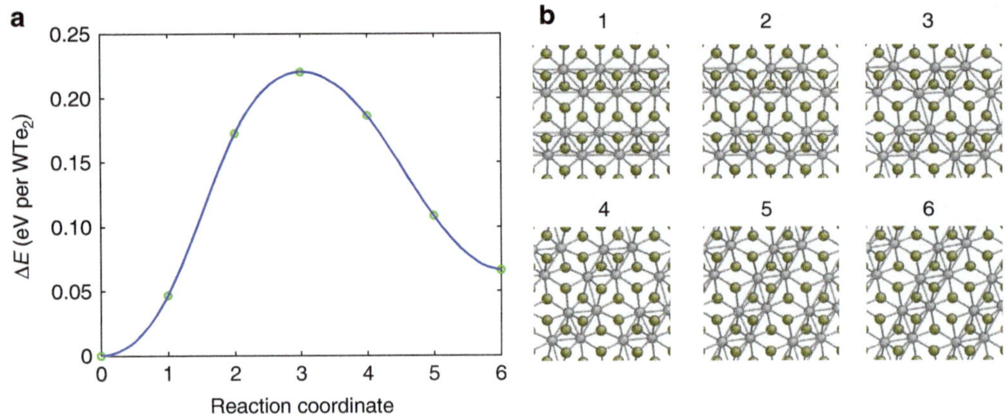

Figure 6 | NEB calculation of transformation barrier and the pathway for orientation switching. The initial configuration of the NEB calculation is the O1 variant of 1T′-WTe_2 monolayers at ground state (zero external stress), while the final state is the strained O2 variant with the same supercell dimensions as those of the O1 variant. (**a**) Change of the system energy per chemical formula unit as a function of reaction coordinate. (**b**) The corresponding atomistic structure of the system along the reaction coordinate. Four different supercells were used to carry out the NEB calculations: 2×2, 2×4, 4×2 and 4×4 supercell of the 1T′ phase, which all give identical results.

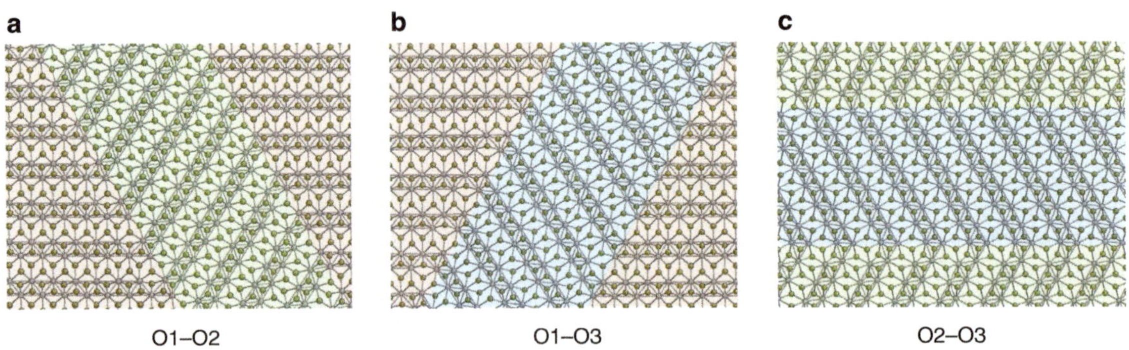

Figure 7 | Ferroelastic domain boundaries between different orientation variants. (**a-c**) Illustrate the DFT-relaxed atomistic structures of domain boundaries formed between the O1 and O2, O1 and O3, and O2 and O3 variants of 1T′-WTe_2 monolayers, respectively. To help guide the eyes, domains of different variants are shaded in distinct colours: O1 variant in orange, O2 variant in green and O3 variant in blue.

$MoSe_2$, $MoTe_2$, WS_2, WSe_2 and WTe_2 monolayers to be 27, 46, 40, 22, 51 and 52 meV · Å^{-1}, respectively. In comparison, the formation energy of another type of 1D defects in 3D crystals, dislocations, is in the order of several hundred meV per Angstrom. Such small-domain boundary energies will facilitate the ferroelastic switching between different orientation variants.

Discussion

The thermodynamic and kinetic analysis above have provided strong evidence that strain-induced ferroelastic switching of orientation variants can occur in 1T′-MX_2 monolayers, with a few percent of local strain. Our calculations indicate that variant switching can most easily happen when stretching the 1T′-MX_2 monolayers along the direction of dimerized metal chains. This prediction, if experimentally realized, will render 1T′-MX_2 monolayers as the first class of 2D ferroelastic materials[32]. Signatures of such ferroelastic switching in experiments include hysteresis in stress–strain curves[27], and the existence of a force plateau when the externally applied strain is beyond a critical value that corresponds to the onset of variant coexistence. Direct experimental proof of ferroelastic domain switching may be realized by carrying out *in situ* transmission electron microscopy experiments of mechanical deformation of 1T′-MX_2 monolayers. As the domains of different variants have different crystallographic orientations, the migration of domain walls during strain-induced variant switching can be observed by dark-field transmission electron microscopy, which has been demonstrated for domain imagining in graphene and MoS_2 monolayers[33,34]. Selective area electron diffraction could also reveal the formation of twinning domains, as variant switching results in the rotation of the underlying Bravais lattice of the 1T′ structure, which will manifest in selective area electron diffraction as the rotation of diffraction patterns.

Our prediction of ferroelasticity in the TMD monolayers can be readily tested experimentally in 1T′-WTe_2 and 1T′-$MoTe_2$, for which bulk single crystals have been synthesized on a large scale and exfoliated down to the monolayer or few-layer regime[20,21]. Recently, large-area and high-quality $MoTe_2$ few layers in 1T′ phase have been grown via chemical vapour deposition[23,24]. Local and controlled phase transformation of $MoTe_2$ from the 2H to 1T′ phase can also be realized using laser ablation[35]. In principle, ferroelastic domain switching can be observed not only in monolayers but also in few-layer samples, since dimerized metal chains within different layers of the 1T′ phase orient along the same direction in naturally grown crystals[36].

For other group VI MX_2 that include MoS_2, $MoSe_2$, WS_2 and WSe_2, as the 2H phase is energetically more stable than the 1T′ phase under normal conditions, the 1T′ phase can be realized using a phase engineering approach[37,38]. The 1T or 1T′ phase of these materials are now actively being explored for applications in energy and electronics[38]. Monolayers of WS_2, MoS_2 and $MoSe_2$ in 1T′ phase have been obtained via liquid-phase exfoliation of the bulk crystals intercalated by alkaline metals[18,25,39]. The transformation from the 2H phase to 1T′ phase by alkaline metal intercalation is attributed to charge transfer from the intercalated alkali atoms to the TMDs[37]. We have performed DFT calculations to study the effect of lithium atom adsorption on the relative energetics of 2H-MX_2 and 1T′-MX_2 monolayers, and it is indeed shown that, with increased amount of adsorbed lithium, the 1T′ phase becomes energetically more favourable than the 2H phase for all the MX_2 monolayers, as illustrated in Supplementary Fig. 5. It has also been proposed that the substitutional doping of MX_2 with elements having more valence electrons (for example, Re) than the transition metal ions can be another effective way to stabilize the 1T′ phase[40].

From an application perspective, ferroelastic behaviours have close connection to the shape memory effect, which has been exploited to make actuators in a wide range of industries. In 3D, shape memory alloys (SMAs) is a well known and technologically important class of ferroelastic materials. SMAs can undergo a diffusionless martensitic phase transformation below a critical temperature, from the high-temperature austenite to the low-temperature martensite phase[41]. The martensite phase of SMAs is ferroelastic: it has several equivalent orientations or variants, that can be switched from one to another by an appropriate uniaxial or shear stress. The martensite phase in SMAs can undergo large inelastic deformation through stress-induced migration of twin boundaries between different variants. On heating the deformed crystal above the martensitic phase transformation temperature, the martensite phase can revert to the austenite phase and recover its original shape before deformation. For the ferroelastic 1T′-MX_2 monolayers, if the 1T′ phase can be reversibly transformed to the 1T or 2H phase under external stimuli (which do not have to be thermal but could be other fields), then MX_2 monolayers could be 2D shape memory materials, with operating principles similar to SMAs. As such, 1T′-MX_2 monolayers can be used to make ultrathin actuators for applications in nanoscale-integrated electromechanical systems.

In closing, we would like to make a few additional comments. First, since the 1T′-MX_2 monolayers were predicted to be quantum spin Hall insulators[22], topological effect plays an important role in determining their electronic properties. The twinning domain boundaries formed through ferroelastic switching in the 1T′ phase, which are 1D defects in 2D quantum materials, may possess exotic physics and provide a rich playground for domain boundary engineering[42]. Second, the possibility of ferroelastic switching in 2D materials may not be limited to group VI MX_2 monolayers, considering the rich family of 2D materials[43]. Our initial studies indicate that several other TMD monolayers, including ReS_2, $NbTe_2$ and $TaTe_2$, which have low-symmetry distorted crystal structures similar to 1T′[16], also possess distinct orientation variants and could be ferroelastic as well. Indeed, experimental evidence of local strain-induced orientational switching in ReS_2 and $ReSe_2$ monolayers has recently been reported[44]. Our finding of potential 2D ferroelastic behaviours in monolayer materials could therefore open doors to many exciting discoveries in 2D materials with low-symmetry distorted crystal structures, which may also include ferroelectric, ferromagnetic and multiferroic behaviours in the future.

Methods

First-principles calculations. DFT calculations were performed using the Vienna Ab initio Simulation Package with a plane-wave basis set[45,46] and the projector-augmented wave[47] pseudopotentials. Exchange-correlation effects were treated using the generalized gradient approximation[48] in the Perdew–Burke–Ernzerhof form[49]. The kinetic energy cutoff for wavefunction expansion was fixed to be 350 eV. The TMD monolayers were modelled in supercells with a vacuum region in the direction perpendicular to the 2D planes of the monolayers (the z direction). The length of the supercells along the z direction was chosen to be 20 Å. Brillouin zone integration employed a Gamma point centred $m \times n \times 1$ Monkhorst-Pack[50] **k**-point grid and a Gaussian smearing of 50 meV, where the numbers m and n were chosen such that the **k**-point sampling spacing is <0.1 Å^{-1} along the supercell reciprocal vectors in the x–y plane. The energy convergence thresholds for electronic and ionic relaxations were 10^{-6} and 10^{-5} eV, respectively. The maximum residual forces resulted from these convergence criteria are smaller than 5×10^{-3} eV Å^{-1}. We provide the DFT-relaxed atomistic structures of the O1, O2 and O3 variants of 1T′-MX_2 monolayers in the POSCAR format of Vienna Ab initio Simulation Package, as listed in Supplementary Tables 4–9.

References

1. Novoselov, K. S. *et al.* Two-dimensional atomic crystals. *Proc. Natl Acad. Sci. USA* **102,** 10451–10453 (2005).
2. Castro Neto, A. H., Guinea, F., Peres, N. M. R., Novoselov, K. S. & Geim, A. K. The electronic properties of graphene. *Rev. Mod. Phys.* **81,** 109–162 (2009).
3. Lee, C., Wei, X. D., Kysar, J. W. & Hone, J. Measurement of the elastic properties and intrinsic strength of monolayer graphene. *Science* **321,** 385–388 (2008).
4. Wang, Q. H., Kalantar-Zadeh, K., Kis, A., Coleman, J. N. & Strano, M. S. Electronics and optoelectronics of two-dimensional transition metal dichalcogenides. *Nat. Nanotechnol.* **7,** 699–712 (2012).
5. Xu, X., Yao, W., Xiao, D. & Heinz, T. F. Spin and pseudospins in layered transition metal dichalcogenides. *Nat. Phys.* **10,** 343–350 (2014).
6. Xia, F., Wang, H., Xiao, D., Dubey, M. & Ramasubramaniam, A. Two-dimensional material nanophotonics. *Nat. Photon.* **8,** 899–907 (2014).
7. Mak, K. F., Lee, C., Hone, J., Shan, J. & Heinz, T. F. Atomically thin MoS_2: A new direct-gap semiconductor. *Phys. Rev. Lett.* **105,** 136805 (2010).
8. Mak, K. F., He, K., Shan, J. & Heinz, T. F. Control of valley polarization in monolayer MoS_2 by optical helicity. *Nat. Nanotechnol.* **7,** 494–498 (2012).
9. Zeng, H., Dai, J., Yao, W., Xiao, D. & Cui, X. Valley polarization in MoS_2 monolayers by optical pumping. *Nat. Nanotechnol.* **7,** 490–493 (2012).
10. Bertolazzi, S., Brivio, J. & Kis, A. Stretching and breaking of ultrathin MoS_2. *ACS Nano* **5,** 9703–9709 (2011).
11. Cooper, R. C. *et al.* Nonlinear elastic behavior of two-dimensional molybdenum disulfide. *Phys. Rev. B* **87,** 035423 (2013).
12. Feng, J., Qian, X. F., Huang, C. W. & Li, J. Strain-engineered artificial atom as a broad-spectrum solar energy funnel. *Nat. Photon.* **6,** 866–872 (2012).
13. Radisavljevic, B., Radenovic, A., Brivio, J., Giacometti, V. & Kis, A. Single-layer MoS_2 transistors. *Nat. Nanotechnol.* **6,** 147–150 (2011).
14. Lee, G.-H. *et al.* Flexible and transparent MoS_2 field-effect transistors on hexagonal boron nitride-graphene heterostructures. *ACS Nano* **7,** 7931–7936 (2013).
15. Fiori, G. *et al.* Electronics based on two-dimensional materials. *Nat. Nanotechnol.* **9,** 768–779 (2014).
16. Wilson, J. A. & Yoffe, A. D. Transition metal dichalcogenides discussion and interpretation of observed optical, electrical and structural properties. *Adv. Phys.* **18,** 193–335 (1969).
17. Heising, J. & Kanatzidis, M. G. Structure of restacked MoS_2 and WS_2 elucidated by electron crystallography. *J. Am. Chem. Soc.* **121,** 638–643 (1999).
18. Eda, G. *et al.* Coherent atomic and electronic heterostructures of single-layer MoS_2. *ACS Nano* **6,** 7311–7317 (2012).
19. Duerloo, K. A. N., Li, Y. & Reed, E. J. Structural phase transitions in two-dimensional Mo- and W-dichalcogenide monolayers. *Nat. Commun.* **5,** 4214 (2014).
20. Keum, D. H. *et al.* Bandgap opening in few-layered monoclinic $MoTe_2$. *Nat. Phys.* **11,** 482–486 (2015).
21. Ali, M. N. *et al.* Large, non-saturating magne-toresistance in WTe_2. *Nature* **514,** 205–208 (2014).
22. Qian, X.-F., Liu, J.-W., Fu, L. & Li, J. Quantum spin Hall effect in two-dimensional transition metal dichalcogenides. *Science* **346,** 1344–1347 (2014).
23. Park, J. C. *et al.* Phase-engineered synthesis of centimeter-scale 1T′-and 2H-molybdenum ditelluride thin films. *ACS Nano* **9,** 6548–6554 (2015).
24. Zhou, L. *et al.* Large-area synthesis of high-quality uniform few-layer $MoTe_2$. *J. Am. Chem. Soc.* **137,** 11892–11895 (2015).
25. Voiry, D. *et al.* Enhanced catalytic activity in strained chemically exfoliated WS_2 nanosheets for hydrogen evolution. *Nat. Mater.* **12,** 850–855 (2013).
26. Aizu, K. Possible species of ferroelastic crystals and of simultaneously ferroelectric and fer-roelastic crystals. *J. Phys. Soc. Jpn* **27,** 387–396 (1969).
27. Salje, E. K. H. Ferroelastic materials. *Annu. Rev. Mater. Res.* **42,** 265–283 (2012).
28. Salje, E. K. H. *Phase Transitions in Ferroelastic and Co-Elastic Crystals* (Cambridge Univ. Press, 1990).

29. Kushima, A., Qian, X.-F., Zhao, P., Zhang, S.-L. & Li, J. Ripplocations in van der Waals layers. *Nano Lett.* **15,** 1302–1308 (2015).
30. Henkelman, G., Uberuaga, B. P. & Jonsson, H. A climbing image nudged elastic band method for finding saddle points and minimum energy paths. *J. Chem. Phys.* **113,** 9901–9904 (2000).
31. Sheppard, D., Xiao, P., Chemelewski, W., Johnson, D. D. & Henkelman, G. A generalized solid-state nudged elastic band method. *J. Chem. Phys.* **136,** 074103 (2012).
32. Tang, S., Mahanti, S. D. & Kalia, R. K. Ferroelastic phase transition in two-dimensional molecular solids. *Phys. Rev. Lett.* **56,** 484–487 (1986).
33. Huang, P. Y. *et al.* Grains and grain boundaries in single-layer graphene atomic patchwork quilts. *Nature* **469,** 389–392 (2011).
34. van der Zande, A. M. *et al.* Grains and grain boundaries in highly crystalline monolayer molybdenum disulphide. *Nat. Mater.* **12,** 554–561 (2013).
35. Cho, S. *et al.* Phase patterning for ohmic homojunction contact in $MoTe_2$. *Science* **349,** 625–628 (2015).
36. Brown, B. E. The crystal structures of WTe_2 and high-temperature $MoTe_2$. *Acta Crystallogr.* **20,** 268–274 (1966).
37. Voiry, D., Mohite, A. & Chhowalla, M. Phase engineering of transition metal dichalco-genides. *Chem. Soc. Rev.* **44,** 2702–2712 (2015).
38. Chhowalla, M., Voiry, D., Yang, J., Shin, H. S. & Loh, K. P. Phase-engineered transition-metal dichalcogenides for energy and electronics. *MRS Bull.* **40,** 585–591 (2015).
39. Gordon, R. A., Yang, D., Crozier, E. D., Jiang, D. T. & Frindt, R. F. Structures of exfoliated single layers of WS_2, MoS_2, and $MoSe_2$ in aqueous suspension. *Phys. Rev. B* **65,** 125407 (2002).
40. Enyashin, A. N. *et al.* New route for stabilization of 1T-WS_2 and MoS_2 phases. *J. Phys. Chem. C* **115,** 24586–24591 (2011).
41. Bhattacharya, K. *Microstructure of Martensite: Why It Forms and How IT Gives Rise to the Shape-Memory Effect* (Oxford Univ. Press, 2003).
42. Salje, E. & Zhang, H. Domain boundary engineering. *Phase Transit.* **82,** 452–469 (2009).
43. Lebegue, S., Bjorkman, T., Klintenberg, M., Nieminen, R. M. & Eriksson, O. Two-dimensional materials from data filtering and Ab Initio calculations. *Phys. Rev. X* **3,** 031002 (2013).
44. Lin, Y.-C. *et al.* Single-layer ReS_2: two-dimensional semiconductor with tunable in-plane anisotropy. *ACS Nano* **9,** 11249–11257 (2015).
45. Kresse, G. & Furthmiiller, J. Efficiency of ab-initio total energy calculations for metals and semiconductors using a plane-wave basis set. *Comput. Mater. Sci.* **6,** 15–50 (1996).
46. Kresse, G. & Furthmiiller, J. Efficient iterative schemes for ab initio total-energy calculations using a plane-wave basis set. *Phys. Rev. B* **54,** 11169–11186 (1996).
47. Blochl, P. E. Projector augmented-wave method. *Phys. Rev. B* **50,** 17953–17979 (1994).
48. Perdew, J. P. *et al.* Atoms, molecules, solids, and surfaces: Applications of the generalized gradient approximation for exchange and correlation. *Phys. Rev. B* **46,** 6671–6687 (1992).
49. Perdew, J. P., Burke, K. & Ernzerhof, M. Generalized gradient approximation made simple. *Phys. Rev. Lett.* **77,** 3865–3868 (1996).
50. Monkhorst, H. J. & Pack, J. D. Special points for Brillouin-zone integrations. *Phys. Rev. B* **13,** 5188–5192 (1976).

Acknowledgements

This work was primarily supported by the Center for Excitonics, an Energy Frontier Research Center funded by the US Department of Energy, Office of Science, Basic Energy Sciences under award no. DE-SC0001088. Our work was also supported by National Science Foundation under grant no. DMR-1410636. Computational time on the Extreme Science and Engineering Discovery Environment (XSEDE) under the grant no. TG-DMR130038 is gratefully acknowledged.

Author contributions

J.L. and W.L. designed the research. W.L. carried out the first-principles calculations. W.L. and J.L. wrote the paper.

Additional information

Competing financial interests: The authors declare no competing financial interests.

6

A liquid-crystalline hexagonal columnar phase in highly-dilute suspensions of imogolite nanotubes

Erwan Paineau[1], Marie-Eve M. Krapf[2], Mohamed-Salah Amara[1], Natalia V. Matskova[1], Ivan Dozov[1], Stéphan Rouzière[1], Antoine Thill[2], Pascale Launois[1] & Patrick Davidson[1]

Liquid crystals have found wide applications in many fields ranging from detergents to information displays and they are also increasingly being used in the 'bottom-up' self-assembly approach of material nano-structuration. Moreover, liquid-crystalline organizations are frequently observed by biologists. Here we show that one of the four major lyotropic liquid-crystal phases, the columnar one, is much more stable on dilution than reported so far in literature. Indeed, aqueous suspensions of imogolite nanotubes, at low ionic strength, display the columnar liquid-crystal phase at volume fractions as low as ~0.2%. Consequently, due to its low visco-elasticity, this columnar phase is easily aligned in an alternating current electric field, in contrast with usual columnar liquid-crystal phases. These findings should have important implications for the statistical physics of the suspensions of charged rods and could also be exploited in materials science to prepare ordered nanocomposites and in biophysics to better understand solutions of rod-like biopolymers.

[1] Laboratoire de Physique des Solides, CNRS, Univ. Paris-Sud, Université Paris-Saclay, Orsay 91405, France. [2] LIONS, NIMBE, CEA, CNRS, Université Paris-Saclay, CEA-Saclay, Gif sur Yvette 91191, France. Correspondence and requests for materials should be addressed to E.P. (email: erwan-nicolas.paineau@u-psud.fr).

Lyotropic liquid crystals are obtained when rod-like objects (stiff polymers, detergent micelles, nanoparticles) are dispersed in a solvent. According to their symmetries, liquid-crystalline phases belong to four main classes: nematic, lamellar, columnar and cubic[1]. On dilution, the nematic phase remains stable down to very low concentrations (~0.1%) if its constituents have a large enough aspect ratio[2]. Lamellar and cubic phases stable at low concentrations (~1%) were also reported[3–6]. In contrast, the columnar phase, which occurs in suspensions of DNA, rod-like viruses, surfactant micelles and metal oxide nanorods, was only detected at large concentrations (~10–70%)[7–11].

Imogolite nanotubes (INTs) which were first discovered in soils formed from volcanic ash are the aluminosilicate counterparts of carbon nanotubes (CNT)[12,13]. However, unlike the latter, INT are easily synthesized in large amounts by hydrothermal techniques, without need for post-synthesis purification steps, and they are readily dispersed in aqueous solvents to form stable colloidal suspensions[14,15]. INT are very similar to CNT in terms of dimensions, aspect ratios and rigidity. They could therefore be used for the same kind of applications as CNT, where geometrical features are important, such as nanostructuring agents or for the elaboration of composite materials. Indeed, both types of nanotubes have comparable Young moduli ($2–10 \times 10^8$ Pa) but INTs are much easier to functionalize than CNT, which could help optimizing interfacial interactions with the matrix. Moreover, germanium-based (Ge-INT) INTs, similar to usual silicon-based INTs (Si-INT), are easily synthesized from germanium precursors. Furthermore, like CNT, INT occur as single-wall (SW) or double-wall (DW) nanotubes. As expected for highly elongated nanorods, suspensions of INT form a nematic phase at very low volume fractions[16].

Here we report on the observation of a columnar phase in suspensions of INTs at concentrations around 0.3%. The stability domain of the columnar phase therefore extends to much lower concentrations than previously inferred[17–19]. Moreover, this dilute phase has low visco-elasticity and can easily be aligned in an electric field, unlike other lyotropic columnar mesophases.

Results

Nanotube characterization. Two different types of INT were investigated: SW Si-INT and DW Ge-INT. Wide-angle X-ray scattering measurement (WAXS) (Fig. 1) was carried out to determine the diameter of the nanotubes. For sufficiently small scattering wave-vectors, that is, $< 10\,\mathrm{nm}^{-1}$, the WAXS diagrams can be fitted within the homogeneous approximation: imogolite SW nanotubes can be modelled by a homogeneous hollow cylinder with an internal radius R_i, a wall thickness t_{wall} and an average electronic density ρ_{INT}. The external radius is $R_e = R_i + t_{wall}$ (inset Fig. 1a). DW nanotubes can be modelled by two concentric cylinders with the same wall thickness. The external radius of DW nanotubes is thus $R_e = R_i + 2t_{wall} + t_{intertube}$ with $t_{intertube}$ the gap between the inner and outer tubes (inset Fig. 1b).

The scattered X-ray intensity for individual SW nanotubes well-dispersed in suspension writes:

$$I_{ind}(Q) \propto \frac{1}{Q}\left[(\rho_{INT} - \rho_{water})(F_{R_e} - F_{R_i})\right]^2, \tag{1}$$

where F_R refers to the Fourier transform of the projection of a full cylinder of radius R along its axis:

$$F_R(Q) \propto \frac{RJ_1(QR)}{Q}, \tag{2}$$

with J_1 the cylindrical Bessel function of order 1; ρ_{water} is the water electronic density inside and outside the cylinder, assumed to be the same in first approximation. For individual DW nanotubes in suspension, one finds:

$$I_{ind}(Q) \propto \frac{1}{Q}\left[(\rho_{INT} - \rho_{water})(F_{R_e} - F_{R_e - t_{wall}} + F_{R_i + t_{wall}} - F_{R_i})\right]^2. \tag{3}$$

Equations (1) and (3) were used to fit the experimental scattering curves in Fig. 1 and to estimate nanotube diameters. SW Si-INT and DW Ge-INT have, respectively, diameters of 2.7 nm and 4.3 nm and lengths of about 700 nm and 500 nm, the latter being determined by atomic force microscopy (AFM) (Supplementary Fig. 1). The length of the nanotubes is very polydisperse (~55% and 77%, respectively) but their diameter is well-defined (~5%)[15].

Polarized optical microscopy. In polarized-light microscopy, the suspensions of INTs exhibit a clear phase separation at very low volume fractions ϕ (Fig. 2). The bottom phase is birefringent and denser than the isotropic (I) top phase, which is not devoid of INT because it shows strong flow-birefringence. The schlieren texture of the bottom phase proves that it is nematic (N). Indeed, in freshly prepared samples, nematic droplets, called tactoids, form in a few days and sediment, both in test-tubes and in capillaries stored vertically (Fig. 3). This first order I/N transition,

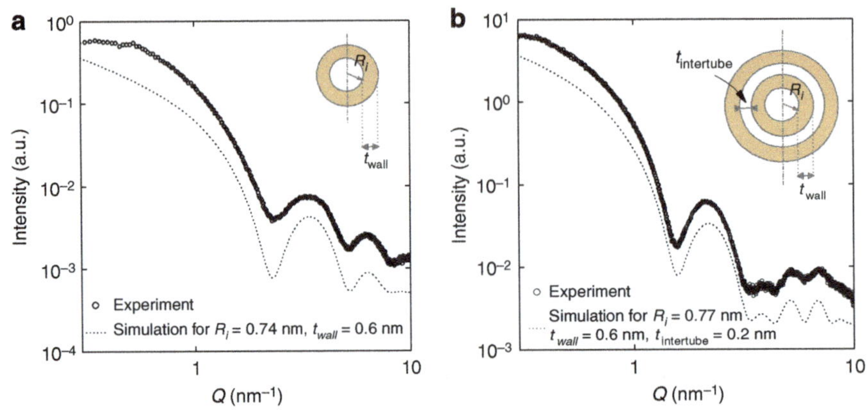

Figure 1 | WAXS measurements of imogolite nanotubes. Experimental (open circles) and calculated (dotted line) WAXS curves of (**a**) SW Si-INT and (**b**) DW Ge-INT in suspension. The insets represent sketches of a single-walled and a double-walled nanotube. For the sake of clarity, calculated curves (using equations (1) and (3)) have been translated vertically.

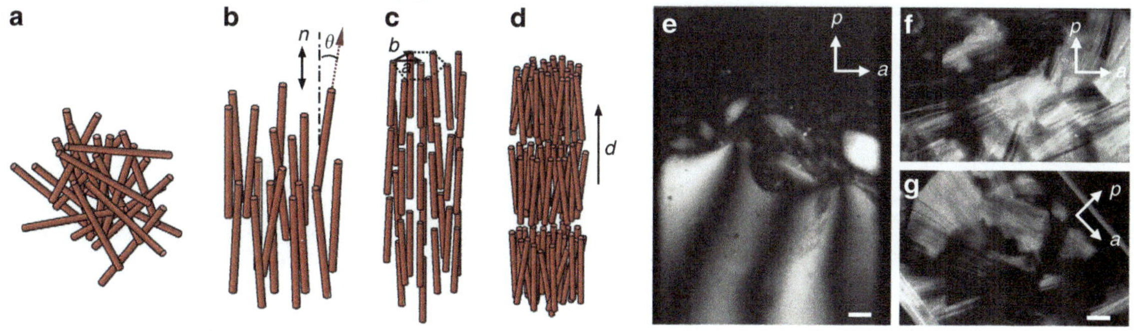

Figure 2 | Liquid-crystal phases in imogolite suspensions. Schematic representations of (**a**) isotropic liquid, (**b**) nematic, (**c**) columnar and (**d**) smectic liquid-crystal phases. Texture photographs of (**e**) isotropic/nematic phase coexistence with sedimenting spindle-shaped nematic droplets ($\phi = 0.07\%$) and (**f**,**g**) columnar phase ($\phi = 0.31\%$) obtained with SW Si-INT suspensions. Scale bar, 200 μm.

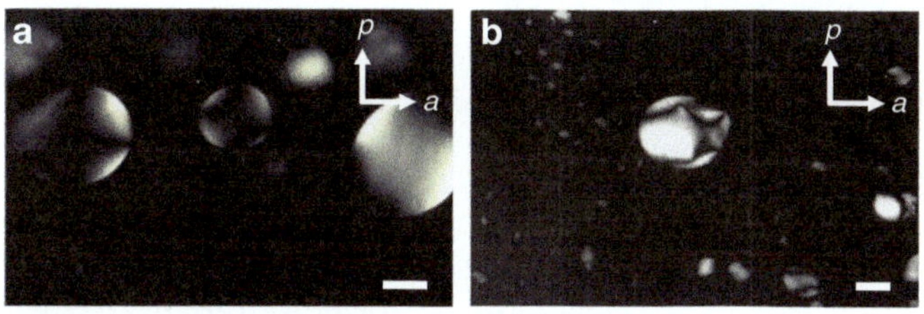

Figure 3 | Nematic tactoids in imogolite suspensions. Texture photographs of (**a**) SW Si-INT ($\phi = 0.07\%$) and (**b**) DW Ge-INT ($\phi = 0.12\%$) suspensions. Scale bar, 200 μm.

already reported for other suspensions of INT, is well-described by the Onsager model[2,15,16,20]. However, at slightly larger volume fractions (Fig. 2b,c), very different textures suggest the existence of another liquid-crystalline phase but do not allow unambiguous identification of its nature.

Small-angle X-ray scattering experiments. The small-angle X-ray scattering (SAXS) patterns of the nematic phase (Figs 4a,b and 5a,b) do not show any sharp reflection, as expected. They display several broad scattering rings (arrows in Figs 4b and 5b) whose positions depend on volume fraction. These rings are due to a liquid-like structure factor arising from the local positional order of the INT in the plane perpendicular to their main axis.

In stark contrast, the scattering patterns of the second mesophase display many thin diffraction rings that can all be indexed with a two-dimensional (2D) hexagonal lattice (Fig. 4c,d and Supplementary Fig. 2). Therefore, the phase is of the columnar hexagonal (Col_H) type where the rods organize on a hexagonal lattice perpendicular to the INT average direction (C_6 axis). The dilute columnar phase was discovered in suspensions of both SW Si-INT (Fig. 5c,d) and DW Ge-INT, which shows that its occurrence is not related to either nanotube composition or wall structure. Altogether, on increasing volume fraction ϕ, the phase sequence of SW Si-INT is I 0.06% N 0.24% Col_H and that of DW Ge-INT is I 0.08% N 0.13% Col_H. At larger ϕ ($\sim$0.7%), a sol/gel transition occurs that prevents samples from reaching full thermodynamic equilibrium. No lamellar mesophase was observed, most probably because of the large length polydispersity of the INT.

The sharpness of the X-ray reflections allows us to measure accurately the lattice parameter, a (Supplementary Fig. 2). Assuming a homogeneous hexagonal columnar phase of nanotubes, one can calculate its volume fraction using the

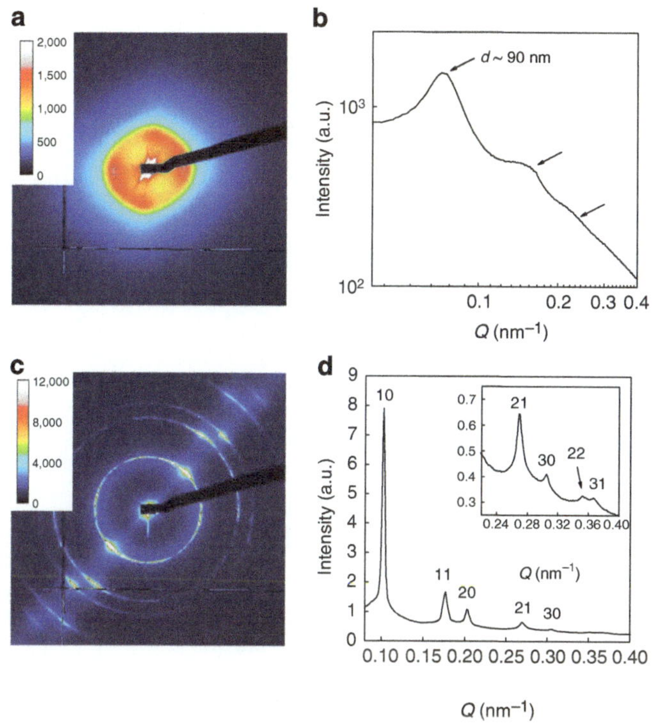

Figure 4 | SAXS measurements of DW Ge-INT suspensions. SAXS patterns of (**a**) nematic ($\phi = 0.12\%$) and (**c**) columnar ($\phi = 0.22\%$) DW Ge-INT suspensions and (**b**,**d**) their corresponding $I(Q)$ curves. (The hk indices in **d** refer to the 2D hexagonal reciprocal lattice peaks.)

formula: $\phi = (2\pi/3^{1/2})(R/a)^2$ where R is the nanotube radius. For example, for SW Si-INT, we obtain $\phi = 0.28\%$ to be compared with the volume fraction of 0.31% measured by slow drying. Similarly, for DW Ge-INT, we obtain 0.25% to be compared with 0.23%. This confirms both the high dilution of the columnar phase and its homogeneity.

Due to the high dilution, the lattice parameter a of the columnar phase is quite large; for example, it is 71.3 nm for DW Ge-INT at $\phi = 0.22\%$. This value is more than 15 times larger than the nanorod outer diameter (4.3 nm). Moreover, numerical simulations of the scattering patterns from INT on a 2D hexagonal lattice allowed us to determine the columnar domain size from the widths of the diffraction lines (see Methods section). The typical domain size obtained, $\sim$1.6 μm, in the plane perpendicular to the C_6 axis, is about 20 times the lattice spacing in a domain, meaning that some 450 INT diffract X-rays

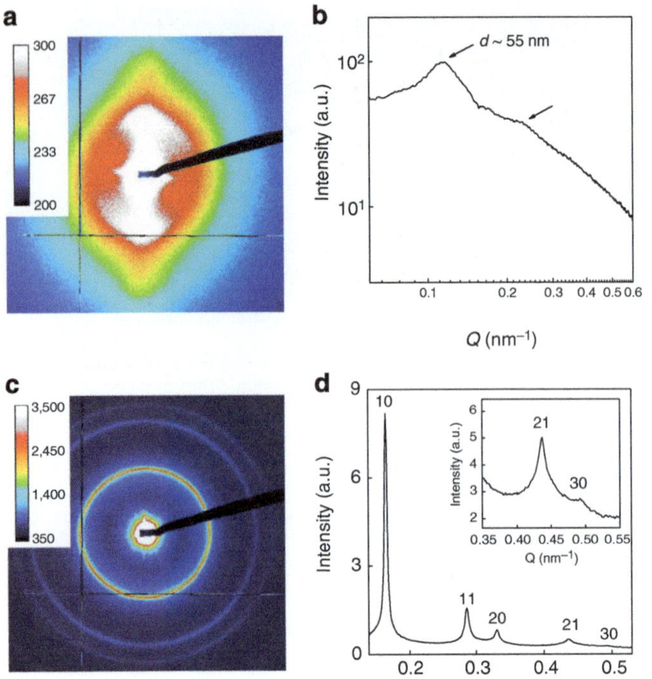

Figure 5 | SAXS measurements of SW Si-INT suspensions. SAXS patterns of (**a**) nematic ($\phi = 0.14\%$) and (**c**) columnar ($\phi = 0.31\%$) SW Si-INT suspensions and (**b**,**d**) their corresponding $I(Q)$ curves. (The hk indices in **d** refer to the 2D hexagonal reciprocal lattice peaks.)

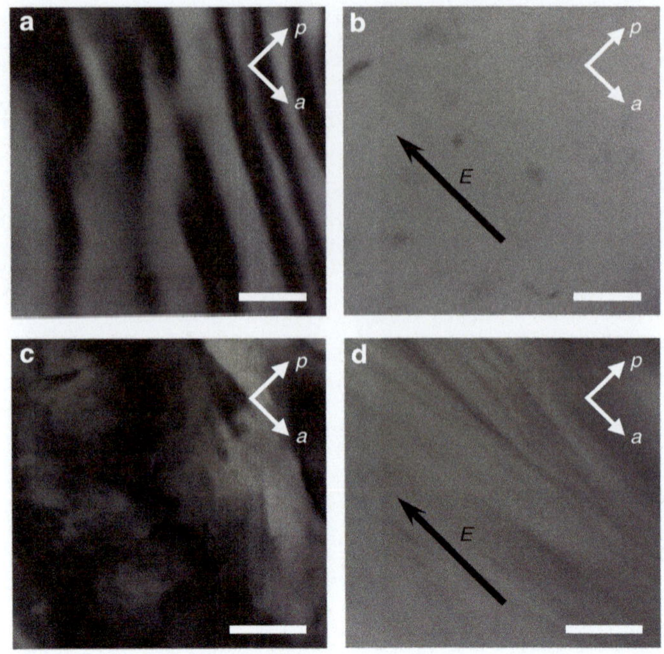

Figure 6 | Texture photographs of aligned imogolite suspensions. (**a**,**b**) nematic ($\phi = 0.12\%$) and (**c**,**d**) columnar ($\phi = 0.22\%$) DW Ge-INT suspensions before (**a**,**c**) and after (**b**,**d**) application of an electric field $E = (100\,\mathrm{V\,mm^{-1}}, 500\,\mathrm{kHz})$. Scale bar, 100 μm.

coherently in that plane. Correlatively, the observation of 7 orders of diffraction shows that the positional fluctuations of the INT are very low, only a few % of the lattice spacing.

Alignment under electric field. INTs in aqueous suspensions readily respond to a high-frequency alternating current electric field because they are highly charged anisotropic objects surrounded by ionic clouds. Indeed, both nematic and columnar phases are easily aligned by applying a moderate field ($\sim 100\,\mathrm{V\,mm^{-1}}$) for about 10 s only (Fig. 6).

The birefringence Δn of these columnar monodomains was measured ($\Delta n = 8.2 \times 10^{-5}$ for SW Si-INT and 1.7×10^{-4} for DW Ge-INT), which allows cross-checking the low value of the volume fraction (see Methods section). Scattering patterns of well-aligned INT were recorded from suspensions submitted to an electric field *in situ*, both in the nematic and columnar phases (Figs 7 and 8). These patterns prove that the INT align parallel to the field. The nematic order parameter, $S = 0.90 \pm 0.05$, is very large, which agrees with the strongly first order character of the I/N transition. In the columnar phase, the dependence of scattered intensity versus azimuthal angle should rather be fitted by a gaussian distribution of columnar crystallites. The small width of this distribution ($\sim 12°$ full width at half maximum) illustrates the good alignment of the columnar phase in the field, which is very original for such colloidal phases that are usually rather stiff.

Discussion

The occurrence at large volume fractions of a columnar mesophase is firmly predicted for suspensions of rigid hard rods of polydispersity in length larger than 18% (ref. 18). However, the calculation of the phase diagrams of charged nanorods in polar solvents is still an active research field, due to the intrinsic complexity of the electrostatic interactions in these systems[19,21,22]. The influence of particle polydispersity is probably less important in these dilute systems than in dense ones but it still needs to be investigated theoretically. Experimentally, the columnar phases of rod-like particles (DNA, viruses, polypeptides, polysaccharides, metallic or mineral nanoparticles...) reported so far occur at volume fractions of ~10–50% and the lattice spacing is close to the rod diameter. In comparison, the columnar phase of INT occurs at volume fractions that are 2 orders of magnitude lower. Such difference of behaviour is probably related to the huge aspect ratio (100–300) of INTs, which also favours the nematic phase[16]. Moreover, the ionic strength of INT suspensions was kept low on purpose ($\sim 10^{-4}\,\mathrm{M}$) to ensure a large Debye length ($\kappa^{-1} \sim 30\,\mathrm{nm}$) extending the range of electrostatic repulsions between the charged INT. Indeed, adding salt to the suspensions brings about their flocculation. To account for the very low volume fraction of the columnar phase of charged INT, it is tempting to use a simple renormalization of the phase diagram of hard rods by introducing an effective diameter, $D_{\mathrm{eff}} = D + 2\kappa^{-1}$. Starting from a typical volume fraction of 0.5 for the onset of the columnar phase of hard rods[18], with $\kappa^{-1} = 30\,\mathrm{nm}$ at an ionic strength of $10^{-4}\,\mathrm{M}$, volume fractions of 0.093% and 0.22% are, respectively, obtained for SW Si-INT and DW Ge-INT. These predicted values are in fair agreement with the experimental ones (0.24% and 0.13%, respectively). However, this renormalization procedure is quite dubious because the interaction potential of charged particles is always anisotropic, even at long distance and intrinsically much more complex than that of hard particles, as recently demonstrated in the case of disks[22]. The absence of strong enough electrostatic repulsions in most suspensions of CNT, which could balance the van der Waals attractions, is probably the reason why a similar dilute columnar phase was not yet reported in these systems[23].

The very low concentration of the columnar phase of INT makes this phase reminiscent of the highly ‘swollen’ lamellar

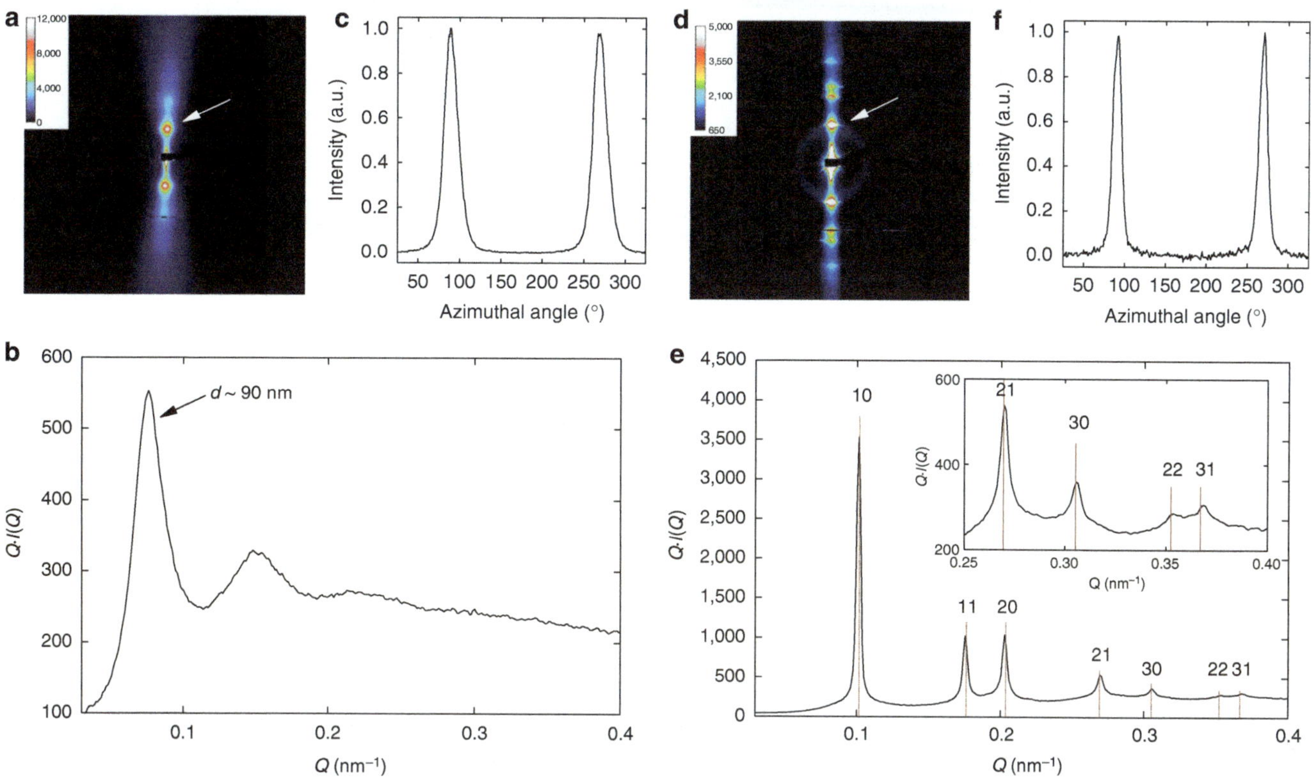

Figure 7 | SAXS measurements of aligned imogolite nanotube suspensions. (**a-c**) nematic ($\phi = 0.12\%$) and (**d-f**) columnar ($\phi = 0.22\%$) DW Ge-INT suspensions submitted to an electric field $E = 100\ \mathrm{V\,mm^{-1}}$ at 500 kHz. (**a,d**) 2D-SAXS patterns, (**b,e**) structure factor $Q\ I(Q)$ and (**c,f**) azimuthal profiles recorded along a circle going through the diffuse scattering maximum (white arrow in **a** and **d**). In **e**, the hk indices refer to the 2D hexagonal reciprocal lattice peaks (red lines) of Ge-INT in the columnar phase with lattice parameter $a = 71.3$ nm.

phase of surfactant membranes and of the nematic phase of 'swelling clays'[3,24,25]. However, the case of nanotubes is even more surprising because they are 1D objects that can fluctuate more easily in orientation and position without colliding than 2D objects like membranes and nanosheets. Moreover, the orientational order in this system is somewhat comparable to that of gold nanorods diluted in a liquid crystal although the interactions in that case are mediated by the matrix[26,27]. Furthermore, the 2D positional long-range order of the nanotubes in the columnar phase reminds of the 3D crystalline order of both charged spherical particles in highly dilute colloidal crystals[28] and charged nanorods in cubic plastic crystals[29]. However, the columnar liquid-crystalline phase is more intricate than the two other systems because of the onset of long-range orientational order.

The remarkable positional ordering of the charged INT is strong evidence of the large intensity of the electrostatic repulsions between charged linear objects, which stabilize the columnar phase down to very low volume fractions. Therefore, the phase diagrams of suspensions of rod-like objects in polar solvents clearly deserve more thorough theoretical investigations. From an applied point of view, the appearance of the columnar phase at high dilution and its easy alignment in an electric field could be the first step towards the elaboration of anisotropic nanocomposite materials with well-dispersed nanotube filler. Indeed, such hybrid materials keeping liquid-crystalline organization can easily be produced by polycondensation of silica precursors under field, as illustrated in the case of chitin nanorods[30]. Moreover, this system that is optically transparent in a wide spectrum and that can be manipulated by moderate electric fields is somewhat similar to a strongly anisotropic 2D photonic crystal. Although the optical contrast is presently too low to really expect photonic bandgap and other spectacular optical properties, it could be enhanced by doping the nanotubes with dyes[31]. This will result in a periodic anisotropic 2D structure with a strong spatial variation of optical properties, such as dichroism and birefringence, with potential photonic applications.

Methods

Synthesis of aluminosilicate Si-INT nanotubes. Aluminosilicate nanotubes were synthesized by adding tetraethoxysilane to a dilute ($C = 10^{-3}\ \mathrm{mol\,l^{-1}}$) aluminium perchlorate solution in a Teflon beaker. Then, the mixture was slowly hydrolysed by the addition of a $0.1\ \mathrm{mol\,l^{-1}}$ NaOH solution to reach a hydrolysis ratio ([OH]/[Al]) of 2. Solutions were stirred overnight at room temperature and then aged for 5 days into an oven at 90 °C.

Synthesis of aluminogermanate Ge-INT nanotubes. We recently described the simple single-step synthesis of micron-long germanium imogolite-like nanotubes[15]. Briefly, tetraethoxygermanium was mixed in a Teflon beaker with a $0.2\ \mathrm{mol\,l^{-1}}$ aqueous solution of aluminium perchlorate using the imogolite stoichiometric ratio [Al]/[Ge] = 2. Then, a urea ($CO(NH_2)_2$) solution was added at room temperature up to a [urea]/[Al] ratio of 1. Indeed, the thermal decomposition of urea produces *in situ* hydroxyl ions (two hydroxyls per urea molecule) homogeneously distributed in the solution. Immediately after mixing, the solution was transferred in an autoclave and placed in an oven at 140 °C for 5 days.

Sample preparation. At the end of the aging process, the Si- and Ge-INT mixtures were cooled down and dialysed against ultrapure water using 10-kDa membranes (Visking) until the conductivity fell below $5\ \mu\mathrm{S\,cm^{-1}}$. Si-INT suspensions, due to their low concentration of aluminium precursor, were concentrated with an ultrafiltration set-up. In both cases (Si and Ge-INT), the pH of the suspensions was close to 6. Samples with different concentrations were then prepared by dilution with ultrapure water and stored in air-tight vials before measurements.
For all samples, the mass concentration was determined by weight loss on drying.

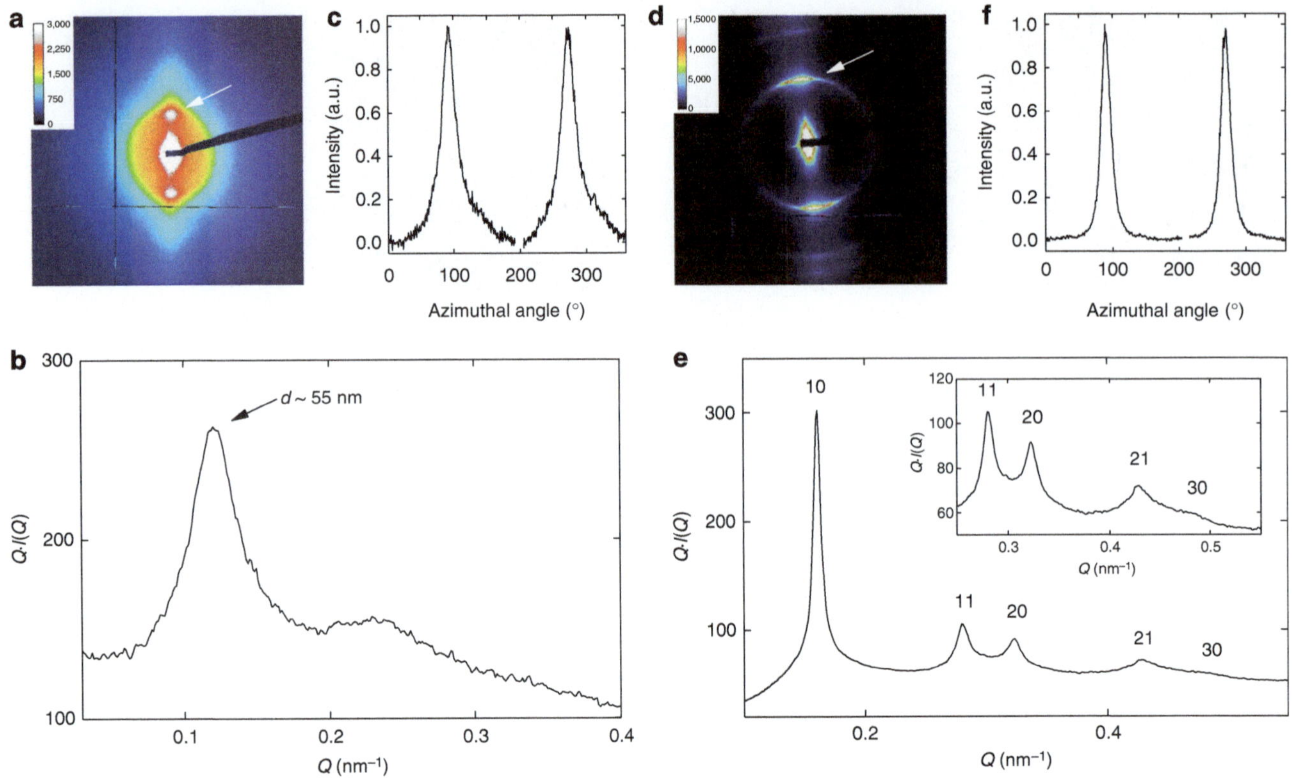

Figure 8 | SAXS measurements of aligned Si-INT suspensions. (**a–c**) Nematic ($\phi = 0.14\%$) and (**d–f**) columnar ($\phi = 0.31\%$) SW Si-INT suspensions submitted to an electric field $E = 200\ \mathrm{V\,mm^{-1}}$ at 500 kHz. (**a,d**) 2D-SAXS patterns, (**b,e**) structure factor $Q.I(Q)$ and (**c,f**) azimuthal profiles recorded along a circle going through the scattering maximum (white arrow in **a** and **d**). In **e**, the hk indices refer to the 2D hexagonal reciprocal lattice peaks.

Atomic force microscopy. The average length of Si and Ge-INT samples were measured using AFM (Digital Instruments, Nanoscope V). The tapping mode was employed for imaging nanotubes. Freshly cleaved mica sheets were dipped into very dilute suspensions ($\sim 20\ \mathrm{mg\,l^{-1}}$) for 1 min followed by three cleanings in ultrapure water to remove unabsorbed nanotubes. The mica sheets were then dried at 60 °C overnight before performing measurements.

Electrostatic parameters of INTs. The point of zero charge (PZC) of synthetic INTs was either taken from literature for SW Si-INT (PZC ~ 9.5)[32] or measured by electrophoretic mobility for DW Ge-INT (PZC = 11). This means that INTs are positively charged at pH 6. Their ζ-potential, deduced from the measurements of electrophoretic mobility, is + 45 mV for SW Si-INT and + 50 mV for DW Ge-INT, at this pH. A bare linear charge density of ~ $0.7e^-$ per nm was estimated for INTs from the measurements of electrophoretic mobility and the nanotube dimensions[33].

Polarized-light microscopy. The samples were transferred into thin flat optical glass capillaries (VitroCom, 0.2 × 2 mm), which were flame sealed to avoid water evaporation and stored vertically to follow the phase separation. Optical textures were recorded with a polarizing microscope (BX51-P, Olympus) equipped with a CCD camera.

Birefringence measurements. The birefringence of the samples was measured directly under the microscope using an Olympus Berek optical compensator. Moreover, using the formula: $\Delta n = \Delta n_p \phi S$, with $S \sim 0.9$ and where Δn_p is the estimated specific birefringence of the particle ($\Delta n_p = 0.019$ for SW Si-INT and 0.063 for DW Ge-INT), the volume fraction of the columnar phase can be retrieved and compares very well with those determined by weight loss on drying, all the more when one considers the approximations involved in the calculation of Δn_p. For example, we obtain $\phi = 0.42\%$ instead of 0.31% for SW Si-INT and $\phi = 0.27\%$ instead of 0.23% for DW Ge-INT.

Electric field set-up. The electric field is applied to the samples along the capillary axis, perpendicular to the X-ray beam or to the direction of optical observation. This geometry presents an important advantage for the *in situ* application of the field in the SAXS experiments, as the electrodes are outside of the observation window and only the thin 10-μm glass walls of the capillary are in the beam path. The electric field set-up is described in details in ref. 34. Briefly, a pair of external electrodes made of aluminium foil, encircles completely the capillary, 2 mm apart along the capillary axis. The electrodes are pressed onto the capillary protected by a cushion of soft foam to optimize the electrode/glass wall contact without damaging the capillary tube. High-frequency (500 kHz) sinusoidal alternating current voltage, either continuous or pulsed, was applied to the electrodes, with amplitudes varying from 0 to 1,100 V. The electric field penetrating into the suspension is moderately strong, up to about $200\ \mathrm{V\,mm^{-1}}$ root mean squared (r.m.s.) value (corrected for the attenuation due to the screening by the mobile charges and the dielectric mismatch between suspension and capillary). The field is parallel to the capillary axis, with highly uniform strength and orientation in the whole inter-electrode area accessible for observation. This electric-field holder was fitted onto either the stage of the polarizing microscope for electro-optical measurements or the experimental table of the SWING beamline at synchrotron SOLEIL, with the capillary tubes (and the electric field) held horizontal for electro-SAXS experiments. We report the experimental results as a function of the applied voltage. The r.m.s. value of the electric field E inside the sample was estimated from numerical simulations and frequency-dependent electric-birefringence experiments. Under the present experimental conditions, we obtained $E = 180 \times U_{\mathrm{r.m.s.}}\ (\mathrm{V\,m^{-1}})$, giving $E = 200\ \mathrm{V\,mm^{-1}}$ for the highest applied voltage, $U_{\mathrm{r.m.s.}} = 1{,}100$ V.

Small and wide-angle X-ray scattering. SAXS experiments were carried out at the SWING beamline of the synchrotron SOLEIL (Orsay, France). Measurements were performed at a wavelength of $\lambda = 0.138$ nm with an incident X-ray beam of $\sim 350 \times 100\ \mu\mathrm{m}^2$ in the horizontal and vertical directions, respectively. 2D scattering patterns were acquired on an AVIEX CCD detector formed by four detectors with 167 μm pixel size and placed in a vacuum detection tunnel, at a sample to detector distance of 6.57 m. The typical accessible range of scattering vector modulus Q was $0.01 - 0.5\ \mathrm{nm^{-1}}$ ($Q = 4\pi \sin\theta/\lambda$, where 2θ is the scattering angle).

Samples of suspensions were filled by mild centrifugation into borosilicate capillary tubes (GLAS, Schönwalde bei Berlin, Germany) of 1 mm diameter and stored vertically after flame-sealing. (Other samples were prepared in 1.5 mm diameter glass tubes and in Kapton tubes as well and showed exactly the same results.)

The sample size in all directions is at least 1,000 times larger than any particle dimension (or spacing). Such ratio rules out any confinement effect[35] in the appearance of the columnar phase, which must then be of thermodynamic bulk origin. SAXS patterns, in the absence of field, are most often anisotropic because of the spurious partial alignment of the phase and/or the presence of several large domains in the volume illuminated by the X-ray beam.

WAXS measurements were performed using a rotating anode generator ($\lambda_{MoK\alpha} = 0.0711$ nm, Rigaku), a multilayer W/Si mirror (Osmic) and home-made hybrid slits, providing a monochromatic beam of 1×1 mm^2 at the sample position. A vacuum chamber behind the sample allows minimization of the X-ray scattering signal of air. The transmitted flux was measured continuously with a photodiode placed in the beam stop. 2D scattering patterns were collected on a MAR research X-ray-sensitive 345-mm plate detector with 100 µm pixel size, placed behind the output window of the vacuum chamber, at a distance of 720 mm from the sample. The typical accessible range of scattering vector modulus Q was $0.3 - 30\,\mathrm{nm}^{-1}$. Samples of suspensions were transferred into kapton cylindrical capillaries of 2.6 mm diameter.

In both cases, the curves of scattering intensity versus Q were deduced from angular integration over $[0, 2\pi]$ of the X-ray scattering patterns previously corrected for water and glass or kapton scattering. For anisotropic patterns, the angular integration was reduced around the direction of maximum intensity. Furthermore, in the case of samples aligned under electric field, angular profiles of scattered intensity at constant Q were also performed. The nematic order parameter, S, of the nematic samples was retrieved from a fit of the scattered intensity versus azimuthal angle[36].

SAXS diagram of INT organized on a 2D hexagonal lattice. Evaluation of the coherence domain size in the columnar phase was made by comparing the measured peak widths with those obtained by calculating the intensity $I(Q)$ scattered by domains of columnar phase of extension D_{domain}, perpendicular to the long axes of INT. One easily obtains[37]:

$$I(Q) \propto \frac{(F_{R_e} - F_{R_i})^2}{Q} \sum_{i,j} J_0(QR_{ij}), \tag{4}$$

for SW-INT and

$$I(Q) \propto \frac{(F_{R_e} - F_{R_e - t_{\mathrm{wall}}} + F_{R_i + t_{\mathrm{wall}}} - F_{R_i})^2}{Q} \sum_{i,j} J_0(QR_{ij}), \tag{5}$$

for DW-INT, where J_0 is the cylindrical Bessel function of order zero and the sum runs over all nanotubes i and j within the columnar domain, R_{ij} is the distance between nanotubes i and j perpendicularly to their axis ($R_{ij} \leq D_{\mathrm{domain}}$). Calculated peak widths decrease with increasing domain extension. Small measured peak widths thus lead to large domain sizes.

References

1. De Gennes, P.-G. & Prost, J. *The Physics of Liquid Crystals* Vol. 23 (Clarendon press Oxford, 1993).
2. Vroege, G. J. & Lekkerkerker, H. N. W. Phase-transitions in lyotropic colloidal and polymer liquid-crystals. *Rep. Prog. Phys.* **55,** 1241–1309 (1992).
3. Safinya, C. R., Sirota, E. B., Roux, D. & Smith, G. S. Universality in interacting membranes—the effect of cosurfactants on the interfacial rigidity. *Phys. Rev. Lett.* **62,** 1134–1137 (1989).
4. Peter, U., Konig, S., Roux, D. & Bellocq, A. M. Extremely dilute lyotropic 3D crystalline phase in a water/oil/surfactant/cosurfactant mixture. *Phys. Rev. Lett.* **76,** 3866–3869 (1996).
5. Gabriel, J. C. P. *et al.* Swollen liquid-crystalline lamellar phase based on extended solid-like sheets. *Nature* **413,** 504–508 (2001).
6. Barriga, H. M. G. *et al.* Temperature and pressure tuneable swollen bicontinuous cubic phases approaching nature's length scales. *Soft Matter* **11,** 600–607 (2015).
7. Bawden, F. C., Pirie, N. W., Bernal, J. D. & Fankuchen, I. Liquid crystalline substances from virusinfected plants. *Nature* **138,** 1051–1052 (1936).
8. Livolant, F. & Leforestier, A. Condensed phases of DNA: structures and phase transitions. *Prog. Polym. Sci.* **21,** 1115–1164 (1996).
9. Vroege, G. J., Thies-Weesie, D. M. E., Petukhov, A. V., Lemaire, B. J. & Davidson, P. Smectic liquid-crystalline order in suspensions of highly polydisperse goethite nanorods. *Adv. Mater.* **18,** 2565–2568 (2006).
10. Grelet, E. Hard-rod behaviour in dense mesophases of semiflexible and rigid charged viruses. *Phys. Rev. X* **4,** 021053 (2014).
11. Ramos, L. & Fabre, P. Swelling of a lyotropic hexagonal phase by monitoring the radius of the cylinders. *Langmuir* **13,** 682–686 (1997).
12. Yoshinaga, N. & Aomine, S. Imogolite in some Ando soils. *Soil Sci. Plant Nutr.* **8,** 22–29 (1962).
13. Cradwick, P. D. *et al.* Imogolite, a hydrated aluminium silicate of tubular structure. *Nat. Phys. Sci.* **240,** 187–189 (1972).
14. Thill, A. *et al.* Physico-chemical control over the single- or double-wall structure of aluminogermanate imogolite-like nanotubes. *J. Am. Chem. Soc.* **134,** 3780–3786 (2012).
15. Amara, M. S. *et al.* Single-step formation of micron long (OH)(3)Al2O3Ge (OH) imogolite-like nanotubes. *Chem. Commun.* **49,** 11284–11286 (2013).
16. Kajiwara, K., Donkai, N., Hiragi, Y. & Inagaki, H. Lyotropic mesophase of imogolite.1. Effect of polydispersity on phase-diagram. *Makromol. Chem.* **187,** 2883–2893 (1986).
17. Hentschke, R. & Herzfeld, J. Isotropic, nematic, and columnar ordering in systems of persistent flexible hard-rods. *Phys. Rev. A* **44,** 1148–1155 (1991).
18. Bates, M. A. & Frenkel, D. Influence of polydispersity on the phase behaviour of colloidal liquid crystals: a Monte Carlo simulation study. *J. Chem. Phys.* **109,** 6193–6199 (1998).
19. Wensink, H. H. Columnar versus smectic order in systems of charged colloidal rods. *J. Chem. Phys.* **126,** 194901 (2007).
20. Onsager, L. The effects of shape on the interaction of colloidal particles. *Ann. NY Acad. Sci.* **51,** 627–659 (1949).
21. Trizac, E., Bocquet, L. & Aubouy, M. Simple approach for charge renormalization in highly charged macroions. *Phys. Rev. Lett.* **89,** 248301 (2002).
22. Jabbari-Farouji, S., Weis, J. J., Davidson, P., Levitz, P. & Trizac, E. On phase behaviour and dynamical signatures of charged colloidal platelets. *Sci. Rep.* **3,** 3559 (2013).
23. Bisoyi, H. K. & Kumar, S. Carbon-based liquid crystals: art and science. *Liq. Cryst.* **38,** 1427–1449 (2011).
24. Safinya, C. R. *et al.* Steric interactions in a model multimembrane system—a synchrotron x-ray study. *Phys. Rev. Lett.* **57,** 2718–2721 (1986).
25. Paineau, E. *et al.* Liquid-crystalline nematic phase in aqueous suspensions of a disk-shaped natural beidellite clay. *J. Phys. Chem. B* **113,** 15858–15869 (2009).
26. Liu, Q. K. *et al.* Self-alignment of plasmonic gold nanorods in reconfigurable anisotropic fluids for tunable bulk metamaterial applications. *Nano Lett.* **10,** 1347–1353 (2010).
27. Liu, Q. K., Yuan, Y. & Smalyukh, I. I. electrically and optically tunable plasmonic guest-host liquid crystals with long-range ordered nanoparticles. *Nano Lett.* **14,** 4071–4077 (2014).
28. Okubo, T. Giant colloidal single-crystals of polystyrene and silica spheres in deionized suspension. *Langmuir* **10,** 1695–1702 (1994).
29. Liu, B. *et al.* Switching plastic crystals of colloidal rods with electric fields. *Nat. Commun.* **5,** 3092 (2014).
30. Boltoeva, M. Y. *et al.* Electric-field alignment of chitin nanorod-siloxane oligomer reactive suspensions. *Langmuir* **29,** 8208–8212 (2013).
31. Amara, M. S. *et al.* Hybrid, tunable-diameter, metal oxide nanotubes for trapping of organic molecules. *Chem. Mater.* **27,** 1488–1494 (2015).
32. Rotoli, B. M. *et al.* Imogolite: an aluminosilicate nanotube endowed with low cytotoxicity and genotoxicity. *Chem. Res. Toxicol.* **27,** 1142–1154 (2014).
33. Gustafsson, J. P. The surface chemistry of imogolite. *Clay Clay Miner.* **49,** 73–80 (2001).
34. Dozov, I. *et al.* Electric-field-induced perfect anti-nematic order in isotropic aqueous suspensions of a natural beidellite clay. *J. Phys. Chem. B* **115,** 7751–7765 (2011).
35. de las Heras, D., Velasco, E. & Mederos, L. Capillary smectization and layering in a confined liquid crystal. *Phys. Rev. Lett.* **94,** 017801 (2005).
36. Davidson, P., Petermann, D. & Levelut, A. M. The measurement of the nematic order-parameter by X-ray-scattering reconsidered. *J. Phys. II* **5,** 113–131 (1995).
37. Launois, P. *et al.* Transformation of C-70 peapods into double walled carbon nanotubes. *Carbon* **48,** 89–98 (2010).

Acknowledgements

This work was funded by C'Nano Ile de France and by the French ANR agency (contract no. ANR-11-BS08-02, HIMO2). We thank the Soleil Synchrotron (Saint-Aubin, France) for the award of beam-time 20130644 and Javier Perez for local contact during the SAXS experiments. We thank V.Peyre, O.Taché, P.Picot, B.Sarrazin and D.Petermann for technical assistance, and A.Fernandez-Martinez, J.C.Gabriel, O.Poncelet and E.Trizac for helpful discussions.

Author contributions

E.P. and P.D. designed the study. E.P., M.-E.M.K., M.S.A., I.D., A.T., P.L. and P.D. performed the syntheses and experiments. E.P., M.S.A., N.V.M., I.D., S.R., P.L. and P.D. contributed to the data interpretation. E.P. and P.D. wrote the manuscript and all authors participated in manuscript editing.

Additional information

Competing financial interests: The authors declare no competing financial interests.

7

Nonmagnetic band gap at the Dirac point of the magnetic topological insulator $(Bi_{1-x}Mn_x)_2Se_3$

J. Sánchez-Barriga[1], A. Varykhalov[1], G. Springholz[2], H. Steiner[2], R. Kirchschlager[2], G. Bauer[2], O. Caha[3], E. Schierle[1], E. Weschke[1], A.A. Ünal[1], S. Valencia[1], M. Dunst[4], J. Braun[4], H. Ebert[4], J. Minár[4,5], E. Golias[1], L.V. Yashina[6], A. Ney[2], V. Holý[7] & O. Rader[1]

Magnetic doping is expected to open a band gap at the Dirac point of topological insulators by breaking time-reversal symmetry and to enable novel topological phases. Epitaxial $(Bi_{1-x}Mn_x)_2Se_3$ is a prototypical magnetic topological insulator with a pronounced surface band gap of ~100 meV. We show that this gap is neither due to ferromagnetic order in the bulk or at the surface nor to the local magnetic moment of the Mn, making the system unsuitable for realizing the novel phases. We further show that Mn doping does not affect the inverted bulk band gap and the system remains topologically nontrivial. We suggest that strong resonant scattering processes cause the gap at the Dirac point and support this by the observation of in-gap states using resonant photoemission. Our findings establish a mechanism for gap opening in topological surface states which challenges the currently known conditions for topological protection.

[1] Helmholtz-Zentrum Berlin für Materialien und Energie, Elektronenspeicherring BESSY II, Albert-Einstein-Straße 15, 12489 Berlin, Germany. [2] Institut für Halbleiter und Festkörperphysik, Johannes Kepler Universität, Altenbergerstr. 69, 4040 Linz, Austria. [3] Department of Condensed Matter Physics, CEITEC, Masaryk University, Kotlarska 2, 61137 Brno, Czech Republic. [4] Department Chemie, Ludwig-Maximilians-Universität München, Butenandtstr. 5-13, 81377 München, Germany. [5] New Technologies Research Centre, University of West Bohemia, Univerzitni 8, 306 14 Pilsen, Czech Republic. [6] Department of Chemistry, Moscow State University, Leninskie Gory 1/3, 119991 Moscow, Russia. [7] Department of Condensed Matter Physics, Charles University, Ke Karlovu 5, 12116 Prague, Czech Republic. Correspondence and requests for materials should be addressed to J.S.-B. (email: jaime.sanchez-barriga@helmholtz-berlin.de) or to O.R. (email: rader@helmholtz-berlin.de).

The topological surface state (TSS) of the three-dimensional topological insulator (3D TI) Bi_2Se_3 forms a Dirac cone in the band dispersion $E(k_{||})$ of electron energy E versus wave vector component $k_{||}$ parallel to the surface[1]. Differently from graphene, the electron spin is nondegenerate and locked to $k_{||}$, and the Dirac crossing point is protected by time-reversal symmetry (TRS) against distortions of the system[1,2]. This stability has been theoretically investigated demonstrating that nonmagnetic impurities at the surface and in the bulk can form a resonance in the surface-state density of states (DOS) without opening a band gap[3–5]. A magnetic field breaks TRS lifting the Kramers degeneracy $E(k_{||}, \uparrow) = E(-k_{||}, \downarrow)$ between up ($\uparrow$) and down ($\downarrow$) spins, and can open a band gap at the Dirac point[1]. For the two-dimensional quantum-spin-Hall system HgTe, the effect of a perpendicular magnetic field on the topologically protected edge state has been demonstrated successfully[6], and similar effects are expected from magnetic impurities in this system[7]. In subsequent studies on HgTe, the effect of the magnetic field was much smaller, and recently in an inverted electron-hole bilayer from InAs/GaSb, helical edge states proved robust in perpendicular magnetic fields of 8 T (ref. 8). At the surface of a 3D TI, calculations show that magnetic impurities can open a gap at the Dirac point and exhibit ferromagnetic order with perpendicular anisotropy mediated by the TSS through Ruderman–Kittel–Kasuya–Yosida (RKKY) exchange coupling[4,9,10].

3D TIs with magnetic impurities have been studied by angle-resolved photoelectron spectroscopy (ARPES). The magnetic impurities have been incorporated into the bulk[11] and at the surface of Bi_2Se_3 (ref. 10). For Fe impurities, the opening of large surface band gaps of $\sim$100 meV was reported in both cases[10,11]. In a previous work, however, we found that Fe deposited on Bi_2Se_3 does not open a gap for a wide range of preparation conditions, revealing a surprising robustness of the TSS towards magnetic moments[12]. This conclusion also extends to Gd/ Bi_2Se_3 (ref. 13) and Fe/Bi_2Te_3 (ref. 14).

The absence of a surface band gap is consistent with the recent finding that Fe on Bi_2Se_3 does not favour a magnetic anisotropy perpendicular to the surface, at least in the dilute limit[15]. In this respect, bulk impurities in TIs have an advantage over surface impurities in that ferromagnetic order and perpendicular magnetic anisotropies have been achieved in the bulk systems[11,16–21]. Fe incorporated in bulk Bi_2Te_3 is known to order ferromagnetically with a Curie temperature (T_C) of $\sim$12 K for concentrations of $x = 0.04$ showing an easy axis perpendicular to the base plane[16,17]. On the other hand, $(Bi_{1-x}Fe_x)_2Se_3$ has not been found to be ferromagnetic at temperatures above 2 K (ref. 17). For $x = 0.16$ and 0.25, ferromagnetic order was found at 2 K (ref. 11). By ARPES[22], surface-state band gaps in $(Bi_{1-x}Fe_x)_2Se_3$ were reported and assigned to a magnetic origin for Fe concentrations from $x = 0.05$ to 0.25 including nonferromagnetic concentrations such as $x = 0.12$ with a band gap of 45 meV (ref. 11). These band gaps have been attributed to short-range magnetic order. Concerning Mn incorporation, bulk $(Bi_{1-x}Mn_x)_2Se_3$ with $x = 0.02$ has been shown to exhibit a surface band gap with an occupied width of 7 meV (ref. 11), which was suggested as an indication of ferromagnetic order induced by the TSS[11]. However, much larger surface band gaps were observed for n-doped $(Bi_{1-x}Mn_x)_2Se_3$ films[20], where ferromagnetic order of Mn at the surface was found to be strongly enhanced with T_C up to $\sim$45 K. The ferromagnetic order resulted in an unusual spin texture of the TSS. The T_C at the surface was much higher than in the bulk[20], which was partially attributed to Mn surface accumulation. A strong enhancement of the surface T_C was also predicted by mean-field theory for this system (for example, from 73 K (bulk) to 103 K (surface))[23]. Magnetically doped TIs with ferromagnetic order are important as realizations of novel topological phases. When spin degeneracy is lifted by the exchange splitting, bulk band inversion can occur selectively for one spin. If also the Fermi level is in the band gap, as predicted for Cr and Fe in Bi_2Se_3 (ref. 24), this gives rise to an integer quantized Hall conductance σ_{xy} in thin films termed the quantized anomalous Hall effect[7,24]. This has recently been reported for Cr in $(Bi,Sb)_2Te_3$ (ref. 25). When a perpendicular magnetization breaks TRS at the surface of a bulk TI, the resulting mass and surface band gap give rise to quantized edge states. In this case σ_{xy} is half-integer quantized yielding a half quantum Hall effect[26] which can be probed at ferromagnetic domain boundaries at the surface[27] and leads to exotic topological magnetoelectric effects such as the point-charge-induced image magnetic monopole[28,29]. Another topological phase predicted for magnetically doped Bi_2Se_3 is the realization of a 3D Weyl fermion system in which the Dirac-like dispersions become a property of the bulk[30].

Here we present a detailed investigation on the correlation between the TSS and magnetic properties in the Bi-Mn-Se system, where the magnetic impurity Mn is introduced in the bulk of $\sim$0.4 μm thick $(Bi_{1-x}Mn_x)_2Se_3$ epilayers with concentrations up to $x = 0.08$. By ARPES we find remarkably large band gaps of $\sim$200 meV persisting up to temperatures of 300 K far above T_C, which is found to be $\sim$10 K at the surface showing only a limited enhancement by $\leq$4 K over the bulk T_C. We find that the magnitude of the band gaps by far exceeds that expected from TRS breaking due to magnetic order. By ARPES from quantum-well states we exclude that Mn changes the inverted bulk band gap. Using resonant photoemission and *ab initio* calculations we conclude that, instead, impurity-induced resonance states destroy the Dirac point of the TSS. We further support our conclusion by showing that extremely low-bulk concentrations of nonmagnetic In do also open a surface band gap in Bi_2Se_3. Our findings have profound implications for the understanding of the conditions for topological protection.

Results

Concentration and temperature dependence of surface band gap. Figure 1 shows Mn-concentration-dependent high-resolution ARPES dispersions of the TSS and bulk valence-band (BVB) states of $(Bi_{1-x}Mn_x)_2Se_3$ with x up to 0.08 measured at 12 K and 50 eV photon energy. At this energy, contributions from the bulk-conduction band (BCB) do not appear due to the dependence of the photoemission transitions on the component of the electron wave vector perpendicular to the surface $k_\perp$ (ref. 22), but the BCB is partially occupied as data at 21 eV photon energy reveal (Supplementary Fig. 1). The pure Bi_2Se_3 film (Fig. 1a) is n-doped and exhibits a well-resolved Dirac point with high photoemission intensity at a binding energy of $E_D \sim 0.4$ eV that is seen as an intense peak in the energy-distribution curve at zero momentum superimposed as red curve on the right-hand side of the panel. The intact and bright Dirac point marked by a horizontal dashed line in Fig. 1a demonstrates that the TSS is gapless in Bi_2Se_3. At larger binding energies, the lower half of the Dirac cone overlaps with the BVB, which is observed for all $(Bi_{1-x}Mn_x)_2Se_3$ samples (Figs 1b–d). Increasing Mn concentration, we find a gradual upward shift of the band edges in energy, revealing a progressive p-type doping (hole doping). Most notably, a surface band gap opens at the Dirac point with increasing Mn content. In each energy-distribution curve, the opening of the gap is also evident from the development of an intensity dip at the binding energy of the original Dirac point. Thus, the energy separation between the upper Dirac band minimum and the energy position of the intensity dip is about

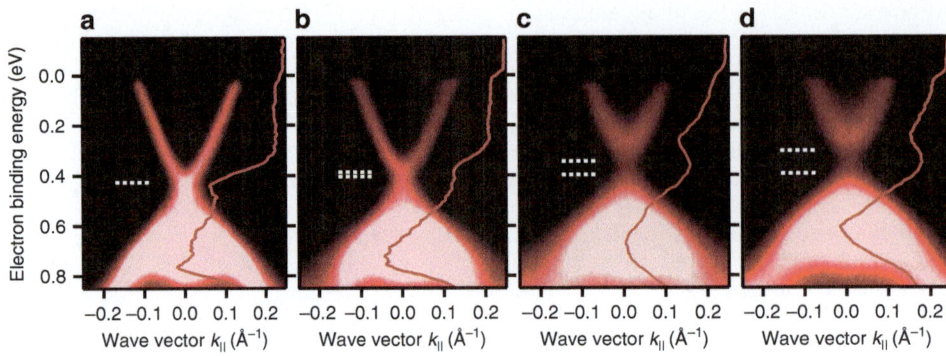

Figure 1 | Effect of Mn doping on the electronic band dispersions of $(Bi_{1-x}Mn_x)_2Se_3$. (**a–d**) Mn-concentration-dependent high-resolution ARPES dispersions measured at 50 eV photon energy and a temperature of 12 K for x values of (**a**) 0, (**b**) 0.02, (**c**) 0.04 and (**d**) 0.08. The red lines represent EDCs obtained in normal emission ($k_{||}=0$). A surface band gap opens at the Dirac point of the TSS which increases in size with increasing Mn content. The horizontal white-dashed lines highlight the minimum limit of the gap size after taking into account the contribution from the linewidth broadening. The opening of the gap leads to an intensity dip in the EDCs around the energy region of the original Dirac point. For $x=0.02$, the suppression of intensity at the Dirac point and the more parabolic surface-state dispersion are signatures of a small gap. Besides the increased linewidth broadening due to the Mn impurities, these effects become more pronounced with increasing Mn content.

half of the gap size. Note that the horizontal dashed lines in Figs 1b–d highlight the minimum limit of the gap size after taking into account the contribution from the linewidth broadening to the ARPES spectra (see Supplementary Note 1 and Supplementary Fig. 2). The surface band gap rapidly increases with Mn content and exceeds 200 meV for $x=0.08$. For $x=0.02$, the suppression of the Dirac point intensity as compared with the undoped Bi_2Se_3 film indicates the existence of a small gap. This is consistent with the fact that the TSS dispersion becomes slightly parabolic and is characterized by an increased effective mass of $m^* \approx 0.09m_e$ for $x=0.02$. This value increases further to $\approx 0.15m_e$ for $x=0.08$, where m_e is the free-electron mass.

The observed surface band gaps are similar to those previously reported[20], which have been attributed to long-range ferromagnetism and TRS breaking of the topologically protected surface state[20]. Figures 2a–d show high-resolution ARPES dispersions of the TSS, as well as the normal-emission spectra through the Dirac point at temperatures of 12 and 300 K for $x=0.08$ (see also Supplementary Fig. 3). Strikingly, there is no significant change as temperature is raised, and very clearly the band gap in the TSS persists up to room temperature. Moreover, we find a similar behaviour for the whole sample series independently of the Mn content, which challenges the dominant role of ferromagnetic order in inducing the band gap in the TSS, unless the surface T_C is above room temperature.

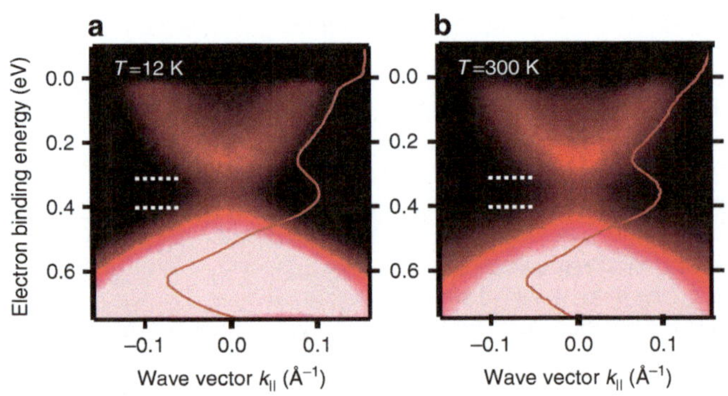

Figure 2 | Temperature dependence of the large energy scale surface band gap. (**a**,**b**) Energy-momentum ARPES dispersions obtained for Mn-doped Bi_2Se_3 epilayers with 8% Mn concentration. The red lines are the energy-distribution curves obtained in normal emission ($k_{||}=0$). The horizontal white-dashed lines highlight the surface gap. Measurements are taken at 50 eV photon energy and a temperature of (**a**) 12 K and (**b**) 300 K. The surface band gap does not show a remarkable temperature dependence.

Bulk and surface magnetism. We begin in Fig. 3 with characterizing the bulk magnetic properties using superconducting quantum-interference device (SQUID) magnetometry. The comparison of measurements with in-plane and out-of-plane applied magnetic field in Fig. 3a shows that at a temperature of 2 K the bulk easy axis lies in the surface plane. This holds irrespective of the Mn concentration. Since the opening of a gap at the Dirac point requires a magnetization perpendicular to the surface[9], we concentrate in the following on the out-of-plane component of the magnetization. In the hysteresis loops displayed in Fig. 3b measured with an out-of-plane applied magnetic field at different temperatures, we observe a narrow magnetic hysteresis at 4.2 K, whereas paramagnetic behaviour emerges at 7 K once the ferromagnetic transition has been crossed (inset in Fig. 3b). For a better determination of T_C we present in Fig. 3c modified Arrott plots according to a 3D Heisenberg ferromagnet, normalized to the mass of the sample, with exponents $\beta=0.348$ and $\gamma=1.41$, from which we deduce $T_C=5.5$ K in the bulk (see also Supplementary Fig. 4). This is well below the temperature of our ARPES measurements shown in Figs 1 and 2.

The only remaining possibility that long-range ferromagnetism opens the gap at the Dirac point is an enhanced surface magnetism with magnetization perpendicular to the surface[9,20,23]. This cannot be verified directly by the bulk-sensitive SQUID magnetometry. Therefore, we perform X-ray magnetic circular dichroism (XMCD) measurements at the Mn $L_{2,3}$-edges to probe the near-surface ferromagnetic order. The detection by total electron yield leads to a probing depth in the nanometre range. Figure 4a shows for $x=0.04$ the normalized intensity of the Mn $L_{2,3}$ absorption edges obtained upon reversal of the photon helicity at a temperature of 5 K with an out-of-plane applied magnetic field of 3 T. The XMCD difference spectrum is shown in Fig. 4b, with the normalized XMCD difference signal following one part of the out-of-plane hysteresis as a function of the applied magnetic field as inset. The temperature dependence of the XMCD signal measured in remanence (0 T applied magnetic field) is presented in Fig. 4c for Mn concentrations of $x=0.04$ and 0.08. This signal is obtained after switching off an

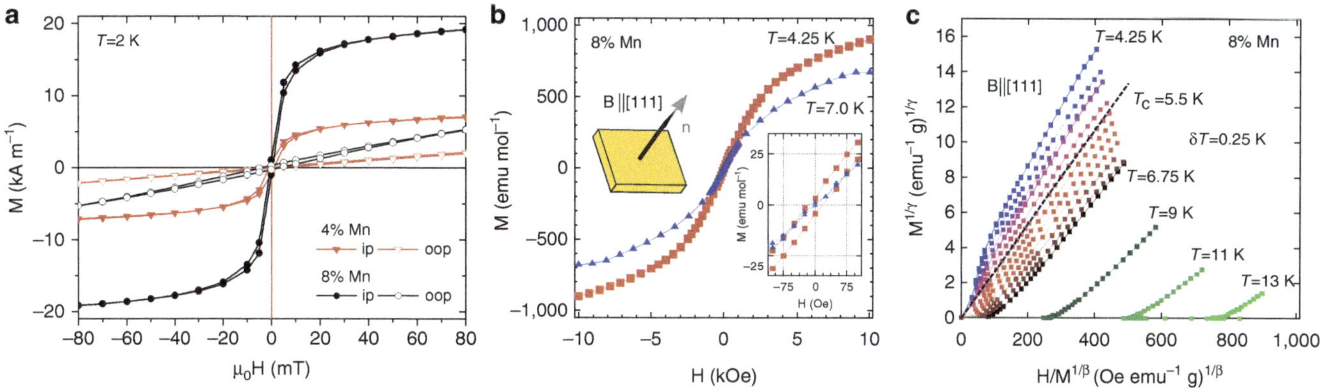

Figure 3 | Characterization of the bulk magnetic properties of Mn-doped Bi_2Se_3 films. (**a**) In-plane (ip) and out-of-plane (oop) magnetization curves for Mn concentrations of $x = 0.04$ and $x = 0.08$, measured at a temperature of 2 K using SQUID magnetometry. The magnetic anisotropy lies in the surface plane irrespective of the Mn concentration. (**b**) Corresponding hysteresis loops for $x = 0.08$ measured at a temperature of 4.25 K (squares) and 7 K (triangles), respectively. The applied magnetic field is perpendicular to the (111) sample surface (see sketch). Inset: a zoom-in around zero magnetic field showing hysteresis at 4.25 K and a paramagnetic state at 7 K. (**c**) Arrott plots at various temperatures from which a Curie temperature T_C of 5.5 K is deduced. Between 4.25 K and 6.75 K, data are shown for temperature increments of $\delta T = 0.25$ K.

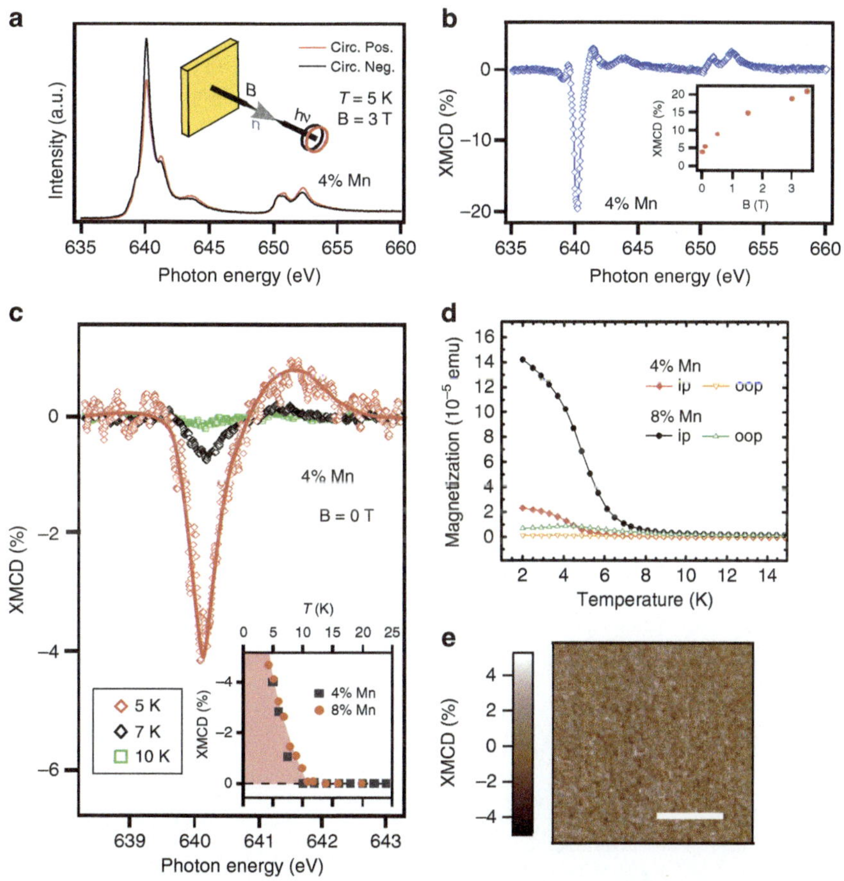

Figure 4 | XMCD in total electron yield at the Mn $L_{2,3}$-edges. (**a**) X-ray absorption spectra measured for opposite helicities of the circularly polarized incident light at 5 K with an out-of-plane applied magnetic field of 3 T. A sketch of the experimental geometry is also shown. (**b**) Corresponding XMCD difference spectrum. The inset shows XMCD measurements for different applied magnetic fields. (**c**) Temperature dependence of the remanent XMCD for a Mn concentration of $x = 0.04$. Inset: detailed temperature dependence for $x = 0.04$ and 0.08. (**d**) To re-examine the bulk magnetic properties, field-cooling measurements are obtained under an in-plane (ip) and out-of-plane (oop) applied magnetic field of 10 mT using SQUID magnetometry. The temperature dependence for $x = 0.04$ and 0.08 compares well with the surface-sensitive measurements shown in the inset of (**c**). (**e**) XMCD-PEEM image obtained using photoelectron microscopy for $x = 0.08$ at room temperature, revealing the absence of magnetic domains with partial or full out-of-plane magnetic order. Scale bar, 500 μm (horizontal white-solid line).

applied magnetic field of 5 T perpendicular to the surface, and represents the remanent XMCD, which is proportional to the remanent magnetization of the film surface. We clearly observe a ferromagnetic Mn component in the remanent XMCD appearing around 640 eV which becomes increasingly more pronounced at lower temperatures. The lineshape of the XMCD spectrum is similar to that of Mn in GaAs[31], indicating a predominant d^5 configuration. More specifically, the lineshape compares rather well with Mn in Bi_2Te_3 for which the comparison with an Anderson impurity model recently gave a superposition of 16% d^4, 58% d^5 and 26% d^6 character[32]. The ferromagnetic Mn component shows surface ferromagnetic order in the remanent XMCD only below 10 K for Mn concentrations of $x = 0.04$ and 0.08 (inset in Fig. 4c). On one hand, this result is not inconsistent with the predicted strong-surface enhancement of T_C since the enhancement effect vanishes already when chemical potential and Dirac point differ by 0.2 eV (ref. 23). On the other hand, our XMCD result means that the magnetically induced gap would have to close above 10 K.

Having established that the bulk and surface magnetic properties of our samples are very similar, as a next step, we performed field-cooling experiments to re-examine the bulk magnetic properties. Figure 4d shows field-cooled SQUID data for an applied magnetic field of 10 mT parallel and perpendicular to the surface plane. Above T_C, there is no preferential orientation of the anisotropy axis perpendicular to the surface. Additional zero-field-cooled SQUID data compare rather well with these results (Supplementary Fig. 5). Further XMCD measurements under an in-plane and out-of-plane applied magnetic field are consistent with the SQUID data and are shown in Supplementary Fig. 6. These XMCD and SQUID results strongly suggest that above T_C also short-range static magnetic order does not play a role at the surface or in the bulk, respectively (see Supplementary Note 2), and that domains with partial or full out-of-plane magnetic order are absent. We further support this conclusion by additional XMCD measurements carried out by the means of X-ray photoelectron emission microscopy (XMCD-PEEM) at room temperature, with a lateral resolution of ~20 nm. Figure 4e shows the XMCD image which due to the light incidence (16° grazing) is sensitve to both in- and out-of-plane components of the magnetization, with additional data given in Supplementary Figs 7,8. The XMCD image displays no magnetic domains at room temperature thus also excluding short-range static inhomogeneous magnetic order with magnetization partially or fully in or out of the surface plane, at least within our lateral resolution (see Supplementary Note 3 for details). A similar conclusion can be drawn regarding out-of-plane magnetic short-range fluctuations (see discussion in Supplementary Note 4 and Supplementary Fig. 9). All our observations taken together lead us to the important conclusion that the surface band gap is not due to magnetism and, thus, the novel topological phases cannot be realized with $(Bi_{1-x}Mn_x)_2Se_3$.

Absence of topological phase transition with concentration. There exists an alternative nonmagnetic explanation for the surface band gap. It is in principle possible that the incorporation of Mn changes the bulk band structure and even reverts the bulk band inversion, rendering Mn-doped Bi_2Se_3 topologically trivial. This topological phase transition with concentration is, for example, the basis for the HgTe quantum spin-Hall insulator[6]. Interestingly, it has been argued that a gap in the TSS can be a precursor of a reversed bulk band inversion, as discussed for $TlBi(S_{1-x}Se_x)_2$ (ref. 33). Indium substitution in Bi_2Se_3 behaves similarly, and leads to a topological-to-trivial quantum-phase transition with an inversion point between 3 and 7% In in thin films[34]. Figure 5 shows that the bulk band gap stays constant within 4% of its value for 8% Mn incorporation. This statement is possible because at low photon energies changes in the bulk band gap are traced most precisely from quantum-well states in normal emission ($k_{||} = 0$). The simultaneous quantization of BCB and BVB is created by surface band bending after adsorption of small amounts of residual gas[35]. This means that the scenario of reversed bulk band inversion does not hold here.

Another remarkable feature is the measured size of the surface band gap. A perpendicular magnetic anisotropy has recently been predicted by density-functional theory (DFT) for Co in Bi_2Se_3 (ref. 36) as well as for 0.25 monolayer Co adsorbed on the surface in substitutional sites, leading to a surface band gap of ~9 meV (ref. 36), which, however, is one order of magnitude smaller than the gaps observed here. For $(Bi_{1-x}Mn_x)_2Te_3$, a surface band gap of 16 meV has been calculated under the condition of perpendicular magnetic anisotropy[37]. For Mn in Bi_2Se_3 at the energetically favoured Bi-substitutional sites, an in-plane magnetic anisotropy is predicted[38]. The size of the gap at the Dirac point is only 4 meV (ref. 38). Perpendicular anisotropy and ferromagnetic order are absent in our samples at the temperature of the ARPES measurements.

Role of the local magnetic moment and intercalation scenarios. If the large local magnetic moment of the Mn is responsible for the measured band gaps via TRS breaking, the effect should be equal or larger when Mn is deposited directly on the surface. Here we argue in the following way: If TRS is broken only due to the Mn magnetic moment, we do not require hybridization to open a gap. By depositing Mn we create a completely different environment for the Mn at the surface, allowing us to understand whether the Mn magnetic moment is solely responsible for the opening of large surface band gaps via TRS breaking and, ideally, without hybridization playing a role. Mn is very well suited for such comparison, as due to the high stability of its half-filled d^5 configuration, its high magnetic moment is to a large extent independent of the atomic environment and the resulting hybridization. DFT obtains 4.0 μ_B for most substitutional sites (except for the hypothetical Se-substitutional site with still 3.0 μ_B) and only for interstitial Mn the moment peaks with 5.0 μ_B (ref. 38). Mn deposition was performed at ~30 K in order to keep the Mn atoms isolated from each other and on the surface. As the XMCD showed a predominant d^5 configuration for Mn bulk impurities, which is most stable, we can assume the same magnetic moment for Mn impurities deposited on the surface. Figure 6 shows that a similar p-doping occurs as with the Mn in the bulk. In another respect, the magnetic moments do not act in a similar way as in the bulk. Even 30% of a Mn monolayer on Bi_2Se_3 does not open a band gap at the Dirac point. This result is similar to what we have observed for Fe on Bi_2Se_3 (ref. 12) and Bi_2Te_3 (ref. 14).

Most recently, the appearance of surface band gaps in ARPES at the Dirac point of the TSS has been discussed based on different mechanisms. One is momentum fluctuations of the surface Dirac fermions in real space as observed by scanning tunnelling microscopy measurements of Mn-doped Bi_2Te_3 and Bi_2Se_3 bulk single crystals[39]. However, these fluctuations amount to 16 meV (ref. 39), much less than the present band gaps. Another mechanism is top-layer relaxation, that is, breaking of the van-der-Waals bond between quintuple layers during cleavage of bulk crystals as proposed in Xu *et al.*[20] This would open a gap due to hybridization of TSS's through the layer[40] but such an effect does not occur for epitaxial layers where no sample cleavage is used for surface preparation.

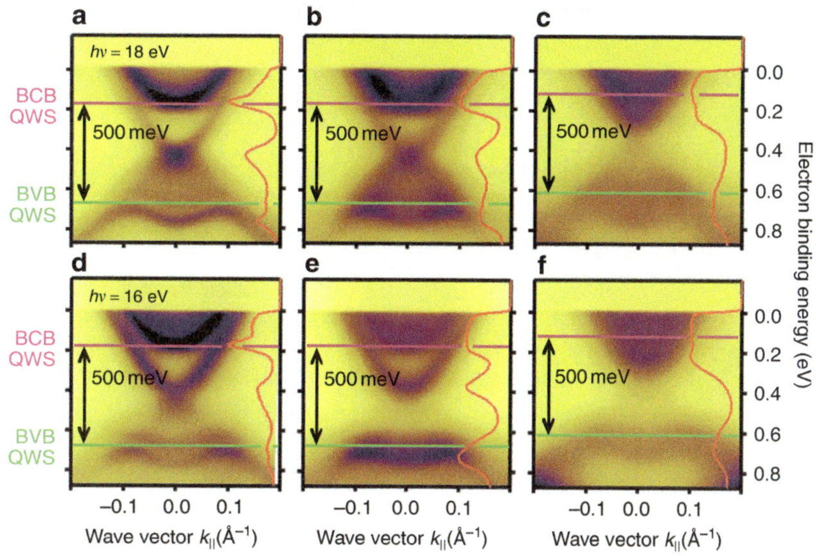

Figure 5 | Tracing changes in the bulk band gap of Mn-doped Bi_2Se_3 films. Changes in the bulk band gap are traced most precisely from quantum-well states (QWS) in normal emission (red curves on the right-hand side of each panel).(**a-c**) Energy-momentum dispersions acquired at a photon energy of $h\nu = 18$ eV for (**a**) Bi_2Se_3, (**b**) 2% and (**c**) 8% Mn doping. The simultaneous quantization of bulk-conduction band (BCB) and valence-band (BVB) is created by surface band bending after adsorption of small amounts of residual gas[35]. It is found that Mn doping does not change the bulk band gap of Bi_2Se_3. (**d-f**) Similar results as in **a-c**, respectively, but at a photon energy of $h\nu = 16$ eV, where large apparent surface band gaps are observed. This unexpected photon-energy dependence of a surface state is interpreted as coupling to bulk states mediated by the Mn impurities.

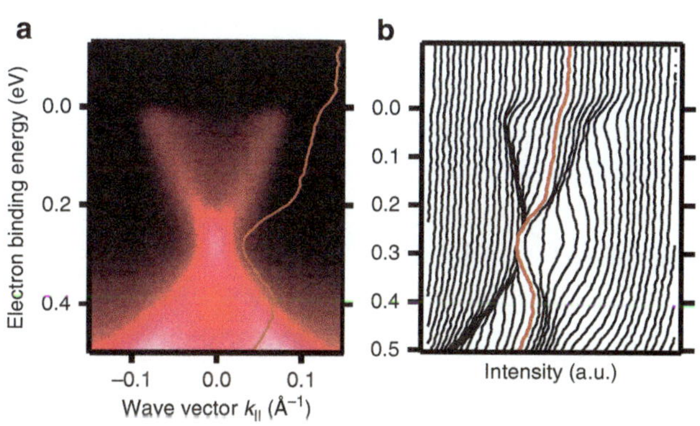

Figure 6 | Effect of Mn deposition on the surface of Bi_2Se_3 films. (**a**) Energy-momentum ARPES dispersion showing a gapless Dirac cone in pure Bi_2Se_3 films after deposition of 0.3 monolayer Mn at 30 K on the surface. Measurements are performed at the same temperature using 50 eV photon energy. (**b**) EDCs extracted from the measurements shown in **a**. The red curves in **a** and **b** emphasize the EDC in normal emission ($k_{||} = 0$).

If such a separation of quintuple layers occurs, it is more likely caused by the Mn. Principally, intercalated Mn in the van-der-Waals gaps could separate quintuple layers electronically, although such effect has not been seen in experiments yet. At these new interfaces TSS's could form and if they would behave like in ultra-thin Bi_2Se_3 films, they would hybridize and open a band gap[40]. However, band gaps of the order of ~100 meV would correspond to an unrealistic Mn intercalation pattern grouping three quintuple layers when compared with the films of Zhang *et al.*[40] In order to estimate the amount of Mn intercalation, we have analysed the change of the bulk lattice constant in our samples by X-ray diffraction (see Supplementary Note 5, Supplementary Figs 10–13 and Supplementary Tables 1,2). We find that the *c*-lattice constant increases by ~0.15% for 4% Mn and by ~0.45% for 8% Mn. If we would assume that this is completely due to an expansion of the van-der-Waals gaps, the interlayer distance would increase by only ~1.8% for 4% Mn and by less than ~5% for 8% Mn. According to DFT, intercalation of transition metals into the first van-der-Waals gap of Bi_2Se_3 leads to an expansion between 10 and 20% (ref. 41). New surface-state features can appear inside the bulk band gap but in order to be split remarkably off the BVB and BCB edges, expansions between 20 and 50% are required[41]. In addition, we point out that, despite our observation of a large surface band gap at room temperature, we observe a nearly zero spin polarization perpendicular to the surface (Supplementary Fig. 14). The absence of a hedgehog spin texture pinpoints a non-TRS breaking mechanism underlying the origin of the gap, a fact which is consistent with our conclusions derived from the magnetic properties. Moreover, it is understood that we cannot verify a small magnetic band gap opening of few meV as calculated in Schmidt *et al.*[36] and Henk *et al.*[37] because much larger gaps are present in the whole temperature range.

Topological protection beyond the continuum model. As we find that neither ferromagnetic order in the bulk or at the surface, nor the local magnetic moment of the Mn are causing the large band gaps that we observe, as we can exclude the nonmagnetic explanation of a reversal of the bulk band inversion and as we do not find sufficient indications for the nonmagnetic explanation of surface-state hybridization by van-der-Waals-gap expansion, we point out a different mechanism based on impurity-induced resonance states that locally destroy the Dirac point.

In fact, recently the treatment of the topological protection in the continuum model[3–5] has been extended for finite bulk band gaps[42,43]. As a result, surface[42] and bulk impurities[43] can mediate scattering processes via bulk states and the localized impurity-induced resonance states emerging at and around the Dirac point lead to a local destruction of the topological protection of the Dirac point. The resulting band gap opening depends on the resonance energy, impurity strength U as well as spatial location of the impurities[42,43]. A typical gap size is of the order of 100 meV (ref. 42). Moreover, the gap does not rely on a particular magnetic property of the impurity and, therefore, the mechanism should apply for magnetic impurities as well. The model can also explain the absence of the gap when Mn impurities are placed on the surface, since bulk-like resonance states do not form when Mn is placed only on the surface. A definite assignment, however, requires realistic electronic structure calculations since a simple atomic Mn d^5 configuration does not lead to states at the Dirac point.

To identify impurity resonances, we calculated for 8% Mn in Bi_2Se_3 at the Bi-substitutional sites the corresponding DOS by *ab initio* theory. The model structure used in the calculations is shown in Fig. 7a, where Bi atoms (yellow) acquire Mn character (blue wedges). The results of the calculation are shown in Fig. 7b, where it is seen that Mn impurity states (blue) strongly contribute to the total DOS (red) near the BVB maximum. The Mn impurity states in the gap are clearly identified by assuming ferromagnetic order and subtracting the minority from majority spin DOS. In experiment, impurity-band states are difficult to observe as the example of $Ga_{1-x}Mn_xAs$ shows[44], where—similarly to the present case—impurity states emerge near the BVB maximum. There are, in fact, indications for the impurity resonances in our data. If we look at the photon-energy dependence from 16 to 21 eV in the ARPES results shown in Supplementary Fig. 15, we see that the surface band gap is only well defined with respect to the minimum of the upper Dirac cone and consistent for 16–21 eV as well as 50 eV (for 50 eV see Figs 1,2, where the BCB does not appear). The lower Dirac cone exhibits despite its

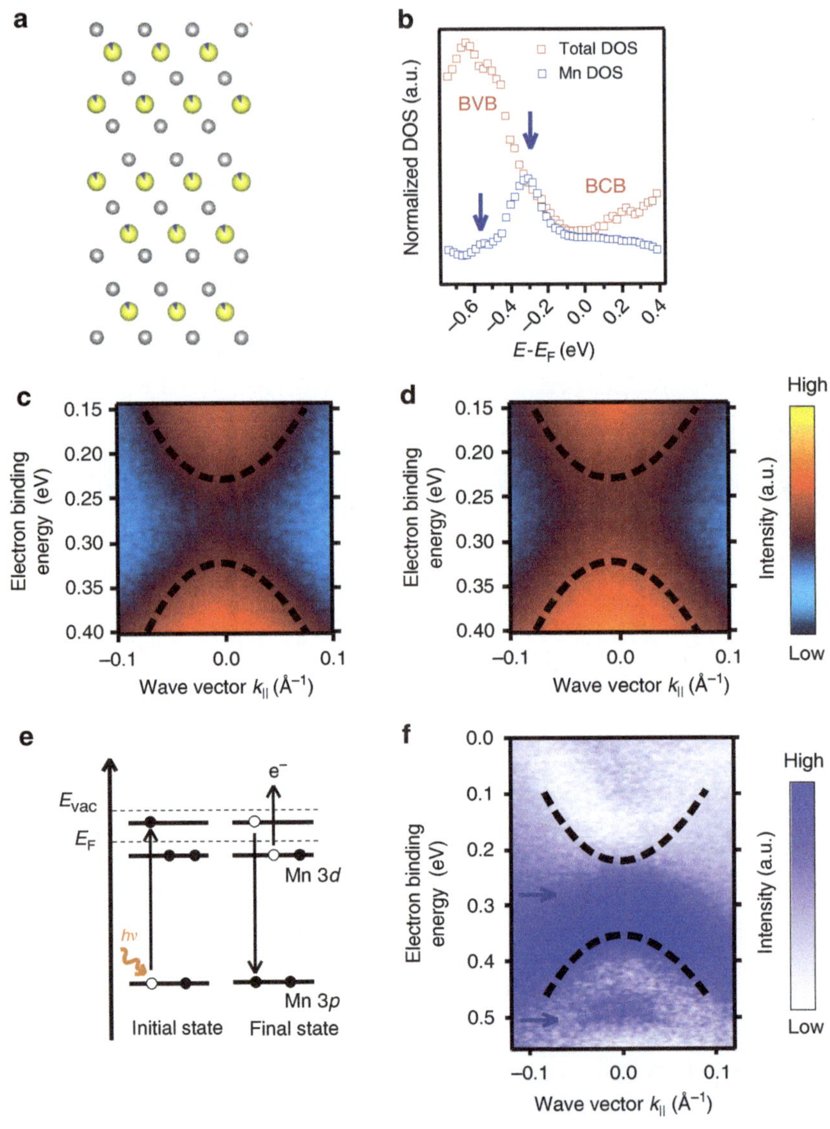

Figure 7 | Resonant photoemission and *ab initio* calculations. (**a**) The structure of Bi_2Se_3 doped with 8% Mn at the Bi-substitutional sites used in the calculations. Within the coherent-potential approximation, the Bi atoms (yellow) acquire Mn character (blue wedges). Se atoms are shown in grey colour. (**b**) Calculated density of states (DOS). The total DOS (red) contains contributions from impurity resonances (vertical blue arrows) of strong d-character, a seen in the partial Mn DOS (blue). For the purpose of the calculation, ferromagnetic order at $T=0$ K was assumed. (**c**,**d**) Resonant photoemission at the Mn *M*-edge, focusing on the region of the surface gap (8% Mn). (**c**) Off-resonant (48 eV photon energy) and (**d**) on-resonant data (50 eV). (**e**) Schematics of the resonant process. At the resonance energy, the excitation occurs via 3*p*-3*d* transitions. The exciting photon energy (*hν*) corresponds to a transition from an occupied core level to a valence-band empty state between the Fermi (E_F) and vacuum (E_{vac}) levels. The relaxation of excited electrons leads to enhanced photoemission from *d*-like Mn states. (**f**) Intensity difference obtained after subtracting off-resonant from on-resonant data. The resonances are seen in blue contrast and are marked with arrows, similar to the calculations shown in **b**.

two-dimensional nature a photon-energy (wave vector perpendicular to the surface, $k_{\perp}$) dependence when Mn is incorporated. The reason for this difference is that the lower Dirac cone strongly overlaps with the BVB while the upper Dirac cone does not overlap with the BCB. This can also be seen in Fig. 5, where an apparent band gap of $\sim$200 meV occurs for 2% Mn at $h\nu = 16$ eV. At this small concentration, such an apparent gap seems to be nearly closed at 18 eV, where the suppression of the Dirac point intensity and the more parabolic TSS dispersion as compared with the undoped Bi_2Se_3 film indicates the existence of a small gap, in agreement with the results of Fig. 1. Note that the minimum band gap in the photon-energy dependence defines the gap size (see Supplementary Fig. 16). The minimum gap size observed at low photon energies agrees well with the one obtained at 50 eV. At 8% Mn, the surface band gap opens at all photon energies and reaches a minimum of $\sim$200 meV, that is, it is determined rather unambiguously, but some intensity appears in the surface band gap. Such intensity also appears in Xu *et al.*[20] The role of the impurity resonance is exactly to couple the TSS to bulk states. This 3D coupling is naturally different for the upper and lower Dirac cone due to the different bulk overlap and thus seen as the photon-energy dependence of the lower half of the Dirac cone.

We applied resonant photoemission which fortunately is comparatively strong for Mn due to its half-filled *d*-shell. Figures 7c and d show for 8% Mn off-resonant ($h\nu = 48$ eV) and on-resonant photoemission (50 eV) measurements via the Mn 3*p* core level, focusing on the region of the surface gap, respectively. The spectra were normalized to the photon flux after taking into account the photon-energy dependence of the photoionization cross sections of Bi 6*p* and Se 4*p*, respectively. Subtracting off-resonant from on-resonant data allows us to visualize directly the contribution from impurity resonances in the ARPES spectrum, as there is enhanced photoemission from *d*-like Mn states through decay of electrons that are excited via 3*p*–3*d* transitions (Fig. 7e). The difference spectrum shown in Fig. 7f reveals the existence of impurity resonances inside the surface band gap and near the valence-band maximum. The resonances are seen in blue contrast and are marked with arrows, similar to the calculations shown in Fig. 7b. The unexpectedly strong dispersion of the resonant states with wave vector $k_{||}$ is in qualitative agreement with our one-step model photoemission calculations (see Supplementary Fig. 17 and Supplementary Note 6), and additionally supports the physical picture of Mn-induced coupling to the bulk states. We should emphasize that the Mn d^5 configuration confirmed by our XMCD measurements offers much fewer states in the energy range of the Dirac cone than the other magnetic transition metals. The fact that already Mn breaks the Dirac cone indicates that the present result is of general importance for TIs doped with magnetic transition metals.

As nonmagnetic control experiment we have chosen thick In-doped Bi_2Se_3 bulk samples which give rise to a trivial phase close to $\sim$5% In concentration. Supplementary Fig. 18 shows that a large band gap of the order of $\sim$100 meV appears at the Dirac point already for a much smaller In concentration, namely 2%, far away from the inversion point of the topological quantum-phase transition (see Supplementary Note 7 and Supplementary Fig. 19 for more details). In addition, spin-resolved ARPES measurements around $k_{||} = 0$ for In-doped samples (Supplementary Fig. 14) reveal that the out-of-plane spin polarization is zero, similarly to the result for the gapped Dirac cone in Mn-doped samples. Interestingly, the size of the band gap for 2% In is of the same order as the one for 8% Mn whereas no gap appears for 4% Sn (Supplementary Fig. 18). This indicates that the concentration range at which the large surface gap develops varies from dopant to dopant. On the basis of ideas proposed recently[42,43], this might be associated with the impurity-dependent strength *U*, regardless whether the dopant is magnetic or not. Our control experiments demonstrate the existence of a mechanism for surface band gap opening which is not directly connected to long-range or local magnetic properties. Although it is not possible to directly search for In impurity resonances in the photoemission experiment as there is no resonant photoemission condition available, for completeness we point out that our conclusion on the nonmagnetic origin of the surface band gap in Mn-doped Bi_2Se_3 films would not have been possible unless several findings in our experiment differed from previous experiments[20]. This refers to the in-plane magnetic anisotropy, the absence of a temperature dependence of the gap, a much lower T_C, no out-of-plane spin polarization at the gapped Dirac point, no enhanced surface magnetism, a photon-energy dependence of the band gap, and an available resonant condition via the Mn 3*p* core level allowing us to directly observe the contribution from in-gap states.

To summarize, we revealed the opening of large surface band gaps in the TSS of Mn-doped Bi_2Se_3 epilayers that strongly increase with increasing Mn content. We find ferromagnetic hysteresis at low temperatures, with the magnetization oriented parallel to the film surface. Magnetic surface enhancement is absent. Even above the ferromagnetic transition, we observe surface band gaps which are one order of magnitude larger than the magnetic gaps theoretically predicted and, moreover, they do not show notable temperature dependence. No indication for a Mn-induced reversal of the bulk band inversion and no significant enhancement of the surface magnetic ordering transition with respect to the bulk are found. Control experiments with nonmagnetic bulk impurities, conducted in the topological phase, reveal that surface band gaps of the order of $\sim$100 meV can be created without magnetic moments. In line with recent theoretical predictions, we conclude that the band gap opening up to room temperature in Mn-doped films is not induced by ferromagnetic order and that even the presence of magnetic moments is not required. Our results are important in the context of topological protection and provide strong circumstantial evidence that Mn-doped Bi_2Se_3 is not suited for observing the quantized anomalous Hall effect or the half quantum Hall effect.

Methods

Sample growth and structural characterization. The samples were grown by molecular beam epitaxy on (111)-oriented BaF_2 substrates at a substrate temperature of 360 °C using Bi_2Se_3, Mn and Se effusion cells. The Mn concentration was varied between 0 and 8%, and a two-dimensional growth was observed by *in situ* reflection high-energy electron diffraction in all cases (see Supplementary Fig. 20 and Supplementary Note 8). All samples were single phase as indicated by X-ray diffraction. After growth and cooling to room temperature, the $\sim$0.4-μm-thick epilayers were *in situ* capped by an amorphous 50-nm thick Se layer, which was desorbed inside the ARPES and XMCD chambers by carefully annealing at $\sim$230 °C for $\sim$1 h. The Mn concentrations determined by core-level photoemission were in good agreement with those obtained from energy dispersive microanalysis, indicating no significant Mn accumulation at the surface of the samples (see Supplementary Fig. 21 and Supplementary Note 9).

High-resolution and spin-resolved ARPES. Temperature-dependent ARPES measurements were performed at the UE112-PGM2a beamline of BESSY II at pressures better than 1×10^{-10} mbar using p-polarized undulator radiation. Photoelectrons were detected with a Scienta R8000 electron energy analyser and the spin-resolved spectra were obtained with a Mott-type spin polarimeter coupled to the hemispherical analyser. For the spin-resolved measurements of Mn-doped Bi_2Se_3 samples, a magnetic field of +0.3 T was applied perpendicular to the surface plane at 20 K right before the acquisition of the spectra. Overall resolutions of ARPES measurements were 5 meV (energy) and 0.3° (angular). Resolutions for spin-resolved ARPES were 80 meV (energy) and 0.75° (angular).

Magnetic characterization. The characterization of the bulk magnetic properties was done by SQUID magnetometry. The bulk magnetization was recorded as a function of temperature and applied magnetic field applied either perpendicular (out-of-plane) or parallel (in-plane) to the surface of the films. The diamagnetic contribution of the BaF_2 substrate was derived from the field-dependent magnetization curves at room temperature and subtracted from all data. The characterization of the surface magnetic properties was done by XMCD and XMCD-PEEM measurements at the UE46-PGM1 and UE49-PGM1a beamlines of BESSY II, respectively. The experiments were performed using circularly polarized undulator radiation and under the same pressure conditions as the ARPES measurements. The XMCD spectra were taken using a high-field diffractometer as a function of temperature and applied magnetic field, and the XMCD-PEEM measurements were performed at room temperature and under grazing incidence (16°) with a lateral resolution of ~ 20 nm.

Theoretical calculations. The one-step model photoemission calculations were performed using the results of *ab initio* theory as an input, and taking into account wave vector and energy-dependent transition matrix elements. The *ab initio* calculations were performed within the coherent-potential approximation using the Munich SPR-KKR program package[45,46], with spin-orbit coupling included by solving the four-component Dirac equation.

References

1. Hasan, M. Z. & Kane, C. L. Topological insulators. *Rev. Mod. Phys.* **82,** 3045–3067 (2010).
2. Qi, X.-L. & Zhang, S.-C. Topological insulators and superconductors. *Rev. Mod. Phys.* **83,** 1057–1110 (2011).
3. Lee, W. C., Wu, C., Arovas, D. P. & Zhang, S. C. Quasiparticle interference on the surface of the topological insulator Bi_2Te_3. *Phys. Rev. B* **80,** 245439 (2009).
4. Biswas, R. R. & Balatsky, A. V. Impurity-induced states on the surface of three-dimensional topological insulators. *Phys. Rev. B* **81,** 233405 (2010).
5. Lu, J., Shan, W.-Y., Lu, H.-Z. & Shen, S.-Q. Non-magnetic impurities and in-gap bound states in topological insulators. *New J. Phys.* **13,** 103016 (2011).
6. König, M. *et al.* Quantum spin Hall insulator state in HgTe quantum wells. *Science* **318,** 766–770 (2007).
7. Liu, C.-X., Qi, X.-L., Dai, X., Fang, Z. & Zhang, S.-C. Quantum anomalous Hall effect in $Hg_{1-y}Mn_yTe$ quantum wells. *Phys. Rev. Lett.* **101,** 146802 (2008).
8. Du, L., Knez, I., Sullivan, G. & Du, R.-R. Robust helical edge transport in gated InAs/GaSb bilayers. *Phys. Rev. Lett.* **114,** 096802 (2015).
9. Liu, Q., Liu, C. X., Xu, C. K., Qi, X. L. & Zhang, S. C. Magnetic impurities on the surface of a topological insulator. *Phys. Rev. Lett.* **102,** 156603 (2009).
10. Wray, L. A. *et al.* A topological insulator surface under strong Coulomb, magnetic and disorder perturbations. *Nat. Phys.* **7,** 32–37 (2011).
11. Chen, Y. L. *et al.* Massive Dirac fermion on the surface of a magnetically doped topological insulator. *Science* **329,** 659–662 (2010).
12. Scholz, M. R. *et al.* Tolerance of topological surface states towards magnetic moments: Fe on Bi_2Se_3. *Phys. Rev. Lett.* **108,** 256810 (2012).
13. Valla, T., Pan, Z.-H., Gardner, D., Lee, Y. S. & Chu, S. Photoemission Spectroscopy of magnetic and nonmagnetic impurities on the surface of the Bi_2Se_3 topological insulator. *Phys. Rev. Lett.* **108,** 117601 (2012).
14. Scholz, M. R. *et al.* Intact Dirac cone of Bi_2Te_3 covered with a monolayer Fe. *Phys. Status Solidi RRL* **7,** 139–141 (2013).
15. Honolka, J. *et al.* In-plane magnetic anisotropy of Fe atoms on Bi_2Se_3 (111). *Phys. Rev. Lett.* **108,** 256811 (2012).
16. Kulbachinskii, V. A. *et al.* Ferromagnetic transition in the new diluted magnetic semiconductor p-$Bi_{2-x}Fe_xTe_3$. *Phys. Lett. A* **285,** 173–176 (2001).
17. Kulbachinskii, V. A. *et al.* Ferromagnetism in new diluted magnetic semiconductor $Bi_{2-x}Fe_xTe_3$. *Physica B* **311,** 292–297 (2002).
18. Hor, Y. S. *et al.* Development of ferromagnetism in the doped topological insulator $Bi_{2-x}Fe_xTe_3$. *Phys. Rev. B* **81,** 195203 (2010).
19. Dyck, J. S., Hájek, P., Lošťák, P. & Uher, C. Diluted magnetic semiconductors based on $Sb_{2-x}V_xTe_3$ ($0.01<x<0.03$). *Phys. Rev. B* **65,** 115212 (2002).
20. Xu, S.-Y. *et al.* Hedgehog spin texture and Berry's phase tuning in a magnetic topological insulator. *Nat. Phys.* **8,** 616–622 (2012).
21. Checkelsky, J. G., Ye, J., Onose, Y., Iwasa, Y. & Tokura, Y. Dirac-fermion-mediated ferromagnetism in a topological insulator. *Nat. Phys.* **8,** 729–733 (2012).
22. Hüfner, S. *Photoelectron Spectroscopy: Principles and Applications* (Springer, 1995).
23. Rosenberg, G. & Franz, M. Surface magnetic ordering in topological insulators with bulk magnetic dopants. *Phys. Rev. B* **85,** 195119 (2012).
24. Yu, R. *et al.* Quantized anomalous Hall effect in magnetic topological insulators. *Science* **329,** 61–64 (2010).
25. Chang, C.-Z. *et al.* Experimental observation of the quantum anomalous Hall effect in a magnetic topological insulator. *Science* **340,** 167–170 (2013).
26. Qi, X. L., Hughes, T. L. & Zhang, S. C. Topological field theory of time-reversal invariant insulators. *Phys. Rev. B* **78,** 195424 (2008).
27. Pankratov, O. A. Supersymmetric inhomogeneous semiconductor structures and the nature of a parity anomaly in $(2+1)$ electrodynamics. *Phys. Lett. A* **121,** 360–366 (1987).
28. Qi, X.-L., Li, R., Zang, J. & Zhang, S.-C. Inducing a magnetic monopole with topological surface states. *Science* **323,** 1184–1187 (2009).
29. Zang, J. & Nagaosa, N. Monopole current and unconventional Hall response on a topological insulator. *Phys. Rev. B* **81,** 245125 (2010).
30. Cho, G. Y. Possible topological phases of bulk magnetically doped Bi_2Se_3: turning a topological band insulator into the Weyl semimetal. Preprint at http://arxiv.org/abs/1110.1939 (2011).
31. Rader, O. *et al.* Identification of extrinsic Mn contributions in $Ga_{1-x}Mn_xAs$ by field-dependent magnetic circular x-ray dichroism. *J. Electron Spectr. Relat. Phenom.* **144,** 789–792 (2005).
32. Watson, M. D. *et al.* Study of the structural, electric and magnetic properties of Mn-doped Bi_2Te_3 single crystals. *New J. Phys.* **15,** 103016 (2013).
33. Sato, T. *et al.* Unexpected mass acquisition of Dirac fermions at the quantum phase transition of a topological insulator. *Nat. Phys.* **7,** 840–844 (2011).
34. Brahlek, M. *et al.* Topological-metal to band-insulator transition in $(Bi_{1-x}In_x)_2Se_3$ thin films. *Phys. Rev. Lett.* **109,** 186403 (2012).
35. Bianchi, M., Hatch, R. C., Mi, J., Iversen, B. B. & Hofmann, P. Simultaneous quantization of bulk conduction and valence states through adsorption of nonmagnetic impurities on Bi_2Se_3. *Phys. Rev. Lett.* **107,** 086802 (2011).
36. Schmidt, T. M., Miwa, R. H. & Fazzio, A. Spin texture and magnetic anisotropy of Co impurities in Bi_2Se_3 topological insulators. *Phys. Rev. B* **84,** 245418 (2011).
37. Henk, J. *et al.* Topological character and magnetism of the Dirac state in Mn-doped Bi_2Te_3. *Phys. Rev. Lett.* **109,** 076801 (2012).
38. Abdalla, L. B., Seixas, L., Schmidt, T. M., Miwa, R. H. & Fazzio, A. Topological insulator Bi_2Se_3(111) surface doped with transition metals: an ab-initio investigation. *Phys. Rev. B* **88,** 045312 (2013).
39. Beidenkopf, H. *et al.* Spatial fluctuations of helical Dirac fermions on the surface of topological insulators. *Nat. Phys.* **7,** 939–943 (2011).
40. Zhang, Y. *et al.* Crossover of the three-dimensional topological insulator Bi_2Se_3 to the two-dimensional limit. *Nat. Phys.* **6,** 584–588 (2010).
41. Eremeev, S. V., Vergniory, M. G., Menshchikova, T. V., Shaposhnikov, A. A. & Chulkov, E. V. The effect of van der Waal's gap expansions on the surface electronic structure of layered topological insulators. *New J. Phys.* **14,** 113030 (2012).
42. Black-Schaffer, A. M. & Balatsky, A. V. Strong potential impurities on the surface of a topological insulator. *Phys. Rev. B* **85,** 121103(R) (2012).
43. Black-Schaffer, A. M. & Balatsky, A. V. Subsurface impurities and vacancies in a three-dimensional topological insulator. *Phys. Rev. B* **86,** 115433 (2012).
44. Gray, A. X. *et al.* Bulk electronic structure of the dilute magnetic semiconductor $Ga_{1-x}Mn_xAs$ through hard x-ray angle-resolved photoemission. *Nat. Mater.* **11,** 957–962 (2012).
45. Ebert, H., Ködderitzsch, D. & Minár, J. Calculating condensed matter properties using the KKR-Green's function method–recent developments and applications. *Rep. Prog. Phys.* **74,** 096501 (2011).
46. Ebert, H. *et al.* The Munich SPR-KKR package, version 6.3. Available at http://ebert.cup.uni-muenchen.de/SPRKKR.

Acknowledgements

Financial support from the priority program SPP 1666 of the Deutsche Forschungsgemeinschaft (Grant No. EB154/26-1 and RA1041/7-1) and the Impuls-und Vernetzungsfonds der Helmholtz-Gemeinschaft (Grant No. HRJRG-408) is gratefully acknowledged. J.M. is supported by the CENTEM (CZ.1.05/2.1.00/03.0088) and CENTEM PLUS (LO1402) projects, co-funded by the ERDF as part of the Ministry of Education, Youth and Sports OP RDI programme. V.H. acknowledges the support of the Czech Science Foundation (project 14-08124S). G.S., H.S., R.K. and G.B. acknowledge support of the Austrian Science Funds (project SFB 025, IRON).

Author contributions

J.S.-B. and A.V. performed the photoemission experiments; G.S., G.B. and L.V.Y. provided samples and performed the growth and characterization; H.S., R.K. and A.N. performed the SQUID measurements; O.C. and V.H. performed X-ray diffraction measurements; J.S.-B., E.S. and E.W. performed the XMCD measurements; J.S.-B., A.A.Ü. and S.V. performed the XMCD-PEEM measurements; J.M., M.D., J.B. and H.E. carried out the calculations; J.S.-B. and E.G. performed the numerical simulations; J.S.-B. performed data analysis and figure planning; J.S.-B. and O.R. performed draft planning and wrote the manuscript with input from all the authors; J.S.-B. and O.R. were responsible for the conception and the overall direction of the study.

Additional information

Competing financial interests: The authors declare no competing financial interests.

8

Orbital two-channel Kondo effect in epitaxial ferromagnetic $L1_0$-MnAl films

L.J. Zhu[1,2], S.H. Nie[1], P. Xiong[3], P. Schlottmann[3] & J.H. Zhao[1]

The orbital two-channel Kondo effect displaying exotic non-Fermi liquid behaviour arises in the intricate scenario of two conduction electrons compensating a pseudo-spin-1/2 impurity of two-level system. Despite extensive efforts for several decades, no material system has been clearly identified to exhibit all three transport regimes characteristic of the two-channel Kondo effect in the same sample, leaving the interpretation of the experimental results a subject of debate. Here we present a transport study suggestive of a robust orbital two-channel Kondo effect in epitaxial ferromagnetic $L1_0$-MnAl films, as evidenced by a magnetic field-independent resistivity upturn with a clear transition from logarithmic- to square-root temperature dependence and deviation from it in three distinct temperature regimes. Our results also provide an experimental indication of the presence of two-channel Kondo physics in a ferromagnet, pointing to considerable robustness of the orbital two-channel Kondo effect even in the presence of spin polarization of the conduction electrons.

[1] State Key Laboratory of Superlattices and Microstructures, Institute of Semiconductors, Chinese Academy of Sciences, PO Box 912, Beijing 100083, China. [2] Institut für Physik, Martin-Luther-Universität Halle-Wittenberg, von-Danckelmann-Platz 3, Halle 06120, Germany. [3] Department of Physics, Florida State University, Tallahassee, Florida 32306, USA. Correspondence and requests for materials should be addressed to J.H.Z. (email: jhzhao@red.semi.ac.cn).

The overscreened two-channel Kondo (2CK) effect displaying exotic non-Fermi liquid (NFL) physics has attracted widespread interest in recent years, especially due to their potential relevance to a host of current topics including strongly correlated physics, Majorana fermions, high-T_c superconductors, topological matters, carbon nanotubes and quantum dots[1–6]. In spin 2CK effect, a spin-1/2 impurity couples to conduction electrons into two equal orbital channels via an exchange interaction[6–9]. Below Kondo temperature (T_K) the 2CK model gives rise to impurity quantum criticality accompanied by exotic NFL behaviour as the consequence of two conduction electron spins attempting to compensate the spin-1/2 impurity. However, the strict requirements of zero local magnetic field and channel symmetry make a direct observation of the spin 2CK effect difficult. Intriguingly, an analogous orbital 2CK effect was proposed to arise from resonant scattering centres with orbital degrees of freedom, for example, two-level systems (TLSs)[10–12]. As depicted in Fig. 1a, in a TLS, the tunnelling entity (for example, an atom, atom group or localized electron) coherently tunnels at a rate of 10^8–$10^{12}\,s^{-1}$ between two independent quantum wells with asymmetry energy Δ_z, tunnelling matrix element Δ_x and energy splitting $\Delta=(\Delta_z^2+\Delta_x^2)^{1/2}$ between the lowest two eigenstates[11,12]. For the noncommutative model this problem is reduced to the 2CK case, the TLS being represented by a pseudo-spin-1/2 and the spin degeneracy of the conduction electrons being the two channels[10,13]. The orbital 2CK effect from TLSs is manifested in electrical transport by a unique temperature (T) dependence in resistivity with three distinct T regimes (Fig. 1b): a low-T upturn characterized by $\Delta\rho_{xx}\sim\ln T$ for $T>T_K$, followed by NFL behaviour ($\Delta\rho_{xx}\sim T^{1/2}$) for T_D $(=\Delta^2/T_K)<T\ll T_K$ and deviation from $T^{1/2}$ dependence upon further cooling[11,12]. The $T^{1/2}$ dependence is a hallmark of the NFL state in the orbital 2CK effect, in striking contrast to the T^2 scaling of Fermi-liquid (FL) behaviour in fully screened Kondo effect[14]. The stability of the low-T orbital 2CK fixed point is an important issue. Theoretically, the fixed point is unstable to Δ, channel-symmetry breaking in the exchange coupling and a magnetic field changing the electron population of the two channels, but stable to exchange anisotropy in the Kondo coupling[15,16]. Although the TLS model with electron-assisted tunnelling[10] breaks down in the weak coupling limit[17,18], the 2CK effect is theoretically expected in the case of resonant scattering and strong coupling[19].

Despite the intensive studies for almost 30 years, the experimental proof for the existence of the orbital 2CK effect has been far from certain. First, it remains a challenge to unambiguously demonstrate the three-regime T dependence of the resistivity expected from the orbital 2CK effect in a single material system. Although the NFL behaviour was reported in Cu point contacts (PCs) and glassy ThAsSe single crystal[20–23], their small Δ prevented the observation of possible breakdown of the $T^{1/2}$ behaviour at lower T. Upadhyay *et al.*[24] observed the NFL behaviour and the low-energy restoration of the FL in conductance spectroscopy of Ti PCs, but there was no indication of a crossover to logarithmic dependence at high energies. Furthermore, the H dependence and microscopic mechanisms of the orbital 2CK effect have remained unsettled. Although present theories expect an imbalance in the channel population to produce a crossover to FL behaviour at low temperatures[25], a magnetic field of up to 14 (5) T did not affect the NFL behaviour in ThAsSe (ref. 22) (Ti PCs[24]). Microscopically, the dynamic tunnelling centres in TLSs were interpreted as a group of nonmagnetic atoms in PCs[20,21], whereas resonant tunnelling of an electron in a polar bond transformation was argued to be responsible for the enhancement of T_K to a few K in ThAsSe (ref. 22). Therefore, materials with TLSs of larger T_K and Δ are desirable for a thorough study of the orbital 2CK physics, including its T and H dependencies and microscopic mechanisms.

Ferromagnetic correlations are predicted to have a detrimental effect on the spin 2CK, while in the case of the orbital 2CK, a coexistence over a large T range is possible as the electron spin variable does not directly enter into the interaction process. A crossover to FL behaviour is, however, expected at a low finite temperature in the neighbourhood of the $T=0$ 2CK fixed point due to the asymmetry in the channel population. So far, there are two reports on orbital Kondo effect in ferromagnets[26,27]. The ferromagnetic $UAs_{1-x}Se_{1+x}$ and $Mn_5Si_3C_x$ display a logarithmic

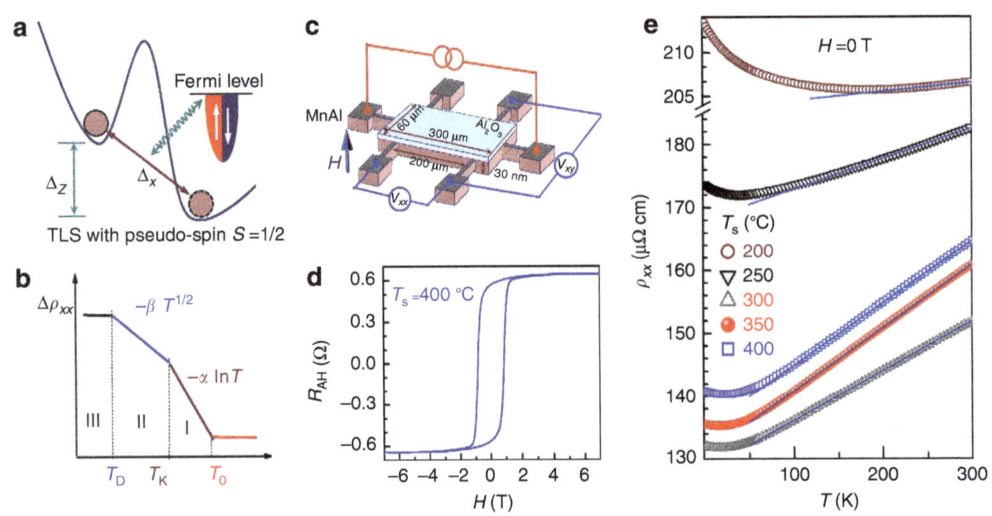

Figure 1 | Orbital 2CK effect and typical transport properties. (**a**) Schematic depiction of a TLS with pseudo-spin $S=1/2$ and strong coupling with conduction electrons. Δ_z and Δ_x represent the asymmetry energy and tunnelling matrix element of the TLS. (**b**) Expected temperature dependence of $\Delta\rho_{xx}$ for TLS-induced 2CK effect: $\Delta\rho_{xx}\sim-\alpha\ln T$ for $T_K<T<T_0$ (I), $\Delta\rho_{xx}\sim-\beta T^{1/2}$ for $T_D<T\ll T_K$ (II) and deviation from $\Delta\rho_{xx}\sim-\beta T^{1/2}$ for $T<T_D$ (III). (**c**) Schematics of the Hall bar device and measurement scheme. The blue arrow represents the direction of the magnetic field. V_{xy} and V_{xx} are Hall and longitudinal voltages, respectively. (**d**) Hysteretic anomalous Hall resistance (R_{AH}) at room temperature for a $L1_0$-MnAl film with $T_s=400$ °C. (**e**) T dependence of ρ_{xx} at zero external magnetic field for $L1_0$-MnAl films with different T_s. The solid lines show the best linear fits at high temperatures, suggesting dominating phonon scattering.

low-T upturn in the resistivity. A crossover to FL behaviour was also observed in $Mn_5Si_3C_x$ at low T. However, neither shows a $T^{1/2}$ dependence characteristic of 2CK effect, possibly because of their low T_K, large T_D and partial polarization of the conduction electrons.

In this work, we report experimental evidence of TLS-induced orbital 2CK effect in a ferromagnetic system, $L1_0$-MnAl epitaxial films with strong perpendicular magnetic anisotropy (PMA). We observed a low-T resistivity upturn with a clear transition from a $\ln T$ dependence to NFL behaviour signified by a $T^{1/2}$ dependence, and deviation from it upon further cooling. The T dependencies are independent of applied magnetic fields up to 8 T. This represents an observation of all three theoretically expected transport regimes from the orbital 2CK effect in the same samples. The greatly enhanced T_K and Δ in this system suggest fast coherent tunnelling of the TLSs and strong coupling with the conduction electrons. Moreover, the structural disorders in $L1_0$-ordered films can be tailored by varying the growth parameters[28–32], offering a convenient pathway of tuning the relevant 2CK parameters.

Results

Sample structure and ferromagnetism. A series of 30-nm-thick $L1_0$-MnAl single-crystalline films (Fig. 1c) were grown at substrate temperatures (T_s) of 200, 250, 300, 350 and 400 °C. The degree of structural disorder decreases with increasing T_s from 200 to 350 °C, and then increases when T_s goes up to 400 °C, which is evidenced by both the intensity and the full width at half maximum of the $L1_0$-MnAl (002) peaks of X-ray diffraction patterns[33]. These films are strongly strained due to the epitaxial growth as indicated by their shorter c axes than bulk ($c = 3.57$ Å)[34], which make them likely candidates to have dynamic defects[24]. These films exhibit strong PMA as revealed by the well-defined hysteretic anomalous Hall resistance (Fig. 1d) and perpendicular magnetization hysteresis[33].

Temperature dependence of the longitudinal resistivity. Figure 1e shows the T dependence of the zero-field longitudinal resistivity (ρ_{xx}) of the $L1_0$-MnAl films. Each film shows a resistivity minimum at a characteristic temperature (T_m) (Supplementary Fig. 1a). In the high T regime ($T > T_m$), ρ_{xx} increases linearly with T due to increasing phonon scattering (Supplementary Fig. 1b). Here we show that the low-T resistivity upturn in our $L1_0$-MnAl films likely arises from the TLS-induced orbital 2CK effect. Figure 2a–c plots the T dependence of resistivity variation at $H = 0$ T for the $L1_0$-MnAl films, which shows distinct signatures associated with the TLS-induced 2CK effect. In the first regime, as displayed in Fig. 2a, the resistivity increase, $\Delta\rho_{xx}$ ($\Delta\rho_{xx} = \rho_{xx} - \rho_1$, with the offset ρ_1 determined from the best linear fit of $\rho_{xx} - \ln T$, see Supplementary Fig. 1c), varies linearly with $\ln T$ below a temperature T_0 for all films with different T_s, similar to the well-known single-channel Kondo (1CK) effect due to static magnetic impurities[14]. $\Delta\rho_{xx}$ deviates from the $\ln T$ dependence and transitions to a $T^{1/2}$ dependence when T drops below T_K (Supplementary Fig. 2). The $T^{1/2}$-dependent resistivity is regarded as a distinct signature of the NFL behaviour from the 2CK effect. Here, an interpretation of localization effects can be excluded, even if a dimensional crossover is considered. Taking into account the resistivity of films (132–214 μΩ cm), which yields a mean-free-path of 2.4–5.5 nm, the film thickness of 30 nm, and the high crossover temperatures (that is, T_K) of up to 23 K, a dimensional crossover seems impossible because neither the thermal length (relevant for e–e interaction) nor the inelastic scattering length (relevant for quantum interference) is likely to approach the film thickness at these temperatures.

As can be seen in Fig. 2b, $\Delta\rho_{xx}$ ($\Delta\rho_{xx} = \rho_{xx} - \rho_2$, where ρ_2 was determined from the best linear fit of $\rho_{xx} - T^{1/2}$; ρ_1 and ρ_2 track each other, see Supplementary Fig. 1c) begins to increase more slowly than $T^{1/2}$ below a characteristic temperature T_D, indicating deviation from the NFL behaviour. The deviation is generic, H-independent (Fig. 3a) and distinct from that induced by quantum corrections in ThAsSe that only appeared at zero field[22,23]. This represents an observation of the TLS theory-expected deviation from the orbital 2CK state below T_D in a diffusive conductor. We emphasize that the deviation is not due to Joule heating because the resistivities measured under a.c. current of 1 μA and d.c. current of 10 μA are virtually identical. Also, such deviation is not expected from conductance corrections due to electron–electron interaction or quantum interference effects. The lower temperature limit for the $T^{1/2}$ regime is given theoretically by $T_D = \Delta^2/T_K$, from which Δ can be determined. In fully screened 1CK systems[14] and frozen slow two-state systems[27,35], ρ_{xx} was observed to saturate following the FL behaviour ($\sim T^2$) at low T. The latter have large tunnelling barrier and negligibly small Δ_x that only allow thermally activated hopping or incoherent tunnelling at a very slow rate ($\ll 10^8$ s^{-1}) (ref. 10). However, in the TLS-induced 2CK effect, it has remained unclear how the system experimentally deviates from the NFL behaviour below T_D. To shed light on this important issue in $L1_0$-MnAl, we measured ρ_{xx} down to 330 mK. As shown in Fig. 2c, none of these films with different T_s shows FL-like saturation (T^2) in the studied temperature range, distinctly different from the expectation of a fully screened Kondo effect[14]. There are three possible reasons for this discrepancy: a broad distribution of T_D of the TLSs; overlapping of the screening clouds of different TLSs; and the spin polarization of the conduction electrons due to the ferromagnetism. A broad distribution of T_D would severely distort the $T^{1/2}$ behaviour and is not a likely scenario. The 2CK displays impurity quantum critical behaviour so that the screening would diverge at $T = 0$ under ideal conditions ($\Delta = 0$), in contrast to the 1CK problem where the screening radius (R) is finite. For orbital 2CK with nonzero Δ, R at $T = 0$ K limit can be estimated by $k_F R \sim D/T_D$, where k_F is Fermi wavevector, D is band width (of the order of Fermi Energy E_F). Using $k_F \sim 1.7$ Å^{-1} ($E_F \sim 11$ eV)[36], $D \sim 10$ eV, $T_D \sim 1$ K for MnAl, one can estimate R to be on the order of $\sim 10^4$ Å at $T = 0$ K, which is much larger than the average distance between TLSs (see below). Hence, with decreasing T the screening clouds will eventually overlap, although this problem has yet to be quantitatively studied. The spin polarization, if homogeneous, should produce a crossover to a T^2 dependence.

Figure 2d plots the slopes $\alpha = -d\rho_{xx}/d(\ln T)$ for $T_K < T < T_0$ and $\beta = -d\rho_{xx}/d(T^{1/2})$ for $T_D < T < T_K$ as a function of T_s. It is evident that α and β have a similar T_s dependence, that is, α (β) first drops quickly from their maximum of 2.5 μΩ cm/lnK (1.0 μΩ cm K$^{-1/2}$) at 200 °C to the minimum of 0.2 μΩ cm/lnK^{-1} (0.15 μΩ cm K$^{-1/2}$) at 300–350 °C, and finally goes up to 0.25 μΩ cm/lnK^{-1} (0.18 μΩ cm K$^{-1/2}$) at 400 °C. As is revealed in the metallic PC experiments, thermal annealing can significantly change the number of TLSs[20,21]. Here, the non-monotonic dependence of α and β on T_s could be ascribed mainly to the thermal tailoring of the density of active TLSs (N_{TLS}). For the strong coupling TLS centres[11], N_{TLS} can be estimated by $N_{TLS} \sim \frac{\Delta\rho_{xxm}}{\rho_{xx}} \frac{N(E_F)}{\tau_e}$ in the diffusive transport regime, where $\Delta\rho_{xxm}$, τ_e and $N(E_F)$ are the maximum resistivity upturn due to the TLSs, electron scattering time and density of states at Fermi level, respectively. τ_e can be determined to be $\sim 10^{-15}$ s by $\rho_{xx} = m^*/ne^2\tau_e$, where m^*, n and e are effective mass, density ($\sim 10^{22}$ cm^{-3}) and charge of conduction electrons, respectively. Using a typical $N(E_F)$ (ref. 36) of $\sim 4 \times 10^{22}$ eV^{-1}cm^{-3} and

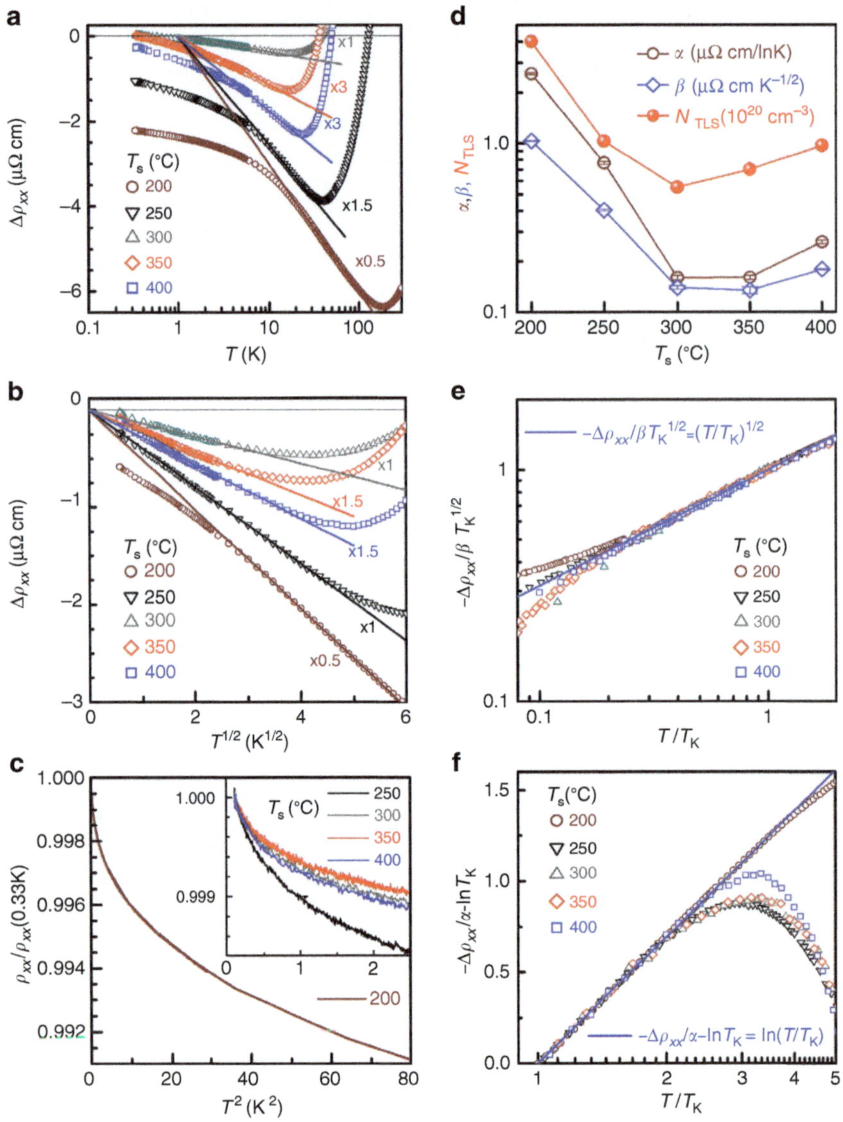

Figure 2 | Temperature dependence of zero-field resistivity. (**a**) Semilog plot of $\Delta\rho_{xx}$ versus T, (**b**) $\Delta\rho_{xx}$ versus $T^{1/2}$, (**c**) ρ_{xx}/ρ_{xx} (0.33 K) versus T^2, (**d**) α, β and N_{TLS} versus T_s, (**e**) log-log plot of $-\Delta\rho_{xx}/\beta T_K^{1/2}$ versus T/T_K and (**f**) semilog plot of $-\Delta\rho_{xx}/\alpha - \ln T_K$ versus T/T_K for $L1_0$-MnAl films grown at different T_s. For clarity, $\Delta\rho_{xx}$ is multiplied by a factor of 0.5 (0.5), 1.5 (1), 1 (1), 3 (1.5) and 3 (1.5) in **a** (**b**) for T_s = 200, 250, 300, 350 and 400 °C, respectively. (**e**) and (**f**) are the scaling plots showing the collapse of the experimental data in the respective temperature regimes; the solid blue lines show the equations $-\Delta\rho_{xx}/\beta T_K^{1/2} = (T/T_K)^{1/2}$ and $-\Delta\rho_{xx}/\alpha\text{-}\ln T_K = \ln(T/T_K)$, respectively.

experimental values of $\Delta\rho_{xxm}$ ($\approx\Delta\rho_{xx}$ at 330 mK), N_{TLS} was calculated and shown in Fig. 1d. The fast TLSs were annealed away[20,21] and its density was reduced quickly as T_s increases to 300 °C; however, the overall population of TLSs increases beyond 350 °C due to the structural deterioration and strain relaxation of the $L1_0$-MnAl films[33]. The large N_{TLS} ($\sim 10^{20}\,\text{cm}^{-3}$) yields an average distance of ~20 Å for the TLSs in these films.

The $T^{1/2}$ and $\ln T$ dependencies of the resistivity upturns in different samples can be visualized more directly by collapsing the T-dependent $\Delta\rho_{xx}$ data of different samples onto single scaling curves of $-\Delta\rho_{xx}/\beta T_K^{1/2} = (T/T_K)^{1/2}$ and $-\Delta\rho_{xx}/\alpha - \ln T_K = \ln(T/T_K)$ for $T_D/T_K < T/T_K < 1$ and $1 < T/T_K < T_0/T_K$, respectively. Figure 2e shows the log-log plot of $-\Delta\rho_{xx}/\beta T_K^{1/2}$ against T/T_K for the $L1_0$-MnAl samples with different T_s, from which $\log(-\Delta\rho_{xx}/\beta T_K^{1/2})$ is shown to scale linearly with $\log(T/T_K)$ with a slope of 1/2 in $T_D/T_K < T/T_K < 1$ for all the samples. Similarly, the data for all the samples in the T range $1 < T/T_K < T_0/T_K$ can be scaled onto a single curve of $-\Delta\rho_{xx}/\alpha - \ln T_K = \ln(T/T_K)$, as demonstrated by the semilog plot of $-\Delta\rho_{xx}/\alpha - \ln T_K$ against T/T_K (Fig. 2f).

Magnetic field effects. To establish more rigorously the orbital 2CK effect in our $L1_0$-MnAl films, we examined the effect of applied perpendicular magnetic fields, H, on the T-dependent resistivity. Here, in these films with strong PMA, anisotropic magnetoresistance (MR) and MR from spin disorder scattering under perpendicular H should be negligible due to the orthogonal magnetization-current relation and the large energy gap in spin wave excitation spectrum. This is highly amenable to study the H dependence of a 2CK effect. As an example, we show ρ_{xx} (T) of the $L1_0$-MnAl film grown at 200 °C under various H from 0 to 8 T in Fig. 3a,b. The magnetic fields have no measurable effects on the T dependence: ρ_{xx} scales linearly with $\ln T$ and $T^{1/2}$ at $T_K < T < T_0$ and $T_D < T < T_K$ ($T_0 \sim 82.5$ K, $T_K \sim 23$ K and $T_D \sim 8.8$ K), respectively. The same features hold for other films

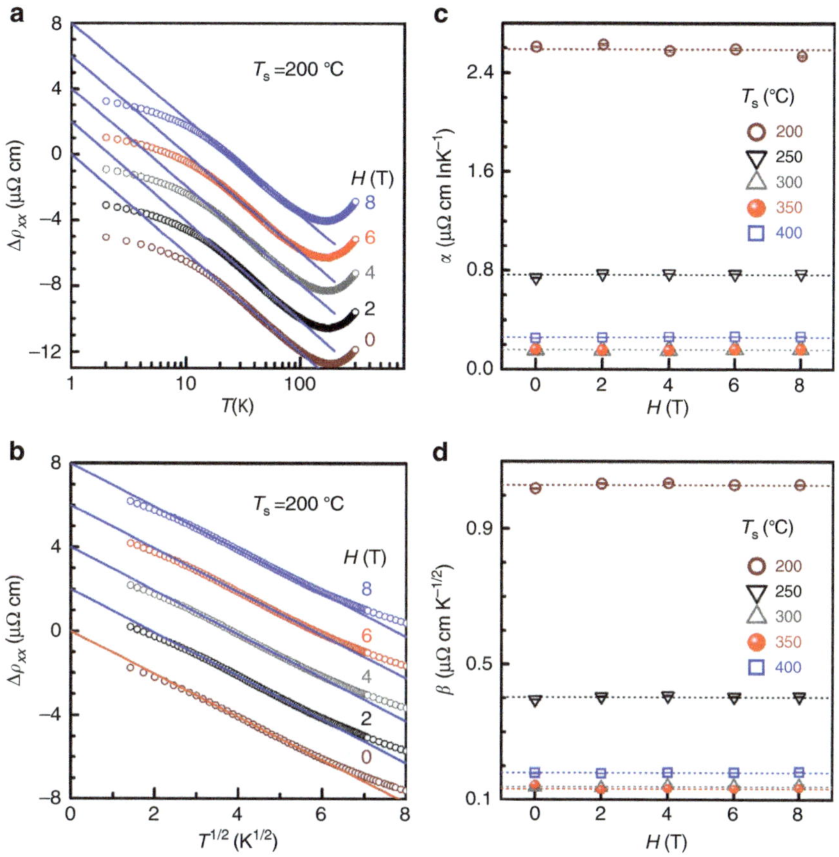

Figure 3 | Magnetic field effects on T dependence of resistivity. (**a**) Semilog plot of $\Delta\rho_{xx}$ versus T and (**b**) $\Delta\rho_{xx}$ versus $T^{1/2}$ for $T_s = 200\,°C$; (**c**) α and (**d**) β plotted as a function of H for $L1_0$-MnAl films with different T_s. The magnetic fields (H) were applied along the film normal. For clarity, the curves in nonzero fields are artificially shifted by steps of 2 μΩ cm in **a** and **b**. The error bars in **c** and **d** correspond to the mean square root of the statistical error for the best linear fits.

with different T_s. Figure 3c,d summarizes the values of α and β as a function of H for the $L1_0$-MnAl films with different T_s. It is clear that both α and β for all films are independent of H, strongly suggesting a nonmagnetic origin of the resistivity upturn scaling in the $L1_0$-MnAl. Specifically, there is no measurable change in T_0, T_K and T_D under different H (Fig. 3a,b), suggesting a negligible effect of H on the coupling strength, the tunnelling symmetry and barrier height of the TLSs. These observations provide strong evidence for the orbital 2CK effect being induced by TLSs of nonmagnetic impurities. Here, it also should be pointed out that the Zeeman energy ($\sim$0.9 meV at $H = 8$ T) is negligibly small in comparison with E_F and ferromagnetic exchange splitting ($E_{exchange} \sim 2$ eV) in $L1_0$-MnAl (ref. 36), hence should not have any measurable effect on the channel asymmetry ($\Delta N = N_\uparrow - N_\downarrow$, where $N_\uparrow$ and $N_\downarrow$ are the numbers of majority and minority spins in the conduction band, respectively) and the three-regime resistivity upturn[16,24,25]. A small negative MR (<0.5%) is observed in the $L1_0$-MnAl at high H in a broad T range from 2 to 300 K, see Supplementary Fig. 3. The MR does not saturate even at 7 T and shows a linear scaling with $H^{1/2}$, which is similar to that in ThAsSe (ref. 20). The MR shows a T dependence, which appears to have no correlation with any characteristic temperatures of the 2CK effect (Fig. 4). Though not yet well understood, such a MR behaviour in materials showing orbital 2CK effect due to TLSs is unlikely to be related to the $T^{1/2}$ dependence of ρ_{xx} at $T_D < T < T_K$.

Characteristic temperatures of orbital 2CK effect. Figure 4 summarizes the relevant characteristic temperatures T_0, T_K and

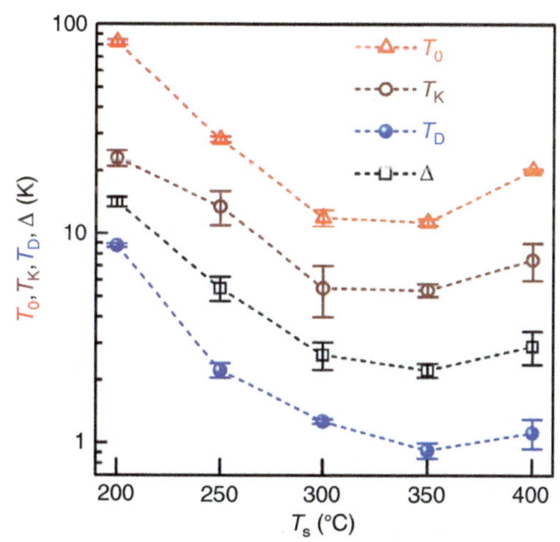

Figure 4 | Characteristic temperatures of orbital 2CK effect. T_0, T_K, T_D and Δ of the $L1_0$-MnAl films are tuned by varying T_s of the epitaxial growth. The error bars correspond to the uncertainty due to the overlapping of the adjacent temperature regimes.

T_D, and Δ of the $L1_0$-MnAl films as a function of T_s. As T_s increases, T_0 drops markedly from 82.5 to 11.4 K, and goes up to 20.0 K as a consequence of the non-monotonic dependence of the TLS population on T_s (Fig. 2d). T_K represents the temperature below which conduction electrons can overscreen the pseudo-spin of the impurity in an orbital 2CK system. Since there

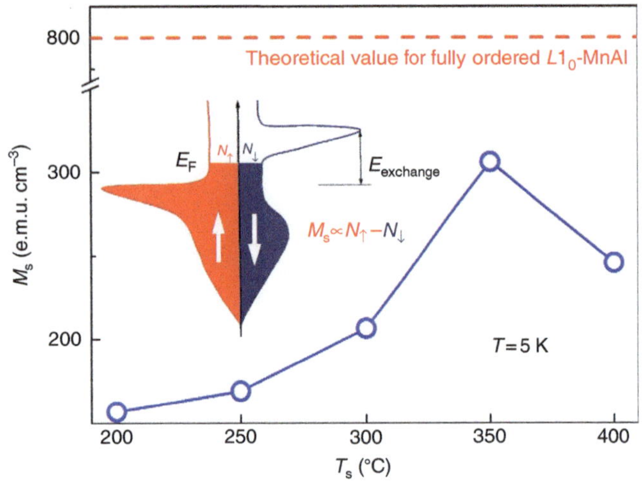

Figure 5 | T_s dependence of saturation magnetization. The red dashed line shows the theoretical value of M_s, that is, 800 e.m.u. cm^{-3}, for fully ordered $L1_0$-MnAl, which is much larger than that at 5 K in the $L1_0$-MnAl films with different T_s. Inset: a schematic of the partial density of states of a ferromagnet, where E_F, $E_{exchange}$ and $N_\uparrow$ ($N_\downarrow$) are the Fermi energy, the exchange splitting and the number of the majority (minority) spin in conduction band, respectively.

is an overlap between the two T regimes of ln$T$ and $T^{1/2}$ dependencies, the values of T_K are defined as the centre of the T overlap. In striking contrast to the small experimental values in Cu PCs (0.1–5 K) and theoretical calculations on amorphous systems ($\ll$1 K), T_K ranges from 23.0 to 5.4 K in the $L1_0$-MnAl. The significant enhancement of T_K in this system could be understood in terms of resonant scattering due to strong coupling of conduction electrons to the scattering centres[19]. Within this scenario, the decrease of T_K at higher T_s may come from a reduced strength of the resonant scattering. Another intriguing observation is that Δ is tuned by as much as a factor of ~5.6 when T_s varies between 200 and 400 °C, which is difficult to achieve in previously studied systems of metallic PCs or glasses using a voltage or magnetic field[20–24]. As T_s goes up, the fastest TLSs are annealed away, leaving the slower ones to dominate the scattering of conduction electrons. Therefore, with T_s increasing from 200 to 350 °C, the value of Δ for active TLSs decreases, leading to a reduction of T_D. The upturn of T_0, T_K and T_D, and Δ at T_s of 400 °C may be attributed to the increased population of fast TLSs and enhanced resonant scattering of conduction electrons due to the structural degradation[31]. Here, it is worth mentioning that the values of Δ observed in $L1_0$-MnAl films are much larger than those reported in Cu PCs (~1 K)[20,21] and ThAsSe (<0.01 K)[22,23]. Besides the small tunnelling barriers and fast scattering rates (that is, intrinsic TLS energy splitting), the adjacent Kondo cloud overlaps suggested by the non-T^2 deviation and the ferromagnetic exchange splitting could also be responsible for the great enhancement in the effective values of Δ in these $L1_0$-MnAl ferromagnetic films. As suggested by the close correlation between the T_s dependence of the characteristic parameters of the TLS-induced 2CK effect, that is, α, β, T_0 and T_K, and Δ with those of structural imperfections and the magnetic properties (for example, magnetization and PMA)[33], electrons as the tunnelling centres of the TLSs in the $L1_0$-MnAl films can be excluded. Similarly, the magnetic Mn atoms can also be ruled out because of the nonmagnetic nature of the orbital Kondo effect. Therefore, we surmise that the nonmagnetic small Al atoms have the role of TLS centres in the present films. In nanoscale tunnelling systems (for example, 3–10 nm-wide Cu PCs[37]), it has been possible to observe individual constituent of the $1/f$ noise, for example, that results from the motion of atomic defects or electrons. However, such constituent is much more difficult to resolve in metallic films[37], for example, our $L1_0$-MnAl epitaxial films. New spectroscopic probe techniques need developing for definitely identifying the microscopic nature of the TLSs in these diffusive metal films.

Coexistence of orbital 2CK with ferromagnetism. Now, we discuss the apparent coexistence of orbital 2CK with ferromagnetism. In a conventional ferromagnet, although the Kondo coupling between the TLS and itinerant electrons is irrelevant to the electron spins, the symmetry of the two spin channels is broken due to ferromagnetic exchange splitting of the d-band. The channel asymmetry should lead to different tunnelling rates of a TLS for two spin channels and thus weaken the NFL behaviours in comparison with its nonmagnetic counterpart ($\Delta N = 0$). If the degree of channel asymmetry ($P = \Delta N/(N_\uparrow + N_\downarrow)$) is large enough, this should be manifested as a decreased magnitude (β) and an enhanced effective breakdown temperature (T_D) of the 2CK effect as a result of the enhanced decoupling of TLS from one spin channel. On the basis of a simple assumption in the Stoner model for itinerant ferromagnetism[38], the saturation magnetization (M_s) can be an index of the channel asymmetry, that is, $M_s \propto \Delta N$. However, this effect is difficult to quantify experimentally as one cannot vary M_s while keeping other parameters (for example, N_{TLS}, coupling strength, Kondo cloud overlap, barrier height and tunnelling rate) of the TLSs constant in a set of samples. For our $L1_0$-MnAl films, the T_s dependence of the measured β and T_D is dominated by the variation of the other parameters, respectively, rather than that of ΔN, as suggested by the quite different T_s dependence of the 2CK parameters (β and T_D) and M_s. The fully ordered $L1_0$-MnAl is theoretically an itinerant magnet with a spin polarization of 23.8% at the Fermi surface[39] and a magnetization M_s of ~800 e.m.u. cm^{-3} (that is, 2.37 μ_B f.u.$^{-1}$)[28,36]. In our disordered $L1_0$-MnAl samples, the measured M_s is much smaller (157–306 e.m.u. cm^{-3}, see Fig. 5), suggesting a very low ΔN ($\ll N_\downarrow$) and P ($\ll 1$). This could be a reason why it is still difficult to establish a definitive quantitative correlation between ΔN and the 2CK parameters (β and T_D).

Discussion

We have observed transport behaviour highly suggestive of a robust orbital 2CK effect from TLSs in epitaxial $L1_0$-MnAl films with strong PMA. The H-independent resistivity upturn scaling with ln$T$ and $T^{1/2}$ in the two T regimes below the resistance minimum, and deviation from the NFL behaviour at the lowest temperatures are consistent with the TLS model. The molecular beam epitaxy growth method afforded unprecedented tunability of the TLS density, T_K and Δ in this material system, leading to the insight on the origin of the 2CK effect. The greatly enhanced T_K and Δ suggest resonant tunnelling of TLSs due to the strong coupling with conduction electrons. The orbital 2CK effect in a material with significant ferromagnetism and conduction spin polarization is an intriguing observation that warrants further theoretical and experimental studies.

Methods

Sample growth and characterization. $L1_0$-MnAl films were grown on 150 nm GaAs-buffered semi-insulating GaAs (001) by molecular beam epitaxy and capped with a ~4 nm-thick Al_2O_3 layer. The thickness and Mn/Al atomic ratio of $L1_0$-MnAl layer is calibrated to be 30 nm and 1.1 by cross-sectional high-resolution transmission microscopy with energy dispersive X-ray spectroscopy[31], respectively. Synchrotron X-ray diffraction and Quantum Design SQUID were used to characterize the structures and magnetization properties of these films.

Device fabrication and transport measurement. These films were patterned into 60 μm wide Hall bars with adjacent electrode distance of 200 μm using photolithography and ion-beam etching for transport measurements (Fig. 1c). The longitudinal resistivity was measured as a function of temperature (T) and magnetic field (H) in a Quantum Design PPMS for $T = 2$–300 K (DC, $I = 10\,\mu A$) and in an Oxford ^{3}He cryostat for $T = 0.33$–5 K (a.c, $I = 1\,\mu A$).

References

1. Béri, B. & Cooper, N. R. Topological Kondo effect with Majorana Fermions. *Phys. Rev. Lett.* **109,** 156803 (2012).
2. Kashuba, O. & Timm, C. Topological Kondo effect in transport through a superconducting wire with multiple Majorana end states. *Phys. Rev. Lett.* **114,** 116801 (2015).
3. Jarillo-Herrero, P. *et al.* Orbital Kondo effect in carbon nanotubes. *Nature* **434,** 484–488 (2005).
4. Mitchell, A. K., Sela, E. & Logan, D. E. Two-channel Kondo physics in two-Impurity Kondo models. *Phys. Rev. Lett.* **108,** 086405 (2012).
5. Delattre, T. *et al.* Noisy Kondo impurities. *Nat. Phys* **5,** 208–212 (2009).
6. Potok, R. M., Rau, I. G., Shtrikman, H., Oreg, Y. & Goldhaber-Gordon, D. Observation of the two-channel Kondo effect. *Nature* **446,** 167–171 (2007).
7. Yeh, S. & Lin, J. Two-channel Kondo effects in Al/AlO$_x$/Sc planar tunnel junctions. *Phys. Rev. B* **79,** 012411 (2009).
8. Nozières, P. & Blandin, A. Kondo effect in real metals. *J. Phys. France* **41,** 193–211 (1980).
9. Schlottmann, P. & Sacramento, P. D. Multichannel Kondo problem and some applications. *Adv. Phys.* **42,** 641–682 (1993).
10. Zawadowski, A. Kondo-like state in a simple model for metallic glasses. *Phys. Rev. Lett.* **45,** 211–214 (1980).
11. Cox, D. L. & Zawadowki, A. Exotic Kondo effects in metals: Magnetic ions in a crystalline electric field and tunneling centres. *Adv. Phys.* **47,** 599–942 (1998).
12. Delft, J. von *et al.* The 2-channel Kondo model: I. Review of experimental evidence for its realization in metal nanoconstrictions. *Annals. Phys.* **263,** 1–55 (1998).
13. Muramatsu, A. & Guinea, F. Low-temperature behavior of a tunneling atom interacting with a degenerate electron gas. *Phys. Rev. Lett.* **57,** 2337 (1986).
14. Hewson, A. C. *The Kondo Problem to Heavy Fermions* (Cambridge Univ. Press, 1993).
15. Pang, H. B. & Cox, D. L. Stability of the fixed point of the two-channel Kondo Hamiltonian. *Phys. Rev. B* **44,** 9454 (1991).
16. Affleck, I. *et al.* Relevance of anisotropy in the multichannel Kondo effect: comparison of conformal field theory and numerical renormalization-group results. *Phys. Rev. B* **45,** 7918 (1992).
17. Aleiner, I. L. *et al.* Kondo temperature for the two-channel Kondo models of tunneling centers. *Phys. Rev. Lett.* **86,** 2629 (2001).
18. Aleiner, I. L. *et al.* Experimental tests for the relevance of two-level systems for electron dephasing. *Phys. Rev. B* **63,** 201401 (R) (2001).
19. Zaránd, G. Existence of a two-channel Kondo regime for tunneling impurities with resonant scattering. *Phys. Rev. B* **72,** 245103 (2005).
20. Ralph, D. C. & Buhrman, R. A. Observations of Kondo scattering without magnetic impurities: a point contact study of two-level tunneling systems in metals. *Phys. Rev. Lett.* **69,** 2118 (1992).
21. Ralph, D. C., Ludwig, A. W. W., Delft, J. v. & Buhrman, R. A. 2-channel Kondo scaling in conductance signals from 2 level tunneling systems. *Phys. Rev. Lett.* **72,** 1064 (1994).
22. Cichorek, T. *et al.* Two-channel Kondo effect in glasslike ThAsSe. *Phys. Rev. Lett.* **94,** 236603 (2005).
23. Cichorek, T. *et al.* TLS Kondo effect in structurally disordered ThAsSe. *J. Mag. Mag. Mater* **272,** 66–67 (2004).
24. Upadhyay, S. H. *et al.* Low-energy restoration of Fermi-liquid behavior for two-channel Kondo scattering. *Phys. Rev. B* **56,** 12033 (1997).
25. Schlottmann, P. & Lee, K. J. B. Quenching of overcompensated Kondo impurities via channel asymmetry. *Physica B* **223–224,** 458–463 (1996).
26. Henkie, Z., Fabrowski, R. & Wojakowski, A. Anisotropies of the electrical resistivity and Hall effect in UAsSe. *J. Alloys Compd.* **219,** 248–251 (1995).
27. Gopalakrishnan, B. *et al.* Electronic transport in magnetically ordered $Mn_5Si_3C_x$ films. *Phys. Rev. B* **77,** 104414 (2008).
28. Zhu, L. J., Nie, S. H. & Zhao, J. H. Recent progress in perpendicularly magnetized Mn-based binary alloy films. *Chin. Phys. B* **22,** 118505 (2013).
29. Zhu, L. J. *et al.* Multifunctional $L1_0$-$Mn_{1.5}Ga$ films with ultrahigh coercivity, giant perpendicular magnetocrystalline anisotropy and large magnetic energy product. *Adv. Mater.* **24,** 4547 (2012).
30. Zhu, L. J., Pan, D., Nie, S. H., Lu, J. & Zhao, J. H. Tailoring magnetism of multifunctional Mn_xGa films with giant perpendicular anisotropy. *Appl. Phys. Lett.* **102,** 132403 (2013).
31. Nie, S. H. *et al.* Perpendicularly magnetized τ-MnAl (001) thin films epitaxied on GaAs. *Appl. Phys. Lett.* **102,** 152405 (2013).
32. Zhu, L. J., Pan, D. & Zhao, J. H. Anomalous Hall effect in epitaxial $L1_0$-$Mn_{1.5}Ga$ films with variable chemical ordering. *Phys. Rev. B* **89,** 220406 (R) (2014).
33. Nie, S. H., Zhu, L. J., Pan, D., Lu, J. & Zhao, J. H. Structural characterization and magnetic properties of perpendicularly magnetized MnAl films grown by molecular-beam epitaxy. *Acta. Phys. Sin* **62,** 178103 (2013).
34. Koch, A. J. J. *et al.* New Material for Permanent Magnets on a Base of Mn and Al. *J. Appl. Phys.* **31,** 75 (1960).
35. Zimmerman, N. M., Golding, B. & Haemmerle, W. H. Magnetic field tuned energy of a single two-level system in a mesoscopic metal. *Phys. Rev. Lett.* **67,** 1322 (1991).
36. Sukuma, A. Electronic structure and magnetocrystalline anisotropy Energy of MnAl. *J. Phys. Soc. Jpn* **63,** 1422 (1994).
37. Ralls, K. S. & Burman, R. A. Microscopic study of 1/*f* noise in metal nanobridges. *Phys. Rev. B* **44,** 5800 (1991).
38. Stoner, E. C. Collective electron ferromagnetism. *Proc. R. Soc. Lond. Ser. A* **165,** 372–414 (1938).
39. Park, J. H. *et al.* Saturation magnetization and crystalline anisotropy calculations for MnAl permanent magnet. *J. Appl. Phys.* **109,** 09A731 (2010).

Acknowledgements

We thank Y.Q. Li and J. Liao for their help on the ultralow temperature measurements. We also thank S. von. Molnár and G. Woltersdorf for discussions. L.J.Z., S.H.N. and J.H.Z. were supported partly by MOST of China (grant no. 2015CB921503), NSFC (grant no. 61334006) and the CAS/SAFEA International Partnership Program for Creative Research Teams. P.X. acknowledges support from NSF grant DMR-1308613. P.S. was supported by the US Department of Energy under grant DE-FG02-98ER45707.

Author contributions

L.J.Z. and J.H.Z. designed the experiment. L.J.Z. and S.H.N. grew the samples and performed the structural and magnetic characterizations. L.J.Z. fabricated the Hall bar devices and performed transport measurements. L.J.Z., P.X., P.S. and J.H.Z. analysed the data and wrote the manuscript.

Additional information

Competing financial interests: The authors declare no competing financial interests.

9

Scrutinizing the double superconducting gaps and strong coupling pairing in $(Li_{1-x}Fe_x)OHFeSe$

Zengyi Du[1,*], Xiong Yang[1,*], Hai Lin[1,*], Delong Fang[1], Guan Du[1], Jie Xing[1], Huan Yang[1], Xiyu Zhu[1] & Hai-Hu Wen[1]

In the field of iron-based superconductors, one of the frontier studies is about the pairing mechanism. The recently discovered $(Li_{1-x}Fe_x)OHFeSe$ superconductor with the transition temperature of about 40 K provides a good platform to check the origin of double superconducting gaps and high transition temperature in the monolayer FeSe thin film. Here we report a scanning tunnelling spectroscopy study on the $(Li_{1-x}Fe_x)OHFeSe$ single crystals. The tunnelling spectrum mimics that of the monolayer FeSe thin film and shows double gaps at about 14.3 and 8.6 meV. Further analysis based on the quasiparticle interference allows us to rule out the *d*-wave gap, and for the first time assign the larger (smaller) gap to the outer (inner) Fermi pockets (after folding) associating with the d_{xy} (d_{xz}/d_{yz}) orbitals, respectively. The gap ratio amounts to 8.7, which demonstrates the strong coupling mechanism in the present superconducting system.

[1]Center for Superconducting Physics and Materials, National Laboratory of Solid State Microstructures and Department of Physics, Collaborative Innovation Center for Advanced Microstructures, Nanjing University, Nanjing 210093, China. * These authors contributed equally to this work. Correspondence and requests for materials should be addressed to H.Y. (email: huanyang@nju.edu.cn) or to H.-H.W. (email: hhwen@nju.edu.cn).

The pairing mechanism and the superconducting gap structure have remained as the core issues in the study on iron-based superconductors. Depending on the atomic structures, the Fermi surface topology changes significantly in different systems, varying from having hole pocket(s) only, to both hole and electron pockets, and to only electron pockets in some FeSe-based systems[1–7]. In many earlier discovered FeAs- or FeSe-based systems[8–10] with both electron and hole pockets, a natural pairing picture was proposed[11,12], which concerns the pairing through the pair-scattering process between the electron and hole pockets leading to an extended *s*-wave gap with the pairing manner denoted by $s\pm$. While in the system with absence of the hole pockets, this weak coupling-based picture is facing a great challenge. The question arising immediately is that whether we have a sign change of the gap function as expected from the repulsive interaction-induced pairing[13], for example, the nodeless *d*-wave with opposite signs between two neighbouring electron pockets. Another puzzling point is that in the scanning tunnelling microscopy/spectroscopy (STM/STS) measurements on the FeSe monolayer thin film[1] on a $SrTiO_3$ substrate, double gaps have been observed on the single tunnelling spectrum with the larger gap ranging from 14 to 22 meV, while in all the angle-resolved photoemission spectroscopy (ARPES) measurements[3,4] there is only one set of electron Fermi pockets near the $\tilde{M}$ point. It is curious to know how to associate the double gaps with only one set of Fermi pockets. In addition, the larger gap discovered either in STM/STS[1] or ARPES[3,4] measurements is very big in a sense of weak coupling-based picture. Therefore it was proposed that some interfacial strong phonon mode may join in the electron-boson coupling leading to the high gap value[2]. However, this scenario needs to be verified in systems without the particular substrate and related phonons. It is thus highly desired to do experiments in the FeSe-based system with the similar electronic structures. The recently discovered $(Li_{1-x}Fe_x)OHFeSe$ phase[5–7] with a superconducting transition temperature $T_c \approx 40$ K shows also only electron pockets around $\tilde{M}$ points in a folded Brillouin zone (BZ)[14,15], which is very similar to the case of monolayer FeSe thin film. This provides us a very good platform to check the fundamental problems, such as the origin of the double gaps and whether the weak coupling-based picture holds still for the superconductivity mechanism.

In this paper, we report the STM/STS measurements on the newly discovered superconductor $(Li_{1-x}Fe_x)OHFeSe$. It is found that the tunnelling spectrum shows a double gap structure and looks very similar to that observed on the FeSe monolayer thin film. The analysis based on the quasiparticle interference (QPI) technique allows us to carefully scrutinize the superconducting gaps on the associated Fermi surfaces near $\tilde{M}$ points in $(Li_{1-x}Fe_x)OHFeSe$. The larger gap (Δ_1) derived here is about 14.3 meV, which yields a huge gap ratio of $2\Delta_1/k_BT_c = 8.7$. This demonstrates the strong coupling mechanism for superconductivity in the present system. Our work shed new light in understanding the puzzling issues in the high-temperature superconducting FeSe monolayer thin films, and will stimulate new efforts in solving the superconductivity mechanism in iron-based superconductors.

Results

Characterization of superconducting properties. Figure 1a shows the temperature-dependent resistivity of the $(Li_{1-x}Fe_x)$OHFeSe single crystal. One can see that the resistive transition is rather sharp near the foot. The rounded shape near the onset transition is induced by the fluctuating superconductivity since the system is highly two-dimensional (2D). In Fig. 1b, we present the temperature dependence of the DC magnetization of the sample. It is measured at an external magnetic field of 20 Oe with the zero-field-cooled (ZFC) and FC modes. The sample shows perfect Meissner shielding effect with the superconducting transition temperature around 38 K, which is comparable to previous reports[5–7]. The magnetization-hysteresis-loops shown in Fig. 1c exhibit a symmetric feature, which suggests the bulk pinning of vortices and thus the bulk superconductivity of the material. The calculated critical current density at 2 K and 0 T from the Bean

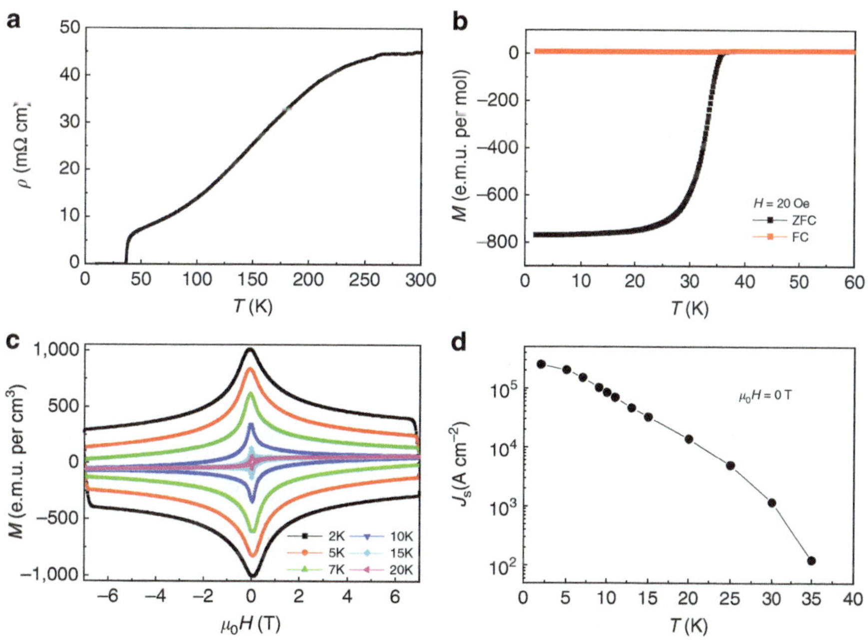

Figure 1 | Superconducting transition and magnetization of the $(Li_{1-x}Fe_x)$OHFeSe crystal. (**a**) Temperature dependence of resistivity of the sample. (**b**) Temperature dependence of DC magnetization measured with the zero-field-cooled (ZFC) and field-cooled (FC) modes at a field of 20 Oe. (**c**) The magnetization-hysteresis-loops measured at temperatures from 2 to 20 K. (**d**) Temperature dependence of the calculated critical current density J_s based on the Bean critical state model at zero magnetic field.

critical state model[16] shown in Fig. 1d is 1–2 orders of magnitude larger than that in $K_xFe_{2-y}Se_2$ and is comparable to that in the optimally doped $Ba(Fe_{1-x}Co_x)_2As_2$ (ref. 17). This indicates that the sample is not phase separated as what occurs in $K_xFe_{2-y}Se_2$. This allows us to have reliable STM/STS measurements.

Topographic STM image and STS spectra. STM/STS measurements can provide essential information on the electronic structure and the gap symmetry of the novel superconductors[18–20]. Figure 2a shows a topographic image of the terminated surface of the cleaved $(Li_{1-x}Fe_x)OHFeSe$ single crystal, and the atomically resolved square lattice is clearly seen. The lattice constants at the perpendicular directions are 3.61 and 3.63 Å, respectively, which are comparable to the lattice constant *a* of 3.79 Å from previous reports[5,7]. The schematic structure of $(Li_{1-x}Fe_x)OHFeSe$ is shown in Fig. 2b. Since the cleaved surface is very stable and the lattice constant is very close to the expected value of Se–Se bond, together with the perfect STS measured (see below), we can reasonably assume that the terminated top layer is the Se-atom layer. Some bright spots with a dumbbell shape on the topography are observed and very similar to those from Cu or Co impurities in our previous works[21,22]. From the structure of the system, it is found that the impurity atom in the centre of a dumbbell spot locates just at the position of an Fe atom in the beneath layer, and it may be induced by the partial substitution of Fe atoms with Li, or some vacancies at the Fe sites in the Fe layer. This can get support from previous results on LiFeAs in which similar defects on Fe sites were proposed[23–25]. The calculated density of the impurities varies from 0.5 to 1.5% of Fe sites on the FeSe layer, which is much lower than the ratio of about 20% of substitution of Li by Fe atoms in the Li(Fe) layers[5–7].

Figure 2c shows a typical tunnelling spectrum measured at 1.5 K. Clearly there are two pairs of coherence peaks on the spectrum, indicating existence of two superconducting gaps. The larger gap is marked by Δ_1 on its peak position of about ± 14.3 meV while the smaller one is marked by Δ_2 at about ± 8.6 meV. The spectrum with two-gap feature is reminiscent of that reported on the monolayer FeSe thin film on $SrTiO_3$ substrate[1]. The recent high-resolution ARPES measurements show only electron pockets as well as one superconducting gap in $(Li_{1-x}Fe_x)OHFeSe$[14,15]. It should be noted that the two-gap feature in the monolayer FeSe thin film was observed only in the STS measurements[1], not in the ARPES. This may be induced by the limited resolution of ARPES, or due to some selection rules in the emission process of photo-electrons. More interestingly, the gaps found by the two ARPES measurements from two different groups on the similar samples are corresponding to the larger[14] and smaller gap[15] maxima of our STS measurements, respectively. The two-gap feature is a common one in our $(Li_{1-x}Fe_x)OHFeSe$ samples, which can get support from the statistical result (Fig. 2d) of the peak positions on 98 spectra collected randomly in the range of 41 nm × 41 nm. By fitting to the Gaussian functions, we obtained the two-gap values of $\Delta_1 = 14.3 \pm 1.3$ and $\Delta_2 = 8.6 \pm 0.9$ meV, and the corresponding gap ratios of $2\Delta_1/k_BT_c = 8.7 \pm 0.8$ and $2\Delta_2/k_BT_c = 5.2 \pm 0.5$. These values are much larger than the one $2\Delta/k_BT_c \sim 3.53$ predicted by the Bardeen–Cooper–Schrieffer (BCS) theory in the weak coupling regime. The huge value $2\Delta_1/k_BT_c = 8.7$ indicates strong coupling mechanism in the system. Worthy of noting is that similar ratio was observed in the cuprate superconductors[26,27].

Temperature-dependent STS spectra and Dynes model fitting. In Fig. 3a, we show the temperature evolution of the tunnelling

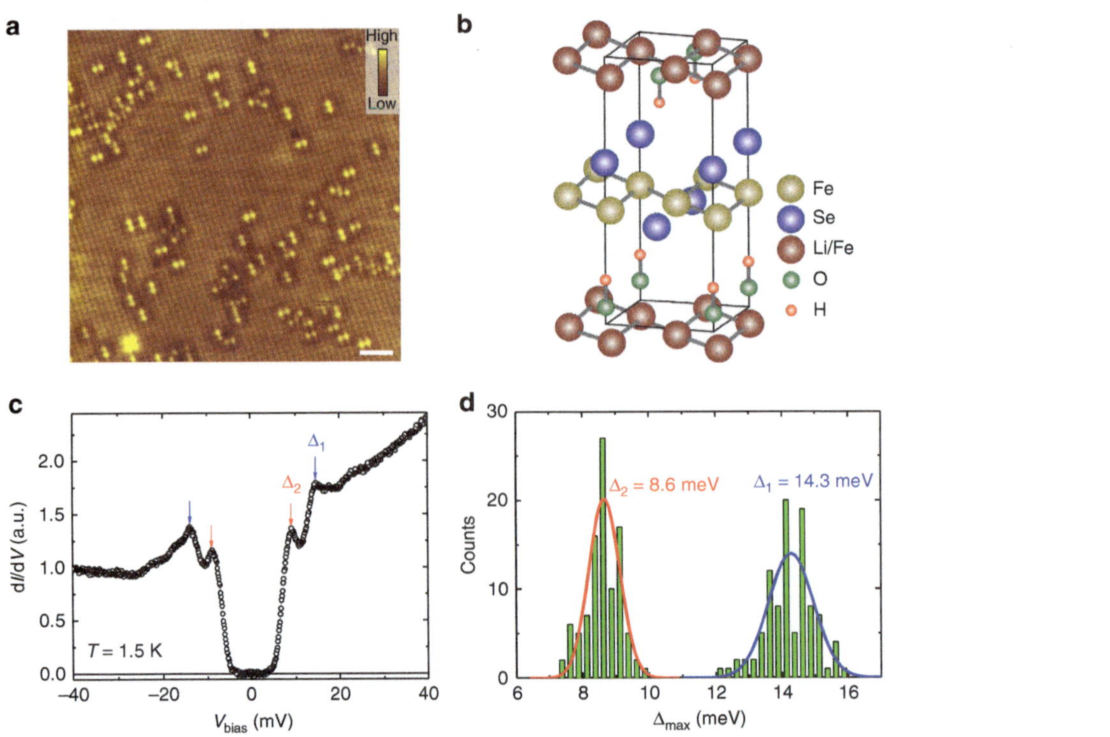

Figure 2 | Topographic STM image and two-gap feature of $(Li_{1-x}Fe_x)OHFeSe$. (**a**) The atomically resolved STM image with a bias voltage of $V_{bias} = 180$ mV and tunnelling current of $I_t = 102$ pA. The defects with the dumbbell shape are observed on the top surface. Scale bar, 2 nm. (**b**) The schematic atomic structure of $(Li_{1-x}Fe_x)OHFeSe$. (**c**) A typical STS spectrum measured at 1.5 K away from the defects. One can clearly see the two-gap feature. Similar STS with double gap structure have been repeated in different areas and different samples. (**d**) Histogram of the superconducting gap values with the fitting results by Gaussian functions (solid lines). Two major gaps can be found.

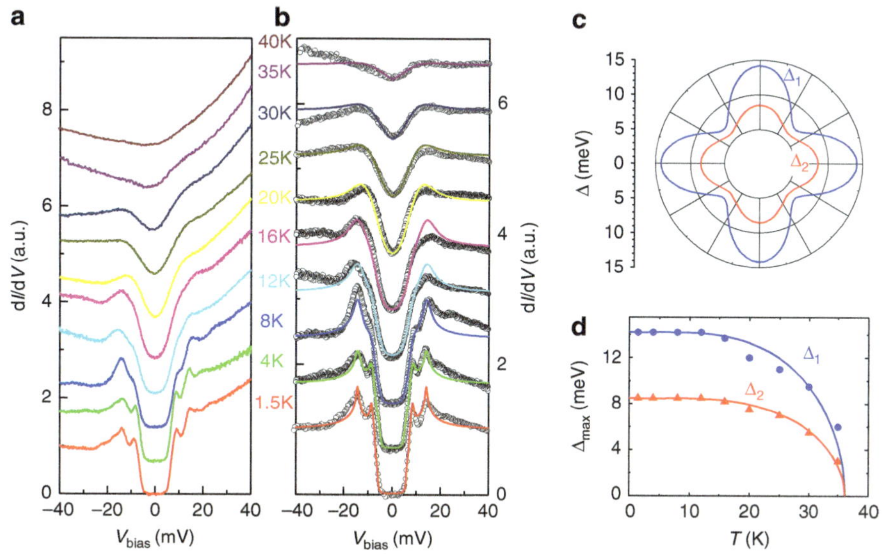

Figure 3 | Temperature-dependent tunnelling spectra and theoretical fits. (**a**) The evolution of the STS spectra measured at temperatures from 1.5 to 40 K. (**b**) Fitting results to the STS spectra normalized by the one measured at 40 K in the normal state. The dark hollow circles represent the experimental data, and the coloured solid lines are the theoretical fits to the data with two anisotropic *s*-wave gaps by the Dynes model. (**c**) The anisotropic-gap functions used in the fitting to the curve measured at 1.5 K. (**d**) Temperature dependence of the two gaps extracted from the Dynes model fitting, and the solid lines denote the theoretical calculations of the superconducting gap from the BCS model.

spectra measured at the same position. The coherence peaks associating with both gaps are suppressed and mix together with increasing temperature, and finally vanish above T_c. To see the possible influence of the electronic bands on the tunnelling spectrum, we show one STS measured in wide energy ranges in Supplementary Fig. 1. The spectra shown in Fig. 3b are normalized (divided by the one measured at 40 K). The two well-resolved coherence peaks together with the fully gapped density of states (DOS) near zero-bias allow us to make reliable fitting using the Dynes model[28]. From the normalized spectrum at 1.5 K, one can see that the bottom of the STS is very flat and has a zero value, however, the coherence peaks are not very sharp. Together with the latter effect is the finite slope of the rising edges of the STS near the bottom. To know whether these contradicting phenomena are induced by the finite scattering rates, we firstly try to use two isotropic *s*-wave gaps and different pairs of scatterings rates $\Gamma_{1(2)}$ to fit the experimental data. Because of the two-gap feature, in the model to fit the normalized spectra, the differential conductivity is constructed as: $G = x\mathrm{d}I_1/\mathrm{d}V + (1-x)\mathrm{d}I_2/\mathrm{d}V$, where $I_{1(2)}(V)$ is the tunnelling current contributed by the larger(smaller) gap, and x is the related spectral weight. The detailed fitting process and parameters are given in Supplementary Note 1. The fitting results with two isotropic *s*-wave gaps under zero and finite scattering rates are presented in Fig. 4. It is clear that, even with the finite scattering rates, the fitting with isotropic gaps fails to catch up the main features of the experimental data. We then use the two anisotropic *s*-wave gaps $\Delta_{1(2)}(\theta) = \Delta_{1(2)\max}\left[1 - p_{1(2)}(1 - \cos 4\theta)\right]$ to fit the experimental data and show the results in Fig. 3b. One can see from Fig. 3b that the theoretical model fits the experimental data very well. The fitting process is detailed in Supplementary Note 1 and the fitting parameters are listed in Supplementary Table 1. The fourfold symmetric gap functions used to fit the data at 1.5 K are presented in Fig. 3c, and the anisotropy p is fixed in all the fittings, that is, 25% for $\Delta_1(\theta)$ and 15% for $\Delta_2(\theta)$, while the weight of the larger gap $\Delta_1(\theta)$ is fixed with the value of 65% at different temperatures. The resultant gap maxima versus temperature are shown in Fig. 3d, and the solid lines are the results of theoretical calculation within the BCS model by fixing $\Delta_{\max}(0)$ and T_c

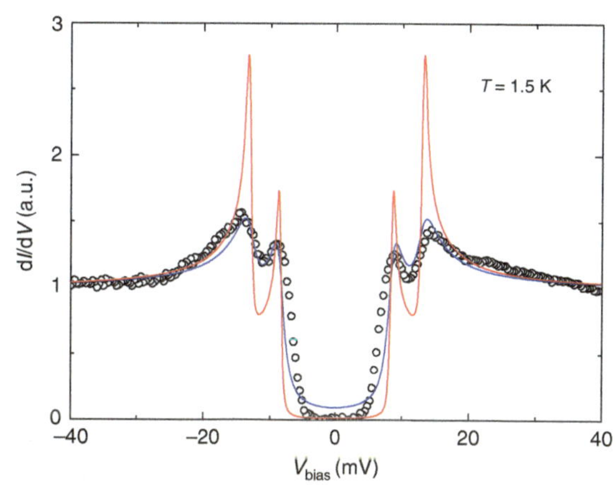

Figure 4 | Theoretical fitting results by two isotropic *s*-wave gaps. The fitting results shown as solid lines with different fitting parameters fail to catch up the main features of the experimental data. For the fittings, $\Delta_1 = 13$ meV and $\Delta_2 = 8.5$ meV, $\Gamma_1 = \Gamma_2 = 0.15$ meV were used for the clean case (red line), while the same gaps and $\Gamma_1 = 1.2$ and $\Gamma_2 = 0.85$ meV for the case of finite scattering rates (blue one). From these fittings, one can see that the tunnelling spectrum cannot be fitted with two isotropic gaps.

derived from our experiment. One can see that the temperature-dependent superconducting gaps can be well-described by the BCS theory. Our results suggest that this material has two nodeless superconducting gaps, each with a significant anisotropy. The two-gap feature is difficult to be understood if only one set of electron Fermi pockets around the $\tilde{\mathrm{M}}$ point exists as observed in the ARPES measurements.

QPI measurements. On the surface of the crystal, we can see some dumbbell-shape impurities. The spectrum measured on the impurity site has no well-defined superconducting coherence peaks as shown in Supplementary Note 2. Instead, we can see the in-gap states with two asymmetric peaks which make the bottom near zero-bias 'V'-shaped (Supplementary Fig. 2). The impurities which act as the scattering centres will produce the standing waves, and such QPI imaging can provide fruitful information on the scattering between the contours of Fermi surfaces with the scattering vector **q** at constant energy *E* in **k**-space[20]. We perform the QPI measurements on two areas with different scanning range as their topographies shown in Supplementary Fig. 3a,b. The details of the measurements of QPI and data treatments are given in Supplementary Note 3. The standing waves can be clearly seen in Fig. 5a measured around the larger energy gap Δ_1. When we do the Fourier transformation (FT) to the QPI image, the resultant $\rho(\mathbf{q}, E)$ image, or called as the FT-QPI image, shown in Fig. 5b,c helps us to investigate the Fermi surfaces and the superconducting gaps of the material. The resultant FT-QPI pattern looks quite similar to the data on the monolayer FeSe thin film on $SrTiO_3$ (001) as recently revealed by the STM measurement[29]. Quite interestingly, the contours near the corner of the FT-QPI pattern corresponding to the large-**q** scattering have clear anisotropic feature, which was also observed on the monolayer FeSe thin film on $SrTiO_3$ (001) and was attributed to the orbital features[29]. The ARPES[14,15] measurements show only one set of electron pockets near $\tilde{M}$ points. However, if taking a closer look at the central part of the FT-QPI images as shown in Fig. 5f,g, we can find that it is unlikely. There are actually two circles, which means that a single set of isotropic ring-shaped Fermi surfaces would not be possible to produce such patterns. Therefore we use elliptic electron Fermi surfaces to simulate the QPI image. Because there are two Fe sites and two distinct Se sites in one unit cell, the two slightly elliptic Fermi pockets will naturally fold into the smaller BZ, as sketched in Fig. 5d. In the picture of folded Fermi surfaces, we can consider the Fermi pockets as inner and outer rings. Due to the possible spin-orbital coupling, the two Fermi pockets may further hybridize into two sets of Fermi pockets with inner and outer contours[30]. We thus do the QPI simulation of this type of Fermi pockets and the resultant pattern (Fig. 5e) is comparable with our experimental data. As shown in Fig. 5b,c, when the bias voltage was increased from $\Delta_{2max} = 8.6$ to $\Delta_{1max} = 14.5$ meV, the peripheral patterns from the inter-pocket scattering have negligible difference while the pattern in the centre corresponding to the intra-pocket scattering changes a lot. The magnitude of the outer circle in the central pattern (Fig. 5f,g) seems to be enhanced obviously when the energy increases from the small energy gap to the larger one. The intensity of the FT-QPI contour reflects the joint DOS (JDOS) between two **k**-points on the Fermi surfaces. The emergence and enhancement of intensity in the contour in **q**-space of the outer circle at a higher energy suggests that the superconducting gap on this Fermi surface is larger than that of the inner one[31–33].

Intraband QPI. We then pay attention to the central pattern by measuring the QPI mapping in a larger real-space scale to get the detailed features of the smaller **q**-space view, and the evolution of the $\rho(\mathbf{q}, E)$ images is shown in Fig. 6. The raw data of the related QPI images are given in Supplementary Fig. 4. One can see that there is almost nothing at the locations of the two interference rings at the energy of 0 and 2 mV. This may rule out the existence of any gap nodes on the Fermi surface, since the FT-QPI images would show an intensity at the **q**-vector connecting two nodal points if they would exist. This is also consistent with the full gap structure as seen from the STS spectrum. At the energy of 6.8 mV, a ring with enhanced intensity on the FT-QPI images can be observed, with the strongest intensity along the $\Gamma\tilde{M}$ direction. At the energy of the small gap, namely 8.6 mV, the FT-QPI pattern looks similar to that at 6.8 mV. However, when the energy is

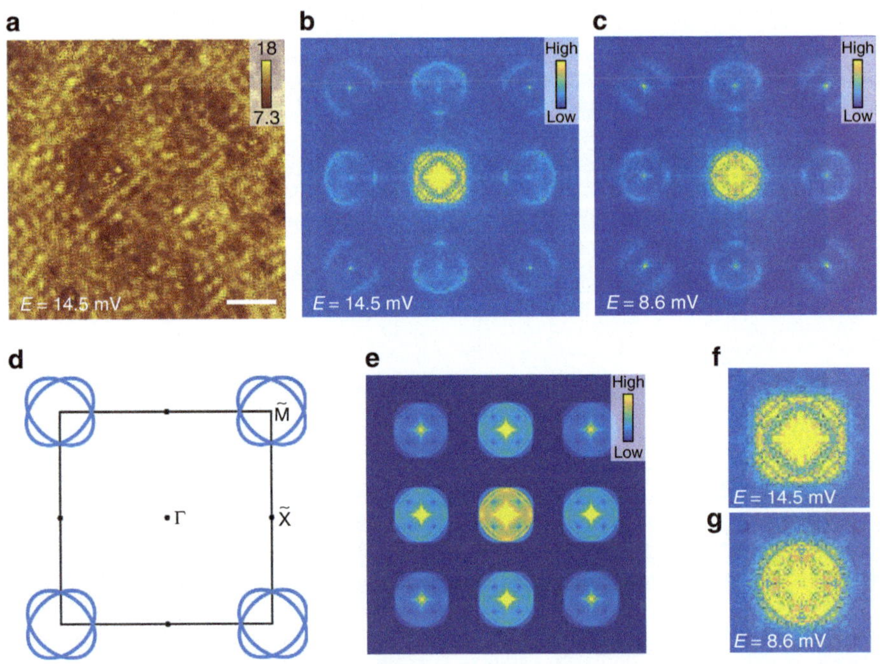

Figure 5 | Quasiparticle interference patterns and theoretical simulation. (**a**) QPI image $g(\mathbf{r}, E = 14.5\,\text{mV})$ in real space measured at 1.7 K near the energy of the coherence peak position of the larger gap. Scale bar, 5 nm. (**b**,**c**) FT-QPI intensity $\rho(\mathbf{q}, E)$ obtained by taking Fourier transformations on the corresponding real-space image measured with energies near the coherence peak position of the two gaps, namely 14.5 and 8.6 mV. The images are fourfold symmetrized to enhance the signal. (**d**) Schematic plot of the elliptic-shaped electron pockets in folded Brillouin zone. (**e**) The theoretical simulation of the QPI scattering intensity by applying autocorrelation to **d**. (**f**,**g**) Zoom-in images of the central parts of **b** and **c**, which contain the information of the small-*q* intra-pocket scattering. The image at 14.5 mV is very similar to the structure of the central pattern in (**e**). However the central pattern at 8.6 mV has a very different shape, and the outer ring seems to be less pronounced.

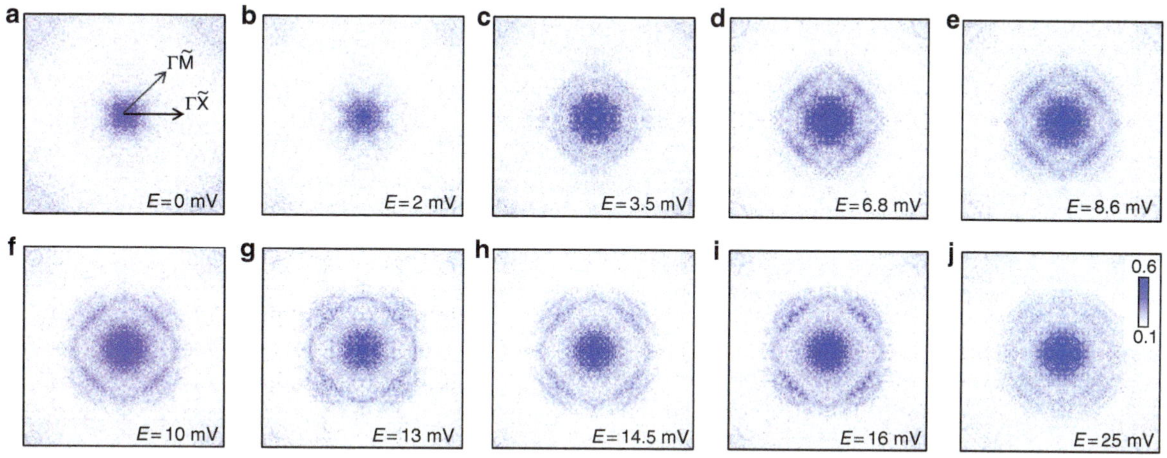

Figure 6 | The evolution of FT-QPI intensity ρ(q, E) at different energies. (**a**-**j**) The FT-QPI ρ(**q**, E) images are derived from Fourier transformation on the QPI images with the real-space scale of 58 nm × 58 nm. A 2D-gaussian-function background was subtracted from the raw FT-QPI image and the fourfold symmetrization was carried out to enhance the signal. The intensity on the inner ring in **q**-space shows up at above 3.5 mV, the ring becomes clear at the energy of the smaller gap, namely 8.6 mV, while the segments corresponding to the outer ring appear gradually with a higher energy above 6.8 mV, and are clear when measuring with the voltage at the larger energy gap 14.5 mV.

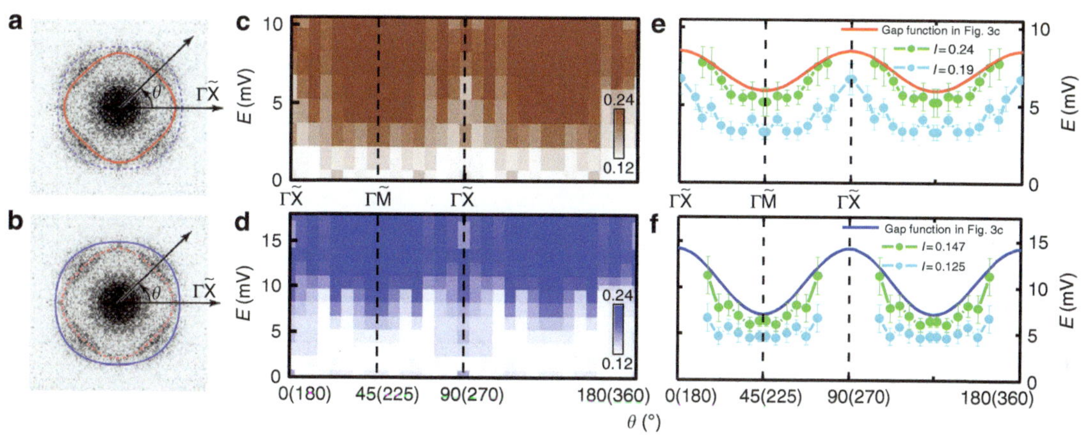

Figure 7 | 2D colour plots of FT-QPI intensity $\rho(\theta, E)$ along the inner and outer contours of the central pattern in q-space. (**a**,**b**) The FT-QPI pattern taken at 14.5 mV. The solid-line contours in **a** and **b** were expressed approximately by $R_1 - r_1\cos4\theta$ and $R_2 + r_2\cos4\theta$ for outer and inner rings, respectively. (**c**,**d**) The FT-QPI intensity along the inner (**a**) and outer (**b**) rings at different angles taken from the fourfold symmetrized FT-QPI patterns in Fig. 6 with various energy values. The initial polar angle starts from the $\Gamma\tilde{X}$ direction. The gap minimum locates in the $\Gamma\tilde{M}$ direction for both contours. (**e**,**f**) The angle-dependent energy thresholds taken from the vertical line-cuts to the colour plots in **c**,**d** by using the intensity $I = 0.19$, 0.24 (0.125, 0.147) for the inner (outer) rings in **a**,**b**, respectively. The gap minimum locates in the $\Gamma\tilde{M}$ direction for both contours. The solid lines in **e**,**f** plots are the gap functions derived from the Dynes model fitting to the spectrum taken at 1.5 K. The error bars were determined from the deviations between the line-cut experimental data and the polynomial fitting curve for each angle.

further increased to the larger gap, that is, 14.5 mV, besides the inner ring, a new set of segments appears at larger **q** vectors, which gradually forms an outer circle. When we take the ρ(**q**, E) intensity along the contours of the inner and outer rings as shown in Fig. 7a,b, an angle-dependent $\rho(\theta, E)$ at various energies can be obtained and presented in Fig. 7c,d as the 2D colour plots. The 2D colour plots of $\rho(\theta, E)$ along the inner and the outer rings both show the fourfold symmetry. The line-cut of QPI intensity versus **q** along the radial direction at several angles is shown in Supplementary Fig. 5. The low or null intensity of $\rho(\theta, E)$ near the $\Gamma\tilde{X}$ direction along each ring may be induced by the band-crossing effect, or suggest that the Fermi surface in this part is still gapped, so both minima of the gaps appear in the $\Gamma\tilde{M}$ directions. To illustrate it more clearly, we take the vertical line-cuts on the colour plot and show in Fig. 7e,f the angle-dependent energy thresholds at two certain FT-QPI intensities along the inner and outer rings. Considering the two anisotropic *s*-wave gap functions derived in the Dynes model fitting to the measured STS spectrum, we put the angle-dependent gap functions, as shown by the solid lines in Fig. 7e,f, onto the vertical line-cut data (shown by the symbols) at two certain $\rho(\theta, E)$ intensities ($I = 0.19$, 0.24 and $I = 0.125$, 0.147 for the inner and outer rings, respectively). If taking the values of $I = 0.24$ and 0.147 as the rough criterion for the gaps on the inner and outer pockets, one can see that the gap functions follow those derived from the Dynes model fitting pretty well. Both gap minima on the two circles appear in the $\Gamma\tilde{M}$ directions, which is consistent with the direction of gap minima of electron pockets in $FeSe_{0.45}Te_{0.55}$ (ref. 34). In the narrow region along the $\Gamma\tilde{X}$ direction, since the intensity is very low and the uncertainty is quite large, we are not able to derive the reliable gap values. Through above analysis and discussions, we can assign the larger (smaller) gap Δ_1 (Δ_2) to the outer (inner) Fermi surfaces, which form two circles in the FT-QPI pattern in the **q**-space.

Discussion

Concerning the two-gap functions with different maximum magnitudes derived from the fitting to both STS spectra and the emergence of two sets of rings in the FT-QPI images, we would like to put forward the picture concerning the band folding (perhaps also hybridization) of the two electron pockets in the folded BZ, as shown in Fig. 5d. In iron-based superconductors, the electron pockets are usually elliptic-shaped and then fold (perhaps also by hybridization) in the folded BZ with two Fe atoms in one unit cell. On the single elliptic Fermi surface, the segments in the pole region are contributed by the d_{xy} orbitals, while the waist areas are derived from the d_{xz}/d_{yz} orbitals[35]. Due to the folding effect, we have two sets of Fermi pockets which correspond well to the inner (d_{xz}/d_{yz}) and outer (d_{xy}) pockets. The crossing area consists of electrons with mixed orbitals of d_{xy} and d_{xz}/d_{yz}. This may be the reason why JDOS is very weak between the overlapped Fermi surfaces and the outer ring seems not to close completely in the $\Gamma\tilde{X}$ directions. The anisotropic FT-QPI intensity has been self-consistently explained as the cause of the gap anisotropy in present study. While it may also be influenced by (or partly related to) the variations in the orbital character of the bands around a given energy contour[25,29,31], and the unclosed part along $\Gamma\tilde{X}$ directions mentioned above may give such suggestion. We cannot rule out the effect of the anisotropic orbital character to the anisotropic FT-QPI patterns. Another possible reason for the anisotropic FT-QPI intensity is the anisotropic self-energy, which can affect the FT-QPI pattern, especially when the measuring energy is above the superconducting gap[36]. In the low energy FT-QPI, such effect may also appear together with the anisotropic superconducting gap[36], although this could be a much weaker effect. Since the anisotropic-gap function is strongly tied with the anisotropic orbital character and self-energy effect, it will be important to figure out which factor is the dominant role for inducing the anisotropic FT-QPI intensity, especially in the energy region above the larger gap. We must emphasize that all these possible alternatives do not change the major results in this study, that is, the identifying of the double electron pockets near $\tilde{M}$ point and the associated double gaps. If the nodeless *d*-wave is applicable in the system, sign reversal of superconducting gaps should exist between two neighbouring elliptic Fermi pockets. In this case, in the band folding (perhaps also hybridized) region, the gap should be zero, which is not consistent with our results. In this picture with electron pocket folding, our data indicate a smaller gap in the inner pocket with the d_{xz}/d_{yz} characteristic, while the larger gap accommodates in the outer pocket with the d_{xy} characteristic. To our knowledge, this is the first time to resolve the two sets of Fermi pockets and associated gaps near the $\tilde{M}$ point. In the situation of *s*-wave gap on the hybridized electron pockets, the theoretically predicted gap maximum locates in the $\Gamma\tilde{X}$ directions and gap minimum locates in the $\Gamma\tilde{M}$ directions[13], which is consistent with our results. Although we cannot recognize that the gap maxima locate in the $\Gamma\tilde{X}$ directions just from the intensity plot of FT-QPI in Fig. 7c–f, however, the tunnelling spectrum shows fully gaped feature near zero-bias together with a fourfold angle dependence of the gap, the FT-QPI shows also a fourfold feature and the gap minimum locates in the $\Gamma\tilde{M}$ directions. All these strongly suggest that the gap distribution may be very similar to the situation shown in Fig. 7c–f. Further ARPES experiments with refined resolution are strongly desired to confirm the conclusions derived here. In addition, our results here resolve two sets of Fermi pockets and associated gaps, which provides a very good stage for further checking whether there are sign reversal gaps between the inner and outer folded (or hybridized) Fermi pockets. This is actually underway by following the recent theoretical predictions[37]. That should be very crucial for a final pinning down of the repulsive interaction-induced pairing. Our observation of two superconducting gaps and the assignment of the two gaps to the folded (or hybridized) two sets of Fermi pockets in the new superconductor $(Li_{1-x}Fe_x)OHFeSe$ is very intriguing. Furthermore, the huge gap ratio $2\Delta_1/k_BT_c = 8.7$ indicates the strong coupling mechanism for superconductivity in the iron-based superconductors.

Methods

Sample synthesis and characterization. The $(Li_{1-x}Fe_x)OHFeSe$ single crystals were synthesized by the hydrothermal ion-exchange method[7]. First, 6 g LiOH (J&K, 99% purity) was dissolved in 15 ml deionized water in a teflon-linked stainless-steel autoclave (volume 50 ml). Then, 0.6 g iron powder (Aladdin Industrial, 99% purity), 0.3 g selenourea (Alfa Aesar, 99% purity) and several pieces of $K_{0.8}Fe_{2-x}Se_2$ single crystals were added to the solution. After that, the autoclave was sealed and heated up to 120 °C followed by staying for 40 to 50 h. Finally, the $(Li_{1-x}Fe_x)OHFeSe$ single crystals can be obtained by leaching. The DC magnetization measurement was carried out with a SQUID-VSM-7T (Quantum Design) with a resolution of 5×10^{-8} e.m.u. The resistivity measurement was done on a PPMS-16T (Quantum Design) with the standard four-probe method.

STM/STS measurements. The STS spectra were measured with an ultra-high vacuum, low-temperature and high-magnetic field scanning tunnelling microscope (USM-1300, Unisoku Co., Ltd.). The samples were cleaved in an ultra-high vacuum with a base pressure about 1×10^{-10} torr. During all STM/STS measurements, tungsten tips were used. To lower down the noise of the differential conductance spectra, a lock-in technique with an ac modulation of 0.8 mV at 987.5 Hz was used.

References

1. Wang, Q. Y. *et al.* Interface-induced high-temperature superconductivity in single unit-cell FeSe films on $SrTiO_3$. *Chin. Phys. Lett.* **29,** 037402 (2012).
2. Lee, J. J. *et al.* Interfacial mode coupling as the origin of the enhancement of T_c in FeSe films on $SrTiO_3$. *Nature* **515,** 245–248 (2014).
3. He, S. L. *et al.* Phase diagram and electronic indication of high-temperature superconductivity at 65 K in single-layer FeSe films. *Nat. Mater.* **12,** 605–610 (2013).
4. Tan, S. Y. *et al.* Interface-induced superconductivity and strain-dependent spin density waves in FeSe/SrTiO3 thin films. *Nat. Mater.* **12,** 634–640 (2013).
5. Lu, X. F. *et al.* Coexistence of superconductivity and antiferromagnetism in $(Li_{0.8}Fe_{0.2})OHFeSe$. *Nat. Mater.* **14,** 325–329 (2015).
6. Pachmayr, U. *et al.* Coexistence of 3d-Ferromagnetism and Superconductivity in $[(Li_{1-x}Fe_x)OH](Fe_{1-y}Li_y)Se$. *Angew. Chem. Int. Ed.* **54,** 293–297 (2015).
7. Dong, X. L. *et al.* $(Li_{0.84}Fe_{0.16})OHFe_{0.98}Se$ superconductor: Ion-exchange synthesis of large single-crystal and highly two-dimensional electron properties. *Phys. Rev. B* **92,** 064515 (2015).
8. Rotter, M., Tegel, M. & Johrendt, D. Superconductivity at 38 K in the Iron Arsenide $(Ba_{1-x}K_x)Fe_2As_2$. *Phys. Rev. Lett.* **101,** 107006 (2008).
9. Chen, G. F., Hu, W. Z., Luo, J. L. & Wang, N. L. Multiple phase transitions in single-crystalline $Na_{1-\delta}FeAs$. *Phys. Rev. Lett.* **102,** 227004 (2009).
10. Hsu, F. C. *et al.* Superconductivity in the PbO-type structure α-FeSe. *Proc. Natl Acad. Sci. USA* **105,** 14262–14264 (2008).
11. Mazin, I. I., Singh, D. J., Johannes, M. D. & Du, M. H. Unconventional superconductivity with a sign reversal in the order parameter of $LaFeAsO_{1-x}F_x$. *Phys. Rev. Lett.* **101,** 057003 (2008).
12. Kuroki, K. *et al.* Unconventional pairing originating from the disconnected Fermi surfaces of superconducting $LaFeAsO_{1-x}F_x$. *Phys. Rev. Lett.* **101,** 087004 (2008).
13. Khodas, M. & Chubukov, A.V. Interpocket pairing and gap symmetry in Fe-based superconductors with only electron pockets. *Phys. Rev. Lett.* **108,** 247003 (2012).
14. Zhao, L. *et al.* Common electronic origin of superconductivity in (Li,Fe)OHFeSe bulk superconductor and single-layer FeSe/SrTiO3 films. Preprint at http://arxiv.org/abs/1505.06361 (2015).
15. Niu, X. H. *et al.* Surface electronic structure and isotropic superconducting gap in $(Li_{0.8}Fe_{0.2})OHFeSe$. *Phys. Rev. B* **92,** 060504 (2015).
16. Bean, C. P. Magnetization of high-field superconductors. *Rev. Mod. Phys.* **36,** 31–39 (1964).
17. Shen, B. *et al.* Intrinsic percolative superconductivity in $K_xFe_{2-y}Se_2$ single crystals. *Europhys. Lett.* **96,** 37010 (2011).

18. Fischer, Ø., Kugler, M., Maggio-Aprile, I., Berthod, C. & Renner, C. Scanning tunneling Spectroscopy of high-temperature superconductors. *Rev. Mod. Phys.* **79,** 353–419 (2007).
19. Hanaguri, T., Niitaka, S., Kuroki, K. & Kakagi, H. Unconventional s-wave superconductivity in Fe(Se,Te). *Science* **328,** 474–476 (2010).
20. Hoffman, J. E. Spectroscopic scanning tunneling microscopy insights into Fe-based superconductors. *Rep. Prog. Phys.* **74,** 124513 (2011).
21. Yang, H. *et al.* In-gap quasiparticle excitations induced by non-magnetic Cu impurities in $Na(Fe_{0.96}Co_{0.03}Cu_{0.01})As$ revealed by scanning tunnelling spectroscopy. *Nat. Commun.* **4,** 2749 (2013).
22. Yang, H. *et al.* Unexpected weak spatial variation in the local density of states induced by individual Co impurity atoms in superconducting $Na(Fe_{1-x}Co_x)As$ crystals revealed by scanning tunneling spectroscopy. *Phys. Rev. B* **86,** 214512 (2012).
23. Chi, S. *et al.* Scanning tunneling spectroscopy of superconducting LiFeAs single crystals: evidence for two nodeless energy gaps and coupling to a bosonic mode. *Phys. Rev. Lett.* **109,** 087002 (2012).
24. Hänke, T. *et al.* Probing the unconventional superconducting state of LiFeAs by quasiparticle interference. *Phys. Rev. Lett.* **108,** 127001 (2012).
25. Chi, S. *et al.* Sign inversion in the superconducting order parameter of LiFeAs inferred from Bogoliubov quasiparticle interference. *Phys. Rev. B* **89,** 104522 (2014).
26. Lang, M. K. *et al.* Imaging the granular structure of high-T_c superconductivity in underdoped $Bi_2Sr_2CaCu_2O_{8+\delta}$. *Nature* **415,** 412–416 (2002).
27. Yazdani, A., Howald, C. M., Lutz, C. P., Kapitulnik, A. & Eigler, D. M. Impurity-induced bound excitations on the surface of $Bi_2Sr_2CaCu_2O_8$. *Phys. Rev. Lett.* **83,** 176–179 (1999).
28. Dynes, R. C., Narayanamurti, V. & Garno, J. Pm. Direct measurement of quasiparticle-lifetime broadening in a strong-coupled superconductor. *Phys. Rev. Lett.* **41,** 1509–1512 (1978).
29. Huang, D. *et al.* Nanoscale imaging of orbital texture in single-layer FeSe/$SrTiO_3$. Preprint at http://arxiv.org/abs/1509.07110 (2015).
30. Lee, D.-H. What makes the T_c of FeSe/$SrTiO_3$ so high? *Chin. Phys. B* **24,** 117405 (2015).
31. Zhang, Y. Y. *et al.* Quasiparticle scattering interference in superconducting iron pnictides. *Phys. Rev. B* **80,** 094528 (2009).
32. Akbari, A., Knolle, J., Eremin, I. & Moessner, R. Quasiparticle interference in iron-based superconductors. *Phys. Rev. B* **82,** 224506 (2010).
33. Allan, M. P. *et al.* Anisotropic energy gaps of iron-based superconductivity from intraband quasiparticle interference in LiFeAs. *Science* **336,** 563–567 (2012).
34. Zeng, B. *et al.* Anisotropic structure of the order parameter in $FeSe_{0.45}Te_{0.55}$ revealed by angle-resolved specific heat. *Nat. Commun.* **1,** 112 (2010).
35. Hirschfeld, P. J., Korshunov, M. M. & Mazin, I. I. Gap symmetry and structure of Fe-based superconductors. *Rep. Prog. Phys.* **74,** 124508 (2011).
36. Allan, M. P. *et al.* Identifying the 'fingerprint' of antiferromagnetic spin fluctuations in iron pnictide superconductors. *Nat. Phys.* **11,** 177–182 (2015).
37. Hirschfeld, P. J., Altenfeld, D., Eremin, I. & Mazin, I. I. Robust determination of superconducting gap sign changes via quasiparticle interference. *Phys. Rev. B* **92,** 184513 (2015).

Acknowledgements

We acknowledge the useful discussions with Qianghua Wang, Jiangping Hu and Dung-hai Lee. This work was supported by the National Natural Science Foundation of China (Grant No. 11534005, No.11190023), the Ministry of Science and Technology of China (973 projects: 2011CBA00100, 2012CB821403) and PAPD.

Author contributions

The low-temperature STS measurements were finished by Z.D., X.Y., D.F., G.D., H.Y. and H.-H.W. The samples were prepared by H.L. and X.Z. The data analysis was done by Z.D., X.Y., D.F. and H.Y. H.-H.W. coordinated the whole work. H.Y. and H.-H.W. wrote the manuscript which was supplemented by others. All authors have discussed the results and the interpretation.

Additional information

Competing financial interests: The authors declare no competing financial interests.

10

Macrospin dynamics in antiferromagnets triggered by sub-20 femtosecond injection of nanomagnons

D. Bossini[1,*,†], S. Dal Conte[2,3,*], Y. Hashimoto[1,†], A. Secchi[1], R.V. Pisarev[4], Th. Rasing[1], G. Cerullo[2,3] & A.V. Kimel[1]

The understanding of how the sub-nanoscale exchange interaction evolves in macroscale correlations and ordered phases of matter, such as magnetism and superconductivity, requires to bridging the quantum and classical worlds. This monumental challenge has so far only been achieved for systems close to their thermodynamical equilibrium. Here we follow in real time the ultrafast dynamics of the macroscale magnetic order parameter in the Heisenberg antiferromagnet $KNiF_3$ triggered by the impulsive optical generation of spin excitations with the shortest possible nanometre wavelength and femtosecond period. Our magneto-optical pump-probe experiments also demonstrate the coherent manipulation of the phase and amplitude of these femtosecond nanomagnons, whose frequencies are defined by the exchange energy. These findings open up opportunities for fundamental research on the role of short-wavelength spin excitations in magnetism and strongly correlated materials; they also suggest that nanospintronics and nanomagnonics can employ coherently controllable spin waves with frequencies in the 20 THz domain.

[1] Institute for Molecules and Materials, Spectroscopy of Solids and Interfaces, Radboud University, Heyendaalseweg 135, 6525 AJ Nijmegen, The Netherlands. [2] Dipartimento di Fisica, Politecnico di Milano, Piazza Leonardo da Vinci 32, I-20133 Milano, Italy. [3] Istituto di Fotonica e Nanotecnologie, Consiglio Nazionale delle Ricerche, Piazza Leonardo da Vinci 32, I-20133 Milano, Italy. [4] Ioffe Physical-Technical Institute, Ferroics Physics Laboratory, Russian Academy of Sciences, St Petersburg 194021, Russia. * These authors contributed equally to this work. † Present addresses: Institute for Photon Science and Technology Graduate School of Science, The University of Tokyo, 7-3-1 Hongo, Bunkyo-ku, Tokyo 113-0033, Japan (D.B.); Spin Quantum Rectification Project, ERATO, Japan Science and Technology Agency, Sendai 980-8577, Japan (Y.H.). Correspondence and requests for materials should be addressed to D.B. (email: davide@gono.phys.s.u-tokyo.ac.jp).

Experimental studies allowing to investigate correlated matter in general, and magnetism in particular, at the sub-nanometre length- and femtosecond timescales of the exchange interaction have recently developed into an exciting research area. The experiments involving ultrashort timescales provided intriguing results, like the femtosecond laser-induced transient ferromagnetic state of a ferrimagnet alloy[1] and even light-induced superconductivity[2]. However, in these cases the wavelengths of the photo-induced excitations lie orders of magnitude above the lengthscale of the exchange interaction. An alternative strategy consists in the investigation of the magnetic order induced by introducing impurities with atomic resolution in space[3–5], but without temporal resolution. A hitherto unexplored approach to the problem, which combines the femtosecond timescale with the nanometre lengthscale, consists in studying the ultrafast dynamics of the macroscale magnetic order parameter triggered by spin excitations with wavelength and period matching the length- and timescales of the exchange interaction.

These spin excitations correspond to magnons (or spin waves) with wavevector near the edges of the first Brillouin zone. In antiferromagnetic materials these magnons can be elegantly excited in the time domain, via a second-order impulsive stimulated Raman scattering process involving pairs of magnons with wavevectors almost equal in magnitude and opposite in sign. It was demonstrated that in the absence of an external magnetic field, such spin excitations can be generated only by a modification of the exchange interaction[6] and cannot be triggered by a change of the spin–orbit coupling. A light-induced modification of the exchange interaction generates pairs of magnons (as shown in the Methods section) and properly describes the Raman spectra[6,7]. The microscopic mechanism of perturbation of the exchange interaction by a laser pulse has been recently described in terms of an effective energy shift of the charge-transfer transition[8,9]. Although pairs of spin waves could in principle be generated throughout the whole Brillouin zone, the magnon density of states is largest in the high-frequency region near the zone edges, where the frequencies are defined mostly by the exchange interaction[6,7] (Supplementary Note 1 and Supplementary Fig. 1). Note that the wavelength of these magnons is comparable to the distance of nearest neighbour spins, whose interactions determine the macroscopic magnetic order. Therefore, studying the photo-induced dynamics of such magnons discloses the response of the magnetic system to excitations on the length- and timescales of the exchange interaction. A bound state of two high-energy, high-wavevector and counter-propagating spin waves, usually denoted as two-magnon (2M) mode, can be induced by a femtosecond light pulse. The frequency and wavevector of this magnetic excitation are the sums of the frequencies and wavevectors of the two magnons involved in the bound state[6,7,10,11]. Although an impulsive excitation of the 2M mode was reported, the subsequent dynamics of the magnetic order parameter has not been discussed[12]. A previous attempt to measure the light-induced evolution of the 2M mode directly in the frequency domain, via the femtosecond-stimulated Raman scattering technique, could reveal a modification of the Raman shift of the 2M peak only during the overlap of the laser pulses[13]. This signal is dominated by nonlinear purely optical effects, originating from the simultaneous interaction of three ultrashort intense laser pulses with the sample. Therefore, disentangling the optical and the magnetic components of the signal was possible only via a sophisticated data analysis and modelling. In femtosecond-stimulated Raman scattering no signal at delays subsequent the instrument response time was observed, which did not allow to disclose the dynamical response of the magnetic order to the 2M excitation.

The present work reports the femtosecond dynamics of the macroscopic magnetic order parameter, by means of an impulsive all-optical injection of the intrinsically highest-frequency spin excitations with wavevectors near the edges of the Brillouin zone. The wavelengths of the corresponding magnons are of the order of 1 nm, and the frequencies in the 20 THz range. We demonstrate the manipulation of the order parameter via the coherent control of the short-range magnons: by changing the conditions of the photo-excitation we arbitrarily reverse the phase of the spin waves and enhance or quench their amplitude. Our time-resolved approach discloses even a dynamical coupling of the high-wavevector magnons with a lattice vibrational mode.

Results

Experimental approach. An excellent system for the all-optical excitation and detection of the dynamics of high-frequency and shortest-wavelength magnons is the cubic Heisenberg antiferromagnet $KNiF_3$, which is ordered below the Néel point $T_N = 246$ K. In this material the Raman cross-section of the 2M mode is so high that it dominates the whole spectrum[6,14]. A time-domain excitation of the 2M mode in $KNiF_3$ ($\nu_{2M} \approx 22$ THz, period ≈ 45 fs, wavevector $\approx 10^7\,cm^{-1}$ and wavelength ≈ 1 nm) can be achieved by the impulsive stimulated Raman scattering mechanism, provided that the duration of the stimulus is shorter than the period of the magnetic mode[15,16]. A successful impulsive excitation of such high-frequency magnons therefore demands laser pulses with a duration significantly shorter than 40 fs. To meet these requirements we used linearly polarized sub-20-fs laser pulses, with a central photon energy of 2.2 eV. For the probe we employed equally short pulses centred around 1.3 eV and with a polarization perpendicular to that of the pump.

Light scattering by a magnon pair can be visualized as two spin-flip events, one on each sublattice, such that the total spin remains unchanged[6,7,14] (Fig. 1a). Consequently, the transient magneto-optical Faraday and Kerr effects, which measure light-induced variations of the total spin, fail to track the dynamics of such a magnetic excitation. On the other hand, the 2M process is expected to be revealed by those second-order magneto-optical effects, which depend quadratically on the transversal spin components (that is, on the spin deviations), so that a pair of spin-flip events can be detected even if the total spin is unaffected (see discussion in Method section and equations (6) and (7)). Such a magneto-optical effect depends on the spin operators via the same spin-correlation function appearing in the Heisenberg term of the Hamiltonian[17] (equation (9)). In particular, we define the antiferromagnetic linear dichroism (ALD) in terms of the symmetric components of the dielectric tensor $\epsilon_s^{\lambda\nu}$ for which $\epsilon_s^{\lambda\nu}=\epsilon_s^{\nu\lambda}$, where ν and λ are indices. The following definition holds:[17,18]

$$\epsilon_s^{\lambda\nu} = \sum_{ij}\sum_{\gamma\delta} \rho^{\lambda\nu\gamma\delta}\langle \hat{S}_i^{\gamma\Uparrow}\hat{S}_j^{\delta\Downarrow}\rangle \qquad (1)$$

where $\rho^{\lambda\nu\gamma\delta}$ is a magneto-optical polar fourth rank tensor and $\lambda\nu\gamma\delta$ are spatial coordinate indices, while $\hat{S}_i^{\Uparrow}$ and $\hat{S}_j^{\Downarrow}$ are the spin operators located on two nearest-neighbour sites (i, j), belonging to oppositely oriented magnetic sublattices ($\Uparrow$ and $\Downarrow$). Note that the ALD induces a rotation of the probe polarization in the experimental configuration shown in Fig. 1b (see Methods section). It is crucial to notice that the two spins in the correlation function belong to two different ionic sites on opposite sublattices. This fact differentiates the ALD from the more conventional magnetic linear birefringence and dichroism (see equation (9) and discussion in the Methods). The

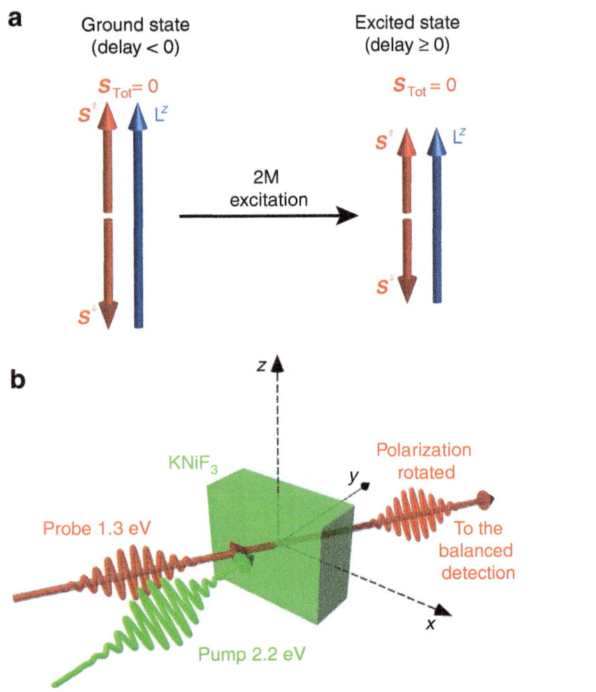

Figure 1 | 2M mode and experimental configuration. (**a**) The 2M excitation is equivalent to a spin-flip event per sublattice. Thus, the magnetization of each sublattice ($\mathbf{S}^{\Uparrow}$ and $\mathbf{S}^{\Downarrow}$, represented by the two red arrows with opposite orientation) and, therefore, the antiferromagnetic vector (L^z, blue arrow) is decreased in the excited state. The sum of the spins of the two sublattices, thus the total magnetization, vanishes both in the ground and in the excited state. (**b**) Schematic representation of the experimental geometry. The pump (green pulse) photon energy was tuned to 2.2 eV in the transparency window of the material. This choice avoided contributions of laser-heated electrons and phonons to the spin dynamics[19]. The central photon energy of the probe beam (red pulse) was 1.3 eV. The arrows indicates the direction of propagation of the two laser beams.

macroscopic magnetic order of an ideal Heisenberg antiferromagnet is conveniently described in terms of the antiferromagnetic vector **L**, which is the order parameter[18] and is defined as

$$\mathbf{L} = \sum_i \langle \hat{S}_i^{\Uparrow} \rangle - \sum_j \langle \hat{S}_j^{\Downarrow} \rangle = \mathbf{S}^{\Uparrow} - \mathbf{S}^{\Downarrow} \tag{2}$$

where $\mathbf{S}^{\Uparrow}$ and $\mathbf{S}^{\Downarrow}$ are the total spins of the two sublattices (Fig. 1a). Our considerations about the dynamics of the spin system triggered by the impulsive excitation of the 2M mode are based on the approximation of non-interacting magnons. In this framework it is straightforward to demonstrate that the z-projection of **L** has the same time dependence as the spin-correlation function, at the leading terms in the magnon operators (equations (11) and (12)). Here z is the direction parallel to the spins in equilibrium. The dynamics of L^z, at the leading order in the excitation intensity, is given by

$$\Delta L^z(t) \propto A \sin(\omega_{2M} t) \tag{3}$$

where ω_{2M} is the frequency of the 2M mode and A is the amplitude. Equation (3) describes a purely longitudinal, non-precessional, dynamics of the antiferromagnetic vector (see Supplementary Note 2 for the complete derivation). Therefore, by detecting the femtosecond dynamics of the ALD, we access the time evolution of the macroscopic order parameter via the spin-correlation function, which ultimately defines also the dynamics of the exchange energy.

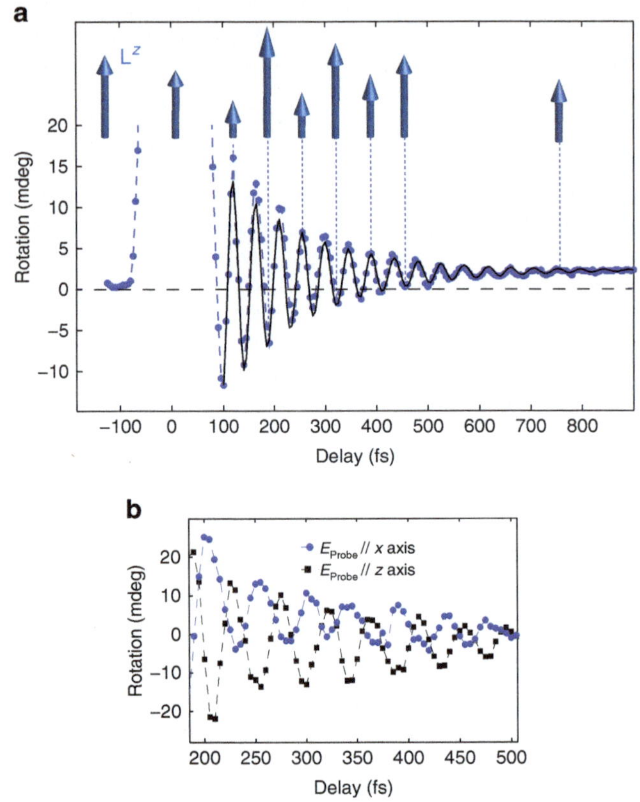

Figure 2 | Laser-induced dynamics of the antiferromagnetic vector. (**a**) The transient rotation of the probe polarization was measured with the electric fields of the pump and the probe beams linearly polarized along the z and x axes, respectively. The pump fluence was set to $\approx 8.6\,\mathrm{mJ\,cm^{-2}}$. The corresponding dynamics of L^z (blue arrows) is schematically represented. When the pump pulses impinge on the sample (0 delay) L^z decreases, due to the generation of magnons. At positive delays, oscillations at the frequency of the 2M mode are visible (equation (3)). The black line is a fit to the data (equation (13)). (**b**) The phase of the oscillation is shifted by π when the orientation of the electric field of the probe ($\mathbf{E}_{\mathrm{Probe}}$) is rotated by 90°. The pump beam was linearly polarized along the z axis, the fluence was set to $\approx 12\,\mathrm{mJ\,cm^{-2}}$. The measurements reported in both panels were performed at 80 K.

Femtosecond dynamics of the magnetic system. Figure 2a shows the typical result of a time-resolved measurement of the laser-induced spin dynamics. The transient rotation of the probe polarization shows oscillations in time with a period of ≈ 45 fs (that is, a frequency of ≈ 22 THz) that are damped on a 500-fs timescale. The oscillatory dynamics is superimposed on an incoherent increase of the background, as it is clear from the difference between the time trace and the zero line at longer delays (> 500 fs). The spectrum of the oscillation measured in the time-domain experiment closely matches the 2M mode as measured by spontaneous Raman. To definitely assess the nature of the 22 THz mode, we performed temperature-dependent measurements (Fig. 3a) and we compared the temperature dependence of the time-domain signal with that of the spontaneous Raman spectra of the 2M bound state (Fig. 3b,c). Figure 3 shows that the frequency and the lifetime of the pump-induced oscillations decrease as the Néel point is approached, in qualitative and quantitative agreement with spontaneous Raman data[6,14]. Thus, Fig. 2a unambiguosly reveals the femtosecond spin dynamics triggered by the impulsive excitation of the 2M mode in $KNiF_3$, which is not accessible with any other experimental approach.

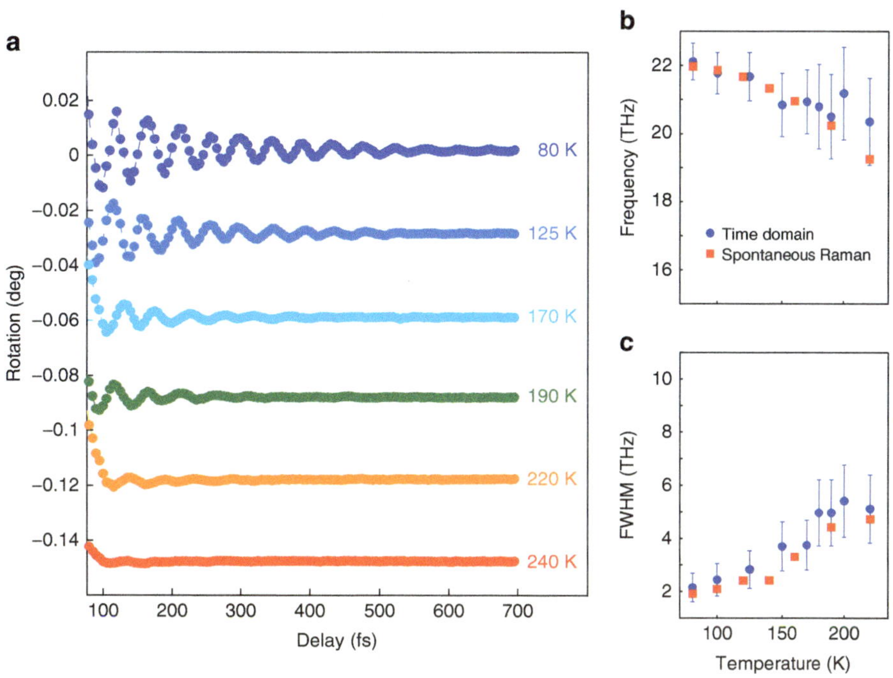

Figure 3 | Temperature dependence of the ultrafast spin dynamics. (**a**) Rotation of the probe polarization as a function of the temperature. The pump and the probe beams are linearly polarized along the *z* and *y* axes, respectively. The fluence is $\approx 8.6\,mJ\,cm^{-2}$. (**b**,**c**) The temperature dependence of the frequency and linewidth of the 2M Raman peak (red squares) compared with the trends obtained by applying the Fourier transform to the time-domain data (blue circles). The error bars in **b** and **c** for the time-domain data are defined as the half width at half maximum of the 2M frequency distribution. The error bars for the Raman data are defined by the instrumental sensitivity ($\approx 2\,cm^{-1}$) and are not visible in this scale.

A fit to the data in Fig. 2a (see Methods section, equation (13)) gave $\tau_d = (167 \pm 4)$ fs (damping of the coherent oscillations) and $\tau_r = (255 \pm 16)$ fs (rise time of the incoherent background response). While τ_d represents the decoherence of the 2M band, we interpret τ_r as the characteristic demagnetization time of the two sublattices, solely driven by magnetic interactions[19]. As a laser pulse excites a continuum of magnons with different frequencies, the damping time τ_d of the oscillations observed in our experiment is actually due to the decoherence of the inhomogeneous ensemble of the coherently excited magnons. This is usually described[20] by means of the characteristic time T_2^*. The demagnetization of the sublattices is a result of the heating of spins, which is caused by the decoherence of single-magnon modes[19] in the ensemble, on a timescale generally indicated with T_2 ($T_2 > T_2^*$)[20]. Measuring the probe polarization dependence of the signal (Fig. 2b) allowed us to demonstrate that the detected rotation of the polarization in our experiment arises from the symmetric components of the $\epsilon^{\lambda,\nu}$ tensor and, in particular, from the ALD. In fact the change of sign and the different amplitude of the two data sets shown in Fig. 2b is predicted by the microscopic theory of the ALD (Supplementary Notes 2 and 3).

Coherent control. The ability to arbitrarily manipulate the properties of spin waves is a necessary requirement for the development of magnon-based devices[21]. Figure 4a demonstrates the control of the phase of the nanomagnons. The upmost time traces were measured exciting the sample with orthogonally polarized pump beams. A clear π shift of the phase of the oscillations is observed, if the polarization of the excitation beam is rotated by 90°. This phase shift is due to the non-equivalence of the excitations of the magnetic system induced by light linearly polarized along and orthogonally to the direction of the spins. This interpretation is substantiated by a symmetry analysis of the two-magnon excitation (Supplementary Note 4), based on the point group of $KNiF_3$. The coherent nature of the photo-induced spin dynamics enables the manipulation of the amplitude of the spin waves, by means of a sequence of laser pulses. We demonstrate this effect in Fig. 4a employing two time-delayed excitation pulses. Tuning the delay we can either amplify or strongly quench the oscillations compared with the case when the spin dynamics is triggered by a single-pump pulse. Figure 4b reveals the periodic dependence of the amplitude of the first magnetic oscillation on the delay between the pump pulses. Such a coherent control of the femto-nanomagnons allows a direct and complete manipulation of the antiferromagnetic order parameter on the femtosecond timescale.

Time-frequency analysis. The real-time measurement of the ultrafast spin dynamics triggered by short-range magnons allows us to disclose another phenomenon not observable with equilibirum techniques. Figure 5a shows the spectrum of the time trace in Fig. 2a obtained via a Fourier transform (blue curve). On the same graph we plot the spectrum of the 2M mode measured via spontaneous Raman scattering at the same temperature (red curve). The zoom in the inset of Fig. 5a shows two sidebands at about 7.5 THz from the central 2M frequency, which appear to originate from a modulation of the 2M mode in the time domain. Figure 5b shows a two-dimensional spectrogram obtained by performing a time-frequency analysis[22,23] of the data in Fig. 2a (Supplementary Note 5). The colour map in Fig. 5b represents the time-dependent spectrum of the 2M mode. The frequencies of the peak of the spectrum at different time delays are traced by a blue dotted line. This curve displays a periodic oscillation of ν_{2M},

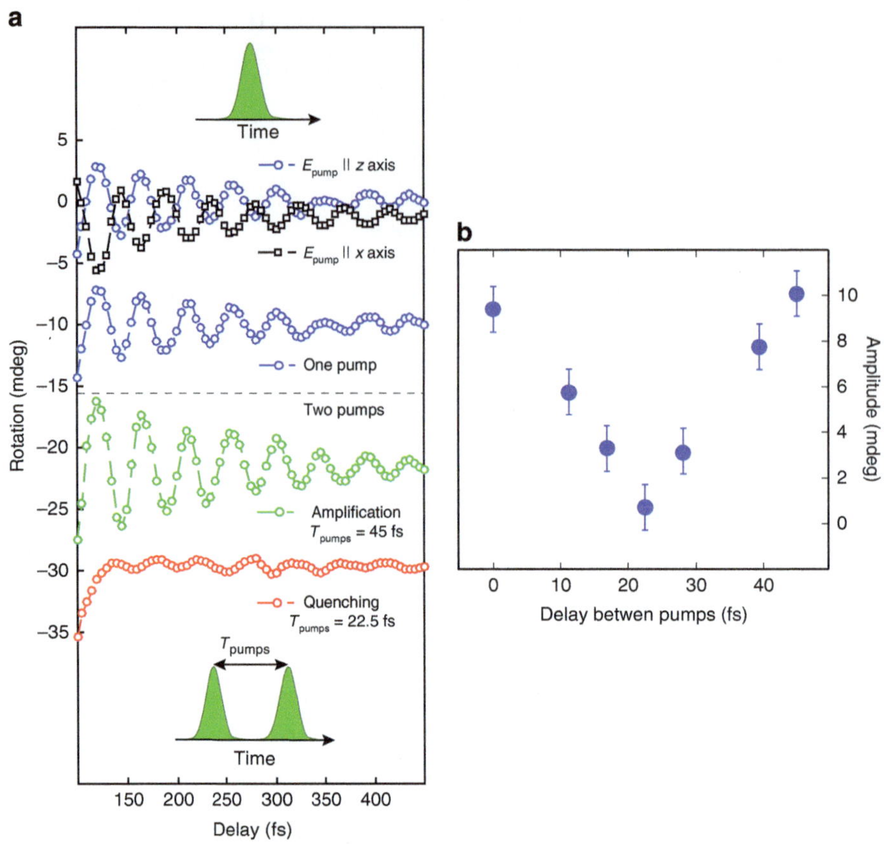

Figure 4 | Coherent control of the spin dynamics. (**a**) The blue and black dotted traces were obtained using a single-pump beam linearly polarized along the *z* axis (blue circles) and the *x* axis (black squares). The phase of the oscillations is shifted by π. The single-pump measurement is repeated for the sake of comparison with the results of the double-pump experiments. The green-circle data set was observed by delaying two pump pulses by $T_{pumps} = 45$ fs, that is, the period of the 2M mode. The coherent amplification of the signal is observed. The coherent oscillations were quenched, when the delay between the two pump pulses was set to $T_{pumps} = 22.5$ fs. The fluence of both pump beams was set to ≈ 1.5 mJ cm^{-2}. The temperature was 80 K. The probe beam was polarized along the *x* axis. (**b**) Amplitude of the first oscillation as a function of the delay between the two pumps (T_{pumps}), the periodic trend is clear. The error bars are defined as twice the sensitivity of our set-up.

which is consistent with the observation of the sidebands in the inset of Fig. 5a. We define the relative frequency shift as

$$\frac{\Delta\nu_{2M}}{\nu_{2M}} = \frac{\nu_{2M} - \langle\nu_{2M}\rangle}{\langle\nu_{2M}\rangle} \qquad (4)$$

where $\langle\nu_{2M}\rangle$ is the average frequency in the temporal interval where the oscillations have a significant amplitude (0–500 fs). We plot the peaks of the spectra obtained by the time-frequency analysis as a function of the delay in the inset of Fig. 5b. The frequency of the modulation of ν_{2M} is $\approx$7.5 THz, which corresponds to the frequency of the infrared-active phonon[24,25] ($\approx$7.7 THz), assigned to the stretching vibration of the Ni–F–Ni bond. Unlike other phonon modes in this material, the frequency of the stretching mode is temperature-independent[24,25], which is consistent with the data (Supplementary Fig. 2). Considering only the term of the polarization of the first order in the electric field of light, it would be concluded that this stretching mode is not Raman-active[26]. However, the light–matter interaction at the next order (hyper-Raman scattering[26,27]) allows to excite lattice vibrations with the symmetry of the stretching mode (F_{1u}) in cubic crystals[27]. It is well established that a frequency modulation as the one shown in Fig. 5b can be modelled in terms of coupled oscillators[23]. Therefore, we assign the modulation of ν_{2M} to the interaction on the femtosecond timescale between the stretching mode and the 2M mode, which are simultaneously and coherently excited by the laser pulse. The 2M–phonon interaction was previously suggested[11], however our time-resolved experiment provides the first evidence of this effect.

Discussion

We would like to underline that the dynamics here reported is totally different from the spin response previously observed after the photo-excitation of the low-energy magnons with wavevectors at the centre of the Brillouin zone in this material[19]. In fact, our previous investigation revealed spin oscillations with the period of 11 ps (that is, a 90-GHz frequency), which are still visible 200 ps after the photo-excitation. This spin dynamics was generated by a single spin-flip event (that is, one-magnon mode), thus the magnetic linear birefringence, which depends quadratically on the spin of a single atom but only linearly on the spin deviation (equation (9)), was employed to probe the response of the magnetic system. The interaction relevant for the low-energy magnons is the spin–orbit coupling, which defines the frequency of such collective excitations[6]. The use of 100-fs laser pulses prevented the access to high-frequency magnetic modes[19]. The present work describes a completely different regime of spin dynamics, whose characteristic timescale is determined by the exchange energy[6]. Moreover, the measurements of the time evolution of the spin-correlation function (via ALD) provide an access to the dynamics of the exchange energy. Within the approximations of our model, the spin-correlation dynamics

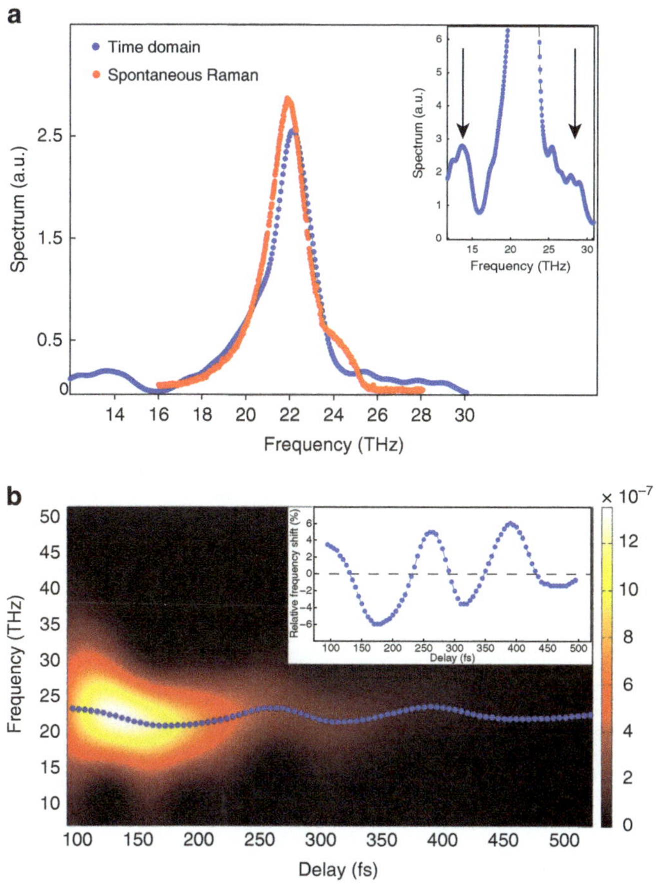

Figure 5 | Spectrum of the ultrafast response of the antiferromagnetic vector. (**a**) The Fourier transform of the time trace measured at 80 K (blue curve) is compared with the spontaneous Raman spectrum obtained at the same temperature (red curve). In the inset a zoom of the Fourier transform reveals two sidebands at $\approx \pm 7.5$ THz away from the peak frequency. (**b**) The squared modulus of the Wigner distribution of the signal is represented by the colour plot. At each time step a blue dotted line highlights the maximum of the spectrum, in which black represents zero intensity. In the inset the relative frequency shift (equation (4)) is plotted as a function of the delay showing ≈ 7.5 THz oscillations.

corresponds to the longitudinal dynamics of the order parameter (equations (11) and (12) and Supplementary Note 2).

In summary, our results reveal the dynamics of the macroscopic order parameter triggered by impulsively excited nanometre-wavelength and femtosecond-period magnons. Moreover, our experiment pushes the coherent control of magnons, previously demonstrated at the centre of the Brillouin zone[28,29], to the edges of the Brillouin zone, laying the foundations for a magnon-based nanotechnology operating in the 20-THz regime. Our investigation concerned an ideal Heisenberg antiferromagnet, however the approach here employed to study the femtosecond dynamics of the macroscopic magnetic order parameter caused by short-range spin excitations can be employed in a broad group of multisublattice systems[6,7]. In particular, it will become possible to monitor the evolution of the exchange energy during a photo-induced phase transition and to probe the femtosecond dynamics of the sub-nanometre range spin correlations in strongly correlated materials. This might even elucidate the dynamical interplay between short-range spin excitations and high-temperature superconductivity in cuprates[30–32].

Methods

Sample. Our sample was a 340-µm-thick (100) plane-parallel plate of $KNiF_3$, which has a perovskite crystal structure (point group $m3m$). Two equivalent Ni^{2+} sublattices are antiferromagnetically coupled below the Néel temperature $T_N = 246$ K (ref. 19). This material is known to be a cubic Heisenberg antiferromagnet because of its very weak anisotropy. The positive sign of the cubic magnetic anisotropy constant determines the alignment of spins along the [001], [010] or [100] axes[18]. The measurements on the $KNiF_3$ sample are carried out at a minimum temperature of 77 K in a liquid nitrogen cryostat. The temperature of the sample is monitored by a thermocouple placed on the sample holder.

Light source. For the pump–probe experiments we used a regeneratively amplified mode-locked Ti:Sapphire laser, providing 150-fs, 500-µJ pulses at 780 nm and 1 kHz repetition rate. The laser drives two non-collinear optical parametric amplifiers (NOPAs) operating in two different spectral ranges[33]. Both NOPAs are pumped by the second harmonic of the laser (that is, 390 nm) and seeded by the white-light continuum produced by focusing the 780-nm beam into a sapphire plate. The amplified pulse from the first NOPA, which initiates the dynamics (pump), has a spectrum spanning the 500–700 nm range and is compressed to nearly transform-limited duration (that is, 8 fs) by a pair of custom-made chirped mirrors. The amplified pulse, generated by the second NOPA (probe), covers the frequency range between 820 and 1050 nm, and is compressed to nearly transform-limited duration (that is, 13 fs) by a couple of fused silica prisms. The temporal resolution of the set-up has been characterized by the cross-correlation frequency-resolved optical gating technique and was below 20 fs (ref. 32). The pump and probe beams were focused on the sample by a spherical mirror down to approximately 100 and 70 µm spot sizes, respectively. The high temporal resolution is preserved by using a very thin (200 µm) fused silica window as optical access to the cryostat. The two pump pulses employed to show the coherent control of the 2M mode were generated in a Michelson interferometer scheme. One of these beams was reflected by mirrors mounted on a stepper-motor delay line, able to introduce a minimum time delay of 0.3 fs. The two orthogonal polarizations were obtained with a polarizer sheet (thickness ≈ 0.3 mm), to avoid the use of thicker waveplates and the consequent broadening of the pulse duration.

Detection scheme. We measured the pump-induced rotation of the probe polarization employing a balanced-detection scheme. The transmitted probe is split by a Wollaston prism into two orthogonal linearly polarized beams, and focused on a couple of balanced photodiodes. The Wollaston prism is rotated to equalize the probe intensities on the two photodiodes. The pump-induced imbalance of the signal registered by the two photodiodes is measured by a lock-in amplifier, which is locked to the modulation frequency of the pump beam (that is, 500 Hz). Our apparatus was able to detect rotations of the polarization on the order of 1 mdeg.

Excitation mechanism. As explained in the main text, a light-induced modification of the exchange interaction generates the 2M mode[6,7]. This concept can be easily verified by taking into account the transient value of the exchange in the energy of the magnetic medium

$$\mathscr{H} = J_0 \sum_{i,j} \hat{S}_i^{\Uparrow} \cdot \hat{S}_j^{\Downarrow} + \sum_{i,j} \Delta J_{ij}(t) \hat{S}_i^{\Uparrow} \cdot \hat{S}_j^{\Downarrow} \quad (5)$$

where $\Delta J_{ij}(t)$ is the laser-induced modification of the exchange interaction, while J_0 is the value of the exchange in the ground state. The ij indices refer to the anisotropic coupling of light with the magnetic system, determined by the polarization of the laser beam, which breaks the isotropic symmetry of J_0 (Supplementary Note 2). The second term of the equation can be rewritten by expanding the dot product in the following way

$$\sum_{i,j} \Delta J_{ij}(t) \left(\hat{S}_i^{x\Uparrow} \hat{S}_j^{x\Downarrow} + \hat{S}_i^{y\Uparrow} \hat{S}_j^{y\Downarrow} + \hat{S}_i^{z\Uparrow} \hat{S}_j^{z\Downarrow} \right) \quad (6)$$

Defining z as the direction along which the spins are oriented at equilibrium, this equation shows that the modification of the exchange generates terms in the energy quadratic in the transversal spin components (that is, x and y). Such contributions describe the process of generation of two magnons[6] reported in Fig. 1a. This becomes evident by introducing the conventional ladder operators $\hat{S}^{\pm} = \hat{S}^x \pm i\hat{S}^y$, in terms of which we can rewrite equation (6) as

$$\sum_{i,j} \Delta J_{ij}(t) \left(\frac{1}{2} \left(\hat{S}_i^{+\Uparrow} \hat{S}_j^{-\Downarrow} + \hat{S}_i^{-\Uparrow} \hat{S}_j^{+\Downarrow} \right) + \hat{S}_i^{z\Uparrow} \hat{S}_j^{z\Downarrow} \right) \quad (7)$$

The physical interpretation of the first two terms of the last equation consists in two spin-flip processes, one on each sublattice. Once a positive direction of the quantization axis is chosen (direction of L^z in Fig. 1a), flipping a spin on the $\Uparrow$ ($\Downarrow$) sublattice results in a decrease(increase) of the total spin, taken into account by $\hat{S}_i^{-\Uparrow}$($\hat{S}_j^{+\Downarrow}$). In conclusion, a transient modification of the exchange interaction, which breaks the isotropic symmetry of J_0, triggers two magnons described by the quadratic dependence of the Hamiltonian on the transversal spin components.

Antiferromagnetic linear dichroism. The Hamiltonian representing the interaction between light and the magnetic system can be written as[6]

$$\hat{\mathscr{H}} = \sum_{\lambda,\nu} \epsilon^{\lambda\nu}(\hat{\mathbf{S}})E^{\lambda}E^{\nu} \quad (8)$$

where $\epsilon^{\lambda\nu}$ is the dielectric tensor that depends on the spin ($\hat{\mathbf{S}}$), while λ, ν are indexes and E is the electric field of light. The tensor $\epsilon^{\lambda\nu}$ can be expanded in powers of the spin operators. Following a well-established approach[6,17], we may express the spin-dependent components of $\epsilon^{\lambda,\nu}$ as

$$\epsilon^{\lambda\nu} = \sum_i \sum_\gamma K^{\lambda\nu\gamma}\langle\hat{S}_i^\gamma\rangle + \sum_i \sum_{\gamma\delta} G^{\lambda\nu\gamma\delta}\langle\hat{S}_i^\gamma\rangle\langle\hat{S}_i^\delta\rangle + \sum_{i,j}\sum_{\gamma\delta} \rho^{\lambda\nu\gamma\delta}\left\langle \hat{S}_i^{\gamma\Uparrow}\hat{S}_j^{\delta\Downarrow}\right\rangle \quad (9)$$

where higher-order terms are omitted. Each term in this expansion represents a magneto-optical effect and a Raman mode, given the intimate connection between the tensors defining the Raman scattering on magnons and the magneto-optical coefficients[6,34]. The first and second terms involve spin operators at a single ionic site i, they describe the linear magneto-optical effects: the complex Faraday effect and the magnetic linear birefringence(and dichroism), respectively. In terms of Raman scattering the first two terms of equation (9) are connected with the lowest energy magnetic excitations, correspondent to a single spin-flip process, the so called one-magnon modes. These spin waves have wavevectors close to the centre of the Brillouin zone and their frequencies are mostly defined by the spin–orbit coupling. Note that although the second term is quadratic in the spin, it describes a single-magnon excitation and a linear magneto-optical effect. In fact this term is linear in the spin deviation, that is, in the spin transversal component, which implies a linear dependence on the spin ladder operators and therefore the generation of a single magnon[6] (see discussion in the previous section). The magnetic linear birefringence was employed to probe the spin dynamics in reference[19], which accordingly reports the excitation and the time evolution of the low-energy one-magnon mode in $KNiF_3$.

In the present study we measure the ALD, which is accounted for by the last term in equation (9): the quadratic dependence on the spin deviations (equations (6) and (7)) of this term allows to unravel the dynamics of the 2M mode, triggered by two spin-flip processes. Moreover, the two spin operators in this term belong to different sublattices, being placed on different ionic sites i, j. This is the pivotal difference between this magneto-optical effect and the others described so far: the ALD probes the correlations between spins on different sites. Since in $KNiF_3$ only the nearest-neighbour interaction is significant, the dynamics of the spin-correlation function in the last term of equation (9) defines also the dynamics of the exchange energy. The transient ALD, which consists of a different absorption for light beams linearly polarized along and orthogonally to the direction of the antiferromagnetic vector, results in the detected rotation of the probe polarization.

Dynamics of the order parameter. The dynamics of the spin-correlation function can be related to the dynamics of the order parameter in a straightforward way. The z-component of the local spins on sites i, j, belonging to the $\Uparrow$ and $\Downarrow$ sublattices respectively, can be written as

$$\hat{S}_i^{z\Uparrow} = S - \hat{b}_i^\dagger\hat{b}_i; \qquad \hat{S}_j^{z\Downarrow} = -S + \hat{a}_j^\dagger\hat{a}_j \quad (10)$$

where the operators $\hat{b}_i^\dagger\hat{b}_i$ and $\hat{a}_j^\dagger\hat{a}_j$ (introduced rigorously in Supplementary Note 2) represent the number of spin deviations from the maximum values of local spins for both sublattices $\pm S$. We can then write the expectation value of the spin-correlation function on a time-dependent state $|\psi(t)\rangle$ as

$$\langle\psi(t)|\hat{S}_i^{z\Uparrow}\hat{S}_j^{z\Downarrow}|\psi(t)\rangle \approx -S^2 + S\langle\psi(t)|\left(\hat{b}_i^\dagger\hat{b}_i + \hat{a}_j^\dagger\hat{a}_j\right)|\psi(t)\rangle, \quad (11)$$

where only the leading terms in the creation and destruction operators have been considered. Recalling the definition of the antiferromagnetic vector in equation (2), the time-dependent z-component of the order parameter is given by

$$\begin{aligned} L^z(t) &= \sum_i \langle\psi(t)|\hat{S}_i^{z\Uparrow}|\psi(t)\rangle - \sum_j \langle\psi(t)|\hat{S}_j^{z\Downarrow}|\psi(t)\rangle \\ &= SN - \frac{2}{N}\sum_{i,j}\langle\psi(t)|\left(\hat{b}_i^\dagger\hat{b}_i + \hat{a}_j^\dagger\hat{a}_j\right)|\psi(t)\rangle, \end{aligned} \quad (12)$$

where N is the total number of spins. It stems from equations (11) and (12) that the time-dependent parts of the spin-correlation function and of the antiferromagnetic vector are proportional.

Fitting procedure. The transient rotation of the probe polarization was fitted employing the following function:

$$f(t) = C\sin[(\tilde{\omega}(t)t + \phi)]e^{(-t/\tau_d)} + H(t)D\left(1 - e^{(-t/\tau_r)}\right) \quad (13)$$

where C and D are amplitude coefficients, ϕ is the phase of the oscillations, $H(t)$ is the Heaviside function, τ_d is the damping time of the oscillations and τ_r is the characteristic rise time of the incoherent contribution to the signal. Considering the outcome of the Wigner analysis in Fig. 5b, we employed a time-dependent frequency $\tilde{\omega}$ in the sinusoidal function, namely

$$\tilde{\omega}(t) = 2\pi\nu_{2M}[1 + G\sin(2\pi\nu_{mod}t + \psi)] \quad (14)$$

where ν_{2M} is the 2M frequency, G is an amplitude coefficient, ν_{mod} is the modulation frequency and ψ is the phase. From the Wigner analysis we set the following parameters: $\nu_{2M} = 22.12$ THz; $G = 0.002$; $\nu_{mod} = 7.5$ THz; $\psi = 45°$; and $D = 2.3 \cdot 10^{-3}$ deg. The fit parameters allowing to reproduction at best the data (Fig. 2) are as follows: $C = (2.5 \pm 0.1) \cdot 10^{-2}$ deg; $\phi = (220 \pm 1)°$; $\tau_d = (167 \pm 4)$ fs; and $\tau_r = (255 \pm 16)$ fs.

References

1. Radu, I. *et al.* Transient ferromagnetic-like state mediating ultrafast reversal of antiferromagnetically coupled spins. *Nature* **472,** 205–208 (2011).
2. Fausti, D. *et al.* Light-induced superconductivity in a stripe-ordered cuprate. *Science* **331,** 189–191 (2011).
3. Loth, S., Baumann, S., Lutz, C. P., Eigler, D. M. & Heinrich, A. J. Bistability in atomic-scale antiferro-magnets. *Science* **335,** 196–199 (2012).
4. Khajetoorians, A. A. *et al.* Atom-by-atom engineering and magnetometry of tailored nanomagnets. *Nat. Phys.* **8,** 497–503 (2012).
5. Khajetoorians, A. A. *et al.* Current-driven spin dynamics of artificially constructed quantum magnets. *Science* **339,** 55–59 (2013).
6. Cottam, M. G. & Lockwood, D. J. *Light Scattering in Magnetic Solids* 1st ed. (Wiley-Interscience, 1986).
7. Fleury, P. & Loudon, R. Scattering of light by one- and two-magnon excitations. *Phys. Rev.* **166,** 514–530 (1968).
8. Secchi, A., Brener, S., Lichtenstein, A. & Katsnelson, M. Non-equilibrium magnetic interactions in strongly correlated systems. *Ann. Phys.* **333,** 221–271 (2013).
9. Mentink, J. H. & Eckstein, M. Ultrafast quenching of the exchange interaction in a mott insulator. *Phys. Rev. Lett.* **113,** 057201 (2014).
10. Balucani, U. & Tognetti, V. Theory of two-magnon raman scattering in the ordered region for cubic antiferromagnets. *Phys. Rev. B* **8,** 4247–4257 (1973).
11. Cottam, M. G. Theory oftwo-magnon Raman scattering in antiferromagnets at finite temperatures. *J. Phys. C Solid State Phys.* **5,** 1461–1474 (1972).
12. Zhao, J., Bragas, A., Lockwood, D. & Merlin, R. Magnon squeezing in an antiferromagnet: reducing the spin noise below the standard quantum limit. *Phys. Rev. Lett.* **93,** 107203 (2004).
13. Batignani, G. *et al.* Probing ultrafast photo-induced dynamics of the exchange energy in a Heisenberg antiferromagnet. *Nat. Photon.* **9,** 506–510 (2015).
14. Chinn, S., Zeiger, H. & O'Connor, J. Two-magnon Raman scattering and exchange interactions in antiferromagnetic $KNiF_3$ and K_2NiF_4 and ferrimagnetic $RbNiF_3$. *Phys. Rev. B* **3,** 1709–1735 (1971).
15. Kimel, A. V. *et al.* Ultrafast non-thermal control of magnetization by instantaneous photomagnetic pulses. *Nature* **435,** 655–657 (2005).
16. Yan, Y.-X., Gamble, E. B. & Nelson, K. A. Impulsive stimulated scattering: general importance in femtosecond laser pulse interactions with matter, and spectroscopic applications. *J. Chem. Phys.* **83,** 5391 (1985).
17. Ferre, J. & Gehring, G. A. Linear optical birefringence ofmagnetic crystals. *Rep. Prog. Phys.* **47,** 513–611 (1984).
18. Landau, L. D. & Lifshitz, E. M. *Electrodynamics of Continuous Media* 2nd edn, Vol. 8 (Elsevier, 1984).
19. Bossini, D., Kalashnikova, A. M., Pisarev, R. V., Rasing, T. & Kimel, A. V. Controlling coherent and incoherent spin dynamics by steering the photoinduced energy flow. *Phys. Rev. B* **89,** 060405 (2014).
20. Yakovlev, D. R. & Bayer, M. in *Spin Physics in Semiconductors,* 1st edn, Ch. 6 135–178 (Springer, 2008).
21. Chumak, A. V., Vasyuchka, V. I., Serga, A. A. & Hillebrands, B. Magnon spintronics. *Nat. Phys.* **11,** 453–461 (2015).
22. Bartelt, H., Brenner, K.-H. & Lohmann, A. The wigner distribution function and its optical production. *Opt. Commun.* **32-38,** 32 (1980).
23. Gambetta, A. *et al.* Real-time observation ofnonlinear coherent phonon dynamics in single-walled carbon nanotubes. *Nat. Phys.* **2,** 515–520 (2006).
24. Sintani, K., Tomono, Y., Tsuchida, A. & Siratori, K. Influence of magnetic ordering on the lattice vibration of $KNiF_3$. *J. Phys. Soc. Jpn* **25,** 99–108 (1968).
25. Tomono, Y., Takaoka, T., Yajima, M., Tanokura, Y. & Jinda, N. Temperature dependence of lattice vibration frequencies of KNiF3 in connection with antiferromagnetism. *J. Phys. Soc. Jpn* **59,** 579–583 (1990).
26. Long, D. A. *The Raman Effect: A Unified Treatment of the Theory of Raman Scattering by Molecules,* 1st edn (Wiley, 2002).
27. Cyvin, S. J., Rauch, J. E. & Decius, J. C. Theory of hyper-Raman effects (nonlinear inelastic light scattering): selection rules and depolarization ratios for the second-order polarizability. *J. Chem. Phys.* **43,** 4083–4095 (1965).
28. Kanda, N. *et al.* The vectorial control of magnetization by light. *Nat. Commun.* **2,** 362 (2011).
29. Kampfrath, T. *et al.* Coherent terahertz control of antiferromagnetic spin waves. *Nat. Photon.* **5,** 31–34 (2010).

30. Le Tacon, M. *et al.* Intense paramagnon excitations in a large family of high-temperature superconductors. *Nat. Phys.* **7,** 725–730 (2011).
31. Dal Conte, S. *et al.* Disentangling the electronic and phononic glue in a high-Tc superconductor. *Science* **335,** 1600–1603 (2012).
32. Dal Conte, S. *et al.* Snapshots of the retarded interaction of charge carriers with ultrafast fluctuations in cuprates. *Nat. Phys.* **11,** 421–426 (2015).
33. Brida, D. *et al.* Few-optical-cycle pulses tunable from the visible to the mid-infrared by optical parametric amplifiers. *J. Opt.* **12,** 013001 (2010).
34. Wettling, W., Cottam, M. G. & Sandercock, J. R. The relation between one-magnon light scattering and the complex magneto-optic effects in YIG. *J. Phys. C Solid State Phys.* **8,** 211–228 (1975).

Acknowledgements

We thank Dr A. Caretta and Prof. P.H.M. van Loosdrecht for the spontaneous Raman measurements. We thank Prof. R. Merlin and Dr J. Mentink for fruitful discussions. This research was supported by LASERLAB-EUROPE (grant agreement no. 284464, EC's Seventh Framework Programme); de Nederlandse Organisatie voor Wetenschappelijk Onderzoek (NWO); de Stichting voor Fundamenteel Onderzoek der Materie (FOM); the European Union's Seventh Framework Program (FP7/2007-2013) Grant No. NMP3-LA-2010-246102 (IFOX), No. 280555 (Go-Fast), No. 281043 (FEMTOSPIN); CNECT-ICT-604391 (Graphene Flagship); the European Research Council under the European Union's Seventh Framework Program (FP7/2007-2013)/ERC Grant Agreement No. 257280 (Femtomagnetism), as well as the programme 'Leading Scientist' of the Russian Ministry of Education of Science (14.Z50.31.0034). R.V.P. acknowledges the partial support by Russian grants 14.B25.31.0025 and 15-02-14222.

Author contributions

D.B. and A.K. conceived the project; S.D.C. and G.C. designed and built the experimental set-up; D.B., S.D.C. and G.C. performed the measurements; A.S. formulated the theory; D.B. and Y.H. analysed the data; R.P. provided the sample; all the authors took part to regular discussions and contributed to the writing of the manuscript.

Additional information

Competing financial interests: The authors declare no competing financial interests.

11

Strain effects on the work function of an organic semiconductor

Yanfei Wu[1], Annabel R. Chew[2], Geoffrey A. Rojas[1], Gjergji Sini[3,4], Greg Haugstad[5], Alex Belianinov[6], Sergei V. Kalinin[6], Hong Li[7], Chad Risko[8], Jean-Luc Brédas[4], Alberto Salleo[2] & C. Daniel Frisbie[1]

Establishing fundamental relationships between strain and work function (WF) in organic semiconductors is important not only for understanding electrical properties of organic thin films, which are subject to both intrinsic and extrinsic strains, but also for developing flexible electronic devices. Here we investigate tensile and compressive strain effects on the WF of rubrene single crystals. Mechanical strain induced by thermal expansion mismatch between the substrate and rubrene is quantified by X-ray diffraction. The corresponding WF change is measured by scanning Kelvin probe microscopy. The WF of rubrene increases (decreases) significantly with in-plane tensile (compressive) strain, which agrees qualitatively with density functional theory calculations. An elastic-to-plastic transition, characterized by a steep rise of the WF, occurs at $\sim$0.05% tensile strain along the rubrene π-stacking direction. The results provide the first concrete link between mechanical strain and WF of an organic semiconductor and have important implications for understanding the connection between structural and electronic disorder in soft organic electronic materials.

[1] Department of Chemical Engineering and Materials Science, University of Minnesota, 421 Washington Ave SE, Minneapolis, 55455 Minnesota, USA. [2] Department of Materials Science and Engineering, Stanford University, 476 Lomita Mall, Stanford, 94305 California, USA. [3] Laboratoire de Physico-chimie des Polymères et des Interfaces, Université de Cergy-Pontoise, 5 Mail Gay Lussac, Neuville sur Oise, Cergy-Pontoise Cedex 95031, France. [4] Physical Science and Engineering Division, Solar & Photovoltaics Engineering Research Center, King Abdullah University of Science and Technology (KAUST), Thuwal 23955-6900, Saudi Arabia. [5] Characterization Facility, University of Minnesota, 100 Union St SE, Minneapolis, 55455 Minnesota, USA. [6] Center for Nanophase Materials Sciences, Oak Ridge National Laboratory, 1 Bethel Valley Rd., Oak Ridge, 37831 Tennessee, USA. [7] School of Chemistry and Biochemistry & Center for Organic Photonics and Electronics, Georgia Institute of Technology, Atlanta, 30332 Georgia, USA. [8] Department of Chemistry & Center for Applied Energy Research, University of Kentucky, Lexington, 40506 Kentucky, USA. Correspondence and requests for materials should be addressed to C.D.F. (email: frisbie@umn.edu).

Strain is known to play a critical role in the physical properties of many conventional inorganic semiconductors. For example, strain modifies the band structure and carrier mobilities of group IV materials including Si and Ge, as well as III–V materials such as GaAs[1–5]. Consequently, controlled introduction of strain is widely employed to engineer the transport and optical properties of these semiconductors for enhanced device performance[6,7]. Similarly, in soft organic semiconductors that serve as the active components in many organic electronic devices[8–12], tensile and compressive strains modify the material electronic properties and function. An intriguing example is the observation by Bao and colleagues that the charge mobility in films of the benchmark organic material 6,13-Bis(triisopropylsilylethynyl)pentacene (TIPS-pentacene) is enhanced under lattice compression[13]. Controlling and understanding strain effects is thus particularly important for device applications of organic semiconductors, especially in the area of flexible electronics, where externally applied strains are routine. Both strain-resistant and strain-sensitive device responses[14–19] are desirable in flexible circuits, and better knowledge of strain–property relationships will facilitate flexible device designs.

Importantly, better understanding of strain–electrical property relationships is also critical for theoretical models of transport in organic semiconductors. Virtually all organic semiconductor films exhibit intrinsic non-uniform strains arising from lattice or thermal expansion mismatch with the substrate, the presence of defects or post deposition treatments such as thermal annealing[20–23]. Non-uniform strains in turn lead to variations in intermolecular electronic coupling and thus to local differences in bandwidths and band gaps[24,25], that is, 'raggedy' valence (conduction) band edges, which can also create shallow charge trap (tail) states. There is, therefore, the intriguing possibility that residual microstrain is a significant cause of charge-carrier trapping in organic semiconductors. Proving this would be a major conceptual step in clarifying the link between structural and electronic disorder, a long-sought goal for organic electronics. However, a challenge is that, as far as we are aware, the quantitative connection between electronic structure and mechanical strain has never been established in these materials.

Here we take a considerable step in this direction by measuring the effect of tensile and compressive strains on the work function (WF) of a prototypical p-type organic semiconductor for the first time. The WF is defined as the energy difference between the vacuum level (E_{vac}) and the Fermi level (E_F), that is, $WF = E_{vac} - E_F$. Strain can modify E_{vac} and E_F of a semiconductor by changing the band-edge positions, the dopant levels in the band gap or the surface dipoles. Thus, strain impacts the WF. In a p-type crystal of π-conjugated organic molecules in which the Fermi level is pinned (fixed doping level), strain effects on E_F can be pictured, to a first approximation, as follows: compressive strains increase the frontier orbital (for example, HOMO) overlap of adjacent molecules, leading to a wider valence band[24], a higher E_F, and a lower WF. Conversely, tensile strains increase the separation between molecules, lowering orbital overlap and decreasing valence bandwidth and E_F, thus increasing the WF. Note that with fixed doping level, the WF change can be directly translated to the change of the ionization potential ($IP = E_{vac} - E_{HOMO}$), which is a more commonly adopted term for describing the electronic structure of intrinsic organic semiconductors.

Our measurements focus on p-type rubrene single crystals, which serve as a model material platform for many fundamental studies of organic semiconductor physics due to their exemplary transport properties, that is, the highest reproducible charge-carrier mobilities to date have been achieved in single crystal rubrene field-effect transistors[26]. By adhering thin rubrene crystals onto substrates with coefficients of thermal expansion (CTEs) distinctly different from rubrene and varying the temperature, we systematically induce large and controlled tensile or compressive strains in rubrene crystals and quantify the elastic portion by X-ray diffraction. The corresponding WF of rubrene, measured by temperature-dependent scanning Kelvin probe microscopy (SKPM), is found to increase (decrease) with the in-plane tensile (compressive) strain. The measured changes in WF (ΔWF) are qualitatively confirmed by density functional theory (DFT) calculations, and verify that indeed small elastic strains $<0.1\%$ can lead to ΔWF surpassing the room temperature thermal energy $kT = 25$ meV. Furthermore, we find that the onset of tensile plastic strain leads to even larger increases of WF with strain. These findings constitute a definitive link between structural deformation and electronic disorder in a model organic semiconductor.

Results

Rubrene single crystals. The crystal structure of rubrene single crystals grown from the vapour phase is shown in Fig. 1a–d. The crystals adopt an orthorhombic structure and slipped π-stack packing motif with the π-stacking direction along the b axis[27]. The optical micrograph (Fig. 1e) shows a lath-like single crystal with the largest facet being the (001) plane and the longest direction aligned with the b axis. Thin rubrene crystals ($\sim 2\,\mu m$) are then laminated on poly(dimethylsiloxane) (PDMS) or silicon (Si) substrates for X-ray diffraction and SKPM measurements. PDMS is chosen for its significantly larger CTE ($\sim 300 \times 10^{-6}\,K^{-1}$) than that of rubrene ($\sim 10$–$80 \times 10^{-6}\,K^{-1}$)[28,29] in order to induce tensile strain in laminated rubrene crystals upon increasing the temperature. Si, on the other hand, is chosen to induce compressive strain since it has a much smaller CTE (3–$4 \times 10^{-6}\,K^{-1}$) than that of rubrene (Supplementary Table 1).

Figure 1f shows the scheme of the SKPM measurement, which operates in a two-pass 'lift mode' to acquire images of topography and the contact potential difference (CPD) between the tip and the sample[30]. The CPD is related to the WF of the tip and the sample by $qCPD = WF_{tip} - WF_{sample}$, where q is the elementary charge. Therefore, with the same tip acting as the reference, the measured CPD change reflects the WF change of the sample, that is, $q\Delta CPD = -\Delta WF_{sample} = -\Delta WF$[31,32]. Typical topography and CPD images of a rubrene (001) surface obtained by SKPM at room temperature are shown in Fig. 1g,h, respectively. The topography image shows a surface with several molecularly flat terraces, which extend over a distance of several micrometres. The height of each terrace obtained from the profile of the white dashed line, as shown in the inset of Fig. 1g, is ~ 13 Å, which coincides with one-half of the c axis unit cell parameter. This is in agreement with previous observations of rubrene molecular steps and confirms that there are two nonequivalent molecular planes in the rubrene crystal unit cell[33]. The CPD map, on the other hand, is almost featureless. The local CPD variations, characterized by the root-mean-square CPD roughness, are less than 3 mV for this $20 \times 20\,\mu m^2$ area. The CPD map thus suggests that the rubrene (001) surface is electrostatically homogeneous, in accordance with the high structural order of rubrene single crystals. It also demonstrates the high voltage resolution of interleave-based SKPM (about 1 mV), which allows small WF changes of the sample to be resolved.

Strain quantification by *in situ* X-ray diffraction studies. To quantify the elastic strain induced in rubrene crystals, temperature-dependent X-ray diffraction measurements were

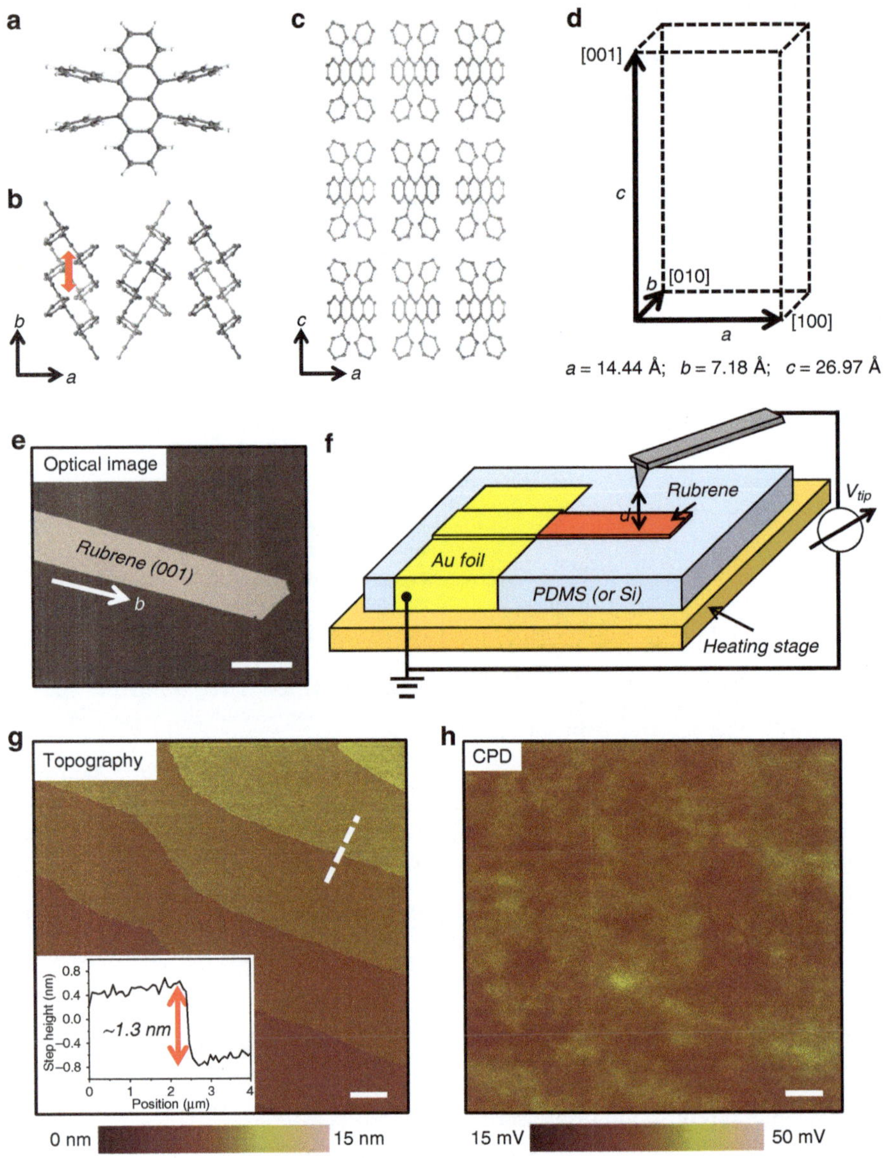

Figure 1 | Crystal structure and SKPM measurement of rubrene single crystals. (**a**) Molecular structure of rubrene. (**b**) Crystal structure in the *a-b* plane; red arrow indicates the π-stacking interaction. (**c**) Crystal structure in the *a-c* plane. (**d**) Orthorhombic structure and lattice parameters of rubrene. (**e**) Optical micrograph of as-grown rubrene crystal. Length of scale bar, 200 μm. (**f**) SKPM setup for CPD measurement. The sample sits on top of a heating stage and is grounded through gold foil. The conducting probe scans in a two-pass 'lift mode' and a constant lift height ($d = 50$ nm) is used in the 'interleave' pass. (**g**) Topography of rubrene single crystal shows typical terrace structure and each terrace has height corresponding to one molecular layer. Length of scale bar, 2 μm. Inset: step height profile of the dashed line. (**h**) CPD image obtained simultaneously with topography shows nearly homogeneous CPD across the surface. Length of scale bar is 2 μm.

carried out. By recording the 2θ shifts of rubrene (0012), (113) and (313) peaks as a function of temperature and calculating the corresponding d-spacings, d_{0012}, d_{113} and d_{313}, the total elastic strains (ε^{total}) along the rubrene *a*, *b* and *c* axes at different temperatures were computed (Supplementary Methods, Supplementary Figs 1-3 and Supplementary Table 2), which are composed of both induced mechanical strain ($\varepsilon^{elastic}$) and thermal strain ($\varepsilon^{thermal}$), that is, $\varepsilon^{total} = \varepsilon^{elastic} + \varepsilon^{thermal}$. Figures 2a,b show the average ε^{total} for rubrene on PDMS and rubrene on Si, respectively, when the samples were heated from room temperature to 75 °C. Also shown are the calculated thermal strains $\varepsilon^{thermal}$ for free crystals using the reported CTEs (Supplementary Table 1)[28]. Clearly, the two types of samples show distinct strain behaviours with increasing temperature. In Fig. 2a, rubrene crystals laminated on PDMS exhibit slightly larger ε^{total} along the *a* axis and much larger ε^{total} along the *b* axis compared with $\varepsilon^{thermal}$ estimated for free crystals. ε^{total} along the *c* axis of rubrene on PDMS, however, is slightly smaller than the estimated $\varepsilon^{thermal}$ for free crystals. On the other hand, crystals on Si show essentially the opposite behaviour, namely much smaller ε^{total} in the *a–b* plane and much larger ε^{total} along the *c* axis than the corresponding $\varepsilon^{thermal}$ predicted for free rubrene.

Such different strain states for rubrene on PDMS and rubrene on Si are explained by the schematics in Fig. 2c,d. For rubrene on PDMS, significant expansion of PDMS due to its large CTE induces tensile strains in the rubrene *a–b* plane, resulting in ε^{total} greater than the estimated $\varepsilon^{thermal}$ of free crystals. The difference in ε^{total} for the *a* and *b* axes with respect to the corresponding $\varepsilon^{thermal}$ may be attributed to the anisotropic CTE of rubrene[28,29], that is, CTE along the *a* axis ($\sim 78 \times 10^{-6}\,K^{-1}$) is significantly larger than that along the *b* axis ($\sim 16 \times 10^{-6}\,K^{-1}$), which leads to anisotropic CTE mismatch between the crystal and PDMS.

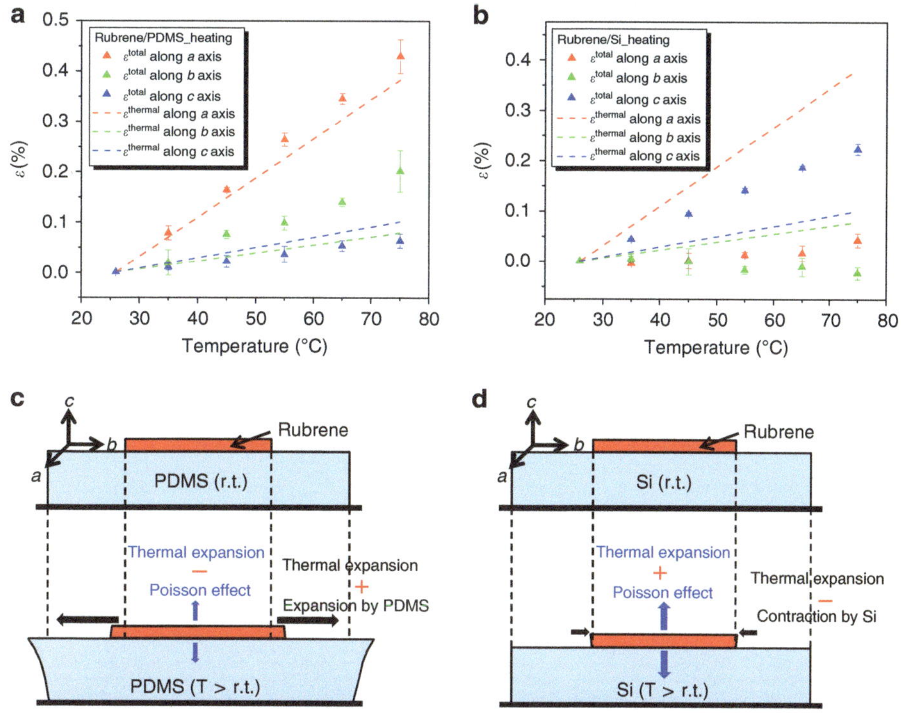

Figure 2 | Total elastic strain in rubrene single crystals. (**a**) Average total elastic strains ε^{total} of rubrene along *a*, *b* and *c* axes as a function of temperature for crystals laminated on PDMS, which are different from the corresponding thermal strains $\varepsilon^{thermal}$ expected for free rubrene crystals (dashed lines). (**b**) Average ε^{total} of rubrene along *a*, *b* and *c* axes as a function of temperature for crystals laminated on Si, different from the corresponding $\varepsilon^{thermal}$ predicted for free rubrene crystals (dashed lines). The error bars in ε^{total} were based on the s.d. calculated as demonstrated in Supplementary Table 2. (**c**) Illustration of strain components for rubrene on PDMS. (**d**) Illustration of strain components for rubrene on Si.

Furthermore, as a result of the Poisson effect, the substrate-induced in-plane tensile strain in rubrene exerts a compressive component in the out-of-plane direction, or the *c* axis. Therefore, ε^{total} measured in the *c* axis is smaller than the predicted $\varepsilon^{thermal}$ if the crystal only undergoes thermal expansion.

In contrast, the Si substrate has negligible expansion ($\sim$0.015%) within the temperature range of interest such that the thermal expansion in the *a*–*b* plane of rubrene is largely constrained. The *a* and *b* axes experience compression induced by the substrate, offsetting thermal expansion and thus leading to very small ε^{total}. Similarly, due to the Poisson effect, the compressive strain creates tension in the *c* axis, making ε^{total} along the *c* axis much larger than the *c* axis thermal expansion expected for free crystals.

The overall tension and compression states of rubrene crystals laminated on PDMS and Si, respectively, are illustrated quantitatively by the percentages of unit cell volume expansion as shown in Fig. 3a. Rubrene on PDMS exhibits a total volume increase of 0.7%, slightly larger than that estimated for free crystals, consistent with net tension, whereas the total volume increases by only 0.25% for rubrene on Si, much smaller compared with the computed free crystal result and consistent with net compression. Though the CTE mismatch between rubrene and PDMS is greater, the magnitude of the tensile strain induced in rubrene by PDMS is actually smaller than the compressive strain induced by Si. This can be understood by the relative stiffness of rubrene compared with the substrates. Rubrene with modulus of $\sim$15 GPa is $\sim$10 times softer than Si whereas it is over 10^4 times stiffer than PDMS[34]. The complexity introduced by the compounding effects of CTE mismatch, relative stiffness and interface adhesion is such that the elastic state of the rubrene crystal cannot be calculated and must be accessed experimentally by *in situ* X-ray diffraction measurements.

The net elastic mechanical strains ($\varepsilon^{elastic} = \varepsilon^{total} - \varepsilon^{thermal}$) along the three principal axes of rubrene on PDMS and Si are shown in Fig. 3b – d. For rubrene on PDMS, tensile strain reaches $\sim$0.05% along the *a* axis and over 0.1% along the *b* axis; the *c* axis compressive strain reaches approximately $-$0.04%. For rubrene on Si, the *a* and *b* axes show large compressive strain, approximately $-$0.35% and $-$0.1%, respectively; the tensile strain along the *c* axis reaches over 0.1%. We will show below that the differences in $\varepsilon^{elastic}$ for the two types of samples correlate with dramatically different WF changes.

WF measurements by temperature-dependent SKPM. To measure the WF change of rubrene as a function of its mechanical states, SKPM as described above was carried out at different temperatures with the same tip. Note that the WF change of the Pt-coated tip with temperature is not significant over the temperature range of interest because the WFs of metals typically exhibit a weak temperature dependence ($\sim 10^{-4}$ eV K^{-1})[35]. Figure 4a,b show the CPD maps of rubrene crystals on PDMS and Si from room temperature to 75 °C. As temperature increases, the CPD of the rubrene (001) surface decreases (bright to dark) on PDMS and increases (dark to bright) on Si. The CPD evolutions are illustrated quantitatively with the histograms extracted from the CPD maps at different temperatures, as shown in Fig. 4c,d. For rubrene on PDMS (Fig. 4c), the CPD changes by more than $-$200 mV from room temperature to 75 °C, whereas for rubrene on Si (Fig. 4d), it changes by about +120 mV across the same temperature range.

By averaging multiple measurements from multiple samples, the average change in WF ($\Delta WF = -q\Delta CPD$) as a function of temperature is plotted for rubrene on PDMS (Fig. 4e) and rubrene on Si (Fig. 4f), respectively, for samples undergoing

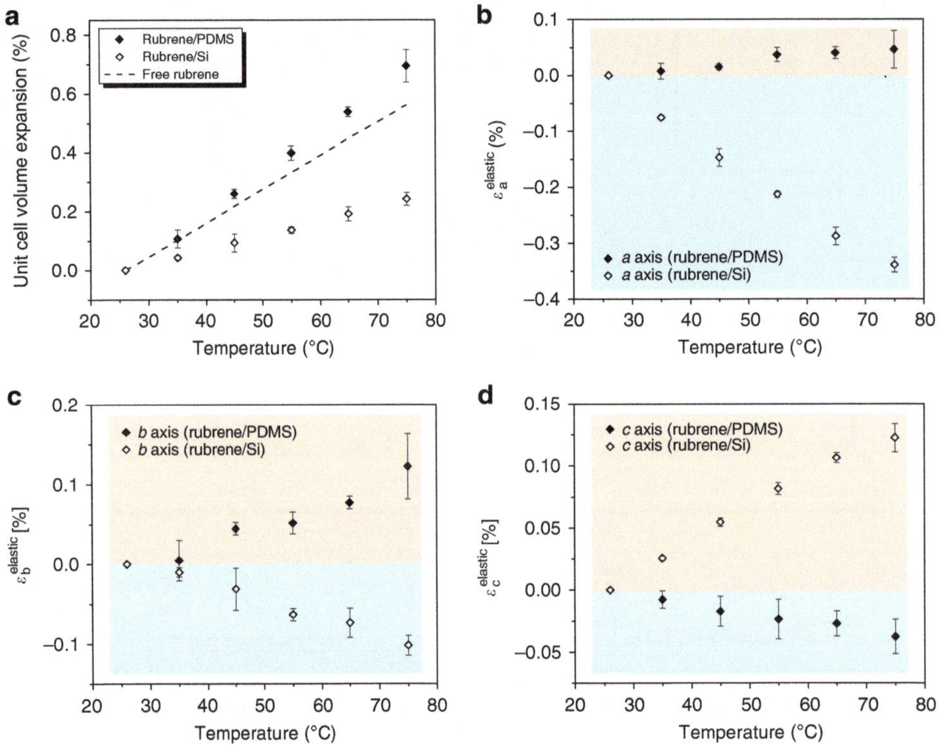

Figure 3 | Unit cell volume expansion and substrate-induced elastic mechanical strain. (**a**) The unit cell volume expansion for rubrene crystals laminated on PDMS and Si as a function of temperature, which are compared with that expected for free crystals. The error bars in unit cell volume expansion were based on the propagation of s.d. errors in ε^{total} of the three axes as shown in Fig. 2a,b. (**b**) Average induced mechanical strain ($\varepsilon^{elastic} = \varepsilon^{total} - \varepsilon^{thermal}$) along rubrene *a* axis as a function of temperature for crystals laminated on PDMS and Si, respectively. (**c**) Average $\varepsilon^{elastic}$ along rubrene *b* axis as a function of temperature for crystals laminated on PDMS and Si, respectively. (**d**) Average $\varepsilon^{elastic}$ along rubrene *c* axis as a function of temperature for crystals laminated on PDMS and Si, respectively. The error bars in $\varepsilon^{elastic}$ were based on the standard deviation errors in ε^{total} as shown in Fig. 2a,b.

consecutive heating and cooling cycles. As shown in Fig. 4e, the WF of rubrene increases ($\Delta WF > 0$) with increasing temperature on PDMS and there is an evident transition when the temperature is close to 50 °C. First, the WF increases only slightly from room temperature to ~50 °C and then above 50 °C it increases much more steeply. As will be discussed later, we attribute the slope change to the elastic-to-plastic transition. Upon cooling, the WF of rubrene decreases continuously ($\Delta WF < 0$) and there remains a large hysteresis of ~100 meV on returning to room temperature. An opposite trend is observed in rubrene crystals on Si, as shown in Fig. 4f. The WF of rubrene decreases ($\Delta WF < 0$) with increasing temperature, and unlike rubrene on PDMS, there is no obvious slope change upon heating and the hysteresis between heating and cooling is almost negligible: the WF comes back to its original value at the end of the cycle. Overall, Fig. 4 demonstrates that changes in WF are significant for both types of samples and there is a qualitative difference in behaviour for rubrene on PDMS and rubrene on Si.

Strain-WF Relationship and DFT calculations. Figure 5 illustrates ΔWF as a function of $\varepsilon^{elastic}$ (tensile and compressive) in rubrene. ΔWF is plotted versus the *b* axis elastic strain for simplicity because it is reasonable to suppose that the relative significance of mechanical strains along different axes is positively correlated with the strength of intermolecular interactions, that is, *b* axis (π-stacking direction) > *a* axis > > *c* axis. It is evident from Fig. 5a that WF increases significantly with tensile strain and decreases with compressive strain, and that the changes can be much greater than kT (25 meV) at room temperature. A similar trend of the WF change has been observed or predicted in strained Si and graphene[36–39]. Figures 5b,c show ΔWF as a function of forward and reverse elastic strains on PDMS and Si substrates, respectively.

We attribute the measured ΔWFs to the substrate-induced mechanical strains $\varepsilon^{elastic}$ in rubrene crystals—instead of intrinsic changes with temperature or surface contamination—for several reasons. First, ΔWF is completely opposite for rubrene on PDMS versus rubrene on Si even though they experience the same temperature change. This rules out temperature as the dominating factor for ΔWF. Second, although in general the WF may be affected by surface contamination, it is unlikely to result in the systematic WF trends we observed, including the hysteresis in rubrene on PDMS. Third, we performed static level ($T = 0$ K) DFT calculations of ΔWF as a function of strain as is shown in Fig. 5b,c. Although they do not match quantitatively, the calculations predict the same signs for ΔWF as observed experimentally and thus qualitatively support our conclusion that ΔWF is dominated by mechanical strain (Supplementary Table 3).

To estimate ΔWF by DFT, the evolution of the valence band maximum (VBM) and the potential energy of an electron at the vacuum level (E_{vac}) were calculated based on the crystal structures of rubrene under the experimentally-attained mechanical strains; we set $WF = E_{vac} - E_F \approx E_{vac} - VBM$ such that $\Delta WF \approx \Delta E_{vac} - \Delta VBM$. This definition of WF is justified because we expect E_F of rubrene to lie closer to the VBM and to track the VBM because of Fermi level pinning or partial pinning, that is, the offset between E_F and VBM is set by the natural p-type doping of as-grown rubrene. From Fig. 5b,c, it is clear that the signs of the calculated ΔWF are determined by ΔVBM. Further, the calculated ΔVBM is consistent with qualitative expectations, namely that tensile strain

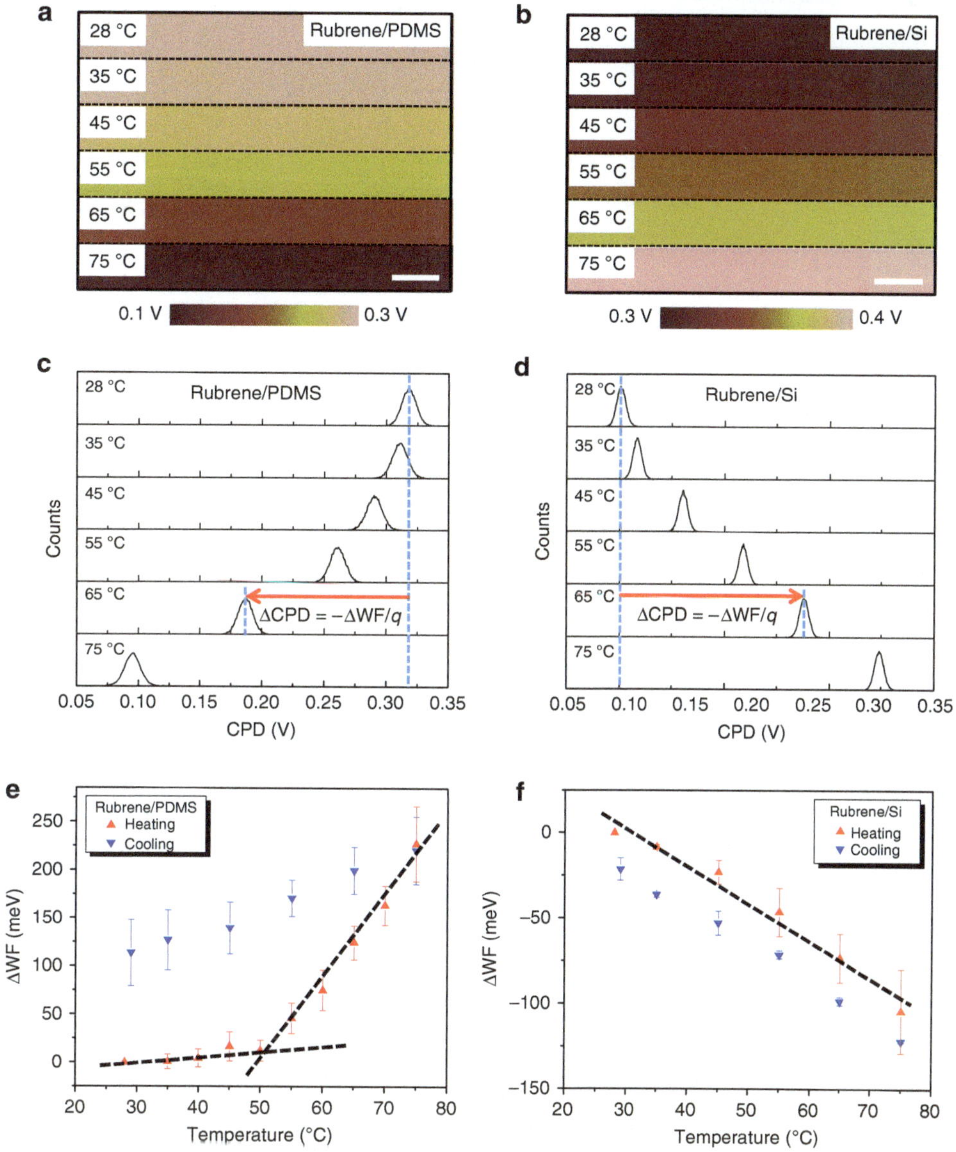

Figure 4 | SKPM measurements of rubrene as a function of temperature. (**a**) CPD of rubrene on PDMS shifts from more positive (bright) to more negative (dark) when the sample is heated from room temperature to 75 °C. Length of scale bar is 500 nm. (**b**) CPD of rubrene on Si shifts from more negative (dark) to more positive (bright) when the sample is heated from room temperature to 75 °C. Length of scale bar, 500 nm. (**c**) CPD histograms extracted from images in **a**. ΔCPD, indicated by the red arrow, is defined as the change of CPD at any elevated temperature from that at room temperature, that is, ΔCPD = CPD(T) − CPD(r.t.). (**d**) CPD histograms extracted from images in **b**. (**e**) Average ΔWF of rubrene on PDMS as a function of temperature with ΔWF defined as $-q\Delta$CPD, where q is the elementary charge. The dashed lines show two distinct regimes of ΔWF at low and high temperatures. The error bars in ΔWF were based on the s.d. of 24 independent measurements from four different samples at each temperature. (**f**) Average ΔWF of rubrene on Si as a function of temperature. The dashed line shows continuous change of ΔWF within the entire temperature range. The error bars in ΔWF were based on s.d. of 12 independent measurements from two different samples at each temperature.

decreases HOMO–HOMO (π–π) interaction, lowering the VBM and thus increasing the WF (ΔWF > 0). Compressive strain, on the other hand, increases the π–π interaction, which increases the VBM and decreases the WF (ΔWF < 0).

The discrepancy between the calculations and experiments may have several origins. First, we note that the calculations did not and cannot take into account plastic deformation that occurs for the case of rubrene on PDMS. So in this case, comparison is only fair at the lowest strains, and here the number of data points is sparse. Second, within the elastic strain regime, the calculations only simulate a static picture, whereas dynamic effects with increasing temperature (for example, local/non-local electron–phonon couplings) are not considered. Such dynamic effects are expected to be more significant in the case of compressive strain. Indeed, a larger quantitative discrepancy with the experimental results is observed for compressive elastic strain (Fig. 5c).

The WF trend in the high-tension regime (Fig. 5a,b) is ascribed primarily to the effect of plastic deformation based on two experimental observations. One is the sudden steep rise in ΔWF at ~0.05% strain, which is not associated with any noticeable signs of surface roughening or structural instability (for example, cracks). It is known from the literature that yield strains for organic crystals are typically <0.1% (ref. 40), consistent with our assignment. A similar WF-strain relationship on yielding has also been reported for metals[41]. The other important observation is the large hysteresis in ΔWF when elastic tensile strain recovers.

This non-recoverable ΔWF is also strongly indicative of plastic strain. In compression, however, ΔWF decreases smoothly as a function of strain without any significant transition point or hysteresis, suggesting that the yield point is not reached. The yield strain is therefore higher for compressive strain than tensile strain, which is also in agreement with observations in other materials and could be due to larger friction between slip layers under compression[41]. The association of plastic deformation with a large non-recoverable ΔWF strongly suggests the existence of a relationship between defects such as dislocations and charge trapping in organic films.

Discussion

Controlled and quantifiable tensile and compressive strains have been induced in rubrene single crystals by utilizing thermal expansion of the substrates. The change of WF as a result of induced strain is measured by SKPM. The WF increases with tensile strain (elastic and plastic) and decreases with compressive strain in the *a*–*b* plane of rubrene crystals, confirmed qualitatively by DFT calculations. In tension, the WF increase is slight but measurable by SKPM in the elastic regime. The WF increase becomes much more significant upon the onset of plastic deformation, which occurs at a relatively small tensile strain (<0.1%). In compression, the WF decreases smoothly with increasing strain and no apparent cross-over from elastic to plastic behaviour is observed within the investigated strain range. We propose that strain-induced WF variations will lead to band-edge fluctuations, which can impact charge transport, and are an important mechanism for the creation of electrostatic disorder (for example, band-tail states) in organic semiconductors. As organic thin films will typically have residual microstrain, the sensitivity of WF to strain in organic semiconductors has great implications for the fundamental electrical properties of these materials and their performance in devices.

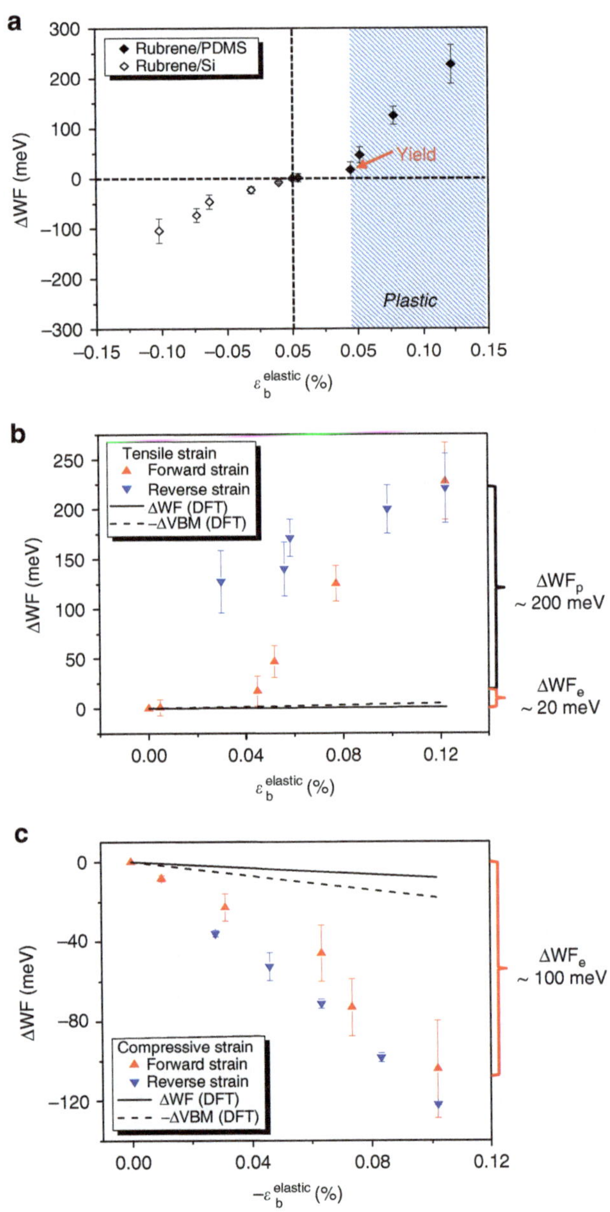

Figure 5 | ΔWF as a function of substrate-induced elastic strain.
(**a**) ΔWF as a function of $\varepsilon^{elastic}$ (tensile and compressive strain) along the *b* axis. The WF increases with tensile strain and decreases with compressive strain. Note that ΔWF in the shaded region is mainly a result of plastic deformation instead of elastic tensile strain. (**b**) ΔWF as a function of elastic tensile strain for forward and reverse strains. There is an elastic-to-plastic transition characterized by a sudden rise in ΔWF with strain. There is a large hysteresis of ΔWF between the forward and reverse tensile strains, indicating the effect of plastic deformation. The DFT calculation results are shown by the dashed line (−ΔVBM) and the solid line (ΔWF). The calculated −ΔVBM and ΔWF slightly increase with elastic tensile strain, a trend that qualitatively agrees with the experimental results. The quantitative disagreement between the calculations and the experimental results in the high-strain regime is attributed to plastic deformation, which is not considered in the calculations. (**c**) ΔWF as a function of elastic compressive strain for forward and reverse strains. The WF decreases with strain smoothly and no apparent elastic-to-plastic transition is observed. There is negligible hysteresis of ΔWF between the forward and reverse strains. The DFT calculation results are shown by the dashed line (−ΔVBM) and the solid line (ΔWF). The calculated −ΔVBM and ΔWF decrease with elastic compressive strain, a trend that qualitatively agrees with the experimental results. The deviation in the ΔWF magnitude of the calculations from the experiments may be due to dynamic effects that are not considered in the theoretical approach. The errors bars in ΔWF were based on standard deviation errors as described in Fig. 4.

Methods

Sample preparation. Rubrene single crystals were grown by physical vapour transport using ultrapure Ar as carrier gas[42]. Commercially available rubrene (sublimed grade, 99.99%, Sigma-Aldrich) was used as received as the source material. The sublimation temperature was ~280 °C. Only thin crystals (<5 μm) with uniform crystalline regions and smooth/clean surfaces were selected for sample preparation. Thicknesses of the crystals were measured by surface profilometry (KLA-Tencor P-16 surface profiler). The average thickness of all crystals used was ~2 μm. Freshly made crystals were laminated onto either PDMS or silicon substrates. Spontaneous adhesion of the crystals to the substrates occurred. In order to electrically ground the sample, vapour deposited gold film with thicknesses around 500 nm was removed from a Si substrate and transferred to cover part of the crystal by tweezers, and silver paint was then used to connect the gold film to the metal SKPM sample puck.

Scanning Kelvin probe microscopy. SKPM measurements were performed with a Cypher ES Environmental AFM (Asylum Research), which works in a two pass 'lift mode'. In the first pass of each line of an image, the conductive probe scans the rubrene surface in attractive-regime dynamic mode to generate the topographic data under conventional amplitude-modulation feedback (also known as 'AC' or 'tapping' mode) while mechanically vibrating the cantilever near resonance. The attractive-regime dynamic mode was used since it better preserves the probe and thereby allows meaningful CPD comparison among different measurements[32]. To stabilize performance in the attractive regime, the cantilever was driven at a drive frequency slightly larger (~150 Hz) than the fundamental resonant frequency, and the setpoint amplitude was about 90% of the free amplitude (~90 nm). In the second 'interleave' pass, the probe was lifted to a constant height above the surface and scanned the topographic trajectory acquired in the first pass. A tip-applied AC bias resonantly excited the cantilever (via a time-varying electrostatic force gradient between tip and sample) while a DC bias was adjusted under feedback so as to null the AC excitation by matching (and thus measuring) the CPD point-to-point across the surface[30]. The samples were heated with an integrated heating stage at a heating rate of $\sim 2\,°C\,min^{-1}$. The samples were held for ~ 10 min at the target temperature and the CPDs of the same area were measured multiple times. Note that the cantilever vibrational tuning was repeated at each temperature to account for changes in resonant frequency with temperature. The typical probes were from Nanosensors (PPP-EFM-W, Pt/Ir coated, resonant frequency ~ 75 kHz, spring constant $\sim 2.8\,N\,m^{-1}$). The lift height during the second pass was 50 nm, which was beyond the range where van der Waals forces come into play. The applied AC voltage in SKPM was 0 ± 3 V in amplitude. The SKPM images were analysed using the freeware Gwyddion.

X-ray diffraction. *In situ* X-ray diffraction measurements were carried out with a PANalytical X'Pert Pro X-ray diffractometer with a Cu Kα source operated at 45 kV and 40 mA filament current. The samples were slowly heated ($\sim 2\,°C\,min^{-1}$) to target temperatures using a thermal stage connected to an Anton Paar temperature control unit (TCU 150) within the diffractometer. The samples were allowed to stabilize at the target temperature for ~ 10 min before the diffraction peaks of interest were measured. To more accurately study the weakly diffracting off-axis peaks, two-dimensional reciprocal space maps were measured. Detailed calculations of strain based X-ray diffraction measurements are shown in Supplementary Methods, Supplementary Figs 1–3 and Supplementary Table 2.

DFT calculations. The geometric and electronic properties of strained rubrene single crystals were computed at the DFT level with the VASP code[43]. The Perdew-Burke-Ernzerhof (PBE) functional were used with a plane-wave basis set (300 eV cutoff) and projector augmented wave potentials[44,45]. The DFT calculations were carried out using on a $2 \times 2 \times 1$ k-point grid and a Gaussian smearing with 0.10 eV width. The calculations considered the effects due to elastic strain in a static sense (that is, the dynamic motions of the molecules at a given temperature were not explicitly considered), with the unit-cell parameters chosen as those modified by mechanical strain obtained experimentally. For each crystal structure, the molecules within a unit cell were allowed to fully relax while the lattice parameters remained fixed to the experimental values. Next, a periodic slab containing two layers of rubrene molecules was extracted from the three-dimensional structure, with 30 Å of vacuum space placed between each slab (to prevent spurious inter-slab interactions). The molecules were then allowed to further relax within the fixed surface (slab) unit cell. The WF was determined by tracking the evolution of the potential energy of an electron at the vacuum level (E_{vac}) and the VBM, as a function of the change in the (experimentally observed) lattice parameters due to mechanical strain.

References

1. Fischetti, M. V. & Laux, S. E. Band structure, deformation potentials, and carrier mobility in strained Si, Ge, and SiGe alloys. *J. Appl. Phys.* **80,** 2234–2252 (1996).
2. Ishikawa, Y. *et al.* Strain-induced band gap shrinkage in Ge grown on Si substrate. *Appl. Phys. Lett.* **82,** 2044–2046 (2003).
3. Kuo, C. P., Vong, S. K., Cohen, R. M. & Stringfellow, G. B. Effect of mismatch strain on band gap in III-V semiconductors. *J. Appl. Phys.* **57,** 5428–5432 (1985).
4. Roberts, M. *et al.* Elastically relaxed free-standing strained-silicon nanomembranes. *Nat. Mater.* **5,** 388–393 (2006).
5. Sun, Y., Thompson, S. E. & Nishida, T. Physics of strain effects in semiconductors and metal-oxide-semiconductor field-effect transistors. *J. Appl. Phys.* **101,** 104503 (2007).
6. Lee, M., Fitzgerald, E., Bulsara, M., Currie, M. & Lochtefeld, A. Strained Si, SiGe, and Ge channels for high-mobility metal-oxide-semiconductor field-effect transistors. *J. Appl. Phys.* **97,** 011101 (2005).
7. Welser, J., Hoyt, J. L. & Gibbons, J. F. Electron mobility enhancement in strained-Si n-type metal-oxide-semiconductor field-effect transistors. *IEEE Electron. Device Lett.* **15,** 100–102 (1994).
8. Kido, J., Kimura, M. & Nagai, K. Multilayer white light-emitting organic electroluminescent device. *Science* **267,** 1332–1334 (1995).
9. Kulkarni, A., Tonzola, C., Babel, A. & Jenekhe, S. Electron transport materials for organic light-emitting diodes. *Chem. Mater.* **16,** 4556–4573 (2004).
10. Dimitrakopoulos, C. D. & Malenfant, P. Organic thin film transistors for large area electronics. *Adv. Mater.* **14,** 99–117 (2002).
11. Hains, A., Liang, Z., Woodhouse, M. & Gregg, B. Molecular semiconductors in organic photovoltaic cells. *Chem. Rev.* **110,** 6689–6735 (2010).
12. Peumans, P., Uchida, S. & Forrest, S. Efficient bulk heterojunction photovoltaic cells using small-molecular-weight organic thin films. *Nature* **425,** 158–162 (2003).
13. Giri, G. *et al.* Tuning charge transport in solution-sheared organic semiconductors using lattice strain. *Nature* **480,** 504–508 (2011).
14. Briseno, A. *et al.* High-performance organic single-crystal transistors on flexible substrates. *Adv. Mater.* **18,** 2320–2324 (2006).
15. Jedaa, A. & Halik, M. Toward strain resistant flexible organic thin film transistors. *Appl. Phys. Lett.* **95,** 103309 (2009).
16. Sekitani, T. *et al.* Bending experiment on pentacene field-effect transistors on plastic films. *Appl. Phys. Lett.* **86,** 073511 (2005).
17. Sekitani, T., Zschieschang, U., Klauk, H. & Someya, T. Flexible organic transistors and circuits with extreme bending stability. *Nat. Mater.* **9,** 1015–1022 (2010).
18. Sokolov, A., Cao, Y., Johnson, O. & Bao, Z. Mechanistic considerations of bending-strain effects within organic semiconductors on polymer dielectrics. *Adv. Func. Mater.* **22,** 175–183 (2012).
19. Yang, C. *et al.* Bending-stress-driven phase transitions in pentacene thin films for flexible organic field-effect transistors. *Appl. Phys. Lett.* **92,** 243305 (2008).
20. Cour, I. *et al.* Origin of stress and enhanced carrier transport in solution-cast organic semiconductor films. *J. Appl. Phys.* **114,** 093501 (2013).
21. Kowarik, S. *et al.* Real-time observation of structural and orientational transitions during growth of organic thin films. *Phys. Rev. Lett.* **96,** 125504 (2006).
22. Mativetsky, J., Burke, S., Fostner, S. & Grutter, P. Templated growth of 3,4,9,10-perylenetetracarboxylic dianhydride molecules on a nanostructured insulator. *Nanotechnology* **18,** 105303 (2007).
23. Murakami, Y. *et al.* Microstructural study of the polymorphic transformation in pentacene thin films. *Phys. Rev. Lett.* **103,** 146102 (2009).
24. Okada, Y., Sakai, K., Uemura, T., Nakazawa, Y. & Takeya, J. Charge transport and Hall effect in rubrene single-crystal transistors under high pressure. *Phys. Rev. B* **84,** 245308 (2011).
25. Schatschneider, B., Monaco, S., Tkatchenko, A. & Liang, J. J. Understanding the structure and electronic properties of molecular crystals under pressure: application of dispersion corrected DFT to oligoacenes. *J. Phys. Chem. A* **117,** 8323–8331 (2013).
26. Podzorov, V. *et al.* Intrinsic charge transport on the surface of organic semiconductors. *Phys. Rev. Lett.* **93,** 086602 (2004).
27. Henn, D. E., Williams, W. G. & Gibbons, D. J. Crystallographic data for an orthorhombic form of rubrene. *J. Appl. Cryst.* **4,** 256 (1971).
28. Jurchescu, O. D., Meetsma, A. & Palstra, T. T. M. Low-temperature structure of rubrene single crystals grown by vapor transport. *Acta Cryst* **B62,** 330–334 (2006).
29. Haas, S. *Crystal structure analysis and trap spectroscopy in organic semiconducting crystals.* PhD Doctoral and Habilitation Theses, ETH, Zürich **16657,** 28–33 (2006).
30. Palermo, V., Palma, M. & Samorì, P. Electronic characterization of organic thin films by kelvin probe force microscopy. *Adv. Mater.* **18,** 145–164 (2006).
31. Ellison, D. J., Lee, B., Podzorov, V. & Frisbie, C. D. Surface potential mapping of SAM-functionalized organic semiconductors by Kelvin probe force microscopy. *Adv. Mater.* **23,** 502–507 (2011).
32. Wu, Y., Haugstad, G. & Frisbie, C. D. Electronic polarization at pentacene/polymer dielectric interfaces: imaging surface potentials and contact potential differences as function of substrate type, growth temperature, and pentacene microstructure. *J. Phys. Chem. C* **118,** 2487–2497 (2014).
33. Lee, B., Choi, T., Cheong, S. & Podzorov, V. Nanoscale conducting channels at the surface of organic semiconductors formed by decoration of molecular steps with self-assembled molecules. *Adv. Func. Mater.* **19,** 3726–3730 (2009).
34. Reyes-Martinez, M., Ramasubramaniam, A., Briseno, A. & Crosby, A. The intrinsic mechanical properties of rubrene single crystals. *Adv. Mater.* **24,** 5548–5552 (2012).
35. Ibragimov, K. h. I. & Korol'kov, V. A. Temperature dependence of the work function of metals and binary alloys. *Inorg. Mater.* **37,** 567–572 (2001).
36. Leu, P. W., Svizhenko, A. & Cho, K. *Ab initio* calculations of the mechanical and electronic properties of strained Si nanowires. *Phys. Rev. B* **77,** 235305 (2008).
37. Unal, K. & Wickramasinghe, H. K. Nanoscale quantitative stress mapping with atomic force microscopy. *Appl. Phys. Lett.* **90,** 113111 (2007).
38. Choi, S.-M., Jhi, S.-H. & Son, Y.-W. Effects of strain on electronic properties of graphene. *Phys. Rev. B* **81,** 081407 (2010).
39. Peng, X., Tang, F. & Copple, A. Engineering the work function of armchair graphene nanoribbons using strain and functional species: a first principles study. *J. Phys. Condens. Mat.* **24,** 075501 (2012).
40. Robinson, P. M. & Scott, H. G. Plastic deformation of anthracene single crystals. *Acta Metall.* **15,** 1581–1590 (1967).
41. Li, W. & Li, D. Y. *In situ* measurements of simultaneous electronic behavior of Cu and Al induced by mechanical deformation. *J. Appl. Phys.* **99,** 073502 (2006).
42. Laudise, R. A., Kloc, C., Simpkins, P. G. & Siegrist, T. Physical vapor growth of organic semiconductors. *J. Cryst. Growth* **187,** 449–454 (1998).
43. Kresse, G. & Furthmüller, J. Efficiency of ab-initio total energy calculations for metals and semiconductors using a plane-wave basis set. *Comput. Mater. Sci.* **6,** 15–50 (1996).
44. Blöchl, P. E. Projector augmented-wave method. *Phys. Rev. B* **50,** 17953–17979 (1994).
45. Kresse, G. & Joubert, D. From ultrasoft pseudopotentials to the projector augmented-wave method. *Phys. Rev. B* **59,** 1758–1775 (1999).

Acknowledgements

This work was primarily supported by the National Science Foundation under Grant No. DMR-0706011. Part of this work was carried out in the Characterization Facility, University of Minnesota, which received partial support from NSF through the MRSEC program under Grant No. DMR-1420013. SKPM measurements in this work were conducted at the Center for Nanophase Materials Sciences, which is a DOE Office of Science User Facility. Part of this work was performed at the Stanford Nano Shared Facilities (SNSF). We acknowledge assistance for crystal growth by Dr Wei Xie and Xinglong Ren, and thank them as well as Dr Christopher Sutton for helpful discussions.

Author contributions

Y.W. and C.D.F. conceived the idea and initiated the project. Y.W. and G.A.R. performed the SKPM measurements with the help of G.H., A.B. and S.V.K. A.R.C. and A.S. executed the X-ray diffraction measurements and contributed to the interpretation of data. G.S., H.L., C.R. and J.-L.B. performed the DFT calculations and provided helpful discussions. Y.W. and C.D.F. wrote the manuscript in collaboration with all the authors.

Additional information

Competing financial interests: The authors declare no competing financial interests.

12

Common electronic origin of superconductivity in (Li,Fe)OHFeSe bulk superconductor and single-layer FeSe/$SrTiO_3$ films

Lin Zhao[1,*], Aiji Liang[1,*], Dongna Yuan[1,*], Yong Hu[1,*], Defa Liu[1], Jianwei Huang[1], Shaolong He[1], Bing Shen[1], Yu Xu[1], Xu Liu[1], Li Yu[1], Guodong Liu[1], Huaxue Zhou[1], Yulong Huang[1], Xiaoli Dong[1], Fang Zhou[1], Kai Liu[2], Zhongyi Lu[2], Zhongxian Zhao[1,3], Chuangtian Chen[4], Zuyan Xu[4] & X. J. Zhou[1,3]

The mechanism of high-temperature superconductivity in the iron-based superconductors remains an outstanding issue in condensed matter physics. The electronic structure plays an essential role in dictating superconductivity. Recent revelation of distinct electronic structure and high-temperature superconductivity in the single-layer FeSe/$SrTiO_3$ films provides key information on the role of Fermi surface topology and interface in inducing or enhancing superconductivity. Here we report high-resolution angle-resolved photoemission measurements on the electronic structure and superconducting gap of an FeSe-based superconductor, $(Li_{0.84}Fe_{0.16})OHFe_{0.98}Se$, with a T_c at 41 K. We find that this single-phase bulk superconductor shows remarkably similar electronic behaviours to that of the superconducting single-layer FeSe/$SrTiO_3$ films in terms of Fermi surface topology, band structure and the gap symmetry. These observations provide new insights in understanding high-temperature superconductivity in the single-layer FeSe/$SrTiO_3$ films and the mechanism of superconductivity in the bulk iron-based superconductors.

[1] National Lab for Superconductivity, Beijing National Laboratory for Condensed Matter Physics, Institute of Physics, Chinese Academy of Sciences, Beijing 100190, China. [2] Department of Physics and Beijing Key Laboratory of Opto-electronic Functional Materials and Micro-nano Devices, Renmin University of China, Beijing 100872, China. [3] Collaborative Innovation Center of Quantum Matter, Beijing 100871, China. [4] Technical Institute of Physics and Chemistry, Chinese Academy of Sciences, Beijing 100190, China. * These authors contributed equally to the present work. Correspondence and requests for materials should be addressed to X.J.Z. (email: XJZhou@aphy.iphy.ac.cn) or to X.L.D. (email: dong@aphy.iphy.ac.cn).

Iron-based superconductors[1–3] represent the second class of high-temperature superconductors after the first discovery of high-temperature superconductivity in cuprates[4]. The electronic structure of the iron-based compounds usually consists of hole-like bands near the Brillouin zone centre and electron-like bands near the Brillouin zone corners, leading to proposals that the electron scattering between the hole pockets around the Brillouin zone centre and the electron pockets around the Brillouin zone corners is a viable mechanism for electron pairing in the iron-based superconductors[5–11]. With the discovery of the $A_xFe_{2-y}Se_2$ (A = K, Cs, Rb, Tl and so on) superconductors[3,12–14], such a Fermi surface nesting picture is challenged, because the superconducting phase in $A_xFe_{2-y}Se_2$ superconductors does not contain hole pockets near the Brillouin zone centre[3,14–18]. The latest discovery of possible high-temperature superconductivity above 65 K in the single-layer FeSe film grown on the $SrTiO_3$ substrate (denoted as FeSe/$SrTiO_3$ film hereafter) puts even stronger constraints on the Fermi surface nesting picture, because there is no Fermi surface present near the zone centre at all[19–23]. Although the implications of the Fermi surface topology in the $A_xFe_{2-y}Se_2$ superconductors and superconducting single-layer FeSe/$SrTiO_3$ films are significant, the material-specific complications make the conclusion on the role of Fermi surface topology ambiguous. The $A_xFe_{2-y}Se_2$ superconductors are well-known to have problems of inhomogeneity, phase separation and small superconducting volume fraction. The exact nature of the superconducting phase remains unclear and its coexistence with the insulating phase suggests that interface may play an important role in the superconductivity[24–27]. There are strong indications that interface in the single-layer FeSe/$SrTiO_3$ films plays an important role in giving rise to high-temperature superconductivity[19–23,28–34]. With the complications of the interface effect and phase separation problem involved in $A_xFe_{2-y}Se_2$ superconductors and superconducting single-layer FeSe/$SrTiO_3$ films, it becomes less straightforward to conclude whether the same effect of Fermi surface topology on superconductivity is operative in all the iron-based superconductors.

In this study, we report high-resolution angle-resolved photoemission (ARPES) measurements on a new FeSe-based superconductor, $(Li_{1-x}Fe_x)OHFe_{1-y}Se$, with a superconducting critical temperature at 41 K (refs 35–40). We find that this superconductor contains electron pocket(s) near the Brillouin zone conners without hole pockets near the zone centre. The superconducting gap around the electron-like Fermi surface near the zone corner is nearly isotropic without nodes and its temperature dependence follows the usual Bardeen-Cooper-Schrieffer form. These behaviours are strikingly similar to that of the superconducting single-layer FeSe/$SrTiO_3$ films. This is a single-phase bulk superconductor with a relatively high $T_c \sim 41$ K, free from the complication of phase separation and interface effect, which shows electron-like Fermi pockets only. Our observations provide insights on the superconductivity mechanism in the iron-based bulk superconductors and the origin of high-temperature superconductivity in the single-layer FeSe/$SrTiO_3$ films.

Results

Fermi surface and band structure. Figure 1a shows the crystal structure of the new FeSe-based superconductor: $(Li_{1-x}Fe_x)OHFe_{1-y}Se$ (refs 36,39). It consists of FeSe layers

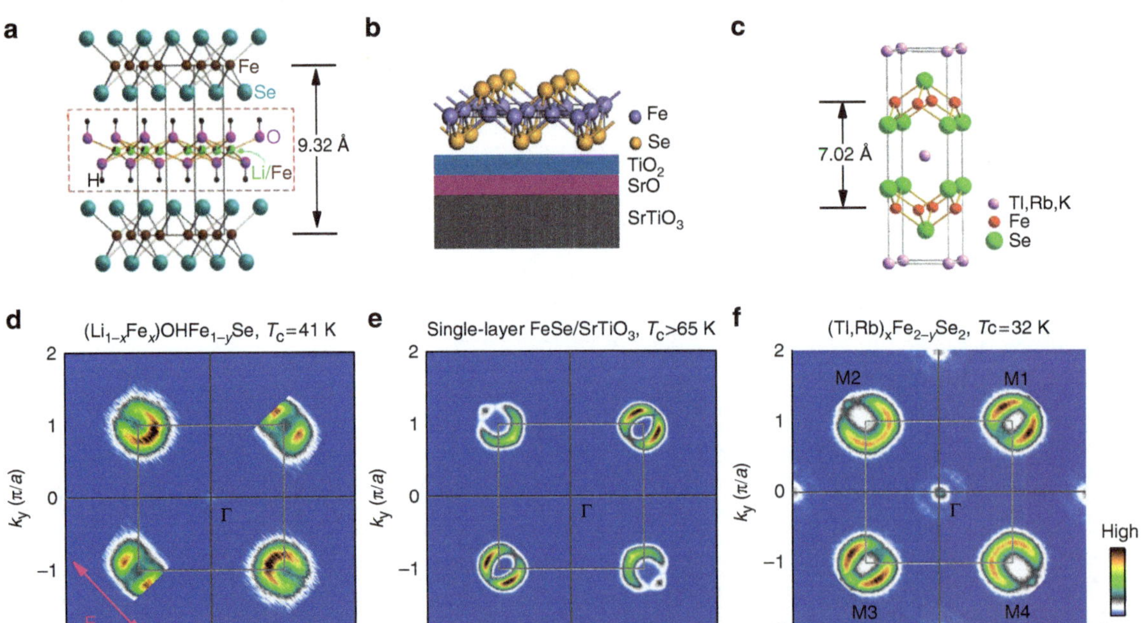

Figure 1 | Crystal structure and Fermi surface of three FeSe-based superconductors. (**a**) The crystal structure of $(Li_{1-x}Fe_x)OHFe_{1-y}Se$. The *c* lattice constant is 9.318 Å for optimally doped sample with a T_c at 42 K (refs 36,39). (**b**) Schematic structure of single-layer FeSe film deposited on a $SrTiO_3$ substrate[19]. The thickness of the FeSe layer is ~5.5 Å. (**c**) Schematic structure of $(Tl,Rb)_xFe_{2-y}Se_2$ (ref. 12). The distance of the adjacent FeSe layers is ~7.02 Å. (**d**) Fermi surface mapping of $(Li_{0.84}Fe_{0.16})OHFe_{0.98}Se$ superconductor measured at 20 K. The spectral weight distribution was obtained by integrating the measured photoemission spectra (EDCs) within [−10, +10] meV energy window with respect to the Fermi level as a function of k_x and k_y. (**e**,**f**) Fermi surface of a superconducting single-layer FeSe/$SrTiO_3$ film[20] and $(Tl,Rb)_xFe_{2-y}Se_2$ (ref. 17). For convenience, the four equivalent M points are labelled as M1(π,π), M2($-\pi$,π), M3($-\pi$, $-\pi$) and M4(π, $-\pi$). The light-grey arrow in **d** marks the main electric field direction on the sample surface from the light source. In **d**-**f**, the data around M1 and M4 are obtained from the data around M3 and M2, respectively, by applying the inversion symmetry.

sandwiched in between the (Li,Fe)OH layers along the c direction. The in-plane lattice constant a or b is 3.78 Å, which is close to that (3.76 Å) in the bulk FeSe superconductor ($T_c \sim 8.5$ K)[41], but the distance of the two adjacent FeSe layers in $(Li_{1-x}Fe_x)OHFe_{1-y}Se$ (9.32 Å) is much larger than that in bulk FeSe (5.5 Å)[41]. This indicates a weak interaction between the two adjacent FeSe layers in $(Li_{1-x}Fe_x)OHFe_{1-y}Se$ and its enhanced two-dimensional nature of the electronic structure compared with bulk FeSe. In comparison, the single-layer FeSe/$SrTiO_3$ film consists of a strictly two-dimensional Se-Fe-Se layer on the top of the $SrTiO_3$ substrate; the thickness of the FeSe layer is 5.5 Å (Fig. 1b)[19]. The distance between the two adjacent FeSe layers in $A_xFe_{2-y}Se_2$ is 7.02 Å; however, there is an in-plane displacement between the two adjacent FeSe layers (Fig. 1c)[12]. The latest success of growing high-quality large-sized $(Li_{1-x}Fe_x)OHFe_{1-y}Se$ single crystals[39] makes it possible to carry out ARPES measurements on the new superconductors. The sample we have measured in this study has a composition of $(Li_{0.84}Fe_{0.16})OHFe_{0.98}Se$ (abbreviated as FeSe11111 hereafter)[39] with a superconducting transition temperature $T_c \sim 41$ K (Fig. 3d).

Figure 1d shows the Fermi surface mapping of FeSe11111 measured at 20 K. The spectral weight at each momentum is obtained by integrating the EDC (energy distribution curve) spectral weight within [−10, +10] meV energy window with respect to the Fermi level (E_F). The measured Fermi surface contains a nearly circular electron-like Fermi surface around M points (Fig. 1d). There is no indication of Fermi crossing around the Γ (0,0) point, except that there is a very weak residual signal near the Fermi level. Such a Fermi surface topology is very similar to that found in the superconducting single-layer FeSe/$SrTiO_3$ film (Fig. 1e) with a superconducting temperature at ~55 K (ref. 20). In comparison, there are electron pockets around the Brillouin centre in the $A_xFe_{2-y}Se_2$ superconductor (Fig. 1f)[17,18]. The area of the electron pocket near M gives a good measure of the electron doping level for the strongly two-dimensional systems. Considering that the Fermi surface around M consists of two degenerate Fermi surface sheets, the estimated electron counting in FeSe11111 (Fig. 1d) is ~0.08 electrons per Fe. The electron counting of ~0.08 electrons per Fe is smaller than 0.10 electrons per Fe in single-layer FeSe film (Fig. 1e) with a $T_c \sim 55$ K (ref. 20) and 0.12 electrons per Fe in vacuum-annealed single-layer FeSe/$SrTiO_3$ film with a $T_c \sim 65$ K (ref. 21). We note that the electron counting in both FeSe11111 and superconducting single-layer FeSe/$SrTiO_3$ film is much smaller than 0.18 electrons per Fe in $A_xFe_{2-y}Se_2$ superconductor (Fig. 1f)[15–18] when only considering the M point electron Fermi surface sheet.

FeSe11111 consists of FeSe layers and (Li,Fe)OH layers; therefore, the natural cleavage surface can be either FeSe surface or (Li,Fe)OH surface. By simple electron counting, both the FeSe surface and the LiOH surface are charge neutral. Substitution in the (Li,Fe)OH layers or off-stoichiometry in the FeSe layers would lead to carrier doping in the FeSe layers. In this case, no charge self-doping is expected for this sample. The composition of $(Li_{0.84}Fe_{0.16})OHFe_{0.98}Se$ is obtained by structure refinement of the single-crystal X-ray diffraction data[39]. By simple electron counting, the electron doping level would be 0.12 electrons per Fe. This is slightly larger than the ~0.08 electrons per Fe we obtained from the Fermi surface measurement. We think this difference may be caused by the composition variation among different samples[38] or uncertainty in the composition determination. For example, neutron scattering on related sample with comparable T_c gives a composition of $(Li_{0.8}Fe_{0.2})OHFeSe$ (ref. 36). It is also possible that

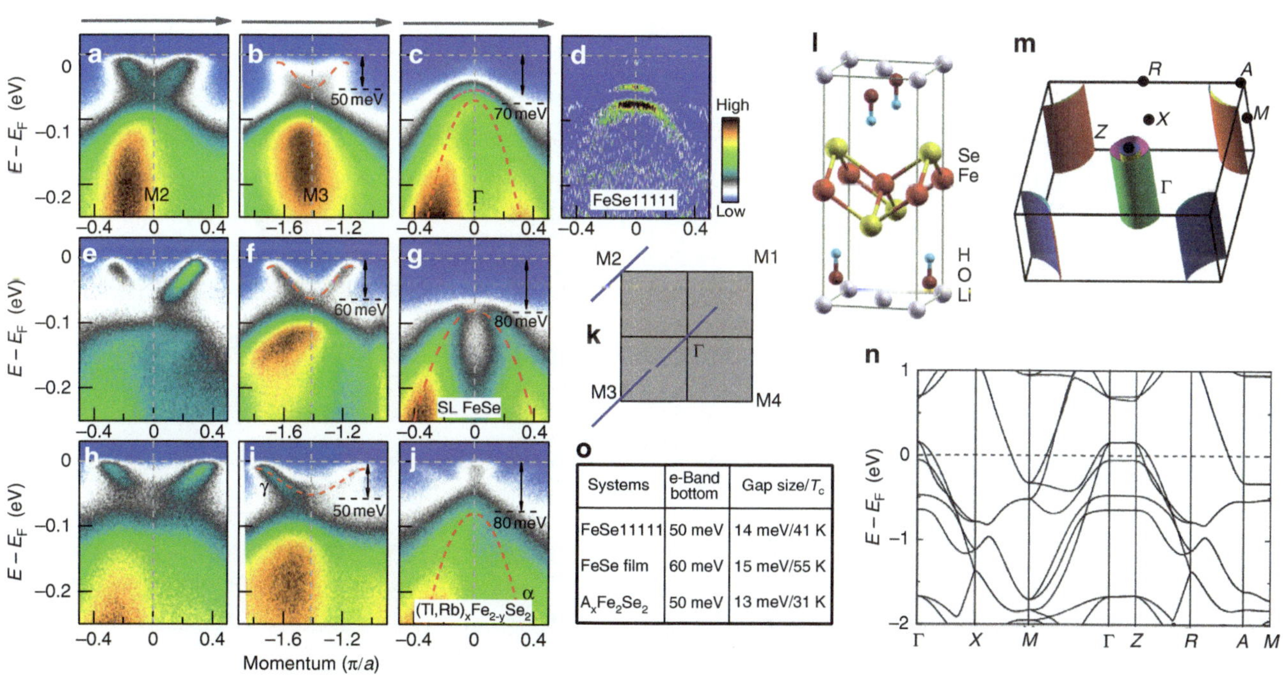

Figure 2 | Band structure and photoemission spectra of $(Li_{0.84}Fe_{0.16})OHFe_{0.98}Se$ superconductor measured along high-symmetry momentum cuts. (**a-c**) Photoemission images of $(Li_{0.84}Fe_{0.16})OHFe_{0.98}Se$ measured at a temperature of 20 K along three high-symmetry cuts, Γ cut, M2 cut and M3 cut, respectively. (**d**) The second-derivative image of **c** with respect to the energy, to reveal the band structure better. For comparison, **e-j** show photoemission images of a superconducting single-layer FeSe/$SrTiO_3$ film and $(Tl,Rb)_xFe_{2-y}Se_2$ measured along the same Γ cut, M2 cut and M3 cut, respectively. The location of the three momentum cuts are shown in **k**. The red dashed lines in **a-j** images are the guide to the eyes for the observed band structures. (**l-n**) DFT calculation results. (**l**) The crystal structure for the DFT calculation. (**m**) The Fermi surface and band structures along high symmetry directions are shown in **n**. In **o**, the table lists the bottom position of the electron band near M, the superconducting gap size and corresponding T_c for these three systems. The gap size for FeSe11111 will be obtained in Fig. 3e.

the electrons in the (Li,Fe)OH layers may not be fully transferred into the FeSe layers.

Figure 2a–c shows the band structure of FeSe11111 measured across three momentum cuts together with the density functional theory (DFT) calculated electronic structures ($l-n$). For comparison, the band structure measured on a superconducting single-layer FeSe/$SrTiO_3$ film[20] (Fig. 3e–g) and on $(Tl,Rb)_xFe_{2-y}Se_2$ superconductor[17] (Fig. 3h-j) along the same three momentum cuts are also presented. Overall, the band structure of FeSe11111 is very similar to that of the superconducting single-layer FeSe/$SrTiO_3$ film with slight quantitative difference in the band position. Near the Γ point, two hole-like bands can be resolved for FeSe11111 (Fig. 2c) that are well below the Fermi level, as seen more clearly in the corresponding second-derivative image (Fig. 2d). We note that there is a very weak signal near the Fermi level at Γ point, which may be due to the leakage from a band above the Fermi level. Similar multiple hole-like bands have also been observed near Γ in the superconducting single-layer FeSe/$SrTiO_3$ films[21] and in $K_xFe_{2-y}Se_2$ superconductor[18], with slight variation in the band top position. In comparison, the electronic structure of $(Tl,Rb)_xFe_{2-y}Se_2$ superconductor shows two electron-like bands crossing the Fermi level (Fig. 2j)[17]. Around M2 and M3 points, we observed in FeSe11111 parabolic electron-like bands crossing the Fermi level (Fig. 2a,b), forming the electron pocket around M. These are very similar to those observed in the superconducting single-layer FeSe/$SrTiO_3$ film (Fig. 2e,f). Careful examination indicates that the Fermi momentum (k_F) for the electron pocket of FeSe11111 is $\sim 0.22\ \pi/a$ (lattice constant $a = 3.78$ Å) (Fig. 2a,b), which is slightly smaller than 0.25 π/a in the superconducting single-layer FeSe/$SrTiO_3$ film (Fig. 2e,f)[20], both being significantly smaller than 0.35 π/a in $(Tl,Rb)_xFe_{2-y}Se_2$ superconductor (Fig. 2h,i)[17]. As a summary in Fig. 2o, the bottom of the electron-like band in FeSe11111 is $\sim$50 meV below the Fermi level(Fig. 2b), slightly shallower than the 60 meV in the superconducting single-layer FeSe/$SrTiO_3$ film (Fig. 2f)[20] but similar to the $\sim$50 meV in the $(Tl,Rb)_xFe_{2-y}Se_2$ superconductor (Fig. 2i)[17]. The combined information of the Fermi momentum and the band width gives a good measure of the electron effective mass that is 2.9, 2.7 and 6.1 m_e (m_e is free electron mass) for FeSe11111, superconducting single-layer FeSe/$SrTiO_3$ film and the $(Tl,Rb)_xFe_{2-y}Se_2$ superconductor, respectively[17,20]. It is interesting to note that electron effective mass for the electron-like band near M is similar for FeSe11111 and superconducting single-layer FeSe/$SrTiO_3$ film, but both are much smaller than that in $(Tl,Rb)_xFe_{2-y}Se_2$ superconductor[17].

To further understand the measured electronic structure, the first-principles electronic structure calculations were carried out with the projector augmented wave method[42,43] as implemented in the VASP software[44–46] with a electron doping 0.1e per Fe in the non-magnetic state, which is close to our sample state. The crystal structure (Fig. 2l) and lattice constant are from ref. 36. Overall, the calculated Fermi surface (Fig. 2m) shows similar result with that of the bulk FeSe (ref. 47) but with

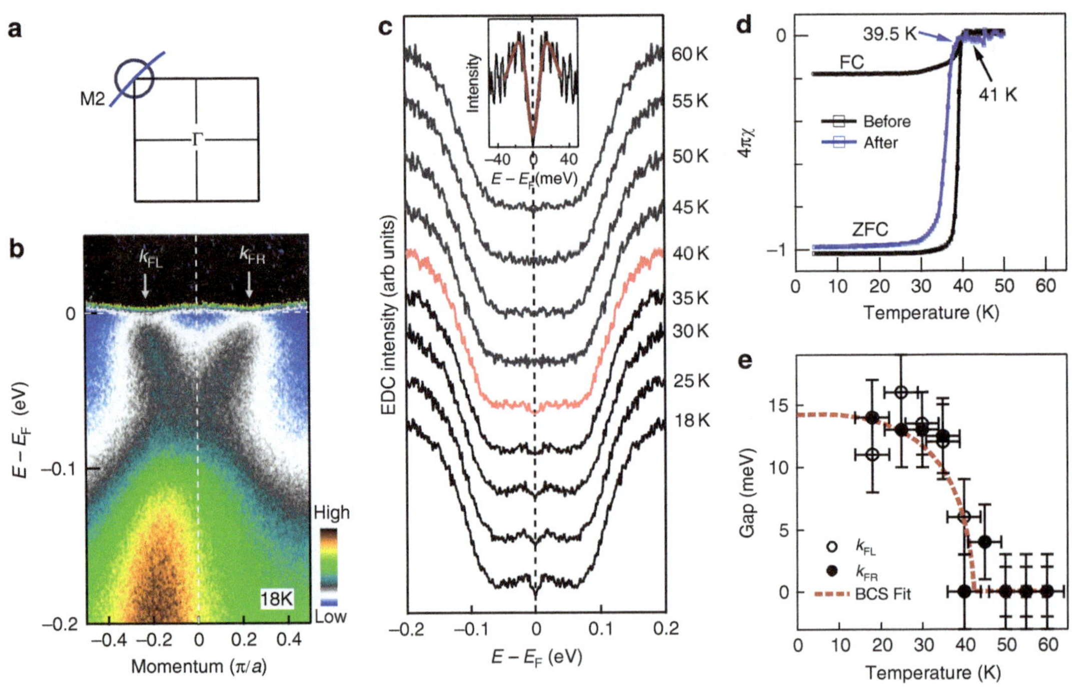

Figure 3 | Temperature dependence of the superconducting gap of the $(Li_{0.84}Fe_{0.16})OHFe_{0.98}Se$ superconductor. (**a**) Schematically, the electron-like Fermi surface near M2 point and the location of the momentum cut (blue line) across the Fermi surface. (**b**) Photoemission image measured at 18 K along the momentum cut near the M2 point (blue line in **a**). The data are divided by the corresponding Fermi distribution function, to highlight the opening of an energy gap around the Fermi level (the original data can be found in Fig. 2a). Two Fermi momenta are marked by arrows as k_{FL} and k_{FR}. (**c**). Symmetrized photoemission spectra (EDCs) at the Fermi momentum k_{FL} measured at different temperatures. The middle-top inset shows the expanded region of the 18 K symmetrized EDC around the Fermi level and the fitting result (red line) using the formula from ref. 50 to get the gap size. (**d**) Magnetic measurement of the superconducting transition temperature (T_c) for the $(Li_{0.84}Fe_{0.16})OHFe_{0.98}Se$ sample we measured. Both field-cooled (FC) and zero-filed-cooled (ZFC) mode measurements show an onset T_c at $\sim$41 K with a sharp transition width of $\sim$1.5 K. (**e**) Temperature dependence of the measured superconducting gap. The gap size is obtained by picking up the peak position from the symmetrized EDCs (**c**) or by fitting the near-E_F symmetrized EDCs using the formula from ref. 50; both approaches give similar results within experimental uncertainty. The error bars are defined as s.d. The gap size from EDCs of both the left Fermi momentum (k_{FL}) and the right Fermi momentum (k_{FR}) are plotted. The red dashed line is a BCS gap form with a gap size of $\sim$14 meV at 0 temperature.

enhanced two-dimensional characteristics, consistent with the larger spacing between the two adjacent FeSe layers in (Li,Fe)OHFeSe. Extra bands around −2.5 and −1.0 eV are present that are from (Li,Fe)OH layers, which are too deep to give contribution around the Fermi level. It is apparent that there is an obvious discrepancy between the band structure calculations and our measurements. For example, the hole pockets near Γ point in the calculations are absent in the measured results. This situation is similar to that of the single-layer FeSe/$SrTiO_3$ films and other FeSe-based materials[23]. More theoretical efforts are needed to understand the experimental results.

The natural cleavage surface of FeSe11111 consists of both the FeSe surface and the (Li,Fe)OH surface; therefore, our ARPES signal should come from both surfaces. We believe our ARPES signal is dominated by the FeSe layers because of the following reasons: (1) our density functional theory band structure calculations (Fig. 2) indicate that the bands from (Li,Fe)OH layers lie ~2.5–1.0 eV below the Fermi level. Thus, the low-energy electronic states near the Fermi level are mainly from the FeSe layers. This is consistent with the electronic structure similarity between (Li,Fe)OHFeSe and the single-layer FeSe/$SrTiO_3$ that has no (Li,Fe)OH layers. (2) Recent scanning tunneling microscope (STM) study on cleaved (Li,Fe)OHFeSe also indicated that the low-energy signal is dominated by the FeSe cleavage surface[48,49]. We note that STM measurement on the (Li,Fe)OH layer reveals that there is a tiny electron pocket around Γ (ref. 49). As we mentioned above, we also observe a very weak signal near the Fermi level at Γ point (Figs 1d and 2c). Whether this weak signal near Γ is due to the tiny electron pocket identified from STM[49] needs further investigation.

Superconducting gap. The clear identification of electron-like Fermi surface sheet near M point makes it possible to investigate the superconducting gap in this new FeSe11111 superconductor. We start by examining the temperature dependence of the superconducting gap. Figure 3b shows the original photoemission image measured at a low temperature of 18 K along the momentum cut (its location is shown in Fig. 3a) crossing the electron-like Fermi surface around M2. The data are divided by the corresponding Fermi distribution function; the spectral weight suppression near the Fermi level indicates the opening of the superconducting gap (Fig. 3b). Following the procedure commonly used in the study of high-temperature cuprate superconductors[50], the symmetrized photoemission spectra (EDCs) measured on one Fermi momentum at different temperatures are shown in Fig. 3c. There is a clear gap opening at low temperatures, as indicated by a dip at the Fermi level in the symmetrized EDCs (Fig. 3c). With increasing temperature, the dip at E_F is gradually filled up and becomes almost invisible above 40 K. We used two methods to extract the gap size in the superconducting state. The first method is to pick the peak position in the symmetrized EDCs, while the second one is to fit the symmetrized EDCs with the phenomenological formula proposed by Norman *et al.*[50]. The top-middle inset of Fig. 3c shows the symmetrized EDC at 18 K and the fitting result. Both methods give consistent gap size within an experimental uncertainty of 2 meV. Figure 3e shows the measured gap size at different temperatures from symmetrized EDCs for both Fermi momenta. The gap size basically follows a Bardeen-Cooper-Schrieffer form with a T_c around (42 ± 2) K and a gap size of ~14 meV at 0 temperature (Fig. 3e). This is in excellent agreement with the T_c~41 K from direct magnetic measurement (Fig. 3d). We note that FeSe11111 is very sensitive to air and processing temperature. To make sure there is no change on the superconductivity of our sample during the ARPES measurements, we carried out magnetic measurements on the measured sample before and after the ARPES experiment. Figure 3d shows the magnetic measurement before ARPES measurement; both the field-cooled and zero-field-cooled modes give similar magnetic onset T_c~41 K with a sharp transition of ~1.5 K. The sharp transition and its 100% diamagnetic shielding demonstrate the high quality of the measured single crystal. After ARPES experiment (Fig. 3d), the magnetic measurement on the cleaved sample shows an onset T_c~40 K with a sharp transition of ~1.5 K. These indicate that there is negligible superconductivity degradation on the sample during the sample handling and ARPES measurements.

Now we come to the momentum dependence of the superconducting gap in the FeSe11111 superconductor. For this purpose, we took high-resolution Fermi surface mapping (energy resolution of 4 meV) of the electron pocket around M2 at 20 K in the superconducting state (Fig. 4a). The corresponding Fermi momenta are identified, marked by red empty circles and labelled by numbers from 1 to 16 in Fig. 4a. Figure 4b shows symmetrized photoemission spectra (EDCs) at the Fermi momenta around the Fermi surface. The extracted superconducting gap is plotted in Fig. 4c. The superconducting gap is nearly isotropic with a gap size at (13 ± 2) meV; there is no gap node (zero gap) around the Fermi surface. For comparison, the gap size around the electron-like Fermi surface at M2 point for the superconducting single-layer FeSe/$SrTiO_3$ film[20] is also plotted (Fig. 4c). It is also nearly isotropic without gap node, although its magnitude (~15 meV) is slightly larger than that in FeSe11111. In FeSe11111, the bonding between the FeSe layer and the adjacent (Li,Fe)OH layer is rather

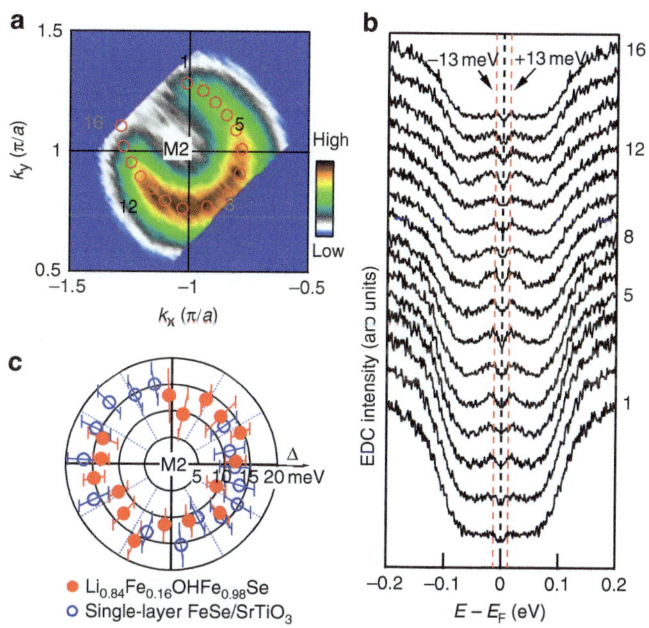

Figure 4 | Momentum dependence of the superconducting gap of ($Li_{0.84}Fe_{0.16}$)$OHFe_{0.98}$Se superconductor. (**a**) Fermi surface mapping near M2 point. The corresponding Fermi crossings are marked by red empty circles with numbers from 1 to 16 (for clarity, the intermittent points are not labelled). (**b**) Symmetrized EDCs measured at 20 K from Fermi crossing 1 to 16 along the measured electron-like Fermi surface around M2. The red dashed lines represent the energy positions of ±13 meV with respect to the Fermi level. (**c**) Momentum dependence of the superconducting gap (red solid circles) along the Fermi surface around M2 for ($Li_{0.84}Fe_{0.16}$)$OHFe_{0.98}$Se superconductor. For comparison, the gap along the electron-like Fermi surface around M2 of the superconducting single-layer FeSe film is also plotted (blue empty circles)[20]. The error bars are defined as s.d.

weak (van der Waals bonding). The separation between the two adjacent FeSe layers is large and their mutual interaction is expected to be rather weak. These make its behaviours close to the single-layer FeSe film that shows strong two dimensionality. It is thus reasonable to conclude that there is no nodes of the superconducting gap in the FeSe11111 superconductor.

Discussion

From the above results, we have found that the FeSe11111 superconductor exhibits similar behaviours to the superconducting single-layer FeSe/$SrTiO_3$ film, in terms of the Fermi surface topology, band structure and the superconducting gap symmetry. Among all the iron-based superconductors discovered so far, the majority of them show common electronic structure with hole-like Fermi surface sheets around the Γ point and electron-like Fermi surface near the M point[1]. The $A_xFe_{2-y}Se_2$ superconductor[3,14–18] and the superconducting single-layer FeSe/$SrTiO_3$ films[20–23] have been the only two exceptions that have distinct Fermi surface topology, that is, no hole-like Fermi surface at the Γ point and there are only electron pockets. The present work has established the FeSe11111 superconductor as a new third exception on the list. It has been proposed that the electron scattering between the hole-like bands near Γ and electron-like bands near M is the cause of electron pairing in the iron-based superconductors[5–11]. The discovery of only electron-like Fermi surface in the $A_xFe_{2-y}Se_2$ superconductor and the superconducting single-layer FeSe/$SrTiO_3$ films challenges such a Fermi surface nesting scenario. However, some material-specific properties prevent us from making a decisive conclusion. The $A_xFe_{2-y}Se_2$ superconductor is well known for its phase separation problem; the superconducting phase occupies only a small fraction of the entire sample volume and the nature of the superconducting phase remains unclear[3,14,24–27]. The coexistence of the superconducting phase and the insulating 245 phase makes people wonder whether the interface between them is responsible for the observed superconductivity. On the other hand, the surprising discovery of high-temperature superconductivity in the single-layer FeSe/$SrTiO_3$ films indicates that interface may play a significant role in inducing or enhancing superconductivity[19–23,28–34]. In this sense, the complications of phase separation and interface effect hinder the direct comparison of the $A_xFe_{2-y}Se_2$ superconductors and superconducting single-layer FeSe/$SrTiO_3$ films with most other iron-based superconductors. The newly discovered FeSe11111 superconductor is advantageous in that it is a single-phase bulk superconductor, free from complications of phase separation and interface effect, thus making it possible to compare directly with other iron-based bulk superconductors. The observation of high-temperature superconductivity with a T_c at 41 K in bulk FeSe11111 superconductor that has electron-pocket-only Fermi surface topology makes the strongest and most decisive case that the hole-like Fermi surface around Γ is not necessary for the superconductivity in the iron-based superconductors. This rules out Fermi surface nesting scenario as the common pairing mechanism for the iron-based superconductors and points to the significant role of the electron pockets near M in governing high-temperature superconductivity in the iron-based superconductors.

The present work also has important implications on the high-temperature superconductivity in the single-layer FeSe/$SrTiO_3$ films. First, our results indicate that the electron-pocket-near-M-only Fermi surface topology is not confined to the superconducting single-layer FeSe/$SrTiO_3$ films only; instead, it can also be present in bulk superconductors. Our experimental results have shown that this form of Fermi surface topology is favourable for achieving high-temperature superconductivity in the iron-based superconductors. Second, it is remarkable that T_c as high as 41 K can be achieved in the bulk FeSe11111 superconductor where there is no strain effect, no interface effect and no strong electron–phonon coupling effect[29] involved as in the superconducting single-layer FeSe/$SrTiO_3$ films. It is known from the superconducting single-layer FeSe/$SrTiO_3$ films that T_c can be tuned by varying the electron doping level[21]. Therefore, it is possible that $T_c = 41$ K is lower in FeSe11111 than that in the superconducting single-layer FeSe films (~65 K or higher), because the electron doping level in FeSe11111 is low. This raises a big possibility that T_c of the FeSe11111 superconductor may be further increased if higher electron doping is made possible and the superconductor–insulator transition[40] can be avoided. The high T_c and its possible enhancement with increasing doping in the FeSe11111 superconductor provide key insight on understanding the superconductivity in the single-layer FeSe/$SrTiO_3$ films. Third, we note that the in-plane lattice constant a or b of the FeSe11111 superconductor (3.78 Å)[36,39] is comparable to the primary bulk FeSe superconductor (3.76 Å), but significantly smaller than the single-layer FeSe film epitaxially grown on the $SrTiO_3$ substrate (3.905 Å). Considering high $T_c = 41$ K and its possible enhancement on further elecron doping in the FeSe11111 superconductor, this indicates that the interface strain does not play primary role in determining T_c. On the other hand, the distance between the two adjacent FeSe layers seems to have strong correlation with superconductivity in the FeSe-based superconductors. As the distance changes from 5.5 Å for bulk FeSe to 9.32 Å in the FeSe11111 superconductor, and to infinity in the single-layer FeSe/$SrTiO_3$ films, T_c increases from 8.5 K to 41 K, to 65 K or higher (Fig. 1). In the meantime, the top of the hole-like band at Γ changes its position from above the Fermi level for bulk FeSe (ref. 51) to 50~70 meV below the Fermi level for FeSe11111, to 80 meV below the Fermi level for the superconducting single-layer FeSe films (Fig. 2). These results again indicate that suppression of the hole-like bands near the Γ point is favourable for achieving high-temperature superconductivity in the iron-based superconductors.

In summary, by performing high-resolution ARPES measurements on the FeSe-based superconductor, FeSe11111, we have discovered that it exhibits strikingly similar electronic behaviours to the superconducting single-layer FeSe/$SrTiO_3$ film, in terms of the Fermi surface topology, band structure and the superconducting gap symmetry. This is the first single-phase bulk superconductor with a relatively high $T_c \sim 41$ K that shows electron-like Fermi-pockets-only Fermi surface topology that offers a direct comparison with other iron-based bulk superconductors.

Methods

Single crystal preparation. High-quality $(Li_{1-x}Fe_x)OHFe_{1-y}Se$ single crystals were grown by hydrothermal approaches[39]. Large crystals of $K_{0.8}Fe_{1.6}Se_2$ (nominal 245 phase) are specially grown and used as a kind of matrix for a hydrothermal ionic exchange reaction. The K ions in $K_{0.8}Fe_{1.6}Se_2$ are completely released into solution after the hydrothermal reaction process. Single-crystal X-ray diffraction (XRD) and powder XRD measurements confirm the high quality and high purity of the FeSe11111 single crystals. In particular, no trace of K ions in the FeSe11111 crystals is detected by energy dispersive X-ray spectrometry and Inductively-coupled plasma atomic emission spectroscopy (ICP-AES) analyses, confirming a complete release of K ions after the hydrothermal ionic exchange. The sample is sensitive to air and treatment temperature due to the existence of hydroxide. All the samples for ARPES measurements were prepared in glove box and vacuum case within a short time and then transferred to ARPES measurement chamber to be kept at low temperature during the ARPES experiments.

High-resolution ARPES measurements. High-resolution ARPES measurements were carried out on our lab system equipped with a Scienta R8000 electron energy analyser[52]. We use helium discharge lamp as the light source, which can provide photon energies of $h\upsilon = 21.218$ eV (helium I). The energy resolution was set at 4 meV for both the Fermi surface mapping and band structure measurements (Figs 1 and 2) and for the superconducting gap measurements (Figs 3 and 4).

The angular resolution is $\sim 0.3°$. The Fermi level is referenced by measuring on a clean polycrystalline gold that is electrically connected to the sample. The sample was cleaved and measured in vacuum with a base pressure better than 5×10^{-11} Torr.

References

1. Stewart, G. R. Superconductivity in iron compounds. *Rev. Mod. Phys.* **83,** 1589–1652 (2011).
2. Wang, F. *et al.* The electron-pairing mechanism of iron-based superconductors. *Science* **332,** 200–204 (2011).
3. Dagotto, E. Colloquium: the unexpected properties of alkali metal iron selenide superconductors. *Rev. Mod. Phys.* **85,** 849–867 (2013).
4. Bednorz, J. G. *et al.* Possible high T_c superconductivity in the BaLaCuO system. *Z. Phys. B* **64,** 189–193 (1986).
5. Kuroki, K. *et al.* Unconventional pairing originating from the disconnected Fermi surfaces of superconducting $LaFeAsO_{1-x}F_x$. *Phys. Rev. Lett.* **101,** 087004 (2008).
6. Mazin, I. I. *et al.* Unconventional superconductivity with a sign reversal in the order parameter of $LaFeAsO_{1-x}F_x$. *Phys. Rev. Lett.* **101,** 057003 (2008).
7. Wang, F. *et al.* Functional renormalization-group study of the pairing symmetry and pairing mechanism of the FeAs-based high-temperature superconductor. *Phys. Rev. Lett.* **102,** 047005 (2009).
8. Chubukov, A. V. *et al.* Magnetism, supercondcutivity, and pairing symmetry in iron-based superconductors. *Phys. Rev. B* **78,** 134512 (2008).
9. Stanev, V. *et al.* Spin fluctuation dynamics and multiband superconductivity in iron pnictides. *Phys. Rev. B* **78,** 184509 (2008).
10. Wang, F. *et al.* Antiferromagnetic correlation and the pairing mechanism of the cuprates and iron pnictides: a view from functional renormalization group studies. *Europhys. Lett.* **85,** 37005 (2009).
11. Mazin, I. I. *et al.* A key role for unusual spin dynamics in ferropnictides. *Nat. Phys.* **5,** 141–145 (2009).
12. Guo, J. G. *et al.* Superconductivity in the iron selenide $K_xFe_2Se_2$ ($0 \leq x \leq 1.0$). *Phys. Rev. B* **82,** 180520(R) (2010).
13. Fang, M. H. *et al.* Fe-based high temperature superconductivity with $Tc = 31$ K bordering an insulating antiferromagnet in $(Tl,K)Fe_xSe_2$ crystals. *Europhys. Lett.* **94,** 27009 (2011).
14. Mou, D. X. *et al.* Structural, magnetic and electronic properties of the iron – chalcogenide $A_xFe_{2-y}Se_2$ (A = K, Cs, Rb, and Tl, etc.) superconductors. *Front. Phys.* **6,** 410–428 (2011).
15. Qian, T. *et al.* Absence of a holelike Fermi surface for the iron-based $K_{0.8}Fe_{1.7}Se_2$ superconductor revealed by angle-resolved photoemission spectroscopy. *Phys. Rev. Lett.* **106,** 187001 (2011).
16. Zhang, Y. *et al.* Nodeless superconducting gap in $A_xFe_2Se_2$ (A = K,Cs) revealed by angle-resolved photoemission spectroscopy. *Nat. Mater.* **10,** 273–277 (2011).
17. Mou, D. X. *et al.* Distinct Fermi surface topology and nodeless superconducting gap in a $(Tl_{0.58}Rb_{0.42})Fe_{1.72}Se_2$ superconductor. *Phys. Rev. Lett.* **106,** 107001 (2011).
18. Zhao, L. *et al.* Common Fermi-surface topology and nodeless superconducting gap of $K_{0.68}Fe_{1.79}Se_2$ and $(Tl_{0.45}K_{0.34})Fe_{1.84}Se_2$ superconductors revealed via angle-resolved photoemission. *Phys. Rev. B* **83,** 140508(R) (2011).
19. Wang, Q. Y. *et al.* Interface induced high temperature superconductivity in single unit-cell FeSe films on $SrTiO_3$. *Chin. Phys. Lett.* **29,** 037402 (2012).
20. Liu, D. F. *et al.* Electronic origin of high-temperature superconductivity in single-layer FeSe superconductor. *Nat. Commun.* **3,** 931 (2012).
21. He, S. L. *et al.* Phase diagram and electronic indication of high-temperature superconductivity at 65 K in single-layer FeSe films. *Nat. Mater.* **12,** 605–610 (2013).
22. Tan, S. Y. *et al.* Interface-induced superconductivity and strain-dependent spin density waves in $FeSe/SrTiO_3$ thin films. *Nat. Mater.* **12,** 634–640 (2013).
23. Liu, X. *et al.* Electronic structure and superconductivity of FeSe-related superconductors. *J. Phys. Condens. Matter* **27,** 183201 (2015).
24. Yan, Y. J. *et al.* Electronic and magnetic phase diagram in $K_xFe_{2-y}Se_2$ superconductors. *Sci. Rep.* **2,** 212 (2012).
25. Bao, W. *et al.* Superconductivity tuned by the iron vacancy order in $K_xFe_{2-y}Se_2$. *Chin. Phys. Lett.* **30,** 027402 (2013).
26. Wang, Z. *et al.* Microstructure and ordering of iron vacancies in the supercondcutor system $K_yFe_xSe_2$ as seen via transmission electron microscopy. *Phys. Rev. B* **83,** 140505(R) (2011).
27. Li, W. *et al.* Phase seperation and magnetic order in K-doped iron selenide superconductor. *Nat. Phys.* **8,** 126–130 (2012).
28. Zhang, W. H. *et al.* Direct observation of high-temperature superconductivity in one-unit-cell FeSe films. *Chin. Phys. Lett.* **31,** 017401 (2014).
29. Lee, J. J. *et al.* Interfacial mode coupling as the origin of the enhancement of T_c in FeSe films on $SrTiO_3$. *Nature* **515,** 245–248 (2014).
30. Liu, X. *et al.* Dichotomy of the electronic structure and superconductivity between single-layer and double-layer $FeSe/SrTiO_3$ film. *Nat. Commun.* **5,** 5047 (2014).
31. He, J. F. *et al.* Electronic evidence of an insulatorâ€“superconductor crossover in single-layer $FeSe/SrTiO_3$ films. *Proc. Natl Acad. Sci.* **111,** 18501–18506 (2014).
32. Peng, R. *et al.* Tuning the band structure and superconductivity in single-layer â€<FeSe by interface engineering. *Nat. Commun.* **5,** 5044 (2014).
33. Peng, R. *et al.* Measurement of an enhanced superconducting phase and a pronounced anisotropy of the energy gap of a strained FeSe single layer in $FeSe/Nb:SrTiO_3/KTaO_3$ heterostructures using photoemission spectroscopy. *Phys. Rev. Lett.* **112,** 107001 (2014).
34. Ge, J. F. *et al.* Superconductivity above 100 K in single-layer FeSe films on doped $SrTiO_3$. *Nat. Mater.* **14,** 285–289 (2015).
35. Lu, X. F. *et al.* Superconductivity in $LiFeO_2Fe_2Se_2$ with anti-PbO-type spacer layers. *Phys. Rev. B* **89,** 020507 (2014).
36. Lu, X. F. *et al.* Coexistence of superconductivity and antiferromagnetism in $(Li_{0.8}Fe_{0.2})OHFeSe$. *Nat. Mater.* **14,** 325–329 (2015).
37. Pachmayr, U. *et al.* Coexistence of 3d-ferromagnetism and superconductivity in $(Li_{1-x}Fe_x)OH(Fe_{1-y}Li_y)Se$. *Angew. Chem. Int. Ed.* **54,** 293–297 (2015).
38. Dong, X. L. *et al.* Phase diagram of $(Li_{1-x}Fe_x)OHFeSe$: a bridge between iron selenide and arsenide superconductors. *J. Am. Chem. Soc.* **137,** 66–69 (2015).
39. Dong, X. L. *et al.* $(Li_{0.84}Fe_{0.16})OHFe_{0.98}Se$ superconductor: ion-exchange synthesis of large single crystal and anomalous normal state properties. *Phys. Rev. B* **92,** 064515 (2015).
40. Lei, B. *et al.*Gate-tuned superconductor-insulator transition in (Li,Fe)OHFeSe. Preprint at http://arxiv.org/abs/1503.02457 (2015).
41. Hsu, F. C. *et al.* Superconductivity in the PbO-type sructure α-FeSe. *Proc. Natl Acad. Sci.* **105,** 14262–14264 (2008).
42. Blöchl, P. E. *et al.* Projector augmented-wave method. *Phys. Rev. B* **50,** 17953 (1994).
43. Kresse, G. & Joubert, D. From ultrasoft pseudopotentials to the projector augmented-wave method. *Phys. Rev. B* **59,** 1758 (1999).
44. Kresse, G. & Hafner, J. *Ab initio* molecular dynamics for liquid metals. *Phys. Rev. B* **47,** 558 (1993).
45. Kresse, G. & Furthmüller, J. Efficiency of *ab-initio* total energy calculations for metals and semiconductors using a plane-wave basis set. *Comp. Mater. Sci.* **6,** 15–50 (1996).
46. Kresse, G. & Furthmüller, J. Efficient iterative schemes for *ab initio* total-energy calculations using a plane-wave basis set. *Phys. Rev. B* **54,** 11169 (1996).
47. Subedi, A. *et al.* Density functional study of FeS, FeSe, and FeTe: electronic structure, magnetism, phonons, and superconductivity. *Phys. Rev. B* **78,** 134514 (2008).
48. Du, Z. *et al.* Scrutinizing the double superconducting gaps and strong coupling pairing in $(Li_{1-x}Fe_xOH)FeSe$, Preprint at http://arxiv.org/abs/1506.04645 (2015).
49. Yan, Y. J. *et al.* Surface electronic structure and plain s-wave superconductivity of $(Li_{0.8}Fe_{0.2})OHFeSe$ revealed by scanning tunneling microscopy. Preprint at http://arxiv.org/abs/1507.02577 (2015).
50. Norman, M. R. *et al.* Phenomenology of the low-energy spectral function in high-T_c superconductors. *Phys. Rev. B* **57,** 11093(R) (1998).
51. Maletz, J. *et al.* Unusual band renormalization in the simplest iron-based superconductor $FeSe_{1-x}$. *Phys. Rev. B* **89,** 220506 (2014).
52. Liu, G. D. *et al.* Development of a vacuum ultraviolet laser-based angle-resolved photoemission system with a superhigh energy resolution better than 1 meV. *Rev. Sci. Instrum.* **79,** 023105 (2008).

Acknowledgements

We thank Dunghai Lee, Tao Xiang and Guangming Zhang for discussions. X.J.Z. thanks financial support from the MOST of China (973 programme numbers 2011CB921703, 2015CB921000 and 2011CBA00110), the NSFC (11190022, 11334010 and 11374335) and the Strategic Priority Research Program (B) of CAS with Grant Number XDB07020300. X.L.D., F. Z. and Z.X.Z. thank financial support from the NSFC(11274358), the MOST of China (973 programme number 2013CB921701) and the Strategic Priority Research Program (B) of CAS with Grant Number XDB07020100. Z.Y.L. thanks financial support from the MOST of China (973 programme number 2011CBA00112). K.L. was supported by the Fundamental Research Funds for the Central Universities and the Research Funds of Renmin University of China (14XNLQ03). Computational resources have been provided by the PLHPC at RUC.

Author contributions

X.J.Z., L.Z., X.L.D. and Z.X.Z. proposed and designed the research. D.N.Y., H.X.Z, Y.L.H., X.L.D., F.Z. and Z.X.Z. contributed to single crystal preparation. L.Z., A.J.L., Y.H., D.F.L., J.W.H., S.L.H., B.S., Y.X., X.L., L.Y., G.D.L., X.L.D., Z.Y.X., C.T.C. and X.J.Z. contributed to the development and maintenance of Laser-ARPES system. L.Z., A.J.L. and Y.H. carried out the experiment with D.F.L., J.W.H., S.L.H., B.S., Y.X. and X.L.D. L.Z., A.J.L., Y.H. and X.J.Z. analysed the data. K.L. and Z.Y.L. did the DFT calculations. X.J.Z. and L.Z. wrote the paper. All authors discussed the results and commented on the manuscript.

Additional information

Competing financial interests: The authors declare no competing financial interests.

13

Seamless growth of a supramolecular carpet

Ju-Hyung Kim[1,2,3,4], Jean-Charles Ribierre[3], Yu Seok Yang[3], Chihaya Adachi[3], Maki Kawai[4], Jaehoon Jung[1,5], Takanori Fukushima[6] & Yousoo Kim[1]

Organic/metal interfaces play crucial roles in the formation of intermolecular networks on metal surfaces and the performance of organic devices. Although their purity and uniformity have profound effects on the operation of organic devices, the formation of organic thin films with high interfacial uniformity on metal surfaces has suffered from the intrinsic limitation of molecular ordering imposed by irregular surface structures. Here we demonstrate a supramolecular carpet with widely uniform interfacial structure and high adaptability on a metal surface via a one-step process. The high uniformity is achieved with well-balanced interfacial interactions and site-specific molecular rearrangements, even on a pre-annealed amorphous gold surface. Co-existing electronic structures show selective availability corresponding to the energy region and the local position of the system. These findings provide not only a deeper insight into organic thin films with high structural integrity, but also a new way to tailor interfacial geometric and electronic structures.

[1] Surface and Interface Science Laboratory, RIKEN, 2-1 Hirosawa, Wako, Saitama 351-0198, Japan. [2] Department of Chemical Engineering, Pukyong National University, 365 Sinseon-ro, Nam-gu, Busan 608-739, Republic of Korea. [3] Center for Organic Photonics and Electronics Research (OPERA), Kyushu University, 744 Motooka, Nishi, Fukuoka 819-0395, Japan. [4] Department of Advanced Materials Science, The University of Tokyo, 5-1-5 Kashiwanoha, Kashiwa, Chiba 277-8561, Japan. [5] Department of Chemistry, University of Ulsan, 93 Daehak-ro, Nam-gu, Ulsan 44610, Republic of Korea. [6] Chemical Resources Laboratory, Tokyo Institute of Technology, 4259 Nagatsuta, Midori-ku, Yokohama 226-8503, Japan. Correspondence and requests for materials should be addressed to J.J. (email: jjung2015@ulsan.ac.kr) or to T.F. (email: fukushima@res.titech.ac.jp) or to Y.K. (email: ykim@riken.jp).

Organic thin films (OTFs) have received much attention for their potential in various electronic and optoelectronic device applications, since they have outstanding advantages (that is, low cost, low weight and mechanical flexibility) in comparison with standard inorganic technologies[1–6]. In particular, it has been recognized in recent years that organic/metal (O/M) interfaces at which charge carriers are injected into OTFs play a crucial role in the operation and performance of organic devices[7–14]. Various interactions at the O/M interface, such as surface–molecule and intermolecular interactions, are of great importance in the formation of intermolecular networks and organic epitaxy, and have a strong correlation with their electronic structures[7,9,13,15–19]. Although the purity and uniformity of OTFs at the O/M interfaces have profound effects on the operating characteristics of organic devices[7], the formation of OTFs with high interfacial uniformity on metal surfaces has suffered from the intrinsic limitation of molecular ordering imposed by surface step edges, due to the relatively weak intermolecular interactions between organic molecules. Extensive research efforts in the development of OTFs with high interfacial uniformity on metal surfaces have mostly focused on examining the formation of covalent bonding and supramolecular complexes between organic precursors via laborious post-treatments, such as epitaxial surface polymerization and metal–organic coordination[20–25]. Recently, one-dimensional (1D) molecular assembly across step edges was achieved on a specific metal surface via hydrogen bonding[26,27], and at the solid–liquid interface, the formation of supramolecular networks with high uniformity on rough graphene has been studied in depth[28]. However, the creation of widely uniform OTFs on metal surfaces that maintain structural integrity based on non-bonding intermolecular interactions remains a challenge for further development and enhancement of organic devices fabricated by one-step deposition of organic molecules.

In this article, we successfully demonstrate the formation of a supramolecular carpet (SMC), with a widely uniform interfacial structure and high adaptability, on a metal surface, via a one-step deposition process. The geometric and electronic structures of the SMC were investigated by means of scanning tunnelling microscopy/spectroscopy (STM/STS) and density functional theory (DFT) calculations. For this work, bis[1,2,5]thiadiazolotetracyanoquinodimethane (BTDA-TCNQ)[29] was used as the key building block with which to realize the SMCs, not only on a single crystal gold (Au) surface but also on a pre-annealed amorphous Au surface. Tetracyanoquinodimethane (TCNQ) is one of the strongest organic electron acceptors to have found wide use in organic devices, and forms a strongly bonded donor–acceptor complex with tetrathiafulvalene (TTF)[17,19,30–32]. Thus, we used a model system of BTDA-TCNQ, which integrates the structural properties of both TCNQ and TTF in a single molecule, for realizing self-organization into an ordered domain. BTDA-TCNQ is a TCNQ derivative fused with 1,2,5-thiadiazole rings, exerting strong intermolecular interactions in a planar fashion to form network structures in the single crystal[29] and even in its charge-transfer crystals[33,34]. The rhombic structure of BTDA-TCNQ exhibits twofold symmetry with two mirror planes, allowing perpendicular alignment of two electrostatically opposite symmetry axes consisting of electronegative and electropositive end groups, respectively. It, therefore, enables equivalent intermolecular interactions along the four sides of the molecule. Such rhombic structure also facilitates access of neighbouring molecules to each other via high electrostatic interactions in the four directions, which leads to topographically favourable intermolecular interactions. The strong non-bonding intermolecular interactions, balanced with surface–molecule interfacial interactions and site-specific rearrangements of the BTDA-TCNQ molecules near surface step edges, enable the step-flow growth mode for the SMC formation[35,36]. This can lead to an extension of the SMC over step edges of the Au surface without loss of its structural integrity, and results in a covering with high interfacial uniformity over multiple surface steps and terraces, even on the pre-annealed amorphous Au surface prepared on a glass substrate. In addition, different types and dimensionalities of the interfacial electronic structures are distinctively observed in the SMC of BTDA-TCNQ on the Au(111) surface, corresponding to the energy region and the local position of the system, implying that various types of electronic structures projected onto the SMC can be selectively accessible. These results suggest that the SMC has great potential for applications in organic electronics, and also provide important guidelines to develop novel materials forming seamless OTFs on various surfaces.

Results

Intermolecular network of BTDA-TCNQ/Au(111). On deposition of BTDA-TCNQ on the Au(111) surface, decoration at the step edge occurs first, and then growth of a highly ordered SMC proceeds from the step edge across the lower terrace without further nucleation occurring on the terrace (see Fig. 1a,b). This step-flow growth mode of the SMC exhibits a quasi-epitaxial nature with an in-plane azimuthal orientation of BTDA-TCNQ (see Supplementary Figs 1 and 2). The reconstructed herringbone structures of Au(111) beneath the SMC are readily observed in

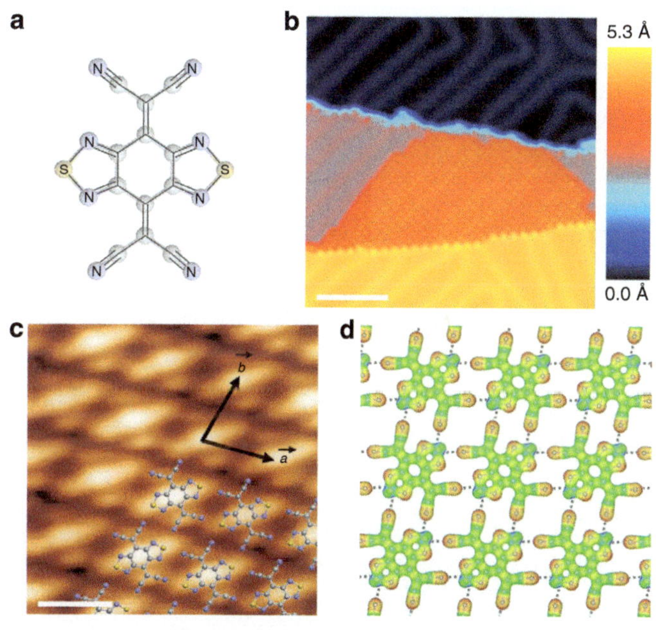

Figure 1 | Network structure of the BTDA-TCNQ molecules. (**a**) The chemical structure of BTDA-TCNQ. (**b**) STM image of BTDA-TCNQ/Au(111), which shows that the molecular network of BTDA-TCNQ grows from the step edge at the lower terrace (sample bias voltage (V_s) = 1,000 mV, tunnelling current (I_t) = 1.00 nA, scale bar (S), 10.0 nm). The reconstructed herringbone structures of Au(111) beneath the BTDA-TCNQ molecules are clearly evident. Height scale is indicated in colour, and the step height of Au(111) is found to be around 2.3 Å. (**c**) Close-up STM image of the network structure formed by BTDA-TCNQ (V_s = 2,000 mV, I_t = 1.0 nA, S = 1.0 nm). Molecular models reveal that the unit cell is noticeably similar to that of the coplanar molecular network in the crystal structure of BTDA-TCNQ. (**d**) The electrostatic potential map of the BTDA-TCNQ network. Blue to red corresponds to positive to negative charges. The distance for the S-N contacts (indicated by grey dotted lines) is 2.80 Å, and the calculated binding energy is ~0.70 eV per molecule.

the STM images, which reveals that the adsorption of BTDA-TCNQ on the Au(111) terraces does not involve quenching of the Au(111) Shockley surface state, as will be discussed below in relation to the STS spectra. These experimental observations indicate that the surface–molecule interfacial interactions of BTDA-TCNQ/Au(111) are sufficiently weak so that lateral diffusion of the molecules across the Au(111) terraces is possible[35,36]. Contrary to the weak adsorption characteristics of BTDA-TCNQ/Au(111), the partial positive and negative charges of the S and N atoms of BTDA-TCNQ, respectively, introduce effective attractive interactions (that is, the S–N intermolecular interactions) between the neighbouring molecules, which determine the array of the molecules in the SMC, as shown in Fig. 1c,d. Detailed analysis of the STM images (see Fig. 1c) gave lattice constants for the SMC unit cell of $a=b=10.0\pm1.0$ Å and $\gamma=80\pm5°$, which are noticeably similar to those achieved by the S–N intermolecular interactions of the coplanar molecular network in the single crystal structure of BTDA-TCNQ[33,34].

DFT calculations were also performed to examine the formation of SMC and to confirm its stability in terms of intermolecular interactions. On the basis of the experimental observations of weak surface–molecule interactions, the substrate was not considered in the calculations. For comparisons among a variety of initial geometries, various molecular orientations rotated by 10° increments were considered with the experimentally observed periodicity of the intermolecular network structure. On the basis of our computational approach, three non-equivalent local potential minima were found corresponding to the varied molecular orientations (see Supplementary Fig. 3). The experimentally observed azimuthal orientation is most favourable with the highest binding energy per molecule of ~0.70 eV (see Fig. 1d), which is more stable than the other two local potential minima by >0.30 eV. Note that the calculated binding energy of ~0.70 eV per molecule is highly competitive with the dipole polarization energy arising from donor–acceptor complexation in the charge-transfer crystal[37,38]. The higher stability of the geometric structure of the SMC, in comparison with the followed second and third local potential minima, mainly originates from the relative orientations of electronegative cyano (–CN) groups between the neighbouring molecules. The electrostatic repulsions between the –CN groups are easily expected in both second and third local potential minima rather than the S–N intermolecular interactions (see Supplementary Fig. 3), which implies that the S–N intermolecular interactions are an impetus for the formation of SMC. The calculated lattice constants of the SMC unit cell ($a=b=9.9$ Å and $\gamma=81.5°$) are almost identical to our STM observations, and the calculated distance for the S–N contacts (~2.80 Å) is also well matched with that of the crystal structure of BTDA-TCNQ (~3.04 Å; ref. 29).

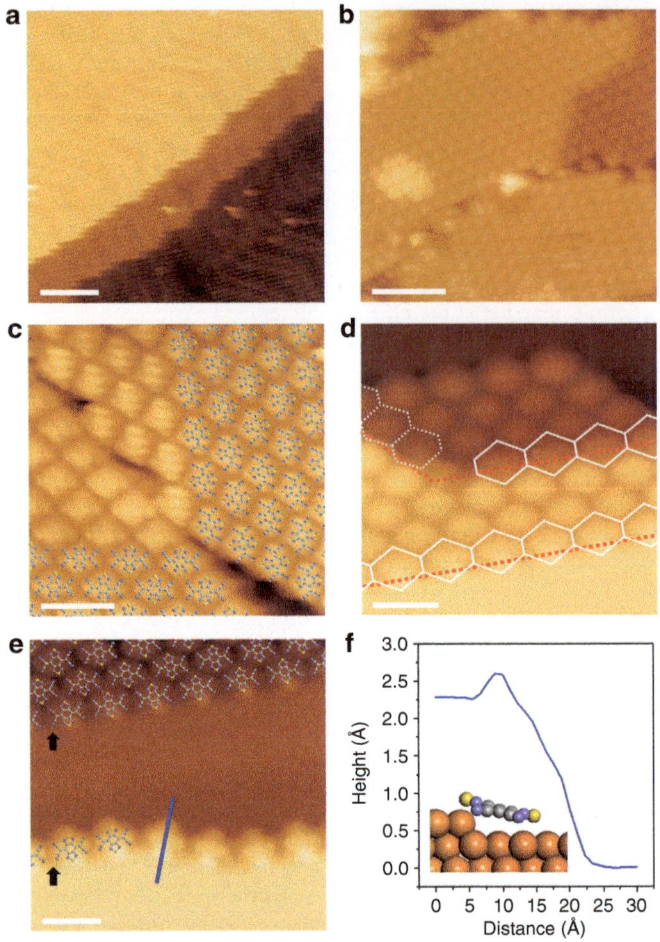

Figure 2 | Highly ordered SMC of BTDA-TCNQ on the Au surfaces. (**a**) STM image of the high-coverage regime of BTDA-TCNQ on Au(111) ($V_s=2{,}000$ mV, $I_t=0.5$ nA, $S=12.0$ nm), showing that the highly ordered SMC of BTDA-TCNQ is extended over multiple surface steps and terraces without losing its structural integrity. (**b**) STM image of an SMC of BTDA-TCNQ on the pre-annealed amorphous Au surface ($V_s=2{,}000$ mV, $I_t=0.3$ nA, $S=5.0$ nm). (**c**) STM image of a structural defect in an SMC of BTDA-TCNQ between two misaligned domains on Au(111) ($V_s=1{,}000$ mV, $I_t=0.1$ nA, $S=2.0$ nm). (**d**) STM image of an SMC of BTDA-TCNQ of which the structural integrity is not interrupted by the surface step edge of Au(111) ($V_s=300$ mV, $I_t=0.5$ nA, $S=2.0$ nm). The BTDA-TCNQ molecules with the slanting adsorption structure (enclosed with white solid lines) and the rearranged flat-lying adsorption structure (enclosed with white dotted lines) are clearly shown near the step edges of Au(111) (indicated by red dotted lines). (**e**) STM image of the BTDA-TCNQ molecules which lean down from the upper terrace to the lower terrace at the step edge ($V_s=500$ mV, $I_t=0.5$ nA, $S=2.0$ nm). The step edges of Au(111) are indicated by black arrows. (**f**) The height profile along the blue line in **e**. corresponding to the slanting adsorption structure of BTDA-TCNQ at the step edge is presented here. The slanting adsorption structure of BTDA-TCNQ is schematically illustrated in the inset of **f**.

Seamless growth of SMC with high structural integrity. In the high-coverage regime of BTDA-TCNQ shown in Fig. 2a, the SMC grows very compactly, covering the whole Au(111) surface, and is extended over multiple surface steps and terraces in a downhill direction, without losing its structural integrity. It is worth noting that single molecular domains of the SMC with a few hundred nanometres in length and width could be achieved on the Au(111) surface, of which sizes are meaningfully large when recent lithography techniques are considered for further applications[39,40]. Interestingly, the outstanding growth characteristics of BTDA-TCNQ are maintained on the pre-annealed amorphous Au surface, as shown in Fig. 2b. The amorphous Au surface was prepared on the cleaned glass substrate by thermal evaporation in a vacuum chamber, and followed by a thermal annealing process without any other preliminary treatments. Note that the amorphous Au surface tends toward the close-packed (111) facet with irregular steps and terraces after annealing, because Au(111) is the most thermodynamically stable facet of Au (see Supplementary Fig. 4). Although structural defects of the SMC, such as domain boundaries and dislocations, are occasionally observed between misaligned domains on Au(111), the structural integrity is not essentially interrupted by the step edges (see Fig. 2c,d). These results imply that the S–N intermolecular interactions in the molecular

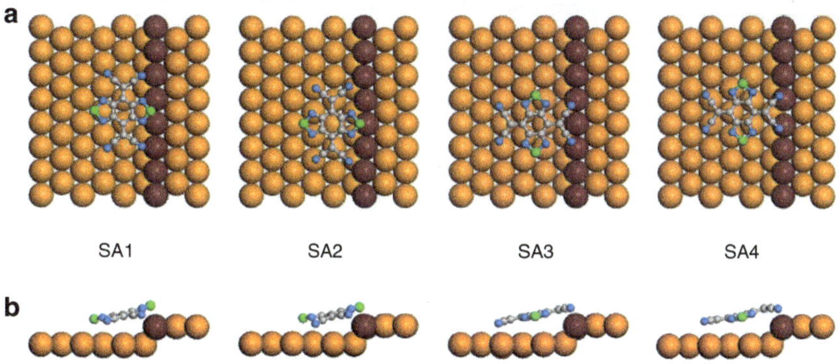

Figure 3 | DFT calculations for the SA structure of BTDA-TCNQ at the step edge of Au(111). (**a**,**b**) Optimized geometries of four different hypothetical model structures of BTDA-TCNQ adsorbed at the step edge of Au(111) (designated 'SA1', 'SA2', 'SA3' and 'SA4', respectively) by DFT calculations. The S-S molecular axes of the molecules are perpendicular to the step line in 'SA1' and 'SA2', and parallel in 'SA3' and 'SA4'. Side views of **a** are shown in **b**.

ordering of BTDA-TCNQ remain intact across the surface steps. This feature distinguishes the behaviour of the BTDA-TCNQ system from that of most other supramolecular assemblies on metal surfaces involving the structural complexities near surface step edges. The adsorption structure of BTDA-TCNQ at the step edges plays a decisive role in maintaining the S–N intermolecular interactions between the two adjacent terraces, thus preserving the structural integrity of the SMC over multiple surface steps and terraces. When the BTDA-TCNQ molecules initially decorate the step edges, they lean down from the upper terrace to the lower terrace (that is, the slanting adsorption (SA) structure), as shown in Fig. 2e,f. The STM observations indicate that the S–S molecular axis (the molecular axis along the two S atoms) of BTDA-TCNQ in this structure is perpendicular to the step line, and thus the two S atoms are located on the upper and lower terraces, respectively (see Fig. 2e,f). In consideration of the symmetry of BTDA-TCNQ, the SA structure enables symmetrical intermolecular interactions on the upper and lower terraces, leading to the intact intermolecular interactions to be valid across the surface steps. Differently from the weak adsorption nature of BTDA-TCNQ on the Au(111) terraces, the surface–molecule interfacial interactions localized near the step edges are the primary factors that determine the SA structure, as will be discussed in relation to the DFT calculations and the STS spectra.

For the BTDA-TCNQ molecules that are aligned along parallel straight lines of the step edges, the SA structure at the step edges is preserved even in the high-coverage regime (see Fig. 2d), thus acting as a bridge for maintaining the S–N intermolecular interactions across the adjacent terraces. However, the BTDA-TCNQ molecules that are adsorbed out of the parallel straight lines at the step edges are displaced from step edges and rearranged onto the lower terraces (that is, the rearranged flat-lying adsorption structure), as shown in Fig. 2d. These phenomena imply that the S–N intermolecular interactions in supramolecular assembly dominate the surface–molecule interactions even at the relatively reactive step edges, and thus result in the SMC formation with structural integrity maintained over wide areas. In addition, the potential energy barrier for the diffusion of molecule crossing the upper step edge is significantly higher than the lower step edge, because the step dipole is mostly formed below the upper step edge[41–43]. The SMC, therefore, grows across the surface steps only in the downhill direction.

Adsorption structure of BTDA-TCNQ at the step edges. To gain deeper insight into the SA structure of BTDA-TCNQ at the step edges, DFT calculations were performed in consideration of four different adsorption models (designated 'SA1', 'SA2', 'SA3' and 'SA4', respectively) as shown in Fig. 3 and Table 1. Among these structure models, the most stable adsorption structure of the BTDA-TCNQ molecule at the step edge was found to be 'SA1' with the adsorption energy of 2.64 eV. In 'SA1', the S–S molecular axis of BTDA-TCNQ is perpendicular to the step line, and the two S atoms are located on the upper and lower terraces, respectively. Compared with 'SA3' that is the most stable adsorption structure with the S–S molecular axis parallel to the step line, 'SA1' shows higher adsorption energy by 0.52 eV, which is consistent with the experimental observations of the SA structure at the step edges.

Table 1 | Relative energies, adsorption energies and optimized nearest atomic distances of S···Au and -CN···Au of four adsorption models, 'SA1', 'SA2', 'SA3' and 'SA4'.

	SA1	SA2	SA3	SA4
E_{rel} (eV)	0.00	0.17	0.52	0.53
E_{ads} (eV)	2.64	2.47	2.12	2.11
Step edge				
S···Au (Å)	3.12 (× 2)	2.54		
-CN···Au (Å)	2.18	2.56	2.31	2.34
Terrace				
S···Au (Å)	3.34	3.33	3.61	3.69
-CN···Au (Å)	3.17	2.36	2.68	3.14

The optimized distances at the step edge and on the terrace are presented, respectively. In 'SA1', a bidentate fashion of S···Au at the step edge is denoted with '× 2'.

The detailed geometric interpretation of the computational results indicates that the –CN groups of the BTDA-TCNQ molecule significantly contribute to determination of the adsorption structure at the step edges. Side views of the molecular adsorption structures clearly reveal that the –CN groups are displaced from the molecular plane and become closer to the surface, compared with the S atoms (see Fig. 3b). In the most stable 'SA1', the nearest atomic distance between the molecule and the reactive Au atoms of the step edge (–CN···Au_{edge}) is 2.18 Å, and is much shorter than the S···Au_{edge} distance of 3.12 Å even though the S atom of BTDA-TCNQ is interacting with the two Au_{edge} atoms. The higher stability of 'SA1' than 'SA3' also can be deduced from the shorter –CN···Au_{edge} distance in 'SA1' than that of 'SA3' by 0.13 Å. Combined with the –CN···Au_{edge} interactions, the relatively weak S···Au_{edge} interactions in 'SA1' also possibly participate in determining the adsorption structure when no existing S···Au_{edge} interaction in 'SA3' is considered. In addition, the charge density difference map of the most stable 'SA1' structure clearly reveals the partial charge transfer from the

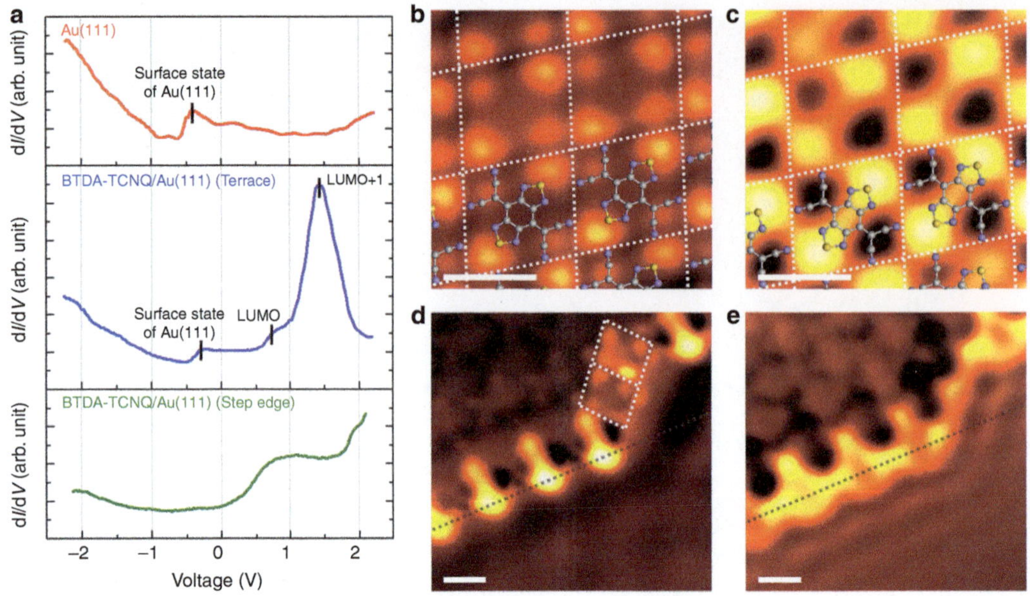

Figure 4 | STS spectra and STS mapping images of BTDA-TCNQ/Au(111). (**a**) STS spectra of (top) the bare Au(111) surface, (middle) the BTDA-TCNQ molecule on the Au(111) terrace and (bottom) the BTDA-TCNQ molecule at the Au(111) step edge. (**b,c**) STS mapping images of BTDA-TCNQ on the Au(111) terrace. The electron probability distributions are clearly observed for the LUMO state of BTDA-TCNQ in **b** (V_s = 800 mV, I_t = 0.8 nA, S = 1.0 nm) and the LUMO + 1 state of BTDA-TCNQ in **c** (V_s = 1,500 mV, I_t = 0.8 nA, S = 1.0 nm). Each BTDA-TCNQ molecule is enclosed with white dotted lines in **b** and **c**. (**d,e**) STS mapping images of BTDA-TCNQ at the Au(111) step edge. Parallel straight lines of the step edge are indicated by grey dotted lines in **d** and **e**, and the BTDA-TCNQ molecules that are rearranged onto the lower terrace from the step edge are enclosed with white dotted lines in **d**. The STS mapping image in **d** (V_s = 500 mV, I_t = 1.0 nA, S = 1.0 nm) reveals the LUMO state for BTDA-TCNQ with the rearranged flat-lying adsorption structure and the LUMO + 1 state for BTDA-TCNQ with the slanting adsorption structure at the lower sample bias than that in **b** and **c**. The STS mapping image in **e** (V_s = 2,000 mV, I_t = 1.0 nA, S = 1.0 nm) also indicates that the unoccupied MO states of BTDA-TCNQ are highly hybridized with the Au surface at the step edge, forming quasi-1D electron dispersion along the step-edge line.

surface to the molecule through the -CN groups at the step edge (see Supplementary Fig. 5), and the corresponding amount of the partial charge transfer evaluated by Bader population analysis is 0.68*e*. The computationally estimated feature of the partial charger transfer strongly supports the surface–molecule interfacial interactions localized at the step edges, resulting from the step dipole as will be discussed in relation to the STS spectra.

Electronic structures of BTDA-TCNQ on the Au surface. To elucidate the electronic structures of BTDA-TCNQ on the Au surface in more detail, intensive STS and STS mapping measurements were performed, as shown in Fig. 4. The Au(111) Shockley surface state, the band edge of which is ~500 meV below the Fermi level (E_F), remains clearly evident in the STS spectrum of the BTDA-TCNQ molecule adsorbed on the Au(111) terrace, indicating the weak adsorption character of BTDA-TCNQ/Au(111) (refs 44,45). The lowest unoccupied molecular orbital (LUMO) and the LUMO + 1 states of BTDA-TCNQ are distinctly observed in the STS spectrum (at ~ +800 mV and ~ +1,500 mV, respectively). STS mapping reveals the electron probability distributions for these unoccupied molecular orbital (MO) states (see Fig. 4b,c), demonstrating that the LUMO state is locally distributed around the -CN groups and the S atoms, forming a rhombic structure, whereas the LUMO + 1 state shows a more conjugated electronic structure along the S–S molecular axis. In general, the unoccupied MO states of BTDA-TCNQ are largely distributed around the S atoms, even though their electronic structures are distinct from each other. The conjugated character of the unoccupied MO states of BTDA-TCNQ results in n-type semiconducting behaviour, which is clarified in a field-effect transistor (FET) configuration (see Supplementary Fig. 6). Note that the distinct electronic structures of the unoccupied MO states lead to strong bias dependence in STM imaging in positive sample bias region (see Supplementary Fig. 1), because STM imaging strongly reflects the integration of the electronic states that can contribute to electron tunnelling[46].

At the step edge of Au(111), however, the STS spectrum indicates a rather strong interaction between BTDA-TCNQ and the Au surface. The unoccupied feature of BTDA-TCNQ at the step edge shows a steep initial rise near the E_F, and then maintains a nearly constant level before another steep rise at ~ +1,700 mV. Such substantial broadening of the unoccupied states is mainly caused by hybridization with the electronic states of the metal surface, and thus the feature for the step-edge state of Au(111) is not distinguishable in the STS spectrum of BTDA-TCNQ adsorbed at the step edge[42]. STS mapping reveals that the unoccupied MO states of the BTDA-TCNQ molecules show a significant downward shift toward the E_F at the step edge. The LUMO + 1 state for the SA structure and the LUMO state for the rearranged flat-lying adsorption structure of BTDA-TCNQ at the step edge are observed at ~ +500 mV, as shown in Fig. 4d. These phenomena mainly originate from the step dipole described as the Smoluchowski effect[41–43]. The step dipole, with an opposite direction to that of the surface dipole, induces a strong localized electric field at the step edge, resulting in Stark shifts of the adsorbed molecules near the step edge[41]. Since the SA structure of BTDA-TCNQ covers the step edge completely, the Stark shift in the SA structure is larger than that of the rearranged flat-lying adsorption structure. On the basis of a detailed analysis of the STS mapping data (see Supplementary Fig. 7), the Stark shifts of the unoccupied MO states of BTDA-TCNQ near the step edge are ~ −900 meV and ~ −300 meV for the SA structure and the rearranged flat-lying adsorption structure, respectively.

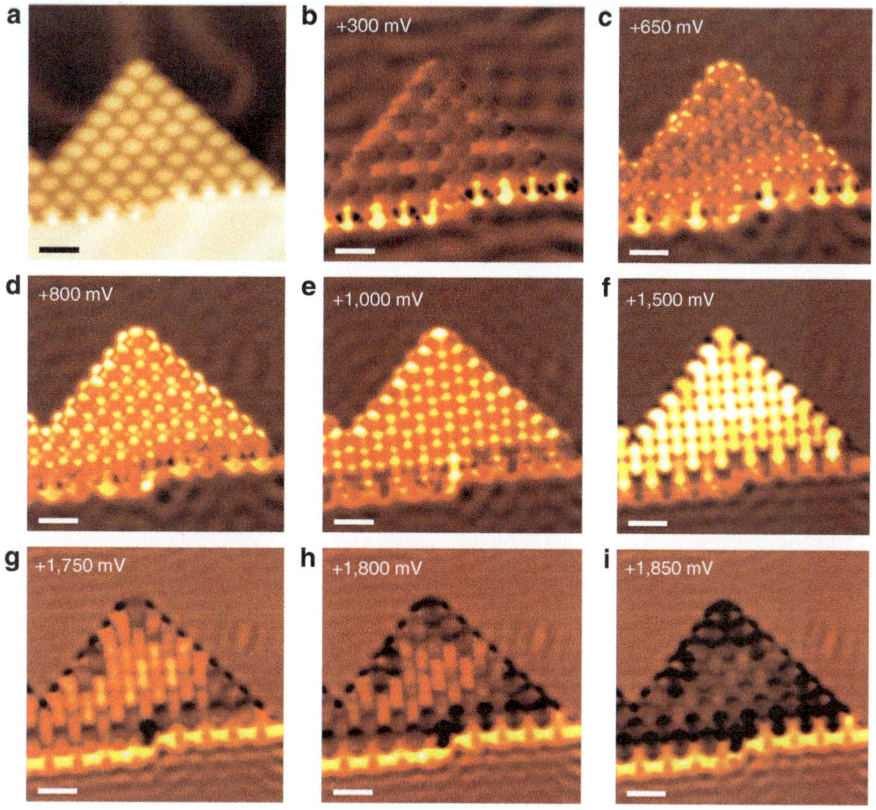

Figure 5 | STS mapping of an SMC of BTDA-TCNQ on the Au(111) surface. (**a**) STM image of an SMC of BTDA-TCNQ on the Au(111) surface ($V_s = 500$ mV, $I_t = 1.0$ nA, $S = 2.0$ nm). (**b**-**i**) STS mapping images of repeating measurements over the same surface area as that in **a**, with varying sample bias ($I_t = 1.0$ nA, $S = 2.0$ nm). The Au(111) surface states are transparently observed through the SMC of BTDA-TCNQ in **b** ($V_s = 300$ mV) and **c** ($V_s = 650$ mV). The electron probability distributions of the MO states of BTDA-TCNQ are the most prominent features in **d** ($V_s = 800$ mV), **e** ($V_s = 1{,}000$ mV) and **f** ($V_s = 1{,}500$ mV). The SMC of BTDA-TCNQ exhibits a highly delocalized electron distribution which is confined within the boundaries of the SMC in **g** ($V_s = 1{,}750$ mV), **h** ($V_s = 1{,}800$ mV) and **i** ($V_s = 1{,}850$ mV). The quasi-1D electron dispersion along the step edge resulting from the hybridization of BTDA-TCNQ and Au at the step is also evident in **g**-**i**.

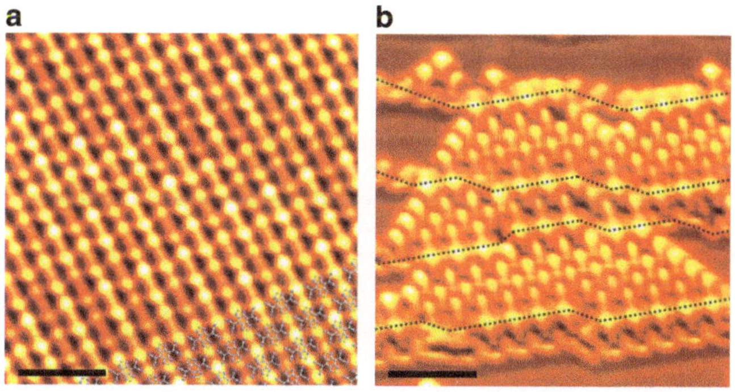

Figure 6 | Widely uniform anisotropic MO states of BTDA-TCNQ/Au(111). (**a**,**b**) STS mapping images of an SMC of BTDA-TCNQ on the Au(111) surface in the energy range corresponding to the LUMO + 1 state ($V_s = 1{,}500$ mV, $I_t = 1.0$ nA). Widely uniform anisotropic MO states of the SMC are clearly observed in **a** ($S = 3.0$ nm), which can be extended over multiple surface steps and terraces as in **b** ($S = 5.0$ nm). Step-edge lines are indicated with black dotted lines in **b**, revealing small sizes (~5 nm in width) of the (111) facets.

These results strongly suggest that the localized electric fields induced by the step dipoles exert influences on the partial charge transfers from the step edges to the unoccupied MO states of BTDA-TCNQ. Considering the LUMO state of BTDA-TCNQ on the Au(111) terrace (at $\sim +800$ meV relative to the E_F), the Stark shift of ~ -900 meV possibly induces the partial charge transfer to the LUMO state for the SA structure from the step edge, and the LUMO + 1 state consequentially represents the nearest unoccupied MO state with respect to the E_F at the step edge (see Supplementary Fig. 7). The computationally estimated charge density difference map also strongly supports the partial charge transfer from the step edge to the molecule (see Supplementary Fig. 5). In addition, the STS spectrum and mapping images show that the unoccupied MO states for the SA structure of BTDA-TCNQ are highly hybridized with the Au surface at the step edge above $\sim +1{,}800$ meV with respect to the E_F, forming quasi-1D electron dispersion along the step-edge line (see Fig. 4e).

Electronic structures between the SMC and the Au surface. The interfacial electronic structures between the SMC of BTDA-TCNQ and the Au(111) surface were also investigated as a function of energy and of local position, by repeating STS mapping over the same area of surface with the sample bias voltage varied between +300 mV and +2,000 mV (see Fig. 5). The interfacial electronic structures of the SMC on the Au(111) terrace are characterized into three main categories according to energy ranges. First, since the MO states of BTDA-TCNQ do not appear near the E_F as indicated in the STS spectrum (see Fig. 4a), the Au(111) surface states are transparently observed through the SMC in the lower energy region near +300 meV with respect to the E_F (see Fig. 5b,c). In this energy range, the STS mapping images exhibit strong standing-wave patterns of the surface-state electrons, which are continuous between two media (that is, the

SMC and the Au(111) terrace) without scattering at the boundaries of the SMC[47]. Second, in the energy range corresponding to the MO states of BTDA-TCNQ, the electronic structures of the SMC can be attributed to the electron probability distributions of the MO states (see Fig. 5d–f). Thus, the STS mapping data reveal the localized MO states of the SMC independently of the Au(111) surface states, which also substantiates the weak adsorption nature of BTDA-TCNQ/Au(111) (ref. 48). Strongly scattered standing-wave patterns, which are caused by the reflection of the surface-state electrons at the boundaries of the SMC, are observed on the Au(111) terrace in this energy region. Compared with the lower energy region (near +300 meV), the observation of the scattering of surface-state electrons indicates that it is the localized MO states of the SMC that serve as potential walls for reflecting the surface-state electrons rather than the geometric footprint of the molecule[49,50]. Finally, above the highly scattering region of the surface-state electrons, the contribution of the MO states of BTDA-TCNQ decreases, as indicated in the STS spectrum (see Fig. 4a). In this higher energy region (near +1,900 meV with respect to the E_F), the SMC exhibits a highly delocalized electron distribution, which is confined within the boundaries of the SMC (see Fig. 5g–i). The delocalized electron distribution shows a pattern of two-dimensional dispersive electronic states, and faintly scattered standing-wave patterns appear on the Au(111) terrace near the boundaries of the SMC. Considering the weak adsorption character of BTDA-TCNQ on the Au(111) terraces and the dissipation of the MO states in this higher energy region, it is the electronic states of Au(111) underneath the SMC that essentially contribute to the dispersive electronic states observed in the SMC[16,17]. On the other hand, the quasi-1D electron dispersion along the step edge resulting from the hybridization of BTDA-TCNQ and Au at the step is observed in this higher energy region separately from the terrace (see Fig. 5g–i), which implies that the surface steps modulate the dimensionality of the electron dispersion. Interestingly, these different types and dimensionalities of the interfacial electronic structures co-exist in the SMC of BTDA-TCNQ on the Au(111) surface. These results suggest that various types of electronic structures projected onto the SMC can be selectively available corresponding to the energy region and the local position of the system. Note that the SMC of BTDA-TCNQ on the Au(111) surface shows widely uniform anisotropic MO states in the energy region corresponding to the LUMO + 1 state (see Fig. 6). Such anisotropic characteristics originate predominantly from the distribution of the LUMO + 1 states along the S atoms and the uniaxial orientation of BTDA-TCNQ via the S–N intermolecular interactions, which are also maintained over multiple surface steps and terraces as shown in Fig. 6b. These results also imply that careful design of favourable intermolecular interactions facilitates the tailoring of direction and alignment of localized MO states in OTFs without the loss of structural integrity.

Discussion

We present a well-designed system, BTDA-TCNQ on the Au surface, to demonstrate the formation of SMC with widely uniform interfacial structure and high adaptability on the metal surface. The strong non-bonding intermolecular interactions induced by the partial positive charge of the S atoms and the partial negative charge of the N atoms of BTDA-TCNQ, and the surface–molecule interactions on both surface steps and terraces enable the step-flow growth mode for the self-organized SMC domain. Owing to the well-balanced interfacial interactions, the SMC of BTDA-TCNQ on the Au surface can extend over multiple surface steps and terraces without losing its structural integrity, accompanying the unidirectional alignment of the MO states. This results in a highly uniform interfacial structure, even on the pre-annealed amorphous Au surface prepared on the glass substrate. In consideration of the typical amorphous Au electrodes widely used in electronics, our findings on the pre-annealed amorphous Au surface may have an impact on the field of organic electronics in terms of molecular orientation, alignment and ordering on the electrodes, which have profound effects on the electrical and optical properties[51,52]. In addition, selective availability of co-existing interfacial electronic structures projected onto the SMC of BTDA-TCNQ is one of the remarkable features of the SMC on the Au surface. We anticipate that the findings of this work will provide not only a deeper insight into the formation of OTFs with high film integrity at the O/M interfaces, but also a new way to tailor the properties of interfacial geometric and electronic structures for future applications of organic devices.

Methods

Low-temperature STM/STS experiments. All the experiments were performed using a low-temperature STM (Omicron GmbH) with an electrochemically etched tungsten tip in an ultrahigh-vacuum (UHV) chamber in which the base pressure was maintained at below 8.0×10^{-11} Torr. The Au(111) and amorphous Au surfaces were cleaned by several cycles of sputtering with argon ions (Ar^+) and annealing at 800 K. Low-temperature STM surface scanning at atomic resolution was used to confirm the cleanliness of each Au surface. BTDA-TCNQ was synthesized and purified in accordance with a previous report[26,30]. Using a Knudsen cell, BTDA-TCNQ (in powdered form at room temperature) was thermally evaporated onto each Au surface under UHV conditions. Evaporation temperature was 473 K for BTDA-TCNQ, which was monitored using a K-type (alumel–chromel) thermocouple. The Au surface was maintained at room temperature during and after each thermal evaporation to confirm a thermodynamically stable phase (~1 h), then cooled to 4.7 K. Note that the domain size of the SMC in the step-flow regime is mainly determined by the amount of adsorbed molecules at sub-monolayer coverage, which can be simply controlled by the deposition rate and time. In the deposition and annealing processes, no intentional temperature control was performed for the Au surface, and the substrate was merely located in the room temperature chamber. When the sample was immediately cooled to 4.7 K (cooling time of ~1 h) even without annealing time after deposition, there were no significant changes observed in the formation of SMC as well following the step-flow growth scheme. The cooled sample was also not damaged if the sample temperature repeatedly increased up to room temperature in the UHV chamber.

All STM and STS measurements were performed at 4.7 K, and STS signals were measured with lock-in detection by applying a modulation of 50 mV (r.m.s.) to the sample bias voltage at 797 Hz. Note that all STS spectra were obtained at a fixed tip-sample separation (feed-back off), whereas STS mapping was performed in the constant current mode, with the sample bias voltage varied from +300 mV to +2,000 mV.

DFT calculations. We used vdW-TS method[53] implemented in the Vienna *Ab initio* Simulation Package (VASP) code[54,55], which can take account of van der Waals interactions, to examine the formation of SMC. The core electrons were replaced by projector-augmented wave pseudopotentials[56], expanded in a basis set of plane waves up to a cutoff energy of 400 eV. The substrate-free DFT calculations were performed, based on the experimental observations of weak surface–molecule interactions. The minimum-size supercell including one molecule was used according to the experimentally observed periodicity of the two-dimensional intermolecular network. For comparisons among a variety of initial geometries, various molecular orientations rotated by 10° increments were prepared with respect to the experimentally proposed molecular arrangement. The calculations using DFT-D2 method[57] were also performed to verify the computational reliability, leading to a good agreement with the results obtained using vdW-TS method. During ionic relaxation, the molecular geometry in a planar fashion was maintained by symmetry constraints, but the volume and shape of the slab were freely adjusted toward a local potential minimum. To ensure two-dimensional molecular architecture, we initially separated the periodically replicated slabs with a large vacuum region of 15 Å. The resultant vacuum regions of the optimized slabs were about 6.0 Å, in which the interactions between the molecular layers are expected to be negligible in obtaining proper geometries and energies of the intermolecular network structures of BTDA-TCNQ. Ionic (electronic) relaxations were performed until atomic forces (energies) were less than $0.05\ \mathrm{eV\ Å^{-1}}$ (10^{-7} eV). The k-point sampling of the Brillouin zone was performed with $8 \times 8 \times 1$ Γ-centred grids in the periodic DFT calculations.

As regards the adsorption structure of BTDA-TCNQ at the step edge of Au(111), we used a (6 × 1)-Au(455) supercell to investigate the adsorption of isolated BTDA-TCNQ molecule, which provides the nearest interatomic distance of longer than 8 Å along the step-edge line and the large distance from the molecule to the next step edge longer than 15 Å. The slab model was composed of six Au layers and vacuum region of ~20 Å, in which the two bottom layers were fixed in their bulk positions. Bader population analysis[58] was also performed to evaluate the partial charge transfer from the surface to the molecule at the step edge. Dipole correction was applied to avoid undesired interactions between periodic slab images. The k-point sampling of the Brillouin zone was performed with $2 \times 2 \times 1$ Γ-centred grids in the periodic DFT calculations.

FET fabrication and characterization. BTDA-TCNQ was incorporated into the FET configuration with a bottom-gate and bottom-contact structure. A highly doped n-type Si wafer with a thermally grown SiO_2 dielectric layer (of ~300 nm thickness) was used as a substrate. The Au source and drain electrodes (of ~40 nm thickness) were deposited onto the pre-cleaned substrate by thermal evaporation in a vacuum chamber, and then annealed at 300 °C for ~1 h in a N_2-filled glove-box. The channel length and width between the Au source and drain electrodes were 50 μm and 975 μm, respectively. After cooling the substrate to room temperature, BTDA-TCNQ (of ~50 nm thickness) was evaporated at the rate of 0.1 Å s^{-1}. The transfer and output characteristics (see Supplementary Fig. 6) were measured using a semiconductor parameter analyser (Agilent B1500A) in the N_2-filled glove-box at room temperature.

References

1. Park, S. H. *et al.* Bulk heterojunction solar cells with internal quantum efficiency approaching 100%. *Nat. Photon.* **3,** 297–302 (2009).
2. Sekitani, T. *et al.* Stretchable active matrix organic light-emitting diode display using printable elastic conductors. *Nat. Mater.* **8,** 494–499 (2009).
3. Sekitani, T., Zschieschang, U., Klauk, H. & Someya, T. Flexible organic transistors and circuits with extreme bending stability. *Nat. Mater.* **9,** 1015–1022 (2010).
4. Zhang, W. Supramolecular linear heterojunction composed of graphite-like semiconducting nanotubular segments. *Science* **334,** 340–343 (2011).
5. Uoyama, H., Goushi, K., Shizu, K., Nomura, H. & Adachi, C. Highly efficient organic light-emitting diodes from delayed fluorescence. *Nature* **492,** 234–238 (2012).
6. Kim, R. H. *et al.* Non-volatile organic memory with sub-millimetre bending radius. *Nat. Commun.* **5,** 3583 (2014).
7. Forrest, S. R. Ultrathin organic films grown by organic molecular beam deposition and related techniques. *Chem. Rev.* **97,** 1793–1896 (1997).
8. Ishii, H., Sugiyama, K., Ito, E. & Seki, K. Energy level alignment and interfacial electronic structures at organic/metal and organic/organic interfaces. *Adv. Mater.* **11,** 605–625 (1999).
9. Hooks, D. E., Fritz, T. & Ward, M. D. Epitaxy and molecular organization on solid substrates. *Adv. Mater.* **13,** 227–241 (2001).
10. Zhu, X.-Y. Electronic structure and electron dynamics at molecule-metal interfaces: implications for molecule-based electronics. *Surf. Sci. Rep.* **56,** 1–83 (2004).
11. Braun, S., Salaneck, W. R. & Fahlman, M. Energy-level alignment at organic/metal and organic/organic interfaces. *Adv. Mater.* **21,** 1450–1472 (2009).
12. Kawano, K. & Adachi, C. Evaluating carrier accumulation in degraded bulk heterojunction organic solar cells by a thermally stimulated current technique. *Adv. Funct. Mater.* **19,** 3934–3940 (2009).
13. Yamane, H., Kanai, K., Ouchi, Y., Ueno, N. & Seki, K. Impact of interface geometric structure on organic–metal interface energetics and subsequent films electronic structure. *J. Electron Spectrosc. Relat. Phenom.* **174,** 28–34 (2009).
14. Ma, H., Yip, H.-L., Huang, F. & Jen, A. K.-Y. Interface engineering for organic electronics. *Adv. Funct. Mater.* **20,** 1371–1388 (2010).
15. Barth, J. V., Costantini, G. & Kern, K. Engineering atomic and molecular nanostructures at surfaces. *Nature* **437,** 671–679 (2005).
16. Temirov, R., Soubatch, S., Luican, A. & Tautz, F. S. Free-electron-like dispersion in an organic monolayer film on a metal substrate. *Nature* **444,** 350–353 (2006).
17. Gonzalez-Lakunza, N. *et al.* Formation of dispersive hybrid bands at an organic-metal interface. *Phys. Rev. Lett.* **100,** 156805 (2008).
18. Bartels, L. Tailoring molecular layers at metal surfaces. *Nat. Chem.* **2,** 87–95 (2010).
19. Tseng, T.-C. *et al.* Charge-transfer-induced structural rearrangements at both sides of organic/metal interfaces. *Nat. Chem.* **2,** 374–379 (2010).
20. Grill, L. *et al.* Nano-architectures by covalent assembly of molecular building blocks. *Nat. Nanotechnol.* **2,** 687–691 (2007).
21. Zwaneveld, N. A. A. *et al.* Organized formation of 2D extended covalent organic frameworks at surfaces. *J. Am. Chem. Soc.* **130,** 6678–6679 (2008).
22. Abdurakhmanova, N. *et al.* Stereoselectivity and electrostatics in charge-transfer Mn- and Cs-$TCNQ_4$ networks on Ag(100). *Nat. Commun.* **3,** 940 (2012).
23. Kley, C. S. *et al.* Highly adaptable two-dimensional metal–organic coordination networks on metal surfaces. *J. Am. Chem. Soc.* **134,** 6072–6075 (2012).
24. Li, Y. *et al.* Coordination and metalation bifunctionality of Cu with 5, 10, 15, 20-tetra (4-pyridyl) porphyrin: toward a mixed-valence two-dimensional coordination network. *J. Am. Chem. Soc.* **134,** 6401–6408 (2012).
25. Floris, A., Comisso, A. & Vita, A. D. Fine-tuning the electrostatic properties of an alkali-linked organic adlayer on a metal substrate. *ACS Nano* **7,** 8059–8065 (2013).
26. Schnadt, J. *et al.* Extended one-dimensional supramolecular assembly on a stepped surface. *Phys. Rev. Lett.* **100,** 046103 (2008).
27. Schnadt, J. *et al.* Interplay of adsorbate-adsorbate and adsorbate-substrate interactions in self-assembled molecular surface nanostructures. *Nano Res.* **3,** 459–471 (2010).
28. Li, B. *et al.* Self-assembled air-stable supramolecular porous networks on graphene. *ACS Nano* **7,** 10764–10772 (2013).
29. Yamashita, Y., Suzuku, T., Mukai, T. & Saito, G. Preparation and properties of a tetracyanoquinodimethane fused with 1,2,5-thiadiazole units. *J. Chem. Soc. Chem. Commun.* **1985,** 1044–1045 (1985).
30. Torrance, J. B. The difference between metallic and insulating salts of tetracyanoquinodimethone (TCNQ): how to design an organic metal. *Acc. Chem. Res.* **12,** 79–86 (1979).
31. Alves, H., Molinari, A. S., Xie, H. & Morpurgo, A. F. Metallic conduction at organic charge-transfer interfaces. *Nat. Mater.* **7,** 574–580 (2008).
32. Kirtley, J. R. & Mannhart, J. Organic electronics: when TTF met TCNQ. *Nat. Mater.* **7,** 520–521 (2008).
33. Suzuki, T. *et al.* Clathrate formation and molecular recognition by novel chalcogen-cyano interactions in tetracyanoquinodimethanes fused with thiadiazole and selenadiazole rings. *J. Am. Chem. Soc.* **114,** 3034–3043 (1992).
34. Suzuki, T., Fukushima, T., Yamashita, Y. & Miyashi, T. An absolute asymmetric synthesis of the [2 + 2] cycloadduct via single crystal-to-single crystal transformation by charge-transfer excitation of solid-state molecular complexes composed of arylolefins and bis[1,2,5]thiadiazolotetracyanoquinodimethane. *J. Am. Chem. Soc.* **116,** 2793–2803 (1994).
35. Vlieg, E. Understanding crystal growth in vacuum and beyond. *Surf. Sci.* **500,** 458–474 (2002).
36. Wagner, S. R., Lunt, R. R. & Zhang, P. Anisotropic crystalline organic step-flow growth on deactivated Si surfaces. *Phys. Rev. Lett.* **110,** 086107 (2013).
37. Aragó, J., Sancho-García, J. C., Ortí, E. & Beljonne, D. *Ab initio* modeling of donor–acceptor interactions and charge-transfer excitations in molecular complexes: the case of terthiophene–tetracyanoquinodimethane. *J. Chem. Theory Comput.* **7,** 2068–2077 (2011).
38. Sini, G., Sears, J. S. & Brédas, J.-L. Evaluating the performance of DFT functionals in assessing the interaction energy and ground-state charge transfer of donor/acceptor complexes: tetrathiafulvalene-tetracyanoquinodimethane (TTF-TCNQ) as a model case. *J. Chem. Theory Comput.* **7,** 602–609 (2011).
39. Beesley, D. J. *et al.* Sub-15-nm patterning of asymmetric metal electrodes and devices by adhesion lithography. *Nat. Commun.* **5,** 3933 (2014).
40. Cho, H. *et al.* Replication of flexible polymer membranes with geometry-controllable nano-apertures via a hierarchical mould-based dewetting. *Nat. Commun.* **5,** 3137 (2014).
41. Wandelt, K. Properties and influence of surface defects. *Surf. Sci.* **251,** 387–395 (1991).
42. Avouris, P., Lyo, I.-W. & Molinàs-Mata, P. STM studies of the interaction of surface state electrons on metals with steps and adsorbates. *Chem. Phys. Lett.* **240,** 423–428 (1995).
43. Kamna, M. M., Stranick, S. J. & Weiss, P. S. Imaging substrate-mediated interactions. *Isr. J. Chem.* **36,** 59–62 (1996).
44. Chen, W., Madhavan, V., Jamneala, T. & Crommie, M. F. Scanning tunneling microscopy observation of an electronic superlattice at the surface of clean gold. *Phys. Rev. Lett.* **80,** 1469–1472 (1998).
45. Kim, J.-H. *et al.* Direct observation of adsorption geometry for the van der Waals adsorption of a single π-conjugated hydrocarbon molecule on Au (111). *J. Chem. Phys.* **140,** 074709 (2014).
46. Wisendanger, R. *Scanning Probe Microscopy and Spectroscopy: Methods and Applications* (Cambridge Univ. Press, 1994).
47. Repp, J., Meyer, G. & Rieder, K.-H. Snell's law for surface electrons: refraction of an electron gas imaged in real space. *Phys. Rev. Lett.* **92,** 036803 (2004).
48. Soe, W.-H., Manzano, C., De Sarkar, A., Chandrasekhar, N. & Joachim, C. Direct observation of molecular orbitals of pentacene physisorbed on Au (111) by scanning tunneling microscope. *Phys. Rev. Lett.* **102,** 176102 (2009).
49. Gross, L. *et al.* Scattering of surface state electrons at large organic molecules. *Phys. Rev. Lett.* **93,** 056103 (2004).
50. Seufert, K. *et al.* Controlled interaction of surface quantum-well electronic states. *Nano Lett.* **13,** 6130–6135 (2013).
51. Dimitrakopoulos, C. D. & Malenfant, P. R. L. Organic thin film transistors for large area electronics. *Adv. Mater.* **14,** 99–117 (2002).
52. Yokoyama, D. Molecular orientation in small-molecule organic light-emitting diodes. *J. Mater. Chem.* **21,** 19187–19202 (2011).
53. Tkatchenko, A. & Scheffler, M. Accurate molecular van der Waals interactions from ground-state electron density and free-atom reference data. *Phys. Rev. Lett.* **102,** 073005 (2009).
54. Kresse, G. & Hafner, J. *Ab initio* molecular dynamics for liquid metals. *Phys. Rev. B* **47,** 558–561 (1993).

55. Kresse, G. & Furthmüller, J. Efficient iterative schemes for *ab initio* total-energy calculations using a plane-wave basis set. *Phys. Rev. B* **54,** 11169–11186 (1996).
56. Kresse, G. & Joubert, D. From ultrasoft pseudopotentials to the projector augmented-wave method. *Phys. Rev. B* **59,** 1758–1775 (1999).
57. Grimme, S. Semiempirical GGA-type density functional constructed with a long-range dispersion correction. *J. Comp. Chem.* **27,** 1787–1799 (2006).
58. Tang, W., Sanville, E. & Henkelman, G. A grid-based Bader analysis algorithm without lattice bias. *J. Phys. Condens. Matter* **21,** 084204 (2009).

Acknowledgements

We thank Dr Eisuke Ohta and Dr Yuki Suna for the synthesis of BTDA-TCNQ. We are grateful for access to the RIKEN Integrated Cluster of Clusters supercomputer system. T.F. acknowledges funding and support from a Grant-in-Aid for Scientific Research on Innovative Areas 'π-Figuration' (No. 26102001) from the Ministry of Education, Culture, Sports, Science and Technology (MEXT). J.J. acknowledges financial support from 2015 Research Fund of University of Ulsan. J.-H.K. also acknowledges support by Leading Foreign Research Institute Recruitment Program through the National Research Foundation of Korea (NRF-2010-00453).

Author contributions

J.-H.K., J.-C.R., C.A., M.K., J.J., T.F. and Y.K. conceived and designed the system, and Y.K. directed the project. J.-H.K. and Y.K. managed the overall experiment. T.F. prepared the BTDA-TCNQ molecules, and J.-H.K. performed the experiment. J.-H.K. acquired and analysed the STM/STS data. J.J. performed and analysed the DFT calculations. J.-H.K., J.-C.R. and Y.S.Y. fabricated the FET devices and analysed the device characteristics. All the authors discussed the results. J.-H.K. wrote the manuscript with contributions from all the other authors.

Additional information

Competing financial interests: The authors declare no competing financial interests.

14

Non-random walk diffusion enhances the sink strength of semicoherent interfaces

A. Vattré[1], T. Jourdan[2], H. Ding[3], M.-C. Marinica[2] & M.J. Demkowicz[3,4]

Clean, safe and economical nuclear energy requires new materials capable of withstanding severe radiation damage. One strategy of imparting radiation resistance to solids is to incorporate into them a high density of solid-phase interfaces capable of absorbing and annihilating radiation-induced defects. Here we show that elastic interactions between point defects and semicoherent interfaces lead to a marked enhancement in interface sink strength. Our conclusions stem from simulations that integrate first principles, object kinetic Monte Carlo and anisotropic elasticity calculations. Surprisingly, the enhancement in sink strength is not due primarily to increased thermodynamic driving forces, but rather to reduced defect migration barriers, which induce a preferential drift of defects towards interfaces. The sink strength enhancement is highly sensitive to the detailed character of interfacial stresses, suggesting that 'super-sink' interfaces may be designed by optimizing interface stress fields. Such interfaces may be used to create materials with unprecedented resistance to radiation-induced damage.

[1] CEA, DAM, DIF, F-91297 Arpajon, France. [2] CEA, DEN, Service de Recherches de Métallurgie Physique, Université Paris-Saclay, F-91191 Gif-sur-Yvette, France. [3] Department of Materials Science and Engineering, Massachusetts Institute of Technology, Cambridge, Massachusetts 02139, USA. [4] Materials Science and Engineering, Texas A&M University, College Station, Texas 77843, USA. Correspondence and requests for materials should be addressed to A.V. (email: aurelien.vattre@cea.fr).

Damage caused by high-energy particles, such as neutrons or ions, severely limits the performance of materials used in nuclear energy applications[1,2]. It arises from the exceedingly high concentrations—many orders of magnitude greater than under thermodynamic equilibrium—of crystal defects created by impinging radiation[3,4]. Preventing radiation-induced damage in engineering solids requires rapid removal of these defects. Materials resistant to radiation damage would markedly improve the safety, efficiency and sustainability of nuclear energy.

One way of removing radiation-induced defects is to provide a high density of sinks, such as grain boundaries or heterophase interfaces[5–7] that continually absorb defects as they are created. This motivation underlies ongoing exploration of the radiation response of nanocrystalline and nanocomposite materials[8–11], because of the large total interface area per unit volume they contain. These investigations have demonstrated wide variations in sink behaviour of different interfaces. Some easily absorb defects, preventing damage in neighbouring material, but become damaged themselves[12]. Others are poor sinks for isolated defects, but excellent sinks for defect clusters[13,14]. The sink behaviour of yet others changes with radiation dose[15–17]. This wide variety of radiation responses prompts us to ask: are some specific interfaces best suited to mitigate radiation damage? Is it possible to identify them without resorting to resource-intensive irradiation experiments?

To answer these questions, we propose an improved computational method for rapidly assessing the vacancy and interstitial sink strength of semicoherent interfaces. This method builds on a reduced order model for elastic fields of heterophase bicrystals[18]. Such interfaces are of particular interest because many of them contain a high density of defect trapping sites[19–21]. Moreover, semicoherent interfaces generate elastic fields that interact directly with radiation-induced defects[18,22,23]. We show that these elastic fields have an unexpectedly large influence on interface sink strength. Unlike previous studies, which highlighted the importance of thermodynamic driving forces for interface sink behaviour[24–26], we find that the principal effect of the elastic fields is to modify defect diffusivities, causing defects to drift preferentially towards the interface through a non-random walk process. Our work also demonstrates that interface sink strength is highly sensitive to the exact distribution of interface elastic fields. These findings motivate a computational search for 'super-sink' interfaces: ones that optimally attract, absorb and annihilate radiation-induced defects.

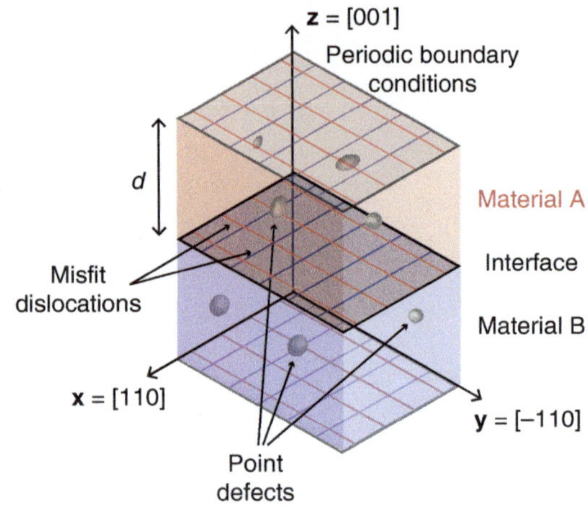

Figure 1 | Schematic illustration of our model. We simulate the diffusion of radiation-induced point defects (illustrated by ovals) to interfaces under the influence of interface elastic fields. In general, materials A and B may be any two crystalline solids. In the present work, they are chosen to be either Cu or Ag (see Table 1 for details).

Results

Kinetic Monte Carlo simulations with elastic interactions. Modelling the removal of radiation-induced point defects at sinks is a challenging task: on one hand, the variety and complexity of defect behaviours call for the flexibility of atomistic modelling. On the other, the relatively slow, thermally activated mechanisms of defect motion require longer simulation times than may be reached using conventional atomistic techniques, such as molecular dynamics. We employ the object kinetic Monte Carlo (OKMC) method[27–30], which is well suited to modelling long-time, thermally activated processes yet is also able to account for nuances of defect behaviour uncovered through atomistic modelling.

Figure 1 illustrates the set up of our simulations. Our models contain two crystalline layers—A and B—separated by semicoherent interfaces. Periodic boundary conditions are applied in all directions, so each model contains two A–B interfaces. Owing to their inherent internal structure, the interfaces create characteristic stress fields in the neighbouring crystalline layers. These stress fields interact with radiation-induced point defects, modifying their diffusion.

We calculate interface stress fields using an efficient, semi-analytical method developed previously[22,31,32]. This method accounts for the elastic anisotropy of solids A and B as well as for differences in elastic constants between them. It generates elastic field solutions consistent with a pre-specified interface crystallographic character (that is, misorientation and interface plane orientation[33]) and with vanishing far-field stresses. The method has been successfully applied to a variety of homo- and hetero-phase interfaces[22,31,32] and has been validated through comparisons with atomistic simulations[18].

Our modelling approach may be used on a wide range of semicoherent interfaces. However, for illustration, we will focus on two specific interfaces in the present work: a low-angle twist grain boundary (GB) on a (001) plane in Ag and a pure misfit (zero misorientation) heterophase interface between (001) planes of Ag and Cu. In our calculations, we use the lattice parameters and elastic constants for Ag and Cu listed in Table 1 (refs 34,35).

Figure 2a shows a plan view of the Ag twist GB, where the adjacent GB planes have been rotated by $\pm\theta/2$ (θ: twist angle). The boundary plane contains two sets of parallel, pure screw dislocations: one aligned with the $\mathbf{x}=[110]$ direction and the other with the $\mathbf{y}=[\bar{1}10]$ direction. For a relative twist angle of $\theta=7.5°$, the spacing between dislocations within each set is $\sim$2.2 nm. Figure 2b shows the interface plane of the Ag–Cu pure misfit interface. Similar to the twist boundary in Fig. 2a, this interface also contains two sets of parallel dislocations aligned with the $\mathbf{x}=[110]$ and $\mathbf{y}=[\bar{1}10]$ directions. Furthermore, the spacing between dislocations in the Ag–Cu interface is the same as in the twist boundary of Fig. 2a: $\sim$2.2 nm. However, unlike in the twist boundary, both sets of dislocations in the misfit interface are of pure edge type.

The two interfaces in Fig. 2 have identical dislocation arrangements, but different dislocation characters. Thus, they contain identical dislocation densities, but have differing stress fields. For instance, all normal stress components for the twist GB are zero throughout the entire bicrystal. This stress field is therefore purely deviatoric. By contrast, owing to symmetry, the shear stress σ_{12} is everywhere zero for the Ag–Cu interface, but all of its other stress components are in general non-zero. In particular, this interface generates significant hydrostatic stresses.

These differences have important implications for interface-defect interactions and defect migration pathways.

We use the force dipole moment approximation to compute elastic interaction energies between point defects and interfaces, $E^{\mathrm{PD/int}}$ (refs 36–38):

$$E^{\mathrm{PD/int}} = -P_{ij}\,\varepsilon_{ij}^{\mathrm{int}}(x,y,z). \quad (1)$$

Here, $\varepsilon_{ij}^{\mathrm{int}}(x,y,z)$ are the components of the previously calculated interface strain field. P_{ij} are the components of the elastic dipole tensor (the '**P**-tensor'), which describes the elastic fields generated by a point defect. $E^{\mathrm{PD/int}}$ values are used to compute stress-dependent energy barriers for defect migration at each location in the simulation cell (see Methods for details). A similar approach has been adopted in previous OKMC studies to describe point defect interactions with dislocations[39,40].

We use density functional theory (DFT) to calculate **P**-tensors for two types of point defects in Ag and Cu (see Methods for details): vacancies and self-interstitials of lowest formation energy, namely ⟨100⟩-split dumbbells[41]. We obtain **P**-tensor values for these defects in their ground states as well as at their saddle point configurations during migration (found using the climbing image nudged elastic band method[42]). Starting from a simulation cell containing a perfect, stress-free crystal, we insert the point defect of interest in the desired location and relax the atom positions while keeping the simulation cell shape fixed. The point defect induces stresses, σ_{ij}, in the simulation cell. They are related to the defect **P**-tensor through

$$P_{ij} = V\sigma_{ij} = P_{ij}^{\mathrm{d}} + p^{\mathrm{h}}\delta_{ij}, \quad (2)$$

where V is the simulation cell volume. P_{ij}^{d} and p^{h} are the deviatoric and hydrostatic (isotropic) **P**-tensor components, respectively. The former is associated with a pure shear (no volume change), whereas the latter is related to isotropic tension (interstitials) or compression (vacancies), which leads to a volume change.

Table 2 lists the **P**-tensors used in the present study. All of them are expressed in the Nye frame, where the **X**, **Y** and **Z** axes are aligned with the [100], [010] and [001] Miller index directions, respectively. The form of the **P**-tensor reflects the symmetry of the corresponding defect. Thus, the **P**-tensor for a vacancy in its ground state is isotropic, whereas that of an interstitial is tetragonal. **P**-tensors for defect orientations other than those given in Table 2 may be calculated using coordinate system rotations. Our **P**-tensors for ⟨100⟩-split dumbbell self-interstitials and vacancies in Cu agree with experimental data[41–44]. Furthermore, the present calculations of relaxation volumes of a vacancy in Ag and Cu are in very good agreement with recent DFT predictions[45].

Figure 3 shows the distribution of ground-state interstitial and vacancy interaction energies with the Ag twist GB and the Ag–Cu misfit interface. A ⟨100⟩-split dumbbell interstitial may take on three different orientations. Figure 3 uses the orientation with lowest $E^{\mathrm{PD/int}}$. For the Ag twist GB, interstitial interaction energies are negative at all locations, as shown in Fig. 3a. Thus, all interstitials in the vicinity of this GB experience a thermodynamic driving force to migrate towards the boundary. The interstitials, however, have nearly isotropic **P**-tensors (see Table 2), so their interaction energies with the Ag twist GB are very small. The interaction energy of vacancies with the Ag twist GB is everywhere zero because of the absence of hydrostatic stresses near this interface. However, the anisotropy of the vacancy saddle

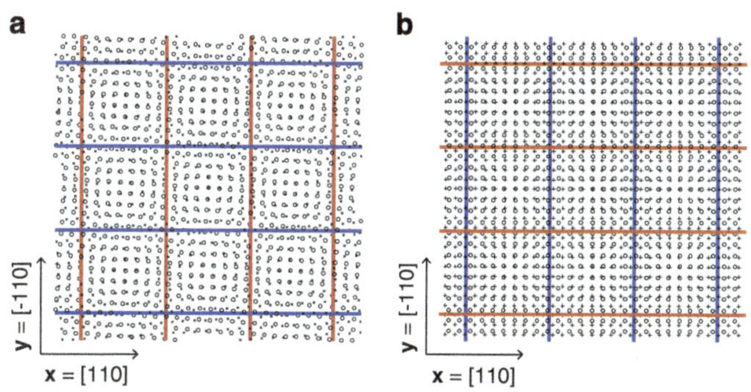

Figure 2 | Planar semicoherent interfaces with identical misfit dislocation arrangements. (**a**) Ag twist GB with pure screw dislocations and (**b**) a Ag–Cu misfit interface with pure edge dislocations.

Table 1 | Materials properties for Silver (Ag) and Copper (Cu).

Element	Cubic lattice parameter (Å)	Elastic constants (GPa)		
		c_{11}	c_{12}	c_{44}
Ag	4.086	124.0	93.4	46.1
Cu	3.615	168.4	121.4	75.5

The lattice parameters and elastic constants of Ag and Cu are used to compute anisotropic elastic fields generated by the semicoherent interfaces with misfit dislocations.

Table 2 | Elastic dipole tensors of point defects from first principles.

Element	Interstitial		Vacancy	
	Ground state	Saddle point	Ground state	Saddle point
Ag	$\begin{bmatrix} 26.80 & 0 & 0 \\ 0 & 26.86 & 0 \\ 0 & 0 & 26.86 \end{bmatrix}$	$\begin{bmatrix} 26.69 & 2.59 & 0 \\ 2.59 & 26.69 & 0 \\ 0 & 0 & 27.74 \end{bmatrix}$	$\begin{bmatrix} -3.04 & 0 & 0 \\ 0 & -3.04 & 0 \\ 0 & 0 & -3.04 \end{bmatrix}$	$\begin{bmatrix} -2.64 & -0.39 & 0 \\ -0.39 & -2.64 & 0 \\ 0 & 0 & 2.15 \end{bmatrix}$
Cu	$\begin{bmatrix} 17.46 & 0 & 0 \\ 0 & 17.66 & 0 \\ 0 & 0 & 17.66 \end{bmatrix}$	$\begin{bmatrix} 18.01 & 1.78 & 0 \\ 1.78 & 18.01 & 0 \\ 0 & 0 & 18.46 \end{bmatrix}$	$\begin{bmatrix} -3.19 & 0 & 0 \\ 0 & -3.19 & 0 \\ 0 & 0 & -3.19 \end{bmatrix}$	$\begin{bmatrix} -3.61 & -0.37 & 0 \\ -0.37 & -3.61 & 0 \\ 0 & 0 & 2.12 \end{bmatrix}$

P-tensors (in eV) for a ⟨100⟩-split dumbbell self-interstitial and a vacancy in Ag and Cu at both the ground-state and saddle point configurations. The ground-state interstitial is oriented in the [100] direction. Its saddle point configuration is for a [100]-to-[010] migration path (see Methods for details). The vacancy saddle point is for migration along the [110] direction.

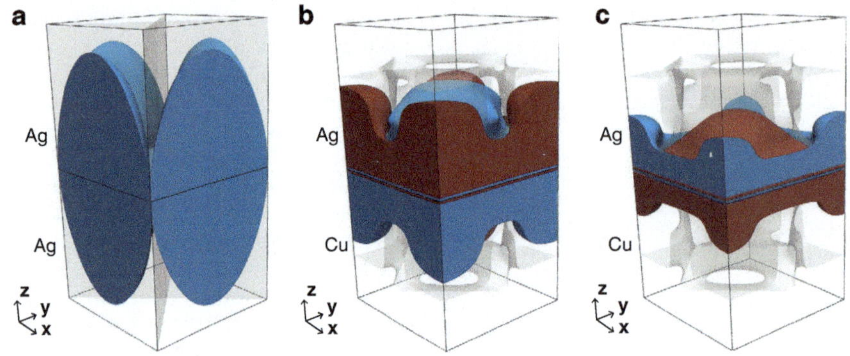

Figure 3 | Elastic energy distributions around modelled interfaces. Elastic interaction energy between (**a**) an interstitial with the Ag twist GB ($E^{PD/int} < -0.002$ eV in the blue isovolume), and between the Ag-Cu misfit interface with (**b**) an interstitial and (**c**) a vacancy ($E^{PD/int} < -0.06$ eV in the blue isovolume; $E^{PD/int} > 0.06$ eV in the red; grey contours are locations with zero interaction energy).

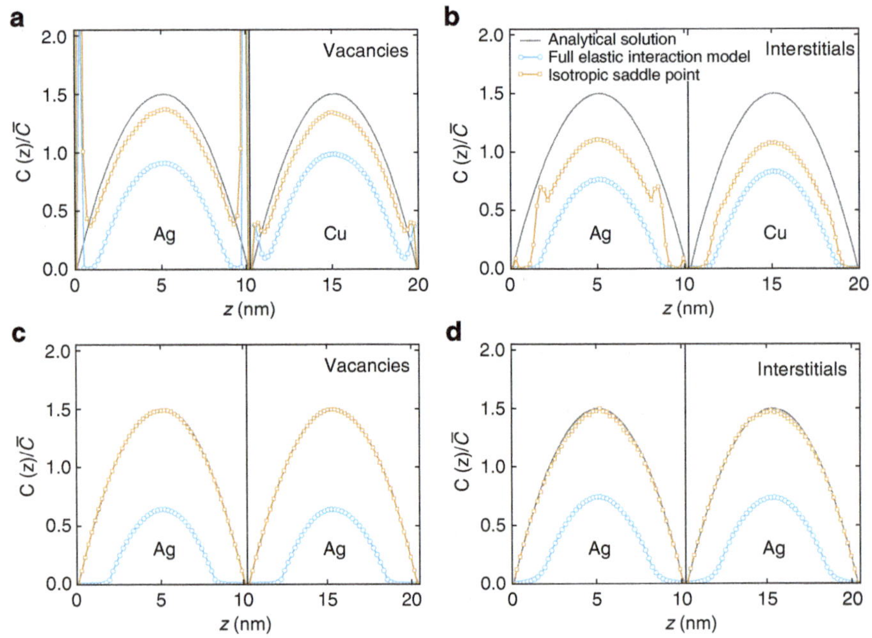

Figure 4 | Steady-state point defect concentrations as a function of location normal to interface planes. The black vertical lines represent the interface planes, whereas the continuous grey lines denote the reference case with no elastic interactions, computed analytically (see Methods for details). OKMC results for both isotropic (orange) and anisotropic (blue) saddle point configurations are shown. (**a**) Vacancy and (**b**) interstitial profiles near Ag-Cu pure misfit interfaces. (**c**) Vacancy and (**d**) interstitial profiles near Ag twist GBs. Concentrations are normalized by the average concentration $\bar{C}$ obtained when no elastic interactions are taken into account (see Methods for details).

point configuration leads to non-zero interaction energies of migrating vacancies with the GB.

Interstitial interaction energies near the Ag–Cu misfit interface, shown in Fig. 3b, may be attractive or repulsive, depending on the location of the defect. Thus, interstitials in Ag are expected to migrate towards the centre of the dislocation pattern, whereas those in Cu are expected to migrate to dislocation cores. Figure 3c shows the interaction energy between vacancies and the Ag–Cu misfit interface. The spatial variation of this interaction energy is similar to that of the interstitials, but with opposite sign.

Our OKMC simulations assume a constant, uniform defect creation rate, *G*. Defects diffuse until they are absorbed by an interface (see Methods for details). Only individual interstitials or vacancies are tracked in our simulations: defect reactions, such as clustering or recombination, are not considered. After a certain simulation time, defect distributions reach a steady state, whereupon the defect concentration is computed as a function of position along the *z*-direction (normal to the interface plane) based on the time spent by each defect on a given atomic site.

Effect of elastic interactions on interface sink strength. Figure 4 shows steady-state vacancy and interstitial concentrations for the two types of interfaces described above for models with 10-nm-thick Ag and Cu layers. In the absence of elastic interactions between defects and interfaces, steady-state defect concentrations may be computed analytically (see Methods for details). We compare our simulation results with these analytical solutions.

Elastic interactions have a dramatic effect on defect concentration profiles. In all cases shown in Fig. 4 except vacancies near Ag–Cu interfaces, there are nearly no defects within ~2-nm-wide zones adjacent to the interfaces. By contrast, without elastic interactions, defect concentrations are zero only at the interfaces themselves. Moreover, even though defect-interface elastic interaction energies are negligible beyond ~2 nm, the zones depleted of defects near the interfaces have a pronounced effect on defect concentrations throughout the entire layer, markedly reducing the average defect concentration. For the simulations in Fig. 4, elastic interactions reduce defect concentrations by about a factor of two even in the middle of the layers. This effect is even more pronounced for thinner layers. For vacancies in Ag–Cu, local traps are responsible for the sharp increase in concentration near the interface.

Our simulations account for numerous aspects of defect-interface elastic interactions, such as defect anisotropy or differences in defect ground-state and saddle point properties. To discover which ones are primarily responsible for the defect concentrations shown in Fig. 4, we artificially 'switch off' some of these characteristics and repeat our OKMC simulations to see whether doing so changes the steady-state defect concentrations. These calculations lead us to conclude that the anisotropy of the **P**-tensor in the saddle point configurations is primarily responsible for the reduced defect concentrations in Fig. 4a,b.

We 'switch off' saddle point anisotropy by replacing the saddle point **P**-tensor with $P_{ij}^{\mathrm{sad}} = p_{\mathrm{sad}}^{\mathrm{h}}\delta_{ij}$, where δ is the Kronecker delta and $p_{\mathrm{sad}}^{\mathrm{h}}$ is one-third of the trace of the true saddle point **P**-tensor. This assumption is tantamount to modelling defects at saddle points as misfitting spherical inclusions in isotropic media. Concentration profiles obtained with this approximation are markedly different from the anisotropic case, as shown in Fig. 4. In the case of the Ag twist GB (Fig. 4c,d), isotropic saddle points yield the same defect concentrations as when there are no defect-interface interactions at all. Indeed, as the twist interface generates no hydrostatic strain field, only the deviatoric components of defect **P**-tensors may interact with these interfaces. Ground-state vacancies have zero deviatoric **P**-tensor components, so the interaction energy with the Ag twist GB vanishes, similar to ground-state interstitials with nearly isotropic

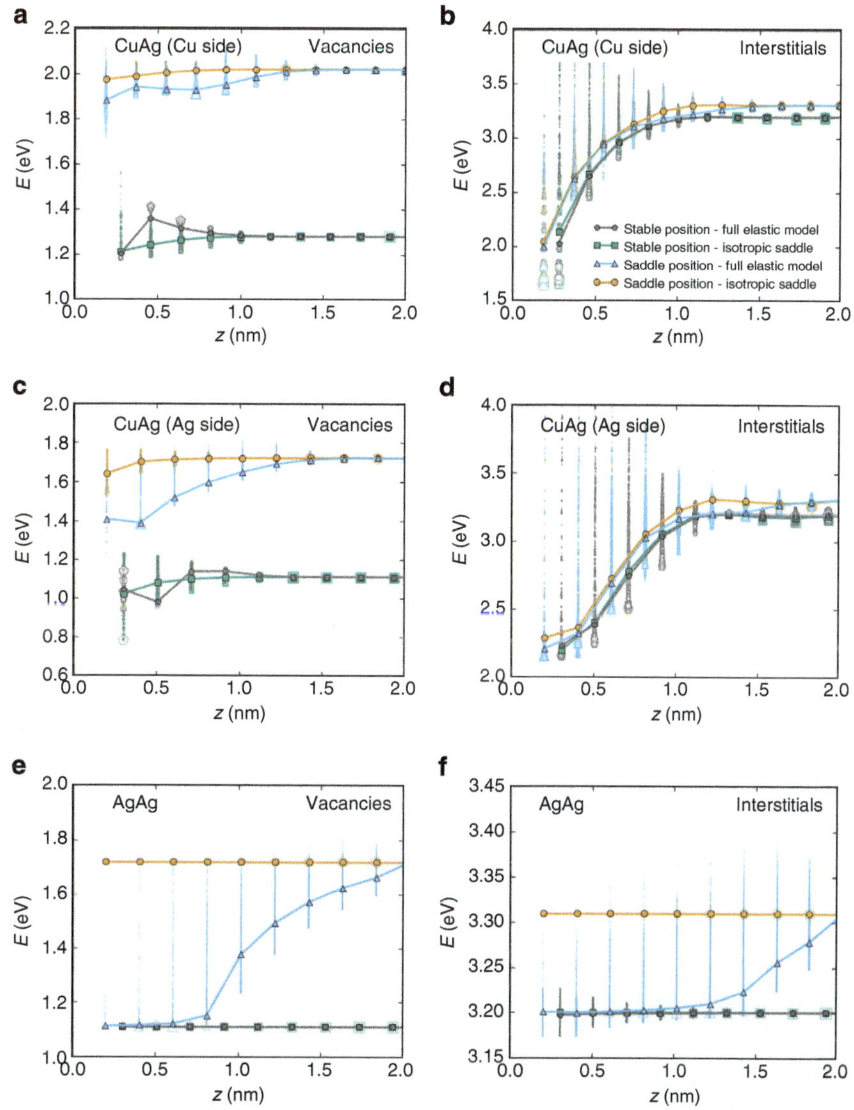

Figure 5 | Effect of saddle point anisotropy on point defect energy landscapes. Mean stable (grey and green lines) and saddle (blue and orange) energy profiles as a function of distance to the interface for **(a)** vacancies and **(b)** interstitials on the Cu side and on the **(c,d)** Ag side of the Ag-Cu interface and for **(e)** vacancies and **(f)** interstitials of the Ag twist GB. Averages are computed over 10^4 trajectories and the dispersion on the energy configurations explored by the defects is highlighted by the open symbols. The size of symbols is proportional to the times a defect passes through a given energy configuration. Orange (green) and blue (grey) lines correspond to simulations without saddle point anisotropy and with the fully anisotropic interaction model, respectively.

P-tensors (Table 2). The same conclusions hold at saddle positions if saddle point anisotropy is 'switched off', as described above. Elastic interactions then do not affect migration energies, explaining why defect concentrations are identical to the case without elastic interactions.

For the Ag–Cu interface, concentration profiles computed without saddle point anisotropy lie between the non-interacting and fully anisotropic cases, as shown in Fig. 4a,b. Vacancy concentrations are only marginally lower than the non-interacting case (Fig. 4a), demonstrating the overriding importance of saddle point anisotropy in their behaviour. Interstitial concentrations obtained without saddle anisotropy lie approximately mid-way between the fully anisotropic and non-interacting cases (Fig. 4b), demonstrating that saddle point anisotropy is at least as important to their behaviour as are $p\Delta V$ interactions, which are more commonly investigated.

To investigate the effect of saddle point anisotropy on saddle point energy, Fig. 5 shows mean stable and saddle energy landscapes for point defects as a function of distance to the interfaces. Although the scatter in saddle point energies can be high at a given distance to interfaces (specially for interstitials near the Ag–Cu misfit interface), the mean saddle point energies (blue lines) are always lower than the mean saddle energies computed with the isotropic approximation (orange lines). For the Ag twist GB, the mean of the energies of saddle positions explored by vacancies and interstitials using the full elastic model decreases dramatically with decreasing distance to the GB. The case of vacancies near the Ag–Cu misfit interface illustrates the same trend, showing that the decrease in migration energy is due entirely to the saddle point configuration. In the case of interstitials near the Ag–Cu misfit interface, both migration energies and thermodynamic driving forces contribute to the enhancement of sink strength. In particular, a reduction of barriers is obtained when $z < 1.5$ nm, thus giving rise to a non-random walk drift towards the Ag–Cu interface. The migration energy effect is especially pronounced in the case of vacancies and for interfaces without hydrostatic stresses.

This analysis therefore shows that, in addition to the classical thermodynamic driving force, reduced migration barriers contribute significantly to sink efficiency and that these reduced barriers are unequivocally due to saddle point anisotropy. The effect of interface elastic fields on defect migration energies is similar to that found in the vicinity of isolated dislocations[46–49]. However, elastodiffusion near interfaces—including semicoherent ones—is not reducible to elastodiffusion near dislocations because the stress fields of the former are in general composed of two contributions: one arising from infinite arrays of discrete dislocations and the other from coherency stresses. These contributions cancel perfectly in the far field, whereas in the near field the cancellation is imperfect[22,31,32]. It is only this imperfect cancellation in the near field that affects defect diffusion near interfaces.

Figure 6 gives a more detailed view of defect concentrations at different locations in the Ag layer of the Ag–Cu interface and in the Ag twist GB. Close to these interfaces, concentrations vary as a function of location parallel to the interface plane, following the strain field pattern created by the interfaces. Indeed, the strain field creates preferential paths for defect migration, as shown by the grey trajectories in Fig. 6. These paths are in general different for interstitials and vacancies. For both the Ag–Cu interface and Ag twist GB, vacancies preferentially migrate to the dislocation lines, whereas interstitials are mostly absorbed between dislocations. This preferential, non-random walk drift of point defects to specific locations is responsible for the enhanced interface sink strengths.

Knowing the steady-state defect concentrations obtained by OKMC, we derive sink strengths for the two interfaces considered above. In the mean field rate theory formalism[50], 'sink strengths' quantify the ability of sinks, such as interfaces, to absorb defects. Within this formalism, the evolution equation for the average defect concentration, $\overline{C}$, follows

$$\frac{d\overline{C}}{dt} = G - k^2 D\overline{C}, \tag{3}$$

where G is the defect creation rate and D is bulk defect diffusivity. The second term on the right hand side is related to the loss of defects at sinks with associated sink strength, k^2. At steady state, the sink strength may be computed from the average concentration:

$$k^2 = \frac{G}{D\overline{C}}. \tag{4}$$

Using the average of the concentration profile computed for defect removal at interfaces in the absence of elastic interactions (see Methods for details), the interface sink strength is analytically found to be $k^2 = 12/d^2$ (ref. 51). When interactions between interfaces and defects are present, we numerically determine the sink strength through equation (4), by using the average steady-state concentration obtained by OKMC simulations and the diffusion coefficient without elastic interactions. The resulting vacancy and interstitial sink strengths for both interfaces are shown in Fig. 7a–f as a function of layer thickness.

In all cases, the sink strength increases significantly when elastic interactions are taken into account. This effect is especially pronounced for thinner layers, as defects undergo elastic interactions with interfaces over a larger fraction of the layer. It is particularly strong for interstitials, whatever the interface type, and for vacancies for the twist interface. These results also confirm the importance of saddle point anisotropy: by comparing with OKMC simulations that use isotropic saddle-point **P**-tensors, we see that it yields order-of-magnitude increases in sink strength, in some cases.

Another quantity of interest for radiation response is the bias factor, B, which expresses the propensity of a given sink to absorb more interstitials than vacancies. It is defined as

$$B = \frac{k_i^2 - k_v^2}{k_i^2}, \tag{5}$$

where k_v^2 and k_i^2 are the sink strengths for vacancies and interstitials, respectively. For example, small interstitial clusters and dislocations exhibit positive bias factors (typically between 0.01 and 0.3 (refs 52,53)) and thus absorb more interstitials than vacancies. The preferential absorption of interstitials by biased sinks leads to an excess of remaining vacancies, which cluster and eventually aggregate into voids[52,54].

Bias factors for the semicoherent interfaces we studied are shown in Fig. 7g–i. Values larger than 0.2 are obtained for the fully anisotropic interaction model in the case of the Ag–Cu interface. Such interfaces would compete for interstitials with dislocations. The presence of two sinks of differing bias magnitude has been given as a possible cause for void swelling suppression in ferritic steels[55]. Interestingly, for the Ag twist GB, the bias factor is negative, meaning that these interfaces tend to absorb more vacancies than interstitials. Similar observations have been made in ref. 56, where the bias factor for single screw dislocations is negative when using anisotropic elasticity theory and zero in the isotropic approximation. Such GBs may therefore deplete excess vacancy concentrations sufficiently to inhibit void nucleation.

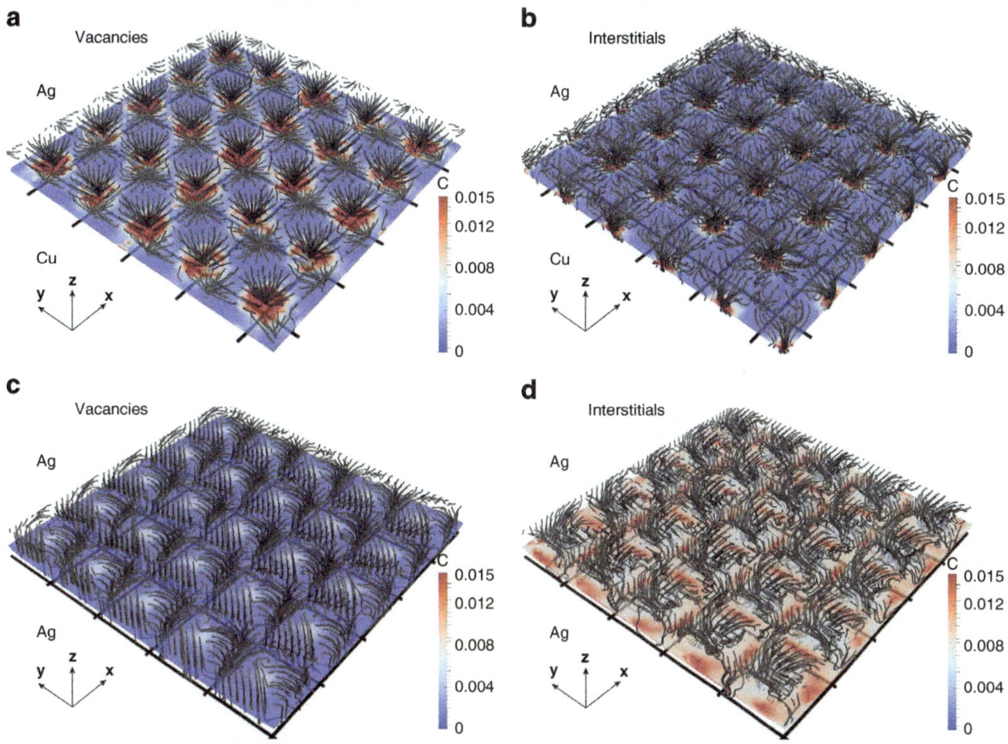

Figure 6 | Preferential drift of point defects towards semicoherent interfaces. Migration paths and local concentrations of (**a**) vacancies and (**b**) interstitials on the Ag side of the Ag-Cu interface and of (**c**) vacancies and (**d**) interstitials in the Ag twist GB. Migration paths are shown as grey lines originating from 1 nm away from the interface. The square grid of black lines represents interface dislocations. Concentrations are plotted in a plane located two atomic distances away from the interface. The concentrations are normalized by $\bar{C}$: the average concentration when no interactions are considered. Any normalized concentration values higher than 0.015 are shown as equal to 0.015.

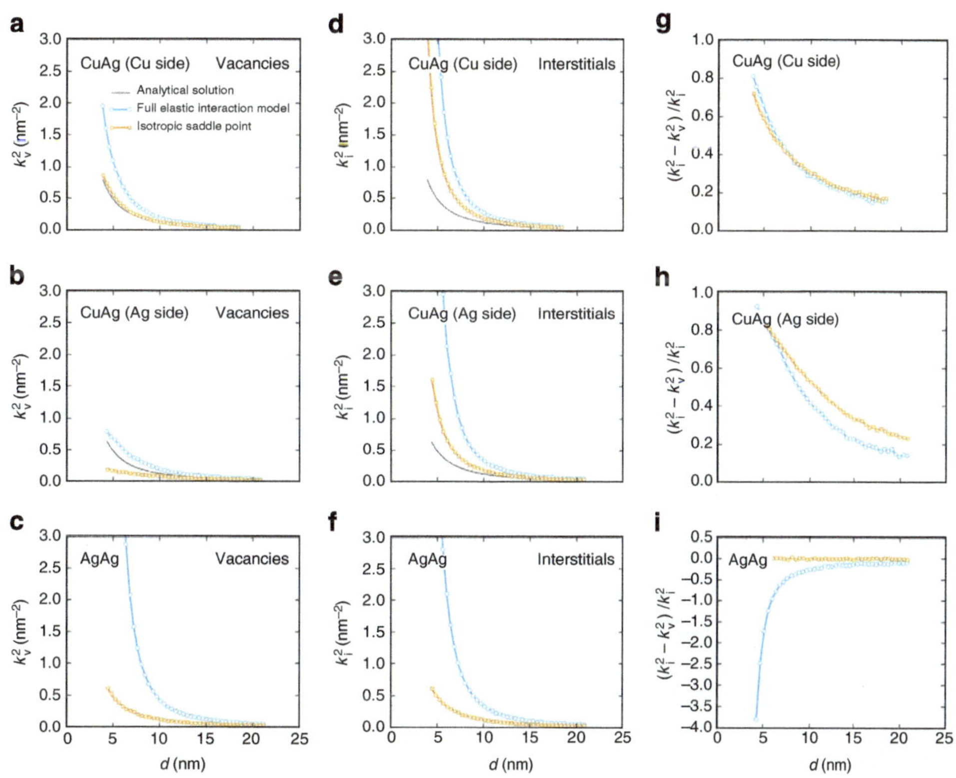

Figure 7 | Enhancement in sink strength of semicoherent interfaces. Sink behaviours of Ag-Cu interfaces and Ag twist GBs for (**a-c**) vacancies (k_v^2) and (**d-f**) interstitials (k_i^2) in a given layer (Ag or Cu), as a function of layer thickness, *d*. (**g-i**) Bias factors of Ag-Cu interface and Ag twist GB. The grey line corresponds to the analytical solution when no interaction is present ($k^2 = 12/d^2$). Orange and blue lines correspond to OKMC calculations without saddle point anisotropy and with the fully anisotropic interaction model, respectively.

Discussion

Our work demonstrates that elastic interactions between radiation-induced point defects and semicoherent interfaces lead to significant increases in interface sink strength, compared with the case with no defect-interface interactions. These conclusions are consistent with other recent simulations, which show that elastic interactions also have a significant influence on defect-dislocation interactions[39,40,56] and on mutual interactions between radiation-induced dislocation loops[57–59].

Our simulations identify non-hydrostatic interactions of point defects in saddle point configurations as the main cause of enhanced interface sink strengths. How might one incorporate such interactions into continuum-level diffusion models? Compared with diffusion in the absence of stress fields, the effect of local variations in defect ground-state and saddle point energies is twofold[44,60]: the diffusivity gains a spatial dependence and the continuum diffusion equation gains a drift term. These modifications are sufficient to model defects that undergo purely hydrostatic, $p\Delta V$ interactions with elastic fields. When non-hydrostatic interactions are considered, local diffusivities are, in general, no longer isotropic and the scalar diffusivity must be replaced by a diffusivity tensor. The effect of saddle point anisotropy on vacancy diffusion may be modelled this way.

Modelling the effect of saddle point anisotropy on the diffusion of $\langle 100 \rangle$-split dumbbell self-interstitials furthermore requires tracking the orientation of the interstitials. Therefore, three separate interstitials concentrations (corresponding to the three different dumbbell orientations) must be defined[44]. As these interstitials re-orient during each migration step (see Methods for details), the evolution of their concentrations is described by three coupled diffusion equations, each with a different location-dependent diffusivity tensor. It may also be necessary to account for thermally activated interstitial re-orientation without migration[61]. Posing and solving a continuum model for such a complex process is likely to remain beyond the scope of most continuum modelling studies in the near future. Thus, OKMC simulations are likely to remain the workhorse method for investigating the effect of elastic interactions on point defect diffusion, especially of self-interstitials.

Our simulations show that the Ag twist GB and the Ag–Cu interface have markedly different sink strengths, even though both have identical dislocation densities. These differences arise from unlike detailed strain distributions, which are due to the different dislocation characters in these interfaces and the unlike coherency strains in the reference states of each interface. The orientations of the adjacent crystals in the twist GB are related through a pure rotation, so the GB has no coherency strains. By contrast, in the Ag–Cu pure misfit interface, unequal stresses of opposite sign are needed to impose coherency[22,31,32].

The sensitivity of interface sink strength to interface elastic fields opens up new opportunities for materials design. For instance, 'super-sink' interfaces with maximal sink strengths may be created by optimizing interface elastic fields. Interface elastic fields may also be designed to yield a desired bias factor. The dependence of interface elastic fields on interface crystallography and the physical properties of the adjoining crystals are well-established[18,22]. Strategies for scanning the interface design space and locating optimal interface crystallographies and compositions have also been developed[18,62].

An advantage of the OKMC-based simulation method presented here is that it may be systematically improved to enhance its accuracy. For example, our simulations did not take into account the interactions of defects with each other. One extension of the model would therefore be to include defect reactions, such as vacancy-interstitial recombination or clustering of like defects[63]. Another simplifying assumption of the current simulations is that they consider only the first-order term in the interaction energy between interfaces and defects[44]. An improved model might account for other interactions, such as higher order moments of the multipole expansion[37] or heterogeneity interactions[64]. The internal structure of interfaces may change because of loading of point defects[15]. Our computational method may be further improved by modelling this structure evolution, for example, following the approach of Uberuaga *et al.*[65]. Finally, all of our simulations assumed perfect trapping of defects that arrive at an interface (see Methods for details), which corresponds to an interface sink efficiency of unity[66]. This assumption may be relaxed to model the sink efficiency of interfaces that are not perfect defect sinks.

Methods

Interface elastic field calculation. We compute the complete elastic strain fields ε^{int} of interface dislocation arrays using the method described in ref. 22. The geometry of the interface dislocation pattern is found by solving the quantized Frank–Bilby equation[33]. Interface dislocations are viewed as Volterra dislocations that have been inserted into a single crystal, coherent reference interface. The complete interface strain is the superposition of the coherency strain ε^{coh} in the reference interface and the strain field of the interface dislocations ε^{int}:

$$\varepsilon^{\text{int}}(x,y,z) = \varepsilon^{\text{coh}} + \varepsilon^{\text{dis}}(x,y,z) = \varepsilon^{\text{coh}} + \sum_k e^{2i\pi k\cdot r(x,y)}\varepsilon_k(z). \quad (6)$$

ε^{dis} is represented using a Fourier expansion on the right hand side of the expression above. The dislocation strain field must satisfy mechanical equilibrium in both of the adjacent crystals:

$$\nabla \cdot (\mathbb{C} : \varepsilon^{\text{int}}(x,y,z)) = \mathbf{0}, \quad (7)$$

where $\mathbb{C}$ the fourth-order anisotropic elasticity tensor. Substituting equation (6) into equation (7), the dislocation strain field is obtained by solving the sextic eigenvalue problem developed by Stroh[67] with specific boundary conditions dedicated to interface dislocations:

- No net far-field strains.
- Consistency of far-field rotations with the prescribed interface misorientation.
- No net tractions along the interface.
- Interface displacement discontinuity matches the disregistry of the desired dislocation pattern.

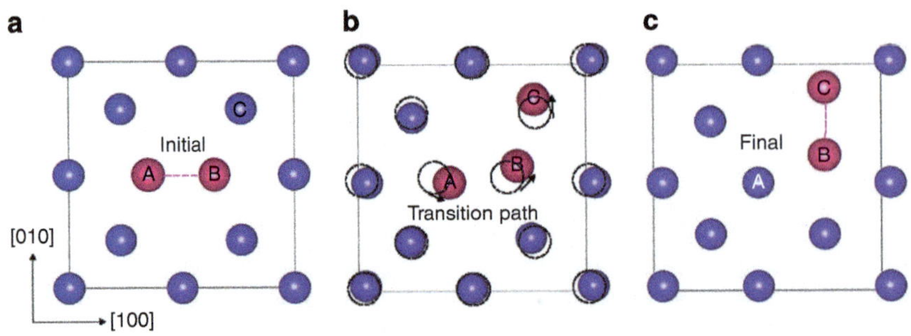

Figure 8 | The migration mechanism of a $\langle 100 \rangle$-split dumbbell interstitial in a face-centred cubic crystal. (**a**) Initial dumbbell configuration, (**b**) saddle point configuration and (**c**) final dumbbell configuration. Initial atom positions are shown by dashed circles in **b**.

Elastic dipole tensor calculation. Defect **P**-tensors are calculated using VASP (the Vienna Ab initio Simulation Package[68]), a plane wave-based, first principles DFT code. A face-centred cubic supercell containing 256 ± 1 atoms (+ 1 and − 1 for interstitial and vacancy, respectively) is used. We also performed LAMMPS (Large-scale Atomic/Molecular Massively Parallel Simulator[69]) classical potential simulations using embedded atom method potentials for Ag (ref. 70) and Cu (ref. 71) to study the convergence of the elastic dipole tensors up to supercell sizes of 2,048 atoms. We find that the discrepancy in **P**-tensor components between the 256-atom supercell and that of 2,048-atom supercell is lower than 4%. This supercell size ensures the convergence of defect formation energies to within few meV, as detailed in the Supplementary Note and Supplementary Figs 1 and 2. We therefore view the 256-atom DFT simulations as well converged with respect to model size.

A 3 × 3 × 3 shifted Monkhorst–Pack *k*-point grid mesh, a Hermite–Gaussian broadening of 0.25 eV (ref. 72) and a plane wave cutoff energy of 400 eV are used. The change of the elastic dipole tensors is less than 0.5% compared with tighter settings. We use the Perdew–Burke–Ernzerhof[73] exchange-correlation functional within the projector-augmented-wave approach[74]. The structures are internally relaxed with a force convergence criterion of 10^{-3} eV Å^{-1}. The climbing image-nudged elastic band method[42] is employed to find the saddle points for defect migration.

In the migration of a vacancy, one of the atoms directly neighbouring the vacancy travels to the vacant site, leaving behind a new vacancy. The more complex migration mechanism of a ⟨100⟩-split dumbbell interstitial is shown in Fig. 8. Here, one atom (B) in the initial dumbbell configuration (A − B, Fig. 8a) migrates to a neighbouring lattice site (C), forming a new ⟨100⟩ dumbbell (B − C, Fig. 8c). Thus, three atoms (A, B and C) are involved in this migration mechanism. The initial and final dumbbells, as well as the transition path, are confined to the same lattice plane. The initial and final dumbbell orientations are orthogonal to each other.

OKMC algorithm. We model defect diffusion using an OKMC code with a residence time algorithm to advance the simulation clock[27,28]. At time *t*, the time step is chosen according to $\Delta t = -(\ln r_1)/w_{tot}$, where r_1 is a random number with $r_1 \in]0,1]$ and w_{tot} is the sum of frequencies of all events $\left(w_{tot} = \sum_i^N w_i\right)$ that may occur at *t*. The chosen event *j* is such that $\sum_i^{j-1} w_i < r_2 w_{tot} \leq \sum_i^j w_i$, where r_2 is another random number with $r_2 \in]0,1]$.

Three kinds of events are considered in the simulations: the jump of a point defect from one stable point to a neighbouring one, the absorption of a defect by an interface and the creation of a new point defect through irradiation. Jump frequencies are given by $w_i = \nu\exp(-\Delta E_i/(kT))$, where ν is an attempt frequency and $\Delta E_i = E_i^{sad} - E_i^{sta}$ is the energy difference between the saddle position and the initial stable position of the jump considered. The stable point energy is

$$E_i^{sta} = -\sum_{k,l} P_{kl,i}^{sta}\, \varepsilon_{kl}^{int}\left(\mathbf{r}_i^{sta}\right), \tag{8}$$

whereas the saddle point energy is

$$E_i^{sad} = E^{m} - \sum_{k,l} P_{kl,i}^{sad}\, \varepsilon_{kl}^{int}\left(\mathbf{r}_i^{sad}\right), \tag{9}$$

with E^m the migration energy in the absence of elastic interactions.

Here, $\mathbf{P}^{sta}$ and $\mathbf{P}^{sad}$ are the defect **P**-tensors in the ground-state and saddle point configurations, respectively. For simplicity, the position of the saddle point $\mathbf{r}_i^{sad}$ is taken mid-way between the two stable points explored by the jump[40].

The defect is considered to have been absorbed by an interface if it reaches the nearest atomic row to the interface. It is then simply removed from the simulation. This absorption condition is used to obtain a first estimate of sink strength, without taking into account the diffusion of point defects along interfaces or their possible reemission. The irradiation rate is fixed at the beginning of each simulation to keep the average number of point defects equal to 200 in the material where the measurements are performed, if no elastic interactions are considered. The actual number of point defects in the system, averaged over the simulation time when steady state is reached, constitutes the basis for our sink strength calculation.

The concentration of defects is recorded every 10^4 iterations, after the concentration has become stationary. At the end of the simulation, an estimate of $\overline{C}$ is computed by averaging over the values C_j $(j = 1,\ldots,n)$:

$$\overline{C}_n = \frac{1}{n}\sum_{j=1}^{n} C_j. \tag{10}$$

The final time is adjusted to obtain sufficient accuracy on $\overline{C}$ and thus on k^2. For this purpose, the estimation of the error on the concentration is given by the standard error of the mean value, that is,

$$\delta\overline{C}_n = \frac{\sigma_n}{\sqrt{n}}, \tag{11}$$

where

$$\sigma_n^2 = \frac{1}{n-1}\sum_{j=1}^{n}\left(C_j - \overline{C}_n\right)^2. \tag{12}$$

The final time for each system is chosen so that the relative error on $\overline{C}$ and k^2 is less than 0.5%.

Analytical solution in the absence of elastic interactions. When there are no defect–interface interactions, the steady-state concentration in a flat crystalline layer may be found analytically. We consider a layer of thickness *d* with interfaces at $z = -d/2$ and $z = d/2$, where zero concentration Dirichlet conditions are imposed. Defects are created by radiation at a constant rate, *G* (per atomic site and per second) and diffuse with a diffusion coefficient *D*. Solving the steady-state diffusion equation leads to

$$C(z) = -\frac{G}{2D}\left(z - \frac{d}{2}\right)\left(z + \frac{d}{2}\right), \tag{13}$$

from which the average concentration $\overline{C}$ per atomic site is readily deduced:

$$\overline{C} = \frac{Gd^2}{12D}. \tag{14}$$

Taking into account admissible jumps for a ⟨100⟩-split dumbbell interstitial in a face-centred cubic lattice, this interstitial diffusion coefficient is

$$D = \tfrac{2}{3}a^2\nu\exp\left(-\frac{E^m}{kT}\right), \tag{15}$$

whereas in the case of vacancies it is

$$D = a^2\nu\exp\left(-\frac{E^m}{kT}\right). \tag{16}$$

References

1. Zinkle, S. J. & Was, G. S. Materials challenges in nuclear energy. *Acta Mater.* **61,** 735–758 (2013).
2. Zinkle, S. J. & Snead, L. L. Designing radiation resistance in materials for fusion energy. *Annu. Rev. Mater. Res.* **44,** 241–267 (2014).
3. Averback, R. S. & Diaz de la Rubia, T. in *Solid State Physics: Advances in Research and Applications* (eds Ehrenreich, H. & Spaepen, F.) (Academic Press Inc, 1998).
4. Bacon, D. J. & Osetsky, Y. N. Modelling atomic scale radiation damage processes and effects in metals. *Int. Mater. Rev.* **47,** 233–241 (2002).
5. Singh, B. N. Effect of grain size on void formation during high-energy electron irradiation of austenitic stainless steel. *Philos. Mag.* **29,** 25–42 (1974).
6. Demkowicz, M. J., Misra, A. & Caro, A. The role of interface structure in controlling high helium concentrations. *Curr. Opin. Solid State Mat. Sci.* **16,** 101–108 (2012).
7. Han, W. Z. *et al.* Design of radiation tolerant materials via interface engineering. *Adv. Mater.* **25,** 6975–6979 (2013).
8. Odette, G. R., Alinger, M. J. & Wirth, B. D. Recent developments in irradiation-resistant steels. *Ann. Rev. Mater. Res.* **38,** 471–503 (2008).
9. Demkowicz, M. J., Bellon, P. & Wirth, B. D. Atomic-scale design of radiation-tolerant nanocomposites. *MRS Bull.* **35,** 992–998 (2010).
10. Beyerlein, I. J. *et al.* Radiation damage tolerant nanomaterials. *Mater. Today* **16,** 443–449 (2013).
11. Chen, Y. *et al.* Damage tolerant nanotwinned metals with nanovoids under radiation environments. *Nat. Commun.* **6,** 7036 (2015).
12. Han, W. Z., Fu, E. G., Demkowicz, M. J., Wang, Y. Q. & Misra, A. Irradiation damage of single crystal, coarse-grained, and nanograined copper under helium bombardment at 450°C. *J. Mater. Res.* **28,** 2763–2770 (2013).
13. Demkowicz, M. J., Anderoglu, O., Zhang, X. & Misra, A. The influence of Σ3 twin boundaries on the formation of radiation-induced defect clusters in nanotwinned Cu. *J. Mater. Res.* **26,** 1666 (2011).
14. Yu, K. Y. *et al.* Removal of stacking-fault tetrahedra by twin boundaries in nanotwinned metals. *Nat. Commun.* **4,** 1377 (2013).
15. Bai, X. M., Voter, A. F., Hoagland, R. G., Nastasi, M. & Uberuaga, B. P. Efficient annealing of radiation damage near grain boundaries via interstitial emission. *Science* **327,** 1631–1634 (2010).
16. Bai, X. M. *et al.* Role of atomic structure on grain boundary-defect interactions in Cu. *Phys. Rev. B* **85,** 214103 (2012).
17. Yu, W. S. & Demkowicz, M. J. Non-coherent Cu grain boundaries driven by continuous vacancy loading. *J. Mater. Sci.* **50,** 4047–4065 (2015).
18. Vattré, A. J., Abdolrahim, N., Kolluri, K. & Demkowicz, M. J. Computational design of patterned interfaces using reduced order models. *Sci. Rep.* **4,** 6231 (2014).
19. Demkowicz, M. J., Wang, J. & Hoagland, R. G. in *Dislocations in Solids* (ed. Hirth, J. P.) (Elsevier, 2008).
20. Demkowicz, M. J., Hoagland, R. G. & Hirth, J. P. Interface structure and radiation damage resistance in Cu-Nb multilayer nanocomposites. *Phys. Rev. Lett.* **100,** 136102 (2008).
21. Shao, S., Wang, J., Misra, A. & Hoagland, R. G. Spiral patterns of dislocations at nodes in (111) semi-coherent FCC interfaces. *Sci. Rep.* **3,** 2448 (2013).
22. Vattré, A. J. & Demkowicz, M. J. Determining the Burgers vectors and elastic strain energies of interface dislocation arrays using anisotropic elasticity theory. *Acta Mater.* **61,** 5172–5187 (2013).

23. Hirth, J. P., Pond, R. C., Hoagland, R. G., Liu, X. Y. & Wang, J. Interface defects, reference spaces and the Frank-Bilby equation. *Prog. Mater. Sci.* **58,** 749–823 (2013).
24. Balluffi, R. W. & Granato, A. V. in *Dislocations in Solids* (ed. Nabarro, F. R. N.) (North-Holland Pub Co, 1979).
25. King, A. H. & Smith, D. A. Calculations of sink strength and bias for point-defect absorption by dislocations in arrays. *Radiat. Eff.* **54,** 169–176 (1981).
26. Jiang, C., Swaminathan, N., Deng, J., Morgan, D. & Szlufarska, I. Effect of grain boundary stresses on sink strength. *Mater. Res. Lett.* **2,** 100–106 (2013).
27. Bortz, A. B., Kalos, M. H. & Lebowitz, J. L. A new algorithm for Monte Carlo simulation of Ising spin systems. *J. Comput. Phys.* **17,** 10–18 (1975).
28. Gillespie, D. T. A general method for numerically simulating the stochastic time evolution of coupled chemical reactions. *J. Comput. Phys.* **22,** 403–434 (1976).
29. Caturla, M. J. *et al.* Comparative study of radiation damage accumulation in Cu and Fe. *J. Nucl. Mater.* **276,** 13–21 (2000).
30. Jourdan, T., Bocquet, J. L. & Soisson, F. Modeling homogeneous precipitation with an event-based Monte Carlo method: application to the case of Fe-Cu. *Acta Mater.* **58,** 3295–3302 (2010).
31. Vattré, A. J. & Demkowicz, M. J. Effect of interface dislocation Burgers vectors on elastic fields in anisotropic bicrystals. *Comput. Mater. Sci.* **88,** 110–115 (2014).
32. Vattré, A. J. & Demkowicz, M. J. Partitioning of elastic distortions at a semicoherent heterophase interface between anisotropic crystals. *Acta Mater.* **82,** 234–243 (2015).
33. Sutton, A. P. & Balluffi, R. W. *Interfaces in Crystalline Materials* (Oxford Univ. Press, 1995).
34. Gray, D. E. *American Institute of Physics Handbook* (McGraw-Hill, 1957).
35. Hirth, J. P. & Lothe, J. *Theory of Dislocations* 2nd edn (Wiley, 1982).
36. Kanzaki, H. Point defects in face-centered cubic lattice-I distortion around defects. *J. Phys. Chem. Solids* **2,** 24–36 (1957).
37. Siems, R. Mechanical interactions of point defects. *Phys. Status Solidi* **30,** 645–658 (1968).
38. Dederichs, P. H., Lehmann, C., Schober, H. R., Scholz, A. & Zeller, R. Lattice-type of point-defects. *J. Nucl. Mater.* **69-70,** 176–199 (1978).
39. Sivak, A. B., Chernov, V. M., Romanov, V. A. & Sivak, P. A. Kinetic Monte-Carlo simulation of self-point defect diffusion in dislocation elastic fields in bcc iron and vanadium. *J. Nucl. Mater.* **417,** 1067–1070 (2011).
40. Subramanian, G., Perez, D., Uberuaga, B. P., Tome, C. N. & Voter, A. F. Method to account for arbitrary strains in kinetic Monte Carlo simulations. *Phys. Rev. B* **87,** 144107 (2013).
41. Ehrhart, P., Jung, P., Schultz, H. & Ullmaier, H. in *Landölt-Bornstein* (ed. Ullmaier, H.) (Springer-Verlag, 1991).
42. Henkelman, G., Uberuaga, B. P. & Jonsson, H. A climbing image nudged elastic band method for finding saddle points and minimum energy paths. *J. Chem. Phys.* **113,** 9901 (2000).
43. Haubold, H. G. & Martinsen, D. Structure determination of self interstitials and investigation of vacancy clustering in copper by diffuse X-ray scattering. *J. Nucl. Mater.* **69-70,** 644–649 (1978).
44. Wolfer, W. G. in *Comprehensive Nuclear Materials* (Elsevier, 2012).
45. Nazarov, R., Hickel, T. & Neugebauer, J. Vacancy formation energies in fcc metals: Influence of exchange-correlation functionals and correction schemes. *Phys. Rev. B* **85,** 144118 (2012).
46. Dederich, P. H. & Schroeder, K. Anisotropic diffusion in stress fields. *Phys. Rev. B* **17,** 2524 (1978).
47. Koehler, J. S. Diffusion of lattice defects in stress field. *Phys. Rev.* **181,** 1015 (1969).
48. Chen, I. W. Anisotropic diffusion of point defects to edge dislocations. *J. Nucl. Mater.* **125,** 52–63 (1984).
49. Borodin, V. A. & Ryazanov, A. I. The effect of diffusion anisotropy on dislocation bias and irradiation creep in cubic lattice materials. *J. Nucl. Mater.* **210,** 258–267 (1994).
50. Brailsford, A. D. & Bullough, R. The theory of sink strengths. *Phil. Trans. R. Soc. A* **302,** 87–137 (1981).
51. Bullough, R., Hayns, M. R. & Wood, M. H. Sink strengths for thin film surfaces and grain boundaries. *J. Nucl. Mater.* **90,** 44–59 (1979).
52. Bullough, R. & Perrin, R. C. The mechanism and kinetics of void growth during neutron irradiation. *ASTM Special Technical Publications* **484,** 317–331 (1970).
53. Heald, P. T. & Speight, M. V. Steady-state irradiation creep. *Philos. Mag.* **29,** 1075–1080 (1974).
54. Mansur, L. K. Void swelling in metals and alloys under irradiation: an assessment of the theory. *Nucl. Technol.* **40,** 5–34 (1978).
55. Little, E. A., Bullough, R. & Wood, M. H. On the swelling resistance of ferritic steel. *Proc. R. Soc. A* **372,** 565–579 (1980).
56. Sivak, A. B. & Sivak, P. A. Efficiency of dislocations as sinks of radiation defects in fcc copper crystal. *Crystallogr. Rep.* **59,** 407–414 (2014).
57. Mason, D. R., Yi, X., Kirk, M. A. & Dudarev, S. L. Elastic trapping of dislocation loops in cascades in ion-irradiated tungsten foils. *J. Phys. Condens. Matter.* **26,** 375701 (2014).
58. Derlet, P. M., Gilbert, M. R. & Dudarev, S. L. Simulating dislocation loop internal dynamics and collective diffusion using stochastic differential equations. *Phys. Rev. B* **84,** 134109 (2011).
59. Dudarev, S. L. *et al.* Langevin model for real-time Brownian dynamics of interacting nanodefects in irradiated metals. *Phys. Rev. B* **81,** 224107 (2010).
60. Schroeder, K. & Dettmann, K. Diffusion reactions in long-range potentials. *Z. Phys. B* **22,** 343–350 (1975).
61. Derlet, P. M., Nguyen-Manh, D. & Dudarev, S. L. Multiscale modeling of crowdion and vacancy defects in body-centered-cubic transition metals. *Phys. Rev. B* **76,** 054107 (2007).
62. Yuryev, D. V. & Demkowicz, M. J. Computational design of solid-state interfaces using O-lattice theory: An application to mitigating helium-induced damage. *Appl. Phys. Lett.* **105,** 221601 (2014).
63. Was, G. S. *Fundamentals of Radiation Materials Science: Metals and Alloys* (Springer, 2007).
64. Eshelby, J. D. The continuum theory of lattice defects. *Solid State Phys. Adv. Res. Appl.* **3,** 79–144 (1956).
65. Uberuaga, B. P., Vernon, L. J., Martinez, E. & Voter, A. F. The relationship between grain boundary structure, defect mobility, and grain boundary sink efficiency. *Sci. Rep.* **5,** 9095 (2015).
66. Beyerlein, I. J., Demkowicz, M. J., Misra, A. & Uberuaga, B. P. Defect-interface interactions. *Prog. Mater. Sci.* **74,** 125–210 (2015).
67. Stroh, A. N. Dislocations and cracks in anisotropic elasticity. *Philos. Mag.* **3,** 625–646 (1958).
68. Kresse, G. & Furthmuller, J. Efficient iterative schemes for *ab initio* total-energy calculations using a plane-wave basis set. *Phys. Rev. B* **54,** 11169 (1996).
69. Plimpton, S. Fast parallel algorithms for short-range molecular-dynamics. *J. Comput. Phys.* **117,** 1–19 (1995).
70. Foiles, S. M., Baskes, M. I. & Daw, M. S. Embedded-atom-method functions for the FCC metals Cu, Ag, Au, Ni, Pd, Pt, and their alloys. *Phys. Rev. B* **33,** 7983 (1986).
71. Mishin, Y., Mehl, M. J., Papaconstantopoulos, D. A., Voter, A. F. & Kress, J. D. Structural stability and lattice defects in copper: *Ab initio*, tight-binding, and embedded-atom calculations. *Phys. Rev. B* **63,** 224106 (2001).
72. Methfessel, M. & Paxton, A. T. High-precision sampling for Brillouin-zone integration in metals. *Phys. Rev. B* **40,** 3616 (1989).
73. Perdew, J. P., Burke, K. & Ernzerhof, M. Generalized gradient approximation made simple. *Phys. Rev. Lett.* **77,** 3865 (1996).
74. Kresse, G. & Joubert, D. From ultrasoft pseudopotentials to the projector augmented-wave method. *Phys. Rev. B* **59,** 1758 (1999).

Acknowledgements

H.D. acknowledges support from the DOE, Office of Nuclear Energy, Nuclear Energy Enabling Technologies, Reactor Materials program, under contract No. DE-NE0000533. H.D. also thanks the DOE-NERSC for computational support. M.C.M. acknowledges support from the GENCI-(CINES/CCRT/IDRIS) computer centre under Grant No. x2015-096973 and from PRACE (Partnership for Advanced Computing in Europe) for access to the resource MareNostrum III at the Barcelona supercomputing centre (TRANSOM project). M.J.D. acknowledges support from the National Science Foundation under Grant No. 1150862.

Author contributions

A.V. carried out elasticity calculations of stress fields and interaction energies. T.J. performed OKMC simulations. H.D. and M.-C.M. computed defect **P**-tensors using DFT simulations. M.J.D. supervised the manuscript. All the authors contributed to data interpretation and to the manuscript preparation and approved the final version of the manuscript.

Additional information

Competing financial interests: The authors declare no competing financial interests.

15

Layer-dependent quantum cooperation of electron and hole states in the anomalous semimetal WTe_2

Pranab Kumar Das[1,2], D. Di Sante[3,4], I. Vobornik[1], J. Fujii[1], T. Okuda[5], E. Bruyer[3], A. Gyenis[6], B.E. Feldman[6], J. Tao[7], R. Ciancio[1], G. Rossi[1,8], M.N. Ali[9], S. Picozzi[3], A. Yadzani[6], G. Panaccione[1,*] & R.J. Cava[9,*]

The behaviour of electrons and holes in a crystal lattice is a fundamental quantum phenomenon, accounting for a rich variety of material properties. Boosted by the remarkable electronic and physical properties of two-dimensional materials such as graphene and topological insulators, transition metal dichalcogenides have recently received renewed attention. In this context, the anomalous bulk properties of semimetallic WTe_2 have attracted considerable interest. Here we report angle- and spin-resolved photoemission spectroscopy of WTe_2 single crystals, through which we disentangle the role of W and Te atoms in the formation of the band structure and identify the interplay of charge, spin and orbital degrees of freedom. Supported by first-principles calculations and high-resolution surface topography, we reveal the existence of a layer-dependent behaviour. The balance of electron and hole states is found only when considering at least three Te-W-Te layers, showing that the behaviour of WTe_2 is not strictly two dimensional.

[1] Istituto Officina dei Materiali (IOM)-CNR, Laboratorio TASC, in Area Science Park, S.S.14, Km 163.5, I-34149 Trieste, Italy. [2] International Centre for Theoretical Physics (ICTP), Strada Costiera 11, I-34100 Trieste, Italy. [3] Consiglio Nazionale delle Ricerche—CNR-SPIN, I-67100 L'Aquila, Italy. [4] Department of Physical and Chemical Sciences, University of L'Aquila, Via Vetoio, I-67100 L'Aquila, Italy. [5] Hiroshima Synchrotron Radiation Center (HSRC), Hiroshima University, 2-313 Kagamiyama, Higashi-Hiroshima 739-0046, Japan. [6] Joseph Henry Laboratories and Department of Physics, Princeton University, Princeton, New Jersey 08544, USA. [7] Department of Condensed Matter Physics and Materials Science, Brookhaven National Laboratory, Upton, New York 11973, USA. [8] Dipartimento di Fisica, Università di Milano, Via Celoria 16, I-20133 Milano, Italy. [9] Department of Chemistry, Princeton University, Princeton, New Jersey 08544, USA. * These authors contributed equally to this work. Correspondence and requests for materials should be addressed to R.J.C. (email: rcava@Princeton.EDU).

Transition metal dichalcogenides (TMDs) are a group of layered materials with chemical formula MX_2, where M is a transition metal and X can be S, Se or Te. Their properties span from pure insulators to good metals, and they also exhibit various low-temperature phenomena such as metal–insulator transitions, superconductivity and charge density waves[1]. An important aspect of TMDs is the presence of anisotropic bonding with different strengths: the X–M–X building blocks are stacked along the crystallographic *c*-direction, and while the inter-layer interaction is mainly of weak van der Waals type, the intra-layer bonding between the atoms is strong and covalent. Dimensionality is thus expected to play a significant role in TMDs, because the transition from a single or a few layers to bulk implies significant change in the symmetry of the orbitals and in quantum confinement, producing important differences in the electronic structure.

Among TMDs, WTe_2 is special because it displays an additional structural distortion: the tungsten atoms form zigzag chains along the crystallographic *a* axis, producing a quasi one-dimensional arrangement. Moreover, WTe_2 is a semimetal with a reduced density of states at the Fermi level coming from a small overlap between valence and conduction bands without a band gap. It exhibits an extremely large uniaxial positive magnetoresistance with no saturation up to a magnetic field as high as 60 T (refs 2,3), which has been attributed to perfect electron and hole compensation. In addition, the presence of two heavy elements points to the importance of spin–orbit coupling (SOC) in determining the details of the Fermi surface and/or the relevant low-energy excitations of the system.

Here, by combining results of scanning tunnelling microscopy (STM), spin- and angle-resolved photoemission spectroscopy (ARPES), and layer-resolved *ab initio* calculations, we explored the details and the evolution of electron and hole states in WTe_2. Our data reveal significant differences between surface and bulk electronic properties, with a clear evolution as a function of depth from the surface. The Fermi surface measured by ARPES is significantly reduced with respect to the one calculated for the bulk, indicating the presence of a reduced, yet still balanced, number of electrons and holes at the surface and near-surface region. The importance of SOC is directly shown by spin-resolved ARPES measurements, and the data are consistent with our theoretical calculations.

Results

STM measurements to characterize the surface properties. To visualize the surface quality of WTe_2, we performed high-resolution scanning tunnelling microscopy (Fig. 1). The STM images reveal that the top layer is composed of two inequivalent Te atoms, labelled Te1 and Te2 in Fig. 1b (see also Supplementary Note 1, and Supplementary Fig. 1). The surface is atomically ordered with extremely high quality and low impurity concentration, with approximately one underlying defect per 3,000 atoms observed (Fig. 1a), corresponding to $\approx 9 \times 10^{11}$ defects cm^{-2}. In addition, the topographic image and its Fourier transform (Fig. 1c) clearly show that the electronic structure of the surface has an angular distortion, similar to that reported previously[4,5]. We observe the distortion at multiple tip-sample biases, in multiple samples, and with different STM tips, indicating that this finding is not related to thermal drift, tip artifacts or a particular energy (see Supplementary Note 2, and Supplementary Figs 2–4). Comparison with TEM data suggest that the distortion is due to surface reconstruction and model calculations for a single WTe_2 monolayer show that it does not significantly affect the band structure (see Supplementary Note 2, and Supplementary Fig. 5).

ARPES results and DFT calculations. We next explore the detailed electronic structure of WTe_2 and the related layer dependence of the electron and hole states. In Fig. 2b we show the experimental *E* versus *k* ARPES spectra overlaid with bulk theoretical band structures calculated using density functional

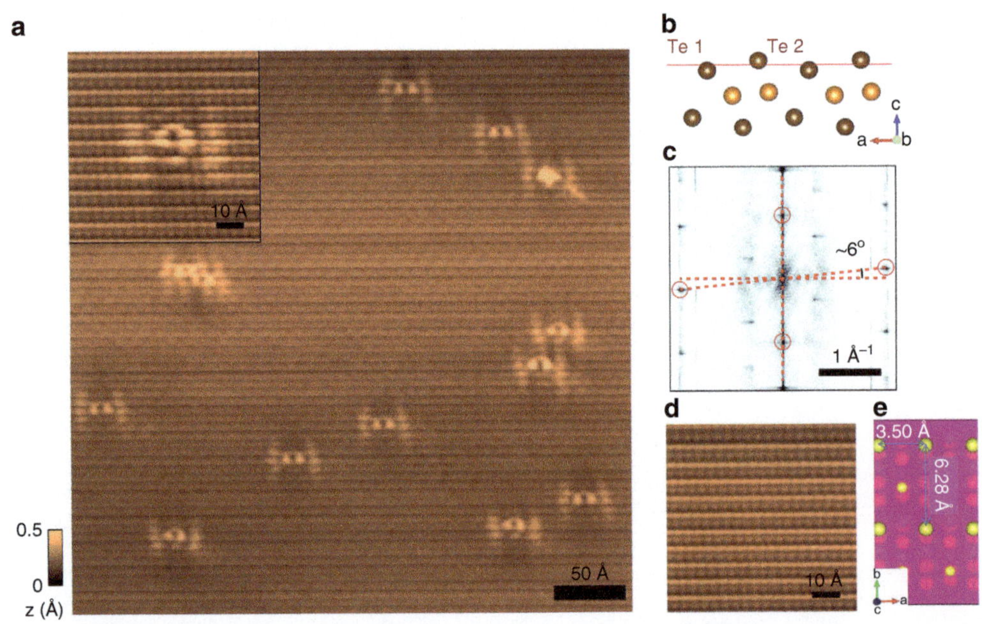

Figure 1 | Surface topography and structure of WTe_2. (**a**) Topographic image of a 400 Å × 400 Å area on the (001) surface of WTe_2 with several 'tie-fighter'-like impurities ($V_{bias} = -80$ mV, $I_{tunneling} = 100$ pA and $T = 30$ K). Inset: Zoom-in on an individual defect. (**b**) Side schematic view of the WTe_2 lattice highlighting the two inequivalent Te atoms. (**c**) Fourier transform of the topograph in **a** reveals Bragg peaks corresponding to the atomic corrugation (red circles indicate the first-order Bragg peaks) and the anomalous angle between the crystallographic axes on the surface (6° in this sample). The degree of the angular distortion varied a few degrees between different samples. (**d**) Zoom-in on a clean area displays the surface Te atoms. (**e**) Top schematic view of the WTe_2 (001) surface highlighting the surface Te atoms and the unit cell dimensions.

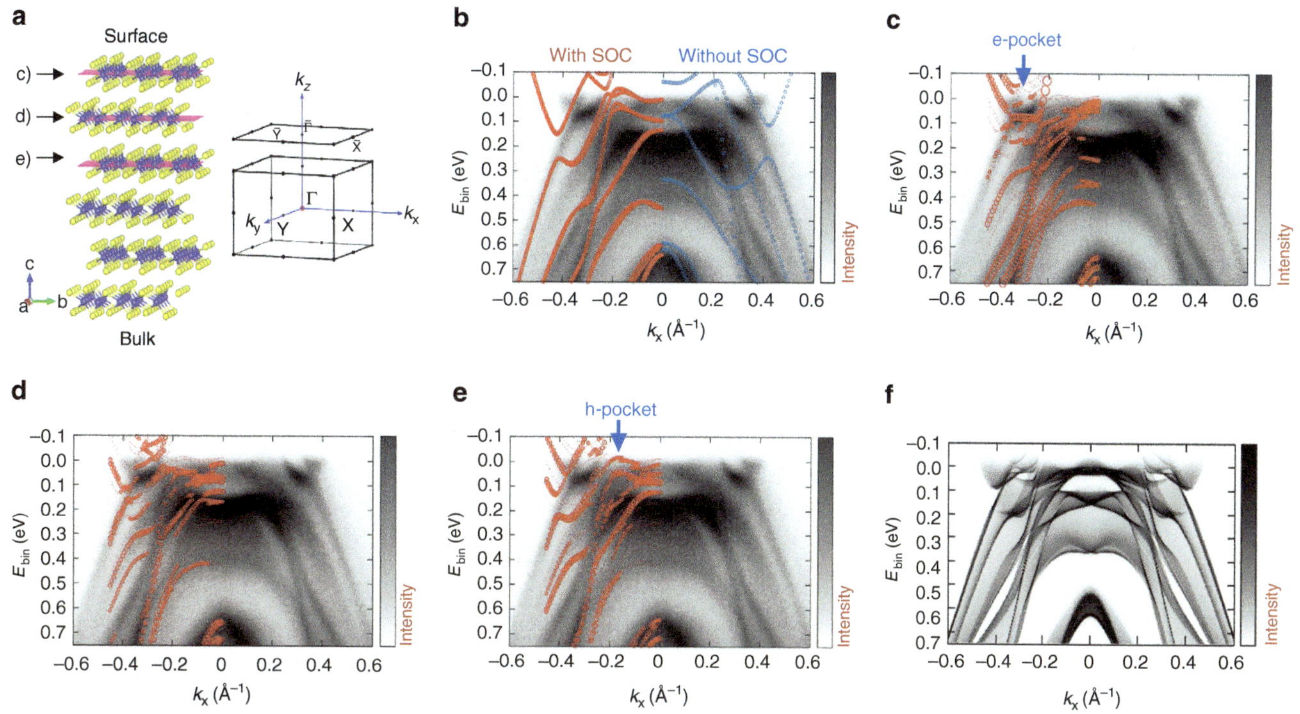

Figure 2 | Evolution of band structure with number of layers. (**a**) Crystal structure of WTe_2 with the bulk and surface Brillouin zones on the right. On the left, the slab model used in the theoretical calculations, consisting of six Te-W-Te layers; each layer is formed by hexagonally packed W sandwiched between two Te layers. (**b–e**) ARPES measurements ($h\nu = 68$ eV, $T = 77$ K) of the electronic structure along the ΓX high symmetry direction (along the W chains); (**b**) bulk electronic structures as calculated with SOC (red bands at negative momenta) and without SOC (blue bands at positive momenta); (**c**) theoretical bands projected on the topmost WTe_2 planes; (**d,e**) theoretical bands projected on second and third plane, respectively. In **c,e**, blue arrows mark the positions of the theoretical electron and hole pockets, respectively. In **b,e**, the size of the circles is proportional to the weight of the layer-resolved orbital character, calculated as the sum of the orbital characters of all the atoms belonging to the respective layer. (**f**) The theoretical surface spectral function $A(\mathbf{k},E)$.

theory (DFT) with (red bands on the left) and without (blue bands on the right) SOC, along the reciprocal space line XΓX. Following the standard procedure and assuming WTe_2 as a pure 2D-layered system, calculations are performed with out-of-plane **k**-vector component $k_z = 0$. We observe well-defined electron and hole pockets at the Fermi energy along the tungsten chain direction (ΓX), confirming previous results[6]. However, a detailed comparison of the observations to bulk theoretical band structure as described above presents some inconsistencies. The theoretical position of the electron pocket is significantly further away from the zone centre and its maximum binding energy is larger than the measured one by more than a factor two. The presence of heavy atoms like W and Te suggests that SOC plays an important role in the interpretation of WTe_2 spectral features. The changes in the theoretical band structure associated with SOC (Fig. 2b, red curves) illustrate that relativistic effects cannot be neglected in WTe_2, and although we obtain somewhat better results, a quantitative agreement is not found in either case. We also note a non-negligible k_z band dispersion[2,6] when comparing experimental spectra with theoretical bands for reciprocal lines parallel to ΓX at different k_z momenta, ranging from the Brillouin zone centre ($k_z = 0$) to the Brillouin zone edge ($k_z = \pi/c$) (see Supplementary Figs 6–9). However, including k_z dispersion does not result in an improved agreement between calculations and experiment, since none of the calculated band structures for any of the given k_z values can fully reproduce the measured dispersion (see Supplementary Fig. 6).

We find a significant improvement, however, when using a more realistic model that takes into account the contribution of individual layers (that is, all the bulk k_z momenta at the same time projected on the surface Brillouin zone). This model is based on a supercell made by van der Waals-bonded WTe_2 planes stacked along the [001] direction (Fig. 2a). From the large number of bands in the supercell calculations, we are able to isolate the individual contribution of each WTe_2 plane in the stack by projecting the electronic band structure onto the atoms belonging to that given plane only. We note that the experimental ARPES spectra correspond to a weighted spectral intensity from the different layers probed, and the contributions of deeper layers are exponentially attenuated by the inelastic mean free path of photoelectrons. Superimposing the theoretical bands on the experimental spectra, and projecting onto the first, second and third WTe_2 planes (Fig. 2c–e, respectively), we filter the layer-dependent information about the electronic structure because we are able to correlate individual features of the layer-resolved calculations in the experimental spectra. In this analysis, the size of the plotted circles is proportional to the contribution of the given WTe_2 plane: the bigger the circle, the larger the contribution of that plane to the spectral features. We thus observe in Fig. 2c that the electron pocket is located at 0.35 Å^{-1} from the zone centre (highlighted by the blue arrow) and is present even when only the first Te–W–Te layer is considered. The hole pocket (blue arrow in panel e) starts appearing only when taking into account the presence of the third layer, and thus has a more bulk-like character. Furthermore, we cannot exclude the existence of a small hole pocket at Γ, as recently proposed by theoretical findings[7] for an isolated WTe_2 monolayer (note that in the same work the electron pockets are closer to the zone centre, similar to our case); a similar hole pocket has also recently been observed in other ARPES experiments[8]. In our case (Fig. 2), the

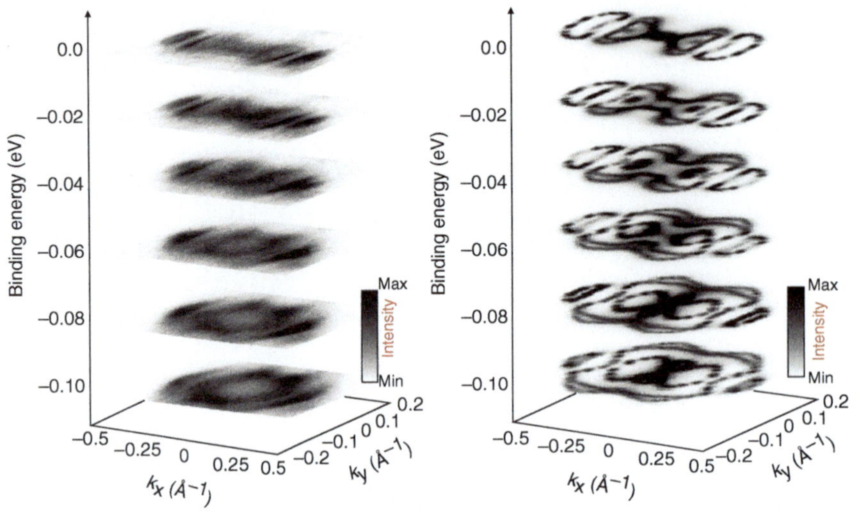

Figure 3 | Fermi surface topography. Constant energy cuts from E_F down to 100 meV binding energy (in steps of 20 meV). Left: experimental spectra measured at a photon energy of 68 eV and at 77 K, right: theoretical calculations for bulk WTe_2 at $k_z = 0$.

hole pocket is fully occupied and the range of explored temperatures is well below the Lifshitz transition recently reported at above 160 K (ref. 9). Given that the hole pocket is very close to E_F, the difference could be due to temperature induced shift of the chemical potential as proposed in ref. 9. Moreover, hole pockets close to each other on either side of the Γ point would justify the unusual magnetic breakdown observed in quantum oscillations[10].

We further stress that our slab calculations do not reveal the existence of distinct surface states. This is supported by the comparison of bulk theoretical band structures for different k_z momenta with the electronic states of the slab (see Supplementary Fig. 7). In fact, all bulk bands lie within the continuum of the slab's bulk band structure projected onto the surface Brillouin zone. Indeed, the bands shown in Fig. 2c–e overlaid to ARPES spectra are bulk states showing substantial spectral weight on the topmost surface layers; this seems in analogy to what is observed in MoS_2 and many other systems, like iron pnictides, with cleaved neutral surfaces[11,12].

The present analysis not only provides a quantitative agreement between experiment and theory in terms of dispersion and momentum space locations of electronic states, but clearly shows the presence of a layer-dependent evolution of electron and hole states: a bulk-like electronic structure characterized by the electron-hole charge balance is obtained only when more than two Te–W–Te layers are taken into account. Our results clearly display some analogies with other 'less than 3D' materials: in the case of topological insulators, for example, it has been theoretically predicted and experimentally confirmed not only that the critical thickness of six quintuple layers is needed to set the topological properties of the surface but also that the spin–orbital texture of a topological insulator evolves in a layer-dependent manner, extending over several nanometers from the surface[13,14].

Given that a nearly equal concentration of electrons and holes is a necessary condition for non-saturating magnetoresistance to be observed in two-component systems[15,16], the present observation has important implications towards the realization of devices based on few-layer TMDs[17].

To further check the robustness and the reliability of our theoretical interpretation, we used two different approaches: (i) an *ab initio* tight-binding model for a 40 layers slab (thickness about 42 nm, see Supplementary Fig. 10); (ii) a renormalization scheme for semi-infinite systems to calculate the surface spectral function, shown in Fig. 2f (ref. 18; see 'Methods' section for details). These checks aimed at excluding spurious effects arising from the interaction between the two extreme surfaces (always present in slab calculations if the number of layers is not large enough). The theoretical bands of Fig. 2f provide a direct link with the measured E versus k intensity maps (except for dipole matrix elements, neglected in our calculations). By using both methods, most of the experimental features are well reproduced, in particular the hole and electron pockets near E_F as well as the bands at higher binding energy. Moreover, no remarkable differences are observed in comparison with the DFT results for the six-layer-thick slab.

Figure 3 shows the experimental (left panel) and bulk theoretical (right panel) constant energy cuts from E_F down to 100 meV binding energy, in steps of 20 meV. The theoretical results qualitatively reproduce the main features of the Fermi surface, including the shape and the distribution of the pockets and a finite intensity at Γ (that is, confirming the presence of a possible hole pocket at Γ). However, the calculated bulk features have a significantly larger area than the measured Fermi surface, providing further evidence that simply recasting bulk calculations is insufficient to explain the experimentally observed ARPES features. The area of the Fermi surface calculated by the bulk theoretical model is larger than the area of the observed Fermi surface, which implies that the total number of carriers is larger in the bulk than on the surface. In spite of this difference, the comparison shows that the balance between electrons and holes is maintained in both cases, to within the sensitivity of our technique (see also Supplementary Fig. 9). This, in turn, indicates that the balancing between electrons and holes, that is, their 'quantum-cooperation' over different layers, is the dominant factor that determines the macroscopic properties of the system such as also the non-saturating magnetoresistance.

Spin-resolved ARPES results. To determine the role of SOC and obtain insight into the spin texture, we performed spin-resolved measurements at a number of **k** points in the Brillouin zone. Spin-resolved ARPES spectra are presented in Fig. 4, as measured at the hole pocket (panels a, c; blue and yellow circles), close to Γ on the ΓX line (panel b, red circle) and close to Γ, but off the ΓX line (panel d, green circle). Panel f shows the calculated spin-resolved band structure along ΓX.

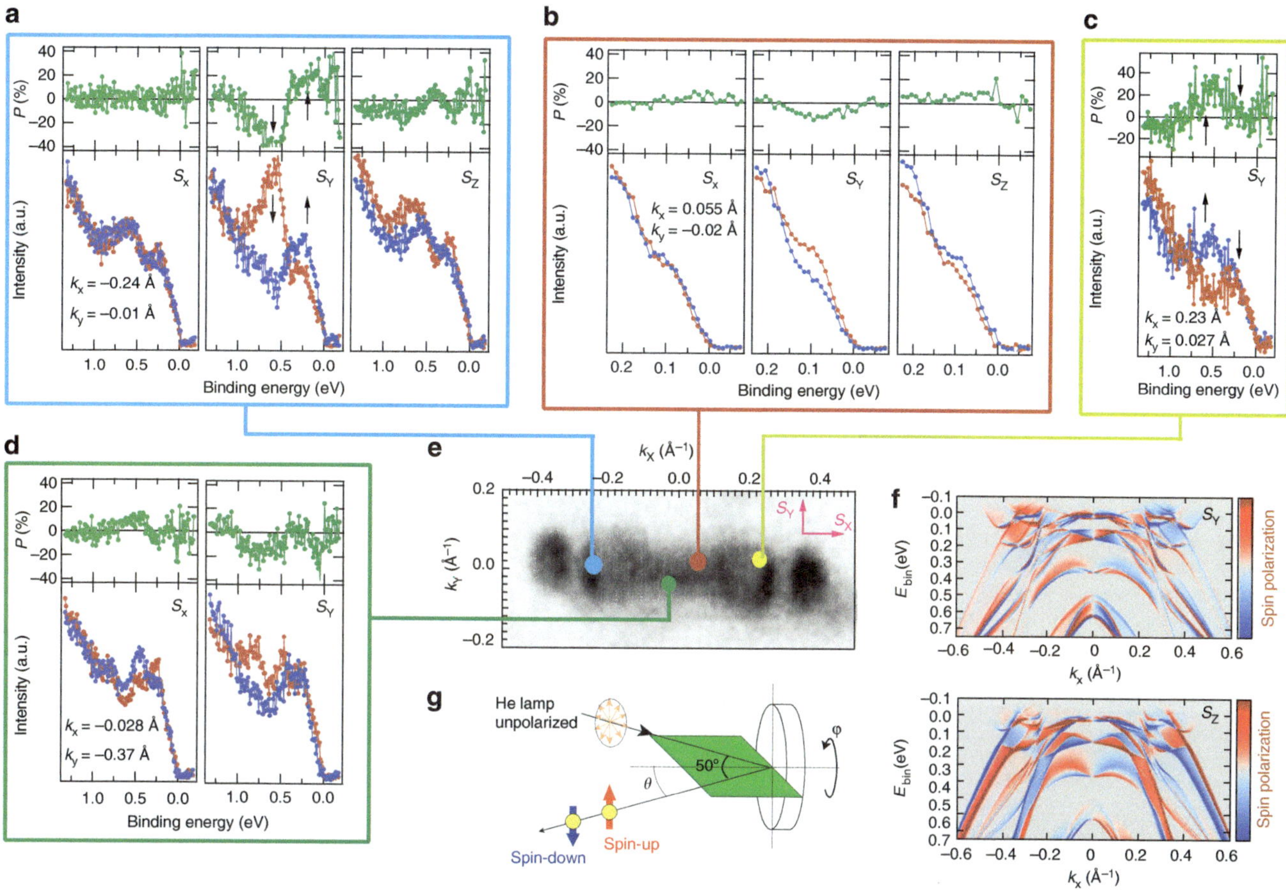

Figure 4 | Experimental and theoretical spin polarized band structure of large SOC and non-centro-symmetric WTe_2. (**a**-**d**) Measured spin-resolved ARPES spectra (red and blue curves) and spin polarization (green curves) at four distinct **k** points as indicated in **e**; (**e**) experimental Fermi surface with the corresponding spin measurement positions; (**f**) calculated band structure along k_x direction, upper/lower panel shows S_y/S_z spin orientation. The colour scales represent the degree of spin polarization; red, blue and white refer to up, down and no-spin polarization, respectively; (**g**) experimental geometry of spin-resolved ARPES setup. In **a**, S_y, the down-spin spectrum (red curve) has a larger peak at 0.6 eV (down arrow) and the up-spin spectrum (blue curve) has a larger peak at 0.2 eV (up arrow), whereas at **c**, on the contrary, the up-spin spectrum has a dominating peak at 0.6 eV and the down-spin spectrum has a dominating peak at 0.2 eV, supporting time-reversal effect.

We measure a sizeable spin polarization (P) for an extended binding energy range. It reaches more than 35% for P_y at a binding energy of 0.55 eV (panel a, black arrows). This indicates that there must not only be a broken space inversion symmetry but also a significant influence of SOC in WTe_2. The spin texture is quite complex, with large oscillations in value and sign of P in a narrow energy window, clearly visible in both P_y and P_z (Fig. 4a). This observation is confirmed by the calculation in Fig. 4f. Experimental error cannot exclude the presence of a finite spin polarization at E_F, but it is negligible with respect to those observed at higher binding energies. The fine spectral structures in the calculation are not visible in the experimental spin polarization measurements, due to the experimental energy and angular resolution. However, we obtain notable qualitative agreement, as detailed below. The P_x component is zero along the ΓX line (Fig. 4a,b), while we observe non-zero P_x component for points away from ΓX (Fig. 4d). This means that the electron spin along ΓX is perpendicular to the W chains, in agreement with calculations that predict only large P_y and P_z spin polarizations (Fig. 4f). The results shown in Fig. 4a,b (at the hole pocket) and Fig. 4f reveal the presence of both in-plane and out-of-plane components of the spin polarization, in contrast with ordinary Rashba systems where P is only in-plane[19]. The spin polarization changes sign upon crossing the Brillouin zone centre, that is, the sign of P is reversed at $\mathbf{k}' = -\mathbf{k}$, as experimentally confirmed in Fig. 4a,c, which show data taken at positive and negative k_x values. This indicates that time reversal symmetry is preserved, that is, $[E(\mathbf{k},\uparrow) = E(-\mathbf{k},\downarrow)]$ and that the observed spin polarization of bands has nonmagnetic origin. A large spin polarization of electronic bands has been recently reported in the semiconducting TMD WSe_2, where P occurs due to the local asymmetry of layers[20] (consecutive Se–W–Se layers have opposite net dipole moment, which modulate the spin texture strongly even though the global inversion symmetry is preserved in the crystal). Unlike WSe_2, the crystal structure of WTe_2 is non-centrosymmetric. Therefore, by symmetry consideration one naturally expects a lifting of the spin degeneracy via spin–orbit coupling. The amplitude of spin polarization depends on many different factors (that is, orbital character, band gap, electric fields and so on), but primarily on the strength of the SOC; therefore, our results indicate that SOC is clearly reflected in the spectral function of WTe_2, as also recently reported by CD-ARPES results[8].

Discussion

Considering our present results from a more general perspective, it is important to underline that exact carrier compensation is a necessary, yet not sufficient, condition for a non-saturating quadratic behaviour of magnetoresistance[15]. Bismuth provides an example where compensation is observed, but quadratic dependence of resistance in a magnetic field and non-saturation are not[15,21]. Other prerequisites for behaviour similar to that observed in WTe_2 are: (a) low density of impurities and defects and (b) a carrier density far from the quantum limit[22,23]. Here we have shown that condition (a) is nearly fulfilled as shown by STM results, and condition (b) is supported by the reported low resistivity and high mobility in WTe_2 (refs 2,7,24). These observations suggest that future experiments exploring the relationship between electronic structure and magnetoresistance in WTe_2 would be worthwhile and that WTe_2 could be a potential candidate to form an excitonic dielectric in the Abrikosov sense[22,25].

In summary, our theoretical and experimental ARPES findings provide clear evidence that the electronic properties of WTe_2 display a layer-dependent evolution from surface to bulk, that is, it cannot be considered a priori as a non-interacting 2D-layered system, in agreement also with recent temperature-dependent magnetoresistance measurements[26]. The balance between the hole and electron states, representing one of the crucial conditions for the non-saturating magnetoresistance in this system, is established only beyond finite number of layers (three) and maintained in the bulk. This consideration provides a fundamental input for future exploitation of TMDs in general, and WTe_2 in particular, in devices and heterogeneous interfaces.

Methods

ARPES experiments. ARPES experiments were performed at APE-IOM beamline[27] at the ELETTRA Sincrotrone Trieste. High quality surfaces were obtained by cleaving the samples in UHV at a base pressure of 1×10^{-10} mbar. The crystallographic orientation was examined by low-energy electron diffraction (LEED) patterns. Core levels, valence-band and Fermi surface measurements were performed using a high-resolution VG-SCIENTA DA30 electron analyzer, in a photon energy range of 20–100 eV, with an angular and energy resolution better than 0.2 deg and 20 meV, respectively. The low temperature data are collected at 16 K using a liquid helium cooled cryostat.

STM. Before the measurements, the samples were cleaved in ultra-high vacuum at room temperature and immediately transferred to our home-built variable temperature STM. Measurements were performed at 28 K.

Spin-resolved ARPES. Spin-resolved ARPES experiments were performed at HiSOR[28]. The spin-resolved spectra were measured by means of VLEED spin detector using Fe-O target; two VLEED detectors positioned orthogonally are able to measure the *x*, *y* (in-plane) and *z* (out of plane) components of the spin polarization. Standard He laboratory light source ($h\nu = 21.22$ eV) was used as incident beam at various temperatures down to 10 K. The asymmetry of the spin polarization was quantified by reversing the current through a coil. The analyser resolution of SR-ARPES was 60 meV and the angular resolution was 1.5 degrees. The spin asymmetry A is given by $A = (I_+ - I_-)/(I_+ + I_-)$. The actual polarization P depends on the instrument and detector setup (target), and is given by $P = A/S_{\text{eff}}$, where S_{eff} is the effective Sherman function value corresponding to the detector and instrumental setup used. Here we have used $S_{\text{eff}} = 0.2$. Next the up and down spin component $S_\uparrow$, $S_\downarrow$ can be calculated following the expression $S_\uparrow(S_\downarrow) = (1 \pm P)(I_+ + I_-)/2$.

Density functional theory calculations. Supercell calculations were performed using the generalized gradient approximation (GGA) as implemented in the DFT code Vienna Ab-Initio Simulation Package (VASP)[29,30]. Atomic positions were fully relaxed starting from data in ref. 31. We used the projector augmented wave method by explicitly treating six valence electrons both for W and Te, while *d* electrons of Te were kept within the core of the PBE pseudopotentials. Integration over the first Brillouin zone was made with a $12 \times 8 \times 1$ Monkhorst-Pack k-mesh centred at Γ ($24 \times 12 \times 2$ for bulk calculations). For all the simulations, a 400 eV plane-wave energy cut-off was used. Spin–orbit coupling has been self-consistently taken into account. Dipole corrections, as implemented in VASP, were applied along the *z* direction to counteract any spurious electric field that might arise from periodic boundary conditions in the presence of a dipole moment normal to the surface in a slab geometry (with 20 Å-thick layer of vacuum). To model surface as well as more bulk features, we considered a slab containing six planes of WTe_2 stacked onto each other along the [001] direction. The effects of van der Waals interactions between WTe_2 planes has been properly taken into account by the DFT-D2 method of Grimme[32], but no significant differences in the electronic properties have been detected with respect to GGA calculations.

To calculate bulk Fermi surfaces reported in Fig. 3 (right panel), we adopt a two step procedure: (i) first, we projected the bulk Hamiltonian onto a basis made of *s* and *d* W-centered, and *s* and *p* Te-centered orbitals, for a total of 112 Wannier functions, by means of the WANNIER90 package; (ii) subsequently, the Wannier Hamiltonian is used to build up the bulk Green's function as $G(\mathbf{k},w) = 1/(w - H(\mathbf{k}) + i\delta)$, which, in turn, gives the Fermi surface maps at the chosen binding energy, *w*. Theoretical spectral functions have been calculated within the framework of the surface renormalization method[18] based on the same Wannier Hamiltonian as described before. The surface spectral function is defined as $A(\mathbf{k},w) = -(1/\pi)ImG_{\text{surf}}(\mathbf{k},w)$, where $G_{\text{surf}}(\mathbf{k},w)$ represents the angular and energy resolved surface Green's function. Major differences between $A(\mathbf{k},w)$ and ARPES data, as, for example, line intensities and high-**k** features, possibly rely in the absence of transition matrix elements in the theoretical description.

References

1. Wang, Q. H. *et al.* Electronics and optoelectronics of two-dimensional transition metal dichalcogenides. *Nat. Nanotechnol.* **7**, 699–712 (2012).
2. Ali, M. N. *et al.* Large, non-saturating magnetoresistance in WTe_2. *Nature* **514**, 205–208 (2014).
3. Crossley, A., Myhra, S. & Sofield, C. J. STM analysis of WTe_2 surfaces—correlation with crystal and electronic structures. *Surf. Sci.* **318**, 39–45 (1994).
4. Hlaa, S. W., Marinković, V., Prodana, A. & Muševiča, I. STM/AFM investigations of β-MoTe2, α-MoTe2 and WTe2. *Surf. Sci.* **352–354**, 105–111 (1996).
5. Pletikosić, I., Ali, M. N., Fedorov, A. V., Cava, R. J. & Valla, T. Electronic structure basis for the extraordinary magnetoresistance in WTe_2. *Phys. Rev. Lett.* **113**, 216601 (2014).
6. Lv, H. Y. *et al.* Perfect charge compensation in WTe2 for the extraordinary magnetoresistance: from bulk to monolayer. *Europhys. Lett.* **110**, 37004 (2015).
7. Jiang, J. *et al.* Signature of strong spin-orbital coupling in the large nonsaturating magnetoresistance material WTe_2. *Phys. Rev. Lett.* **115**, 166601 (2015).
8. Wu, Y. *et al.* Temperature induced Lifshitz transition in WTe_2. *Phys. Rev. Lett.* **115**, 166602 (2015).
9. Zhu, Z. *et al.* Quantum oscillations, termoelectric coefficients, and the Fermi surface of semimetallic WTe_2. *Phys. Rev. Lett.* **114**, 176601 (2015).
10. Gehlmann, M. *et al.* Quasi 2D electronic states with high spin-polarization in centrosymmetric MoS2 bulk crystals. Preprint at http://arxiv.org/abs/1510.04101 (2015).
11. Lankau, A. *et al.* Absence of surface states for LiFeAs investigated using density functional calculations. *Phys. Rev. B* **82**, 184518 (2010).
12. Zhang, Y., He, K. & Chang, C. Crossover of the three-dimensional topological insulator Bi_2Se_3 to the two-dimensional limit. *Nat. Phys.* **6**, 584–588 (2010).
13. Zhu, Z. H. & Levy, G. Rashba spin-splitting control at the surface of the topological insulator Bi_2Se_3. *Phys. Rev. Lett.* **107**, 186405 (2011).
14. Alekseev, P. S. *et al.* Magnetoresistance in two-component systems. *Phys. Rev. Lett.* **114**, 156601 (2015).
15. Parish, M. M. & Littlewood, P. B. Non-saturating magnetoresistance in heavily disordered semiconductors. *Nature* **426**, 162–165 (2003).
16. Wang, L. *et al.* Tuning Magnetotransportin a Compensated Semimetal at the Atomic Scale. Preprint at http://arxiv.org/abs/1510.04827 (2015).
17. Henk, J. & Schattke, W. A subroutine package for computing Green's functions of relaxed surfaces by the renormalization method. *Comput. Phys. Commun.* **77**, 69–83 (1993).
18. Hasan, M. Z. & Kane, C. L. Colloquium: topological insulators. *Rev. Mod. Phys.* **82**, 3045 (2010).
19. Riley, J. M. *et al.* Direct observation of spin-polarized bulk bands in an inversion-symmetric semiconductor. *Nat. Phys.* **10**, 835–839 (2014).
20. Liang, T. *et al.* Evidence for massive bulk Dirac fermions in $Pb_{1-x}Sn_xSe$ from Nernst and thermopower experiments. *Nat. Commun.* **4**, 2696 (2013).
21. Abrikosov, A. A. The transformation of a semimetal into an exciton dielectric in a strong magnetic field. *Sov. Phys. Usp.* **15**, 662–663 (1973).
22. Fenton, E. W. Electrical resistivity of semimetals in the extreme quantum limit. *J. Low Temp. Phys.* **7**, 415–432 (1972).
23. Ali, M. N. *et al.* Correlation of crystal quality and extreme magnetoresistance of WTe_2. *Europhys. Lett.* **110**, 67002 (2015).
24. Abrikosov, A. A. Quantum linear magnetoresistance. *Europhys. Lett.* **49**, 789–793 (2000).
25. Thoutam, L. R. *et al.* Temperature-dependent three-dimensional anisotropy of the magnetoresistance in WTe_2. *Phys. Rev. Lett.* **115**, 046602 (2015).
26. Panaccione, G. *et al.* Advanced photoelectric effect experiment beamline at Elettra: a surface science laboratory coupled with synchrotron radiation. *Rev. Sci. Instrum.* **80**, 043105 (2009).
27. Okuda, T. *et al.* Efficient spin resolved spectroscopy observation machine at Hiroshima Synchrotron Radiation Center. *Rev. Sci. Instrum.* **82**, 103302 (2011).
28. Kresse, G. & Furthmüller, J. Efficient iterative schemes for *ab initio* total-energy calculations using a plane-wave basis set. *Phys. Rev. B* **54**, 11169 (1996).
29. Kresse, G. & Joubert, D. From ultrasoft pseudopotentials to the projector augmented-wave method. *Phys. Rev. B* **59**, 1758 (1999).

30. Mar, A., Jobic, S. & Ibers, A. Metal-metal vs tellurium-tellurium bonding in WTe_2 and its ternary variants $TaIrTe_4$ and $NbIrTe_4$. *J. Am. Chem. Soc.* **114,** 8963 (1992).
31. Grimme, S. Semiempirical GGA-type density functional constructed with a long-range dispersion correction. *Comp. Chem.* **27,** 1787 (2006).
32. Mostofi, A. A. *et al.* wannier90: a tool for obtaining maximally-localised Wannier functions. *Comput. Phys. Commun.* **178,** 685 (2008).

Acknowledgements

This work has been partly performed in the framework of the nanoscience foundry and fine analysis (NFFA-MIUR Italy) project. The electron diffraction study at Brookhaven National Laboratory was supported by the DOE BES, by the Materials Sciences and Engineering Division under contract DE-AC02–98CH10886, and through the use of the Center for Functional Nanomaterials. The work at Princeton was supported by the National Science Foundation MRSEC program grant DMR-1420541, with STM support from NSF-DMR-1104612, ARO-W911NF-1–0262, ARO-MURI program W911NF-12-1-0461 and DARPA-SPWAR Meso program N6601-11-1-4110. D.D.S. and S.P. acknowledge the CARIPLO Foundation through the MAGISTER project Rif. 2013-0726. This work was partly supported by the Italian Ministry of Research through the project PRIN Interfacce di ossidi: nuove proprietà emergenti, multifunzionalità e dispositivi per elettronica e energia (OXIDE).

Author contributions

R.J.C. and G.P. conceived the experiment and wrote the paper with contributions from D.D.S. and I.V. All the authors discussed the results, commented on the manuscript and prepared written contributions. R.J.C. and M.N.A. grew the crystals and characterized the samples. J.T. performed the electron diffraction experiments. A.G., B.E.F. and A.Y. performed STM experiments and analysis. D.D.S., E.B. and S.P. performed the calculations. P.K.D., I.V., G.P., G.R., J.F., R.C. and T.O. performed the synchrotron radiation experiments and analysed the data.

Additional information

Competing financial interests: The authors declare no competing financial interests.

16

Exciton diamagnetic shifts and valley Zeeman effects in monolayer WS_2 and MoS_2 to 65 Tesla

Andreas V. Stier[1], Kathleen M. McCreary[2], Berend T. Jonker[2], Junichiro Kono[3,4,5] & Scott A. Crooker[1]

In bulk and quantum-confined semiconductors, magneto-optical studies have historically played an essential role in determining the fundamental parameters of excitons (size, binding energy, spin, dimensionality and so on). Here we report low-temperature polarized reflection spectroscopy of atomically thin WS_2 and MoS_2 in high magnetic fields to 65 T. Both the A and B excitons exhibit similar Zeeman splittings of approximately $-230\,\mu\mathrm{eV\,T}^{-1}$ (g-factor $\simeq -4$), thereby quantifying the valley Zeeman effect in monolayer transition-metal disulphides. Crucially, these large fields also allow observation of the small quadratic diamagnetic shifts of both A and B excitons in monolayer WS_2, from which radii of ~1.53 and ~1.16 nm are calculated. Further, when analysed within a model of non-local dielectric screening, these diamagnetic shifts also constrain estimates of the A and B exciton binding energies (410 and 470 meV, respectively, using a reduced A exciton mass of 0.16 times the free electron mass). These results highlight the utility of high magnetic fields for understanding new two-dimensional materials.

[1] National High Magnetic Field Laboratory, Los Alamos National Laboratory, Los Alamos, New Mexico 87545, USA. [2] Materials Science and Technology Division, Naval Research Laboratory, Washington, Washington DC 20375, USA. [3] Department of Electrical and Computer Engineering, Rice University, Houston, Texas 77005, USA. [4] Department of Physics and Astronomy, Rice University, Houston, Texas 77005, USA. [5] Department of Materials Science and NanoEngineering, Rice University, Houston, Texas 77005, USA. Correspondence and requests for materials should be addressed to S.A.C. (email: crooker@lanl.gov).

Atomically thin crystals of the transition-metal disulphides (MoS_2 and WS_2) and diselenides ($MoSe_2$ and WSe_2) constitute a novel class of monolayer semiconductors that possess direct optical bandgaps located at the degenerate K and K' valleys of their hexagonal Brillouin zones[1,2]. The considerable recent interest in these two-dimensional (2D) transition-metal dichalcogenides (TMDs) derives from their strong spin–orbit coupling and lack of structural inversion symmetry, which, together with time-reversal symmetry, couples spin and valley degrees of freedom and leads to valley-specific optical selection rules[3–8]: light of σ^+ circular polarization couples to inter-band exciton transitions in the K valley, while the opposite (σ^-) circular polarization couples to transitions in the K' valley. The ability to populate and/or probe electrons and holes in specific valleys using polarized light has renewed long-standing interests[8–11] in understanding and exploiting such 'valley pseudospin' degrees of freedom for both fundamental physics and far-reaching applications in, for example, quantum information processing.

The bands and optical transitions at the K and K' valleys are nominally degenerate in energy and related by time-reversal symmetry. However, in analogy with conventional spin degrees of freedom, this K/K' valley degeneracy can be lifted by external magnetic fields B, which break time-reversal symmetry. Recent photoluminescence studies of the monolayer diselenides $MoSe_2$ and WSe_2 in modest fields have indeed demonstrated this 'valley Zeeman effect', and revealed an energy splitting between σ^+ and σ^- polarized photoluminescence from the lowest-energy 'A' exciton transition[12–17]. In most cases, valley splittings in these monolayer diselenides were found to increase linearly with field at a rate of approximately $-4\mu_B$ ($\equiv -232\,\mu eV\,T^{-1}$), where $\mu_B = 57.9\,\mu eV\,T^{-1}$ is the Bohr magneton. While this value agrees surprisingly well with simple expectations from a two-band tight-binding model (namely, that electron and hole masses are equal, and that the exciton Zeeman shifts derive solely from the hybridized $d_{x^2-y^2} \pm id_{xy}$ atomic orbitals with magnetic moment $\pm 2\mu B$ that comprise the K/K' valence bands[12–15]), it is also widely appreciated that a more complete treatment based on established $\mathbf{k}\cdot\mathbf{p}$ theory should, with proper inclusion of strong excitonic effects, also provide an accurate description. However, initial $\mathbf{k}\cdot\mathbf{p}$ models have so far had limited success accounting for the measured valley Zeeman effect in monolayer TMDs[12,16,18,19].

Regardless of circumstances, magneto-optical studies of these new monolayer semiconductors are still at a relatively early stage, and several outstanding questions remain. In particular, measurements of valley Zeeman effects in the monolayer disulphides WS_2 and MoS_2 have not been reported to date, which would provide a natural complement to the existing data on monolayer WSe_2 and $MoSe_2$. In addition, the valley Zeeman splitting of the higher-energy 'B' exciton has not yet been reported in any of these 2D materials. Both of these studies would provide a more complete experimental basis against which to benchmark new theoretical approaches. And finally, the diamagnetic energy shift of these excitons, which is anticipated to increase quadratically with field and from which the spatial extent of the fundamental (1s) exciton wavefunctions can be directly inferred[20–22], has not yet been observed in any of the monolayer TMDs. Likely, this is because the diamagnetic shift, $\Delta E_{dia} = e^2\langle r^2\rangle_{1s}B^2/8m_r$, is expected to be very small and difficult to spectrally resolve in these materials owing to the small root mean squared (r.m.s.) radius of the 1s exciton ($r_1 = \sqrt{\langle r^2\rangle_{1s}}$), and large reduced mass $m_r = \left(m_e^{-1} + m_h^{-1}\right)^{-1}$. For example, if $r_1 \approx 1.5$ nm and $m_e = m_h \approx m_0/2$ (where m_0 is the bare electron mass and $m_{e/h}$ is the effective electron/hole mass), then ΔE_{dia} is only $\sim 20\,\mu eV$ at $B = 10$ T, clearly motivating the need for large magnetic fields. Crucially, knowledge of ΔE_{dia} can also constrain estimates of the exciton binding energy—a subject of considerable recent interest in the monolayer TMDs[23–35], wherein the effects of non-local dielectric screening and Berry curvature can generate a markedly non-hydrogenic Rydberg series of exciton states and associated binding energies[36–40].

In the following, we address these questions with a systematic study of circularly polarized magneto-reflection from large-area films of monolayer WS_2 and MoS_2 at low temperatures (4 K) and in very high pulsed magnetic fields up to 65 T. Clear valley splittings of about $-230\,\mu eV\,T^{-1}$ are observed for both the A and B excitons, providing measurements of the valley Zeeman effect and associated g-factors in monolayer transition-metal disulphides. Moreover, due to the very large magnetic fields used in these studies, we are also able to resolve the small quadratic diamagnetic shifts of both A and B excitons in monolayer WS_2 (0.32 ± 0.02 and $0.11 \pm 0.02\,\mu eV\,T^{-2}$, respectively), permitting estimates of the r.m.s. exciton radius r_1. These results are compared with similar measurements of bulk WS_2 crystals, and are quantitatively modelled within the context of the non-hydrogenic binding potential[23,36,37] that is believed to exist in 2D semiconductors due to non-local dielectric screening. Within this framework, we estimate A and B exciton binding energies of ~ 410 and ~ 470 meV, respectively, and we show how these values scale with reduced mass m_r.

Results

Samples and experimental set-up. Large-area samples of monolayer WS_2 and MoS_2 were grown by chemical vapour deposition on SiO_2/Si substrates[41,42]. MoO_3 and pure sulphur powder were used as precursor and reactant materials, respectively, and the growth was performed at a reactant temperature of 625 °C. In addition, perylene-3,4,9,10-tetracarboxylic acid tetrapotassium salt was loaded on the SiO_2/Si substrate, which acted as a seeding promoter to achieve uniform large-area monolayer crystals[43]. The monolayer nature and high quality of these samples were confirmed by photoluminescence and Raman studies[42] (Supplementary Figs 1–5 and Supplementary Notes 1–5). In addition, a freshly exfoliated surface of a bulk WS_2 crystal was also prepared.

Magneto-reflectance studies were performed at cryogenic temperatures (down to 4 K) in a capacitor-driven 65 T pulsed magnet at the National High Magnetic Field Laboratory in Los Alamos. Broadband white light from a xenon lamp was coupled to the samples using a 100 μm diameter multimode optical fibre. The light was focused onto the sample at near-normal incidence using a single aspheric lens, and the reflected light was refocused by the lens into a 600 μm diameter collection fibre. A thin-film circular polarizer mounted over the delivery or collection fibre provided σ^+ or σ^- circular polarization sensitivity. The collected light was dispersed in a 300 mm spectrometer and detected with a charge-coupled device detector. Spectra were acquired continuously at a rate of 500 Hz throughout the ~ 50 ms long magnet pulse.

Exciton transitions and Zeeman effects in monolayer TMDs. Figure 1a depicts a single-particle energy diagram of the conduction and valence bands in monolayer TMDs at the K and K' points of the hexagonal Brillouin zone, along with the A and B exciton transitions (wavy lines) and valley-specific optical selection rules. Strong spin–orbit coupling of the valence band splits the spin-up and spin-down components (by $\Delta_v \sim 400$ and 150 meV in WS_2 and MoS_2, respectively), giving rise to the well-separated A and B exciton transitions that are observed in optical absorption or reflection spectra. As depicted, σ^+ circularly polarized light couples to both the A and B exciton transitions in the K valley, while light of the opposite σ^- circular polarization couples to the exciton transitions in the K' valley.

At zero magnetic field, time-reversed pairs of states in the K and K' valleys—for example, spin-up conduction (valence) bands in K and spin-down conduction (valence) bands in K'—necessarily have the same energy and have equal-but-opposite total magnetic moment $\left(\boldsymbol{\mu}_K^{c,v} = -\boldsymbol{\mu}_{K'}^{c,v}\right)$. Therefore, an applied magnetic field, which breaks time-reversal symmetry, will lift the K/K' valley degeneracy by shifting time-reversed pairs of states in opposite directions in accord with the Zeeman energy $-\boldsymbol{\mu}\cdot\mathbf{B}$. This will Zeeman shift the measured exciton energy if the relevant conduction and valence band moments are unequal; viz, $\Delta E_Z = -(\boldsymbol{\mu}^c - \boldsymbol{\mu}^v)\cdot\mathbf{B}$. In the following, we consider strictly out-of-plane fields, $\mathbf{B}\| \pm\hat{z}$.

Figure 1b depicts the field-dependent energy shifts of the conduction and valence bands in the K valley (σ^+ polarized light), for both positive and negative fields. The various contributions to the total Zeeman shift in the monolayer TMDs have been discussed in several recent reports[3,12–15,44], which we summarize as follows. In general, the total magnetic moment $\boldsymbol{\mu}$ of any given conduction or valence band in the K or K' valley contains contributions from three sources: spin (μ_s); atomic orbital (μ_l); and the valley orbital magnetic moment that is associated with the Berry curvature (μ_k). Note that the latter two have been referred to as 'intracellular' and 'intercellular' contributions to the orbital magnetic moment, respectively[12,13]. The spin contribution to the exciton Zeeman shift ΔE_Z is expected to be zero, since the optically allowed transitions couple conduction and valence bands having the same spin $\left(\mu_s^c = \mu_s^v\right)$. In contrast, the conduction and valence bands are comprised of entirely different atomic orbitals: the d_{z^2} orbitals of the conduction bands have azimuthal orbital angular momentum $l_z = 0$ $\left(\mu_l^c = 0\right)$, while the hybridized $d_{x^2-y^2} \pm id_{xy}$ orbitals that comprise the valence bands have $l_z = \pm 2\hbar$ $\left(\mu_l^v = \pm 2\mu_B\right)$ in the K and K' valleys, respectively. This contribution is expected to generate a Zeeman shift of the K and K' exciton of $\mp 2\mu_B B$, respectively, and therefore, a total exciton splitting of $-4\mu_B B$. Finally, the valley orbital (Berry curvature) contributions to the conduction and valence band moments are $\mu_k^c = \pm(m_0/m_e)\mu_B$ and $\mu_k^v = \pm(m_0/m_h)\mu_B$ in the K and K' valleys, respectively. In a simple two-band tight-binding model where $m_e = m_h$, then $\mu_k^c = \mu_k^v$ and shifts due to the valley orbital magnetic moment do not appear in ΔE_Z. In more general models[45] where $m_e \neq m_h$, these Berry curvature contributions may play a role and cause a deviation of the exciton Zeeman splitting away from $-4\mu_B$.

To selectively probe the K and K' transitions in our magneto-reflectivity experiments, we typically fixed the handedness of the circular polarizer to σ^+, and pulsed the magnet in the positive ($+65$ T) and then the negative (-65 T) field direction. The latter case is exactly equivalent (by time-reversal symmetry) to measuring the σ^- optical transitions in positive field (we also explicitly verified this by changing the circular polarizer). Sign conventions were confirmed via magneto-reflectance from a diluted magnetic semiconductor ($Zn_{0.92}Mn_{0.08}Se$)[46].

Valley Zeeman effect in monolayer WS_2. Figure 2a shows the reflection spectrum (raw data) from monolayer WS_2 at 4 K. Both the A and B exciton transitions are clearly visible and are superimposed on a smoothly varying background. Figure 2b shows the well-resolved Zeeman splitting of the A exciton in WS_2 at the maximum ± 65 T applied magnetic field. Red, blue and (dashed) black curves show the normalized reflection spectra, $1 - R/R_0$ (where R_0 is a smooth background), at $+65$, -65 and 0 T respectively. A valley splitting of ~ 15 meV, analysed in detail below, is observed. Moreover, because these measurements are based on magneto-reflectance spectroscopy (rather than photoluminescence), the valley splitting of the higher-energy B exciton in WS_2 can also be observed, as shown in Fig. 2c. For both the A and B excitons, the energy of the exciton transition in positive magnetic fields (hereinafter called E^+) shifts to lower energy, while the exciton energy in negative fields (E^-) shifts to higher energy, as labelled.

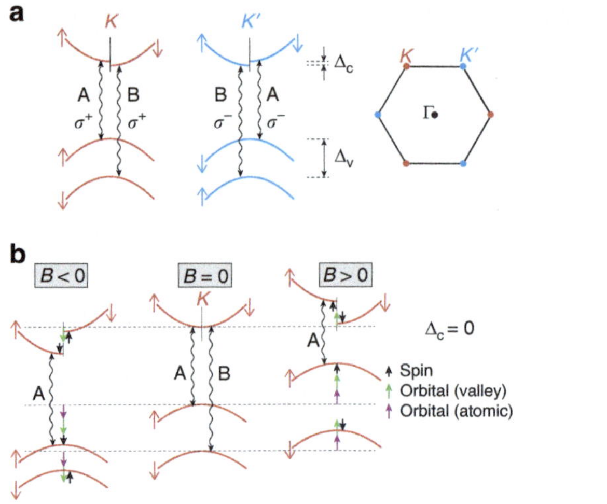

Figure 1 | Excitonic transitions and Zeeman shifts in monolayer TMDs. (**a**) Diagrams of the conduction and valence bands at the K and K' valleys of the monolayer transition-metal dichalcogenides, showing the A and B exciton transitions (wavy lines) and the associated valley-specific optical selection rules for σ^+ and σ^- light. For clarity, the spin-up and spin-down conduction bands are separately drawn on the left and right sides within each valley, respectively. A small spin-orbit splitting in the conduction band, Δ_c, is also depicted for completeness (theory predicts $\Delta_c \sim +30$ and -5 meV for WS_2 and MoS_2, respectively[45]). (**b**) Diagrams depict the relative shifts of the conduction and valence bands in the K valley (that is, σ^+ transitions) due to applied magnetic fields $B\| \pm\hat{z}$. For clarity, $\Delta_c = 0$ here. The contributions to each band's Zeeman shift from spin, atomic orbital and valley orbital (Berry curvature) magnetic moment are depicted separately by black, purple and green arrows. The σ^+ polarized A and B exciton transition energies decrease (increase) in positive (negative) field. By time-reversal symmetry, the shifts depicted here for $B<0$ in the K valley are equivalent to those in the K' valley (σ^- transitions) when $B>0$.

The exciton resonances were fit using complex (absorptive + dispersive) Lorentzian lineshapes to extract the transition energy. The field-dependent energies of the split peaks in monolayer WS_2, $E^+(B)$ and $E^-(B)$, are shown in Fig. 2d,e for the A and B excitons, respectively. The splitting between the two valleys, $E^+ - E^-$, is shown in Fig. 2f for both the A and B excitons. The measured valley Zeeman splitting is negative, and increases in magnitude linearly with applied field, with nearly identical rates of $-228 \pm 2\,\mu\text{eV T}^{-1}$ for the A exciton and $-231 \pm 2\,\mu\text{eV T}^{-1}$ for the B exciton. These values correspond to Landé g-factors of -3.94 ± 0.04 and -3.99 ± 0.04, respectively, thereby quantifying the valley Zeeman effect in the monolayer transition-metal disulphides, and also providing a measurement of the B exciton valley splitting in monolayer TMD materials.

The A exciton valley splitting that we measure in monolayer WS_2 is quite close to that reported recently from magneto-photoluminescence studies of its diselenide counterpart, monolayer WSe_2 (refs 13,16). For comparison, reported g-factors for all the monolayer TMDs are shown in Table 1. As discussed above, our measured values of $g \simeq -4$ agree surprisingly well with a simple two-band tight-binding model, wherein $m_e = m_h$ and

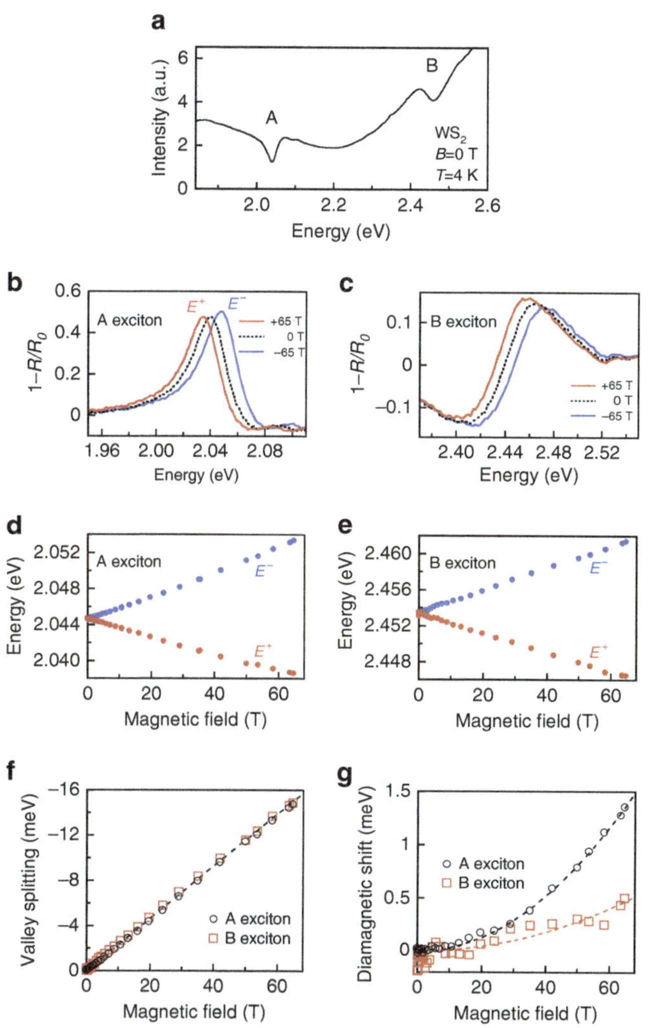

Figure 2 | Valley Zeeman effect and diamagnetic shift in monolayer WS_2. (**a**) Reflection spectrum of monolayer WS_2 at $B=0$ T and $T=4$ K. The A and B exciton resonances are labelled. (**b**) Normalized reflection spectra ($1-R/R_0$) at the A exciton resonance using σ^+ polarized light. The dashed black trace was acquired at $B=0$ T. The red trace was acquired at $+65$ T, and corresponds to the σ^+ transition in the K valley. The blue trace was acquired at -65 T, which is equivalent (by time-reversal symmetry) to the σ^- transition in the K' valley at $+65$ T. The valley Zeeman splitting between these two peaks is clearly resolved. (**c**) Similar reflection spectra and valley Zeeman splitting of the B exciton. (**d**) Energies (E^+ and E^-) of the field-split A exciton versus magnetic field. (**e**) Same, but for the B exciton. (**f**) The measured valley Zeeman splitting ($E^+ - E^-$) versus magnetic field, for both A and B excitons. The dotted line corresponds to a splitting of $-4\mu_B$ ($-232\,\mu$eV T^{-1}). (**g**) The average energy, $(E^+ + E^-)/2$, for both the A and B excitons (the zero-field offset has been subtracted). A small quadratic diamagnetic shift is revealed. The dotted lines show quadratic fits to the data ($\Delta E_{dia} = \sigma B^2$), where σ is the diamagnetic shift coefficient. We find that $\sigma_A = 0.32 \pm 0.02\,\mu$eV T^{-2} and $\sigma_B = 0.11 \pm 0.02\,\mu$eV T^{-2} for the A and B excitons, respectively.

valley moment (Berry curvature) contributions to the exciton magnetic moment cancel out, so that the exciton Zeeman shifts derive solely from atomic orbital magnetic moments of the valence bands. However, Berry curvature contributions to the Zeeman splitting are expected in more general models[45] where $m_e \neq m_h$. Deviations away from $g = -4$, observed, for example, in refs 13,14, have been explained along these lines (although, note that for tightly bound excitons, the total valley moment contribution can vary significantly in magnitude and sign, because this orbital moment must be averaged over a substantial portion of the Brillouin zone[13]).

In view of the above, it is therefore particularly noteworthy that we also measure $g \simeq -4$ for the B exciton in monolayer WS_2, despite the fact that its reduced mass almost certainly differs from that of the A exciton, as shown below from direct measurements of the diamagnetic shift (that is, m_h cannot equal m_e for both spin-up and spin-down valence bands). Note that early studies of bulk MoS_2 (refs 47–49) also indicate that the B exciton mass significantly exceeds that of the A exciton. This suggests that contributions to the orbital moment from Berry curvature effects, expected when $m_e \neq m_h$, may not play a significant role in determining the measured exciton magnetic moment and the valley Zeeman effect.

Non-local dielectric screening in monolayer TMDs. In addition to the reduced mass m_r, the characteristic size of the A and B excitons in monolayer TMDs is an essential parameter for determining material and optical properties. This is especially relevant because of non-local dielectric screening in these and other 2D materials, which fundamentally modifies the functional form of the attractive potential $V(r)$ between electrons and holes[23,36,37]. Rather than a conventional Coulomb potential, $V(r)$ is believed to assume the following form:

$$V(r) = -\frac{e^2}{8\varepsilon_0 r_0}\left[H_0\left(\frac{r}{r_0}\right) - Y_0\left(\frac{r}{r_0}\right)\right] \quad (1)$$

where H_0 and Y_0 are the Struve function and Bessel function of the second kind, respectively, and the characteristic screening length $r_0 = 2\pi\chi_{2D}$, where χ_{2D} is the 2D polarizability of the monolayer material[23,37]. This potential follows a $1/r$ Coulomb-like potential for large electron–hole separations $r \gg r_0$, but diverges weakly as $\log(r)$ for small separations $r \ll r_0$, leading to a markedly different Rydberg series of exciton states with modified wavefunctions and binding energies that cannot be described within a hydrogen-like model[23–25,27].

Diamagnetic shifts in monolayer WS_2. To this end, the use of very large 65 T magnetic fields allows us to measure the small diamagnetic shifts of excitons in monolayer TMDs so that the characteristic size of their wavefunctions can be directly inferred. In general[20–22], an exciton diamagnetic shift ΔE_{dia} is expressed as

$$\Delta E_{dia} = \frac{e^2}{8m_r}\langle r^2\rangle B^2 = \sigma B^2. \quad (2)$$

Here σ is the diamagnetic shift coefficient, m_r is the in-plane reduced mass, r is a radial coordinate in a plane perpendicular to the applied magnetic field B (here for $B\|\hat{z}$, r is in the monolayer plane) and $\langle r^2\rangle_{1s} = \langle\psi_{1s}|(x^2+y^2)|\psi_{1s}\rangle$ is the expectation value of r^2 over the 1s exciton wavefunction $\psi_{1s}(r)$. Equation (2) applies in the 'low-field' limit where the characteristic cyclotron energies $\hbar\omega_c$ (and also ΔE_{dia}) are less than the exciton binding energy, which is the case for excitons in TMDs even at ± 65 T. Given m_r, σ can then be used to determine the r.m.s. radius of the 1s exciton in the monolayer plane, r_1:

$$r_1 \equiv \sqrt{\langle r^2\rangle_{1s}} = \sqrt{8m_r\sigma}/e. \quad (3)$$

This definition is entirely general and independent of $V(r)$. (Note that for a standard Coulomb potential $V(r) = -e^2/(4\pi\varepsilon_r\varepsilon_0 r)$ in two dimensions, $r_1 = \sqrt{\frac{3}{2}}a_{0,2D}$, where $a_{0,2D} = 2\pi\varepsilon_r\varepsilon_0\hbar^2/m_r e^2$ is the classic Bohr radius for hydrogenic 2D excitons.)

Exciton diamagnetic shifts have eluded detection in recent magneto-photoluminescence studies of monolayer $MoSe_2$ and

Table 1 | Summary of effective *g*-factors and diamagnetic shifts in monolayer TMDs.

Material	A exciton *g*-factor	B exciton *g*-factor	σ_A (μeV T^{-2})	σ_B (μeV T^{-2})
Monolayer WS_2	**− 3.94 ± 0.04** **− 3.33 ± 0.04** (bulk) − 3.2 ± 0.3 (bulk)[53]	**− 3.99 ± 0.04**	**0.32 ± 0.02** **0.64 ± 0.02** (bulk)	**0.11 ± 0.02**
Monolayer MoS_2	**− 4.0 ± 0.2** − 4.6 ± 0.08 (bulk)[55] − 3.1 ± 0.1 (bulk)[53]	**− 4.65 ± 0.17**	—	—
Monolayer WSe_2	− 3.7 ± 0.2 (ref. 16) − 4.37 ± 0.15 (ref. 13) − 1.57 to − 2.86 (ref. 14) − 4 ± 0.5 (ref. 17)	—	—	—
Monolayer $MoSe_2$	− 3.8 ± 0.2 (refs 12,16) − 4.1 ± 0.2 (ref. 15)	—	—	

Experimentally measured exciton *g*-factors corresponding to the valley Zeeman effect, and diamagnetic shift coefficients σ, in monolayer transition-metal dichalcogenide materials. Measurements from this work are indicated in boldface. For comparison, also included are selected measurements from bulk crystals.

WSe_2 (refs 12–16), likely due to the limited field range employed ($|B| < 10$ T). Here the diamagnetic shift of the A exciton in monolayer WS_2 can be seen in 65 T fields via the slight positive curvature of both $E^+(B)$ and $E^-(B)$ in Fig. 2d. To directly reveal ΔE_{dia}, Fig. 2g shows the average exciton energy, $(E^+ + E^-)/2$. Overall quadratic shifts are indeed observed, indicating diamagnetic coefficients $\sigma_A = 0.32 \pm 0.02\ \mu\text{eV T}^{-2}$ for the A exciton and a smaller value of $\sigma_B = 0.11 \pm 0.02\ \mu\text{eV T}^{-2}$ for the B exciton. These measurements were repeated on five different regions of the monolayer WS_2 sample, with similar results.

Exciton radii and binding energies. Importantly, knowledge of σ constrains not only the r.m.s. exciton radius r_1 (if the mass is known) but also the exciton binding energy if the potential $V(r)$ is known. Theoretical estimates[3,23,31,32] for the A exciton reduced mass in monolayer WS_2 range from 0.15 to $0.22 m_0$, from which we can then directly calculate $r_{1,A} = 1.48$–1.79 nm via equation (3). These values are in reasonable agreement with recent *ab initio* calculations of the 1*s* exciton wavefunction in monolayer WS_2 (ref. 25), and further support a picture of 2D Wannier-type excitons with lateral extent larger than the monolayer thickness (0.6 nm) and spanning several in-plane lattice constants.

Moreover, σ, m_r and r_1 can then be used to calculate the A exciton wavefunction $\psi_{1s}(r)$ and its binding energy, by numerically solving the 2D Schrödinger equation for describing the relative motion of electrons and holes using the potential $V(r)$ as defined in equation (1), and taking the screening length r_0 as an adjustable parameter to converge on a solution for $\psi_{1s}(r)$ that has the correct r.m.s. radius r_1. For example, using $m_{r,A} = 0.16 m_0$ for the A exciton in WS_2, and using the measured diamagnetic shift σ_A, we find that $r_{1,A} = 1.53$ nm via equation (3). A wavefunction $\psi_{1s}(r)$ with this r.m.s. radius, shown explicitly in Fig. 3a, is calculated if (and only if) the screening length $r_0 = 5.3$ nm, and the binding energy of this state is 410 meV. For comparison, this inferred screening length is somewhat larger than expected for a suspended WS_2 monolayer (where $r_0 = 2\pi\chi_{2D} = 3.8$ nm; ref. 23), but is less than the value of 7.5 nm used recently by Chernikov[24] to fit a non-hydrogenic Rydberg series of excitons in WS_2 from reflectivity data. Similarly, the 410 meV exciton binding energy that we estimate exceeds the value inferred by Chernikov (320 meV), but is less than the 700–830 meV binding energies extracted from two-photon excitation studies[25,26] and reflectivity/absorption studies[28] of monolayer WS_2. We emphasize, however, that the exciton wavefunctions and binding energies that we calculate necessarily depend on the reduced mass m_r and the exact form of the potential $V(r)$, which is sensitive to the details of the dielectric environment and choice of substrate material[33,50].

More generally, Fig. 3b shows a colour-coded surface plot of the exciton binding energy, calculated within the framework of the non-local dielectric screening potential $V(r)$ defined in equation (1), over a range of reduced masses m_r and effective dielectric screening lengths r_0. At each point, the 1*s* exciton wavefunction $\psi_{1s}(r)$, its binding energy, and its r.m.s. radius r_1 were calculated, from which we computed the expected diamagnetic shift coefficient $\sigma = e^2 r_1^2 / 8 m_r$. Importantly, the solid lines on the plot indicate the contours of constant diamagnetic shift that correspond to our experimentally measured values σ_A and σ_B. At intervals along these contours, r_1 is indicated. From this plot, it can be immediately seen that over the range of theoretically calculated masses ($m_{r,A} = 0.15$–$0.22 m_0$), excitons having the appropriate size to give the measured diamagnetic shift σ_A (that is, those lying along the σ_A contour) have binding energies in the range of 480–260 meV. Within this model, excitons with even larger binding energies (but still constrained to exhibit the correct diamagnetic shift) are anticipated if the reduced mass m_r is lighter and the effective screening length r_0 is smaller.

In addition, Fig. 3b also allows us to estimate the mass, binding energy and spatial extent of the B exciton in monolayer WS_2, for which a smaller diamagnetic shift of $\sigma_B = 0.11\ \mu\text{eV T}^{-2}$ was measured. Assuming that the local dielectric environment is similar for A and B excitons (that is, r_0 is unchanged), then parameters for the B exciton lie at a point on the σ_B contour that is directly to the right of those on the σ_A contour. Thus, if $m_{r,A} = 0.16 m_0$ and $r_{1,A} = 1.53$ nm as discussed above, then the B exciton reduced mass is $m_{r,B} = 0.27 m_0$, its r.m.s. radius is $r_{1,B} = 1.16$ nm and its binding energy is 470 meV. These values are qualitatively consistent with trends identified in early optical studies of bulk MoS_2 crystals[48,49], in which B exciton masses and binding energies were found to exceed those of A excitons. These results highlight a further interesting consequence of the potential $V(r)$, which is that exciton binding energies scale only weakly and nonlinearly with m_r, in contrast to the case for hydrogenic potentials.

Zeeman splitting and diamagnetic shifts in bulk WS_2. For direct comparison with monolayer WS_2, circularly polarized magneto-reflectance measurements were also performed on the exfoliated surface of a bulk WS_2 crystal (grown by chemical vapour transport at the Tennessee Crystal Center). Figure 4a shows the well-known[51] A exciton resonance in bulk WS_2, which

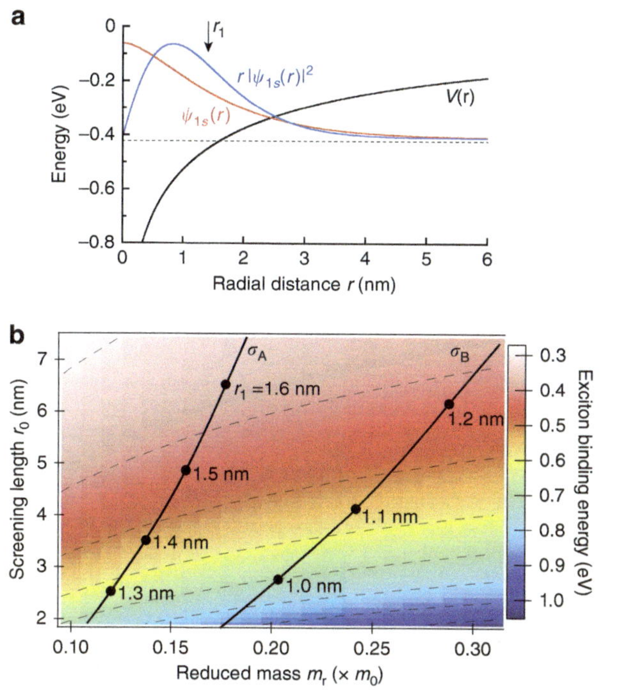

Figure 3 | Constraining exciton binding energies via the diamagnetic shift. (**a**) A plot of the (non-hydrogenic) 1s A exciton wavefunction in monolayer WS_2, $\psi_{1s}(r)$, computed by numerically solving Schrödinger's equation using a reduced mass $m_r = 0.16$ and the non-local dielectric screening potential $V(r)$ defined in equation (1). The screening length r_0 was adjusted to give $\psi_{1s}(r)$ such that the r.m.s. exciton radius $r_1 = \sqrt{\langle\psi_{1s}|r^2|\psi_{1s}\rangle} = 1.53$ nm, which is the value calculated from the measured diamagnetic shift $\sigma_A = 0.32\,\mu eV\,T^{-2}$. (Note that r_1 is an r.m.s. value, and does not correspond to the peak of the 2D radial probability density $r|\psi(r)|^2$.) (**b**) Colour surface plot of the calculated exciton binding energy, using $V(r)$ from equation (1), over a range of reduced mass m_r and screening length r_0. Dashed lines show contours of constant binding energy. Solid lines indicate contours of constant diamagnetic shift corresponding to those measured in Fig. 2g for the A and B excitons in monolayer WS_2. The calculated r.m.s. exciton radius r_1 is indicated at various points along the contours.

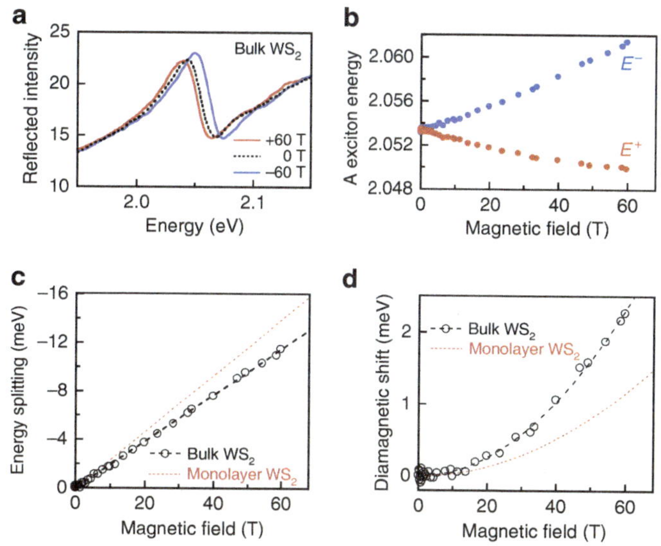

Figure 4 | Zeeman splitting and diamagnetic shift in bulk WS_2. (**a**) Intensity of reflected σ^+ light from the A exciton in bulk WS_2 at 4 K, using $B = 0$, $+60$ and -60 T. (**b**) Energies of the Zeeman-split exciton transitions, E^+ and E^-. (**c**) The measured exciton splitting ($E^+ - E^-$), corresponding to $g = -3.33$. For comparison, the red line shows the valley Zeeman splitting measured in monolayer WS_2 (*cf.* Fig. 2f). (**d**) The average energy $(E^+ + E^-)/2$ showing a diamagnetic shift (0.64 $\mu eV\,T^{-2}$) that is twice as large as that measured in monolayer WS_2 (red line; *cf.* Fig. 2g).

arises from the lowest-energy direct optical transition that is located at the *K* points of the Brillouin zone[52] (this transition, with only slight modification in energy, eventually becomes the lowest overall transition when WS_2 is thinned to a single monolayer and becomes a direct-gap semiconductor[1,2]). At low temperatures and in ± 60 T magnetic fields, the Zeeman splitting of the bulk A exciton is readily resolved (Fig. 4a–c) and is found to increase linearly with field at a rate of $-193\,\mu eV\,T^{-1}$ ($g = -3.33$). This value is in very close agreement with early magnetic circular dichroism measurements of *g*-factors in bulk WS_2 (ref. 53), wherein it was suggested that deviations from $g = -4$ arise from the crystal-field mixing of *p*-type chalcogen atomic orbitals into the predominantly *d*-type character of the conduction and valence bands. Within this context, the value of $g \simeq -4$ that we measured in monolayer WS_2 (Fig. 2f) may suggest that such mixing effects, if present, may be suppressed in atomically thin WS_2.

In addition, Fig. 4d shows that the measured diamagnetic shift of the A exciton in bulk WS_2 is $0.64\,\mu eV\,T^{-2}$, which is twice as large as in monolayer WS_2. Assuming an in-plane reduced mass of $m_r = 0.21 m_0$ in bulk WS_2 (ref. 51), we calculate via equation (3) an in-plane r.m.s. radius of $r_1 = 2.48$ nm for the bulk A exciton, which is substantially larger than that inferred for monolayer WS_2. This large r.m.s. radius indicates an effective dielectric screening constant $\varepsilon_r \simeq 7.0$, in agreement with early work[51], and from which the A exciton binding energy in bulk WS_2 can be estimated via the standard hydrogenic formulation, $m_r/(m_0\varepsilon_r^2) \times 13.6\,eV = 58$ meV. This value agrees extremely well with early work on bulk WS_2 (ref. 51), and is close to that found in other bulk TMDs[54]. Therefore, we find that the binding energy of the A exciton in WS_2 increases by approximately a factor of 7 on reducing the dimensionality of the host crystal from three-dimensional to 2D. Note, however, that these estimates depend on the assumed value of the reduced mass m_r, which has not yet been measured independently by, for example, cyclotron resonance studies.

Valley Zeeman effect in monolayer MoS_2. To complete this study of the monolayer transition-metal disulphides, we also performed high-field magneto-reflectance studies on large-area samples of monolayer MoS_2 (Fig. 5). The A and B exciton linewidths are broader and the optical reflection contrast is lower than for monolayer WS_2 (Fig. 5a). Nonetheless, a clear valley Zeeman splitting of both excitons is observed (Fig. 5b,c). The energies of the field-split exciton peaks are shown in Fig. 5d,e for the A and B excitons, respectively. Although the reduced signals and broader features lead to increased scatter in the fitted data, Fig. 5f shows that the measured valley splitting of the A and B excitons in MoS_2 increases approximately linearly with field at rates of -233 ± 10 and $-270 \pm 10\,\mu eV\,T^{-1}$, corresponding to $g \simeq -4.0 \pm 0.2$ and -4.65 ± 0.17, respectively. For the A exciton, this value is very close to those inferred from low-field magneto-photoluminescence studies[12,15,16] of their diselenide counterpart, monolayer $MoSe_2$ (Table 1). As discussed above for the case of monolayer WS_2, a *g*-factor of -4 for the A exciton agrees surprisingly well with expectations from a simple two-band tight-binding picture, and suggests that the valley Zeeman effect in

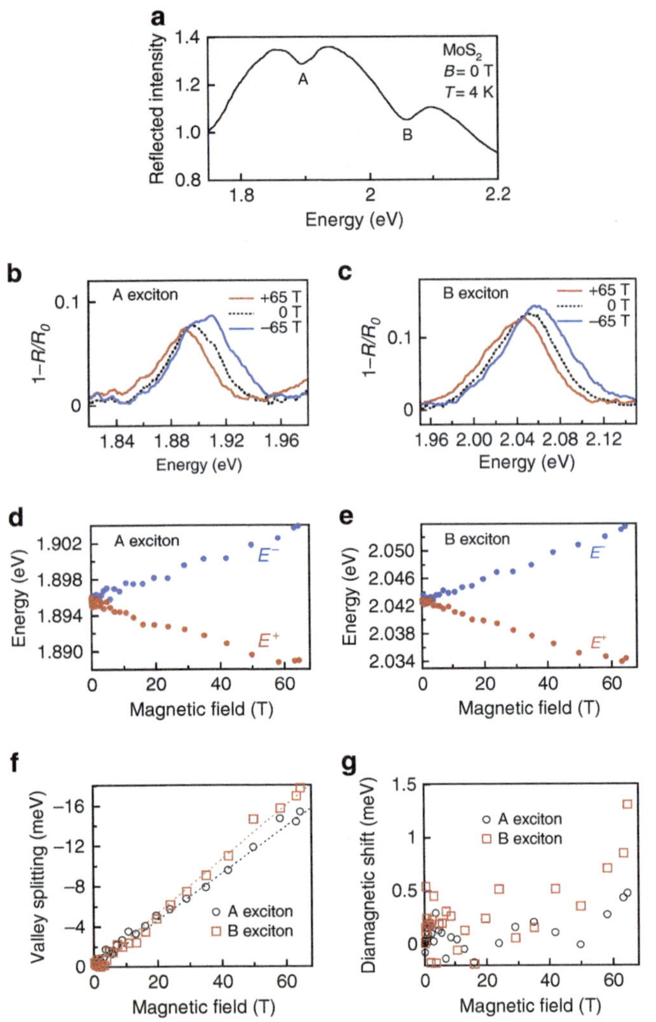

Figure 5 | Valley Zeeman effect in monolayer MoS_2. (**a**) Reflection spectrum of monolayer MoS_2 at $B=0$ T and $T=4$ K. The A and B exciton resonances are labelled. (**b**,**c**) Normalized reflection spectra $(1-R/R_0)$ at the A and B exciton resonances at $B=0$, $+65$ and -65 T. (**d**) Energies of the field-split A exciton transition. (**e**) Same, but for the B exciton transition. (**f**) The measured valley splitting $(E^+ - E^-)$ versus the magnetic field, for both A and B excitons. (**g**) The average energy of the two valley-split resonances, $(E^+ + E^-)/2$, for both A and B excitons. Increased scatter in the data from this MoS_2 sample precludes any clear identification of the diamagnetic shift.

MoS_2, much like $MoSe_2$, is largely uninfluenced by contributions from the valley orbital (Berry curvature) magnetic moment. We note, however, that the measured valley g-factor is somewhat larger for the B exciton in monolayer MoS_2. Unfortunately, the reduced signal levels from these monolayer MoS_2 samples led to correspondingly increased scatter in the fitted exciton energies, precluding an accurate determination of exciton diamagnetic shifts in monolayer MoS_2 (Fig. 5g).

Discussion

In summary, we have presented a comprehensive study of valley Zeeman effect and diamagnetic shifts of excitons in the archetypal monolayer transition-metal disulphides WS_2 and MoS_2. Valley g-factors of the A excitons are approximately -4, which are similar to those obtained from transition-metal diselenides. Unexpectedly, the heavier B exciton in monolayer WS_2 also exhibits $g \simeq -4$, suggesting that the valley Zeeman effect is largely unaffected by the exciton reduced mass. The very large magnetic fields used in these studies also allowed initial measurements of the exciton diamagnetic shifts in a monolayer TMD—specifically, WS_2—from which r.m.s. exciton radii were directly computed ($r_1 = 1.53$ and 1.16 nm for the A and B excitons, respectively). Within a picture of non-local dielectric screening in these 2D semiconductors, these measurements of diamagnetic shifts allowed us to constrain estimates of the exciton binding energies, which we calculate (using a reduced A exciton mass of $0.16m_0$) to be 410 and 470 meV for the A and B excitons, respectively, in monolayer WS_2. These studies highlight the utility of very large magnetic fields for characterizing new 2D material systems.

References

1. Splendiani, A. *et al.* Emerging photoluminescence in monolayer MoS_2. *Nano Lett.* **10,** 1271–1275 (2010).
2. Mak, K. F. *et al.* Atomically thin MoS_2: a new direct-gap semiconductor. *Phys. Rev. Lett.* **105,** 136805 (2010).
3. Xiao, D. *et al.* Coupled spin and valley physics in monolayers of MoS_2 and other group-VI dichalcogenides. *Phys. Rev. Lett.* **108,** 196802 (2012).
4. Mak, K. F. *et al.* Control of valley polarization in MoS_2 by optical helicity. *Nat. Nanotechnol.* **7,** 494–498 (2012).
5. Zeng, H. *et al.* Valley polarization in MoS_2 monolayers by optical pumping. *Nat. Nanotechnol.* **7,** 490–493 (2012).
6. Sallen, G. *et al.* Robust optical emission polarization in MoS_2 monolayers through selective valley excitation. *Phys. Rev. B* **86,** 081301 (R) (2012).
7. Cao, T. *et al.* Valley-selective circular dichroism of monolayer molybdenum disulphides. *Nat. Commun.* **3,** 887 (2012).
8. Xu, X., Yao, W., Xiao, D. & Heinz, T. F. Spin and pseudospin in layered transition metal dichalcogenides. *Nat. Phys.* **10,** 343–350 (2014).
9. Sham, L., Allen, S., Kamgar, A. & Tsui, D. Valley-valley splitting in inversion layers on a high-index surface of silicon. *Phys. Rev. Lett.* **40,** 472–475 (1978).
10. Gunawan, O. *et al.* Valley susceptibility of an interacting two-dimensional electron system. *Phys. Rev. Lett.* **97,** 186404 (2006).
11. Xiao, D., Yao, W. & Niu, Q. Valley contrasting physics in graphene: magnetic moment and topological transport. *Phys. Rev. Lett.* **99,** 236809 (2007).
12. MacNeill, D. *et al.* Breaking of valley degeneracy by magnetic field in monolayer $MoSe_2$. *Phys. Rev. Lett.* **114,** 037401 (2015).
13. Srivastava, A. *et al.* Valley Zeeman effect in elementary optical excitations on monolayer WSe_2. *Nat. Phys.* **11,** 141–147 (2015).
14. Aivazian, G. *et al.* Magnetic control of valley pseudospin in monolayer WSe_2. *Nat. Phys.* **11,** 148–152 (2015).
15. Li, Y. *et al.* Valley splitting and polarization by the Zeeman effect in monolayer $MoSe_2$. *Phys. Rev. Lett.* **113,** 266804 (2014).
16. Wang, G. *et al.* Magneto-optics in transition metal diselenide monolayers. *2D Mater.* **2,** 034002 (2015).
17. Mitioglu, A. A. *et al.* Optical investigation of monolayer and bulk tungsten diselenide (WSe_2) in high magnetic fields. *Nano Lett.* **15,** 4387–4392 (2015).
18. Kormányos, A., Zólyomi, V., Drummond, N. D. & Burkard, G. Spin-orbit coupling, quantum dots, and qubits in monolayer transition metal dichalcogenides. *Phys. Rev. X* **4,** 011034 (2014).
19. Kormányos, A. *et al.* k·p theory of two-dimensional transition metal dichalcogenide semiconductors. *2D Mater.* **4,** 022001 (2015).
20. Knox, R. S. *Theory of Excitons* (Academic, 1963).
21. Miura, N. *Physics of Semiconductors in High Magnetic Fields* (Oxford Univ. Press, 2008).
22. Walck, S. N. & Reinecke, T. L. Exciton diamagnetic shift in semiconductor nanostructures. *Phys. Rev. B* **57,** 9088 (1988).
23. Berkelbach, T. C., Hybertsen, M. S. & Reichman, D. R. Theory of neutral and charged excitons in monolayer transition-metal dichalcogenides. *Phys. Rev. B* **88,** 045318 (2013).
24. Chernikov, A. *et al.* Exciton binding energy and nonhydrogenic Rydberg series in monolayer WS_2. *Phys. Rev. Lett.* **113,** 076802 (2014).
25. Ye, Z. *et al.* Probing excitonic dark states in single-layer tungsten disulphides. *Nature* **513,** 214–218 (2014).
26. Zhu, B., Chen, X. & Cui, X. Exciton binding energy on monolayer WS_2. *Sci. Rep.* **5,** 9218 (2015).
27. He, K. I. *et al.* Tightly bound excitons in monolayer WSe_2. *Phys. Rev. Lett.* **113,** 026803 (2014).
28. Hanbicki, A. T., Currie, M., Kioseoglou, G., Friedman, A. L. & Jonker, B. T. Measurement of high exciton binding energy in the monolayer transition-metal dichalcogenides WS_2 and WSe_2. *Solid State Commun.* **203,** 16–20 (2015).
29. Wang, G. *et al.* Giant enhancement of the optical second-harmonic emission of WSe_2 monolayers by laser excitation at exciton resonances. *Phys. Rev. Lett.* **114,** 097403 (2015).

30. Stroucken, T. & Koch, S. W. Optically bright p-excitons indicating strong Coulomb coupling in transition-metal dichalcogenides. *J. Phys. Condens. Matter* **27,** 345003 (2015).
31. Ramasubramaniam, A. Large excitonic effects in monolayers of molybdenum and tungsten dichalcogenides. *Phys. Rev. B* **86,** 115409 (2012).
32. Shi, H., Pan, H., Zhang, Y.-W. & Yakobson, B. I. Quasiparticle band structures and optical properties of strained monolayer MoS_2 and WS_2. *Phys. Rev. B* **87,** 155304 (2013).
33. Komsa, H.-P. & Krasheninnikov, A. V. Effects of confinement and environment on the electronic structure and exciton binding energy of MoS_2 from first principles. *Phys. Rev. B* **86,** 241201 (2012).
34. Ugeda, M. M. *et al.* Giant bandgap renormalization and excitonic effects in a monolayer transition metal dichalcogenide semiconductor. *Nat. Mater.* **13,** 1091–1095 (2014).
35. Velizhanin, K. A. & Saxena, A. Excitonic effects in 2D semiconductors: path integral Monte Carlo approach. *Phys. Rev. B* **92,** 195305 (2015).
36. Keldysh, L. V. Coulomb interactions in thin semiconductor and semimetal films. *JETP Lett.* **29,** 658–660 (1979).
37. Cudazzo, P., Tokatly, I. V. & Rubio, A. Dielectric screening in two-dimensional insulators: implications for excitonic and impurity states in graphene. *Phys. Rev. B* **84,** 085406 (2011).
38. Berkelbach, T. C., Hybertsen, M. S. & Reichman, D. R. Bright and dark singlet excitons via linear and two-photon spectroscopy in monolayer transition-metal dichalcogenides. *Phys. Rev. B* **92,** 085413 (2015).
39. Srivastava, A. & Imamoğlu, A. Signatures of Bloch-band geometry on excitons: nonhydrogenic spectra in transition-metal dichalcogenides. *Phys. Rev. Lett.* **115,** 166802 (2015).
40. Zhou, J., Shan, W.-Y., Yao, W. & Xiao, D. Berry phase modification to the energy spectrum of excitons. *Phys. Rev. Lett.* **115,** 166803 (2015).
41. McCreary, K. M. *et al.* Large-area synthesis of continuous and uniform MoS_2 monolayer films on graphene. *Adv. Funct. Mater.* **42,** 6449–6454 (2014).
42. McCreary, K. M., Hanbicki, A. T., Jernigan, G. G., Culbertson, J. C. & Jonker, B. T. Synthesis of large-area WS_2 monolayers with exceptional photoluminescence. *Sci. Rep.* **6,** 19159 (2016).
43. Ling, X. *et al.* Role of the seeding promoter in MoS_2 growth by chemical vapor deposition. *Nano Lett.* **14,** 464–472 (2014).
44. Yao, W., Xiao, D. & Niu, Q. Valley-dependent optoelectronics from inversion symmetry breaking. *Phys. Rev. B* **77,** 235406 (2008).
45. Liu, G.-B. *et al.* Three-band tight-binding model for monolayers of group-VIB transition-metal dichalcogenides. *Phys. Rev. B* **88,** 085433 (2013).
46. Furdyna, J. K. Diluted magnetic semiconductors. *J. Appl. Phys.* **64,** R29–R64 (1988).
47. Evans, B. L. & Young, P. A. Exciton spectra in thin crystals. *Proc. R. Soc. Lond. A* **298,** 74–96 (1967).
48. Evans, B. L. & Young, P. A. Delocalized excitons in thin anisotropic crystals. *phys. status solidi b* **25,** 417–425 (1968).
49. Neville, R. A. & Evans, B. L. The band edge excitons in 2H-MoS_2. *phys. status solidi b* **73,** 597–606 (1976).
50. Lin, Y. *et al.* Dielectric screening of excitons and trions in single-layer MoS_2. *Nano Lett.* **14,** 5569–5576 (2014).
51. Beal, A. R. & Liang, W. Y. Excitons in 2H-WSe_2 and 3R-WS_2. *J. Phys. C: Solid State Phys.* **9,** 2459–2466 (1976).
52. Coehoorn, R., Haas, C. & de Groot, R. A. Electronic structure of $MoSe_2$, MoS_2, and WSe_2. II. The nature of the optical band gaps. *Phys. Rev. B* **35,** 6203–6206 (1987).
53. Tanaka, M., Fukutani, H. & Kuwabara, G. Excitons in VI B transition metal dichalcogenides. *J. Phys. Soc. Jpn* **45,** 1899–1904 (1978).
54. Beal, A. R., Knights, J. C. & Liang, W. Y. Transmission spectra of some transition-metal dichalcogenides: II. group VIA: trigonal prismatic coordination. *J. Phys. C: Solid State Phys* **5,** 3540–3551 (1972).
55. Goto, T., Kato, Y., Uchida, K. & Miura, N. Exciton absorption spectra of MoS_2 crystals in high magnetic fields up to 150T. *J. Phys. Condens. Matter* **12,** 6719–6723 (2000).

Acknowledgements

We thank K. Velizhanin and P. Hawrylak for helpful discussions. These optical studies were performed at the National High Magnetic Field Laboratory, which is supported by NSF DMR-1157490 and the State of Florida. Work at NRL was supported by core programs and the NRL Nanoscience Institute, and by AFOSR under contract number AOARD 14IOA018-134141. J.K. was supported by the Air Force Office of Scientific Research under Award Number FA9550-14-1-0268.

Author contributions

S.A.C. and J.K. conceived and directed the experiments; K.M.M. and B.T.J. synthesized and characterized the monolayer samples; A.V.S. built and performed the high-field optical experiments; A.V.S. and S.A.C. analysed the data and wrote the paper in close consultation with all authors.

Additional information

Competing financial interests: The authors declare no competing financial interests.

17

Hexadecapolar colloids

Bohdan Senyuk[1], Owen Puls[1], Oleh M. Tovkach[2,3], Stanislav B. Chernyshuk[4] & Ivan I. Smalyukh[1,5,6]

Outermost occupied electron shells of chemical elements can have symmetries resembling that of monopoles, dipoles, quadrupoles and octupoles corresponding to filled *s*-, *p*-, *d*- and *f*-orbitals. Theoretically, elements with hexadecapolar outer shells could also exist, but none of the known elements have filled *g*-orbitals. On the other hand, the research paradigm of 'colloidal atoms' displays complexity of particle behaviour exceeding that of atomic counterparts, which is driven by DNA functionalization, geometric shape and topology and weak external stimuli. Here we describe elastic hexadecapoles formed by polymer microspheres dispersed in a liquid crystal, a nematic fluid of orientationally ordered molecular rods. Because of conically degenerate boundary conditions, the solid microspheres locally perturb the alignment of the nematic host, inducing hexadecapolar distortions that drive anisotropic colloidal interactions. We uncover physical underpinnings of formation of colloidal elastic hexadecapoles and describe the ensuing bonding inaccessible to elastic dipoles, quadrupoles and other nematic colloids studied previously.

[1] Department of Physics and Soft Materials Research Center, University of Colorado, Boulder, Colorado 80309, USA. [2] Bogolyubov Institute for Theoretical Physics, NAS of Ukraine, Kyiv 03680, Ukraine. [3] Department of Mathematics, University of Akron, Akron, Ohio 44325, USA. [4] Institute of Physics, NAS of Ukraine, Kyiv 03650, Ukraine. [5] Department of Electrical, Computer, and Energy Engineering, Materials Science and Engineering Program, University of Colorado, Boulder, Colorado 80309, USA. [6] Renewable and Sustainable Energy Institute, National Renewable Energy Laboratory and University of Colorado, Boulder, Colorado 80309, USA. Correspondence and requests for materials should be addressed to I.I.S. (email: ivan.smalyukh@colorado.edu).

Colloids form a platform for scalable fabrication of mesostructured composite materials and provide a framework for testing theoretical descriptions of crystals and glasses, albeit they are also commonly encountered in daily life in forms of milk, paints, coffee, fog and so on[1–5]. Nematic liquid crystal (NLC) colloids[6–17] attract particularly strong interest because they can be tuned by weak external stimuli, such as low-voltage fields and light, similar to the NLC host fluids themselves[12,17]. Colloidal particles locally perturb the uniform ground-state alignment of the anisotropic NLC molecules described by the so-called 'director' $\mathbf{n} \equiv -\mathbf{n}$ with non-polar symmetry, producing spatial elastic distortions of the director field $\mathbf{n}(\mathbf{r})$ that resemble electric field configurations around dipolar and quadrupolar charge distributions[7–12]. The type of distortions depends on surface boundary conditions for $\mathbf{n}(\mathbf{r})$, which can be tangential or perpendicular to the particle surfaces and can induce both bulk and surface line and point defects[7–12], topological singularities along which $\mathbf{n}(\mathbf{r})$ and NLC order cannot be defined[1]. Such NLC colloidal particles tend to arrange themselves so that they can share energetically costly $\mathbf{n}(\mathbf{r})$ distortions to minimize free energy, exhibiting highly anisotropic elasticity-mediated interactions that resemble interactions of electrostatic dipoles and quadrupoles[6–12,15,16]. This electrostatics analogy[7] provides a framework for understanding, predicting and engineering the ensuing colloidal self-assembly. However, for many years of very active research efforts[6–12,15,16], only colloidal elastic dipoles and quadrupoles have been found in experiments and theories alike, which limits diversity of the accessible colloidal bonds and self-assembled structures[6–17].

In this work, we describe colloidal elastic hexadecapoles that spontaneously form around solid microspheres immersed in a uniformly aligned NLC. The unusual symmetry of elastic distortions arises from conically degenerate[18] boundary conditions for $\mathbf{n}(\mathbf{r})$ at the colloidal surfaces, which induce surface point and line defects at the same time. Using a combination of holographic optical tweezers (HOT)[19], nonlinear optical imaging[20] and polarizing optical microscopy (POM)[1,6], we probe the $\mathbf{n}(\mathbf{r})$-distortions and quantify colloidal pair interactions by measuring distance and angular dependencies of elastic potentials, demonstrating relations between the director structure and medium-mediated inter-particle forces. Finally, we explain our findings using a model based on elastic multipole expansion and discuss how the experimental framework we have developed may enable colloidal self-assembly into novel forms of tunable pre-engineered matter without known atomic analogues.

Results

Hexadecapolar elastic multipole. When dispersed in a uniformly aligned NLC fluid host, polystyrene microspheres (PSMs) of a radius r_0 locally distort $\mathbf{n}(\mathbf{r})$, which is manifested by eight bright lobes around the particle perimeter seen in POM between the crossed polarizer and analyser (Fig. 1a). These bright lobes are separated by eight dark regions within which $\mathbf{n}(\mathbf{r})$ at the particle's perimeter is parallel to polarizer or analyser. Using a phase retardation plate and interference of polarized light propagating through the particle-distorted structure, we reveal that $\mathbf{n}(\mathbf{r})$ tilting away from the far-field director $\mathbf{n}_0$ switches between clockwise and counterclockwise directions (corresponding to the blue and yellow colours in the micrograph) eight times as one circumnavigates the sphere (Fig. 1b). Bright-field micrographs obtained at different depth of focus reveal presence of weakly scattering surface point defects (boojums) at the particle poles along $\mathbf{n}_0$ as well as a circular loop of a defect line (often called 'Saturn ring')[9,10] at the particle's equator (Fig. 1c,e). Based on POM and three-dimensional nonlinear optical imaging (Fig. 1a–c,e and

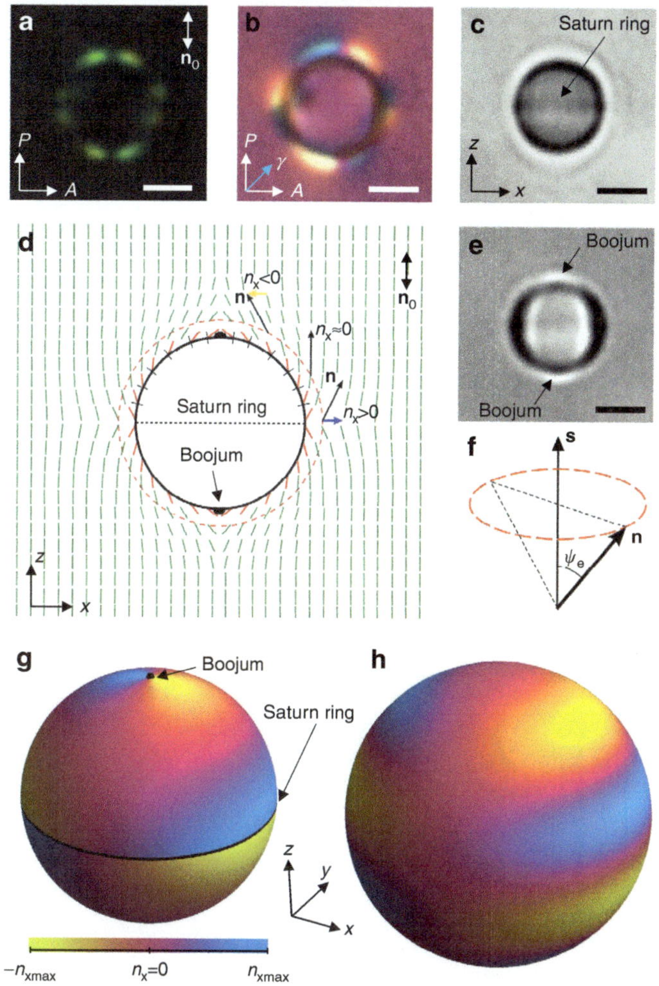

Figure 1 | Elastic hexadecapole induced by a colloidal PSM. (**a**–**d**) Optical micrographs obtained with (**a**,**b**) POM and (**c**,**e**) bright field microscopy. P, A and γ mark the crossed polarizer, analyser and a slow axis of a 530 nm retardation plate (aligned at 45° to P and A), respectively. (**d**) Schematic diagram of induced $\mathbf{n}(\mathbf{r})$ (green rods) satisfying the tilted boundary conditions at the PSM surface (red rods), with the 'easy axis' at a constant angle ψ_e to a local normal $\mathbf{s}$ to the surface (black rods). (**f**) Schematic of conic degenerate surface boundary conditions. (**g**,**h**) Three-dimensional visualization of the x-component n_x of $\mathbf{n}(\mathbf{r})$ (**g**) at the surface of PSM for $\psi_e = 45°$ and (**h**) at a spherical surface of radius $1.2r_0$ shown using a dashed red circle in **d**. Blue, yellow and magenta colours correspond, respectively, to a positive, near-zero and negative n_x according to the colour scheme shown in **g**. Dashed equatorial line in **d** and a black solid line in **g** depict the 'Saturn ring' surface defect loop at the particle's equator visible in **c**. Black hemispheres in **d** and **g** show surface point defects boojums at the poles of the particle visible in **e**. Scale bars, 2 μm.

Supplementary Figs 1–3), we uncover the structure of $\mathbf{n}(\mathbf{r})$ distortions schematically shown in Fig. 1d. This structure is consistent with conically degenerate surface boundary conditions for $\mathbf{n}(\mathbf{r})$ with respect to PSM's local surface normals $\mathbf{s}$, which were previously demonstrated for NLCs at flat polystyrene-coated surfaces[18]. The director's easy axis orientation lies on a cone of equilibrium polar angle ψ_e (Fig. 1f). To minimize the free-energy cost of bulk elastic distortions, interaction of conically degenerate surface boundary conditions on the microsphere with the uniform $\mathbf{n}_0$ lifts this conical degeneracy and yields an axially symmetric $\mathbf{n}(\mathbf{r})$ depicted in Fig. 1d. Extending this analysis to three dimensions, the projection n_x of $\mathbf{n}(\mathbf{r})$ onto the x-axis orthogonal to $\mathbf{n}_0$ can be visualized around PSM using colours that highlight positive, near zero and negative n_x (Fig. 1g). The black points at the poles and ring at the equator of the sphere are regions of discontinuity of $\mathbf{n}(\mathbf{r})$ at the NLC-PSM interface and correspond to the boojum and Saturn ring topological defects, respectively. Away from the particle surface and these topological singularities, the experimentally reconstructed $\mathbf{n}(\mathbf{r})$ is continuous (Fig. 1d) and consistent with the theoretical model for a hexadecapolar director distortions presented in a similar way in Fig. 1h.

Elastic interactions between hexadecapolar colloids. Elasticity-mediated interactions between PSMs (Fig. 2) differ from all NLC colloids studied so far. To get insights into the strength and direction-dependence of these interactions, we utilize HOT to optically trap one 'stationary' PSM and then release another particle at different centre-to-centre vector $\mathbf{R}$ orientations with respect to $\mathbf{n}_0$ as well as at different separation distances. Using video-microscopy, we track the ensuing particle motions, which result from a combination of Brownian jiggling and elasticity-mediated interactions. The submicron waist and relatively low power (~ 10 mW) of a focused trapping beam allow us to avoid the influence of the trapping on the measurements[21]. Furthermore, the laser tweezers are used only to bring the particles to the desired initial conditions and are turned off within the time when the pair-interaction is probed with video microscopy, allowing us to avoid possible artefacts associated with the complex effects of the laser trapping light at small inter-particle distances. The colour-coded time-coordinate trajectories of particles released from optical traps at different $\mathbf{R}$ are shown in Fig. 2a. Unlike in the case of dipolar and quadrupolar NLC colloids[7,8,15,16], elastic forces are relatively short-ranged and exceed the strength of thermal fluctuations only at distances of four-to-five particle radii r_0. However, the angular dependence of these forces is very rich, with eight angular segments of inter-particle attractions separated by eight angular segments of repulsions, with the intermediate angular sectors within which particles move sideways as the inter-particle elastic forces are orthogonal or at large angles to $\mathbf{R}$ (Fig. 2a). These angular sectors of attraction and repulsion correlate with the bright and dark regions of POM micrographs (Fig. 1a) as well as with the predictions of our model based on elastic multipole expansion (Fig. 2).

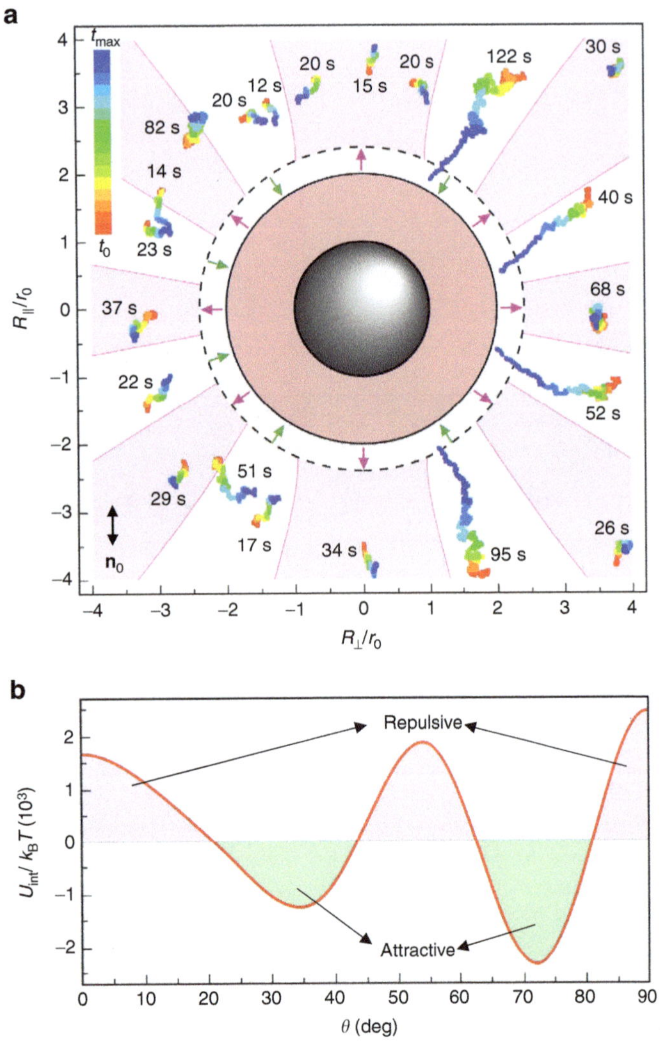

Figure 2 | Elastic interactions of colloidal hexadecapoles. (**a**) Angular dependence of interactions is probed by tracking motion of a particle released from the optical trap and moving with respect to the 'stationary' trapped particle in the centre depending on the orientation of $\mathbf{R}$ with respect to $\mathbf{n}_0$. The elapsed time is coded according to the colour scale (inset) and the maximum elapsed times $t_{max} - t_0$ are marked next to the corresponding trajectories. $R_{||}$ and $R_\perp$ denote the centre-to-centre distance $\mathbf{R}$ components along and perpendicular to $\mathbf{n}_0$, respectively. The PSM of radius r_0 is surrounded by the spherical volume of radius $2r_0$ that is excluded for centres of other PSMs. The nonlinear zone is shown by a dashed circle at $R = 2.4r_0$. (**b**) Pair-potential U_{int} versus θ for two particles at $R = 2.4r_0$. Angular zones of repulsion and attraction are highlighted using magenta and green arrows and colouring, respectively.

To quantify elastic interactions, we first probe the centre-to-centre separation $R = |\mathbf{R}|$ versus time t for particles released at different angles between $\mathbf{R}$ and $\mathbf{n}_0$ within the angular sectors of attraction (Fig. 3a) and then calculate particle velocities $\dot{\mathbf{R}} = d\mathbf{R}/dt$. Because the system is highly over-damped and the inertia effects can be neglected[22], the experimentally measured $\mathbf{R}(t)$ and the simplified equation of motion $0 \approx -\xi\dot{\mathbf{R}} + \mathbf{F}_{int}$ yield the pair interaction potential U_{int} (inset of Fig. 3a), where ξ is a drag coefficient measured separately by characterizing Brownian motion of PSMs (Methods and Supplementary Fig. 4) and $\mathbf{F}_{int} = -\nabla U_{int}$ is the elastic interaction force. The attractively interacting particles stop short of physically touching each other, instead forming stable dimer assemblies with typical $R \approx (2.05\text{-}2.2)r_0$ and stable $\mathbf{R}$ orientations with respect to $\mathbf{n}_0$ within one of the two angular sectors of assembly in each quadrant dependent on ψ_e: $\theta_1 \approx 22°\text{-}26°$ or $\theta_2 \approx 64°\text{-}75°$ (Fig. 3b–f). Multi-particle assemblies with different combinations of angles θ_1 and θ_2 are also observed (Fig. 3d). The inter-particle binding energies are measured to be in the range of hundreds of k_BT (inset of Fig. 3a), making them robust with respect to thermal fluctuations. Although U_{int} versus θ has eight minima, only four of them can be occupied simultaneously in one plane by particles of the same size because of the 'excluded volume' effects, yielding two-dimensional colloidal crystals with rhombic elementary cells (Supplementary Fig. 5 and Supplementary Note 1). Following similar considerations, a large number of low-symmetry three-dimensional colloidal structures can be envisaged too. Since the elastic interactions potential is hundreds of k_BT and the particle assemblies can be entrapped in metastable states, the assembly of two- and three-dimensional colloidal lattices from micrometer-sized particles requires the use of optical tweezers for guiding colloidal particles. Alternatively, the elastic interaction potentials between colloidal particles of smaller size or with weaker surface anchoring can be brought to the order of 10 k_BT and lower, so that the crystal self-assembly can occur without the assistance of optical tweezers, which will be explored elsewhere.

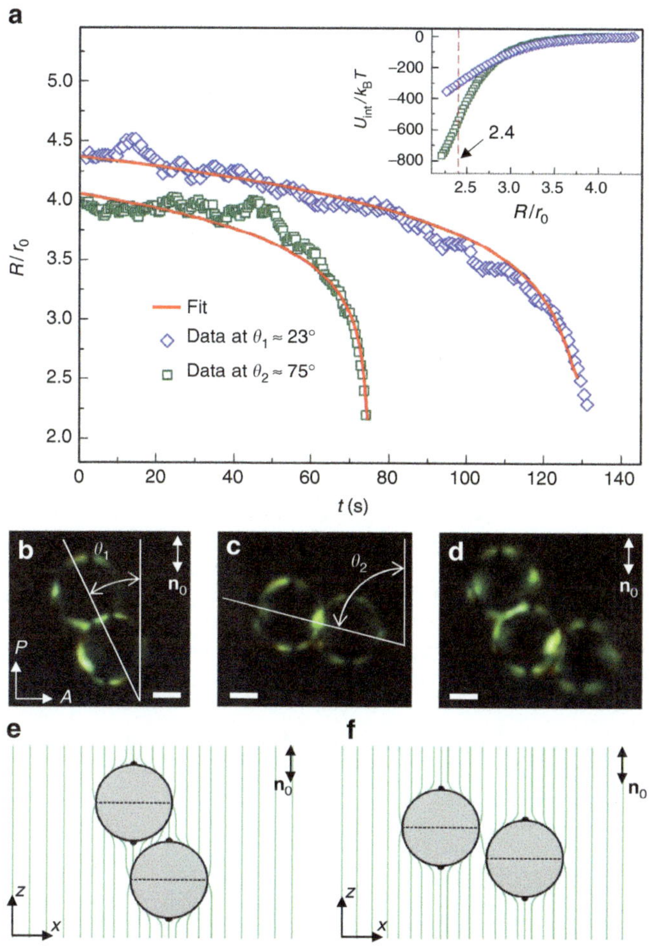

Figure 3 | Self-assembly of hexadecapolar colloids. (**a**) Separation R versus time t for attractive interactions at different θ (measured upon the dimer formation). The solid red line shows a least-squares fit with $R(t)$ obtained from the simplified equation of motion with $(b_2, b_4, b_6) = (-0.017, -0.092, 0.003)$; it is impossible to reproduce such a dependence $R(t)$ with the parameter b_2 only since the quadrupole–quadrupole interaction is repulsive along $\theta \approx 23°$. Inset shows the corresponding U_{int} versus R. (**b**) and (**c**) POM micrographs of self-assembled colloidal dimers of hexadecapoles with **R** aligned at θ_1 and θ_2 to $\mathbf{n}_0$, respectively. (**d**) POM micrograph of a kinked colloidal chain of hexadecapoles. (**e**,**f**) Schematic diagrams of **n(r)** (green lines) for colloidal dimers shown in **c**,**d**, respectively. Dashed lines and black hemispheres in **c** and **d** depict surface defect loops and surface point defects at the poles of PSMs, respectively. Scale bars, 2 µm.

To model experiments, we exploit the electrostatic analogy of the far-field director distortions because of a colloidal sphere that can be represented as elastic multipoles[23], albeit our colloids dramatically differ from elastic dipoles and quadrupoles studied so far[7–12,23–26]. Far from the colloidal sphere, the director deviations n_μ ($\mu = x; y$) from $\mathbf{n}_0 = (0, 0, 1)$ are small. Assuming $\mathbf{n}(\mathbf{r}) \approx (n_x, n_y, 1)$, the NLC elastic free-energy reads[7,8,23]

$$F_{\text{har}} = \frac{K}{2}\sum_{\mu=x,y}\int d\mathbf{r}\nabla n_\mu \cdot \nabla n_\mu, \tag{1}$$

where K is an average Frank elastic constant[1]. Euler–Lagrange equations arising from the functional (1) are of Laplace type, $\Delta n_\mu = 0$, with solutions expanded into multipoles

$$n_\mu(\mathbf{r}) = \sum_{l=1}^{N} a_l(-1)^l \partial_\mu \partial_z^{l-1}\frac{1}{r} \tag{2}$$

where $a_l = b_l {r_0}^{l+1}$ is the elastic multipole moment of the lth order (2^l-pole)[22–26] and one can find coefficients b_l from exact solutions for $\mathbf{n}(\mathbf{r})$ or from relevant experiments (Figs 2a and 3a). Odd moments vanish because $\mathbf{n}(\mathbf{r})$ is symmetric about the particle centre, similar to analogous electrostatic charge distributions that have no dipole or octupole electric moments[22–26]. The multipole expansion of the induced $\mathbf{n}(\mathbf{r})$ shown in Fig. 1d is characterized by coefficients b_2, b_4 and b_6, which also determine the colloidal pair-interaction energy[24]

$$U_{\text{int}} = 4\pi K \sum_{l,l'=2,4,6} a_l a'_{l'}(-1)^{l'}\frac{(l+l')!}{R^{l+l'+1}}P_{l+l'}(\cos\theta), \tag{3}$$

where $P_{l+l'}(\cos\theta)$ are the Legendre polynomials. For colloidal quadrupoles, b_2 dominates and b_4 and b_6 can play a role only at small[25] R. For our particles (Fig. 1), the induced $\mathbf{n}(\mathbf{r})$ can be qualitatively understood as a superposition of configurations of two separate quadrupoles, one with the Saturn-ring and one with boojums[9–11,16,22–26], having opposite signs of quadrupole moments (compare Fig. 4a,d,g,j and 4b,e,h,k). Therefore, the net quadrupole moment is small and the high-order multipoles manifest themselves in a wide range of R. Fitting experimental $R(t)$ within different angular sectors of pair interaction with the corresponding theoretical predictions yields a unique set of parameters b_2, b_4 and b_6 (Fig. 3a). The quadrupole moment $a_2 = -0.017{r_0}^3$ is about two orders of magnitude smaller than that of elastic quadrupoles[16,24], consistent with the hexadecapolar symmetry of $\mathbf{n}(\mathbf{r})$ that is found playing a dominant role, and even the higher-order term (64-pole) plays a detectable role at relatively small R (Supplementary Fig. 6 and Supplementary Note 2). The rich angular dependence of elastic pair-interactions predicted by equation 3 for coefficients b_2, b_4 and b_6 obtained from fitting is consistent with our experimental characterization of the hexadecapolar nature of PSM colloids in the NLC host (Fig. 2).

Discussion

It is instructive to compare hexadecapolar NLC colloids to other known elastic multipoles and their electrostatic analogues. Colloidal spheres with strong tangential anchoring at their surface form $\mathbf{n}(\mathbf{r})$-distortions of quadrupolar configuration[16,24,27,28] and are commonly called elastic quadrupoles (Fig. 4a,d,g,j). They have two surface point defects at the two poles of the spherical particle, which are called 'boojums'. Spherical particles with homeotropic surface anchoring can induce two different elastic multipoles, a dipole[23,28] (Fig. 4c,f,l) and a quadrupole[16,23,27–32] (Fig. 4b,e,h,k). The elastic colloidal dipole has a bulk point defect called 'hedgehog' near one of the poles and lacks mirror symmetry with respect to the plane going through its equator orthogonally to $\mathbf{n}_0$ (Fig. 4f,l). Such a colloidal dipole has a small octupole moment as well[33]. The elastic quadrupole around the particle with a strong homeotropic anchoring has a closed disclination loop around the equator called 'Saturn ring' and, in all four quadrants, has an opposite tilt of the deformed director with respect to $\mathbf{n}_0$ as compared with the tangential elastic quadrupole[16] (compare Fig. 4g,j,h,k). In the case when anchoring is weak, an elastic multipole with weak distortions of quadrupolar configuration can be formed somewhat similar to the elastic quadrupole with 'Saturn ring', but its disclination loop is 'virtual' (within particle's volume) as the director is allowed to deviate away from the easy axis orientation[31,32]. The maps of x-component of the director, n_x, calculated using methods described in refs 23,24 and plotted on

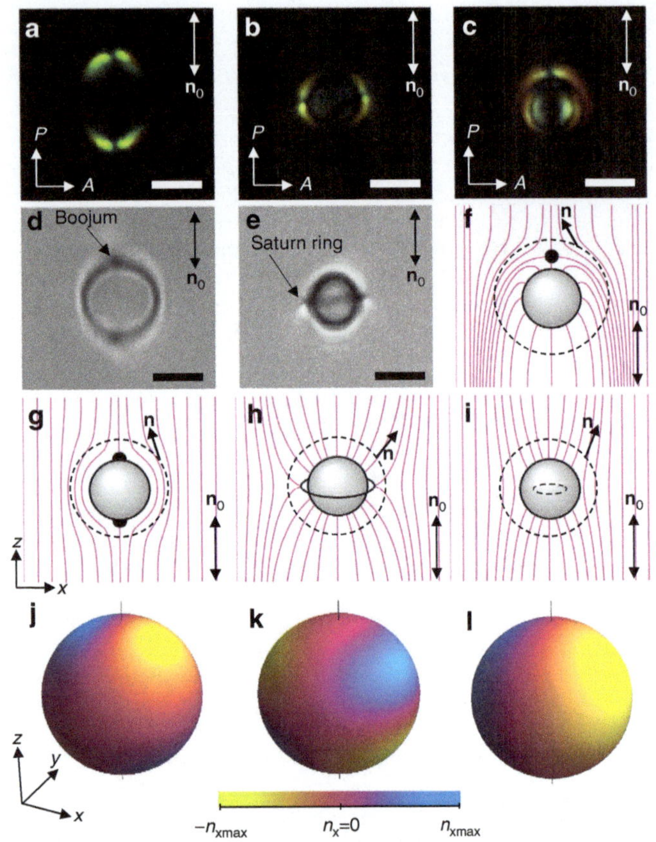

Figure 4 | Elastic multipoles induced by spherical particles with different anchoring. (**a**,**b**) Polarizing and (**d**,**e**) bright field micrographs and (**g**,**h**) schematic diagrams of a director field **n**(**r**) (magenta lines) around elastic quadrupoles, respectively, formed by spheres with a strong (**a**,**d**,**g**) tangential and (**b**,**e**,**h**) homeotropic anchoring. Black filled semicircles at the poles of the particle in **g** show surface point defects dubbed 'boojums.' A line at the particle equator in **h** shows a closed surface disclination loop dubbed 'Saturn ring.' (**c**) Polarizing micrograph and (**f**) a schematic diagram of **n**(**r**) around an elastic dipole formed by a sphere with a strong homeotropic surface anchoring. A black point in the front of the particle shown schematically in **f** depicts a bulk point defect 'hedgehog.' (**i**) Schematic diagram of a quadrupolar **n**(**r**) around a sphere with weak surface anchoring. A dashed loop inside the particle shows a virtual disclination ring. (**j**-**l**) Colour-coded visualizations of the n_x component of **n**(**r**), which is caused by the tilt of **n**(**r**) away from $\mathbf{n}_0$, at the spherical surface enveloping the particles shown, respectively, in **f**,**g**,**h**. The spherical surfaces of colour-coded visualization of director distortions have radius of (1.2-1.4)r_0, as shown by dashed circles in **f**-**h**. Blue, yellow and magenta colours correspond, respectively, to a positive, zero and negative n_x according to a colour scale bar provided in **k**. Scale bars, 4 μm.

spherical surfaces encompassing the particles and the defects, clearly illustrate the multipolar nature of elastic distortions similar to that in the electrostatics analogues[34,35]. A direct comparison of the symmetry of the director distortions around elastic dipoles and quadrupoles with that of hexadecapolar PSMs in NLCs helps to highlight the very different nature of these high-order NLC colloidal multipoles studied in our work (compare Figs 4j-l and 1h). On the other hand, the hexadecapolar NLC colloids further expand the analogy between electrostatic and elastic colloidal multipoles. In fact, the colour presentations of elastic distortions induced by colloidal dipoles (Fig. 4l), two different quadrupoles (Fig. 4j,k) and our hexadecapoles (Fig. 1g,h) resemble very closely the corresponding dipolar, quadrupolar and hexadecapolar electrostatic charge distributions on a sphere described by a spherical harmonic function[35] $\sigma_l^m(\Theta, \phi) = N\cos(m\phi)P_l^m(\cos\Theta)$ with $(l, m) = (1, 1), (2, \pm 1)$ and (4, 1), respectively, where N is a normalization constant, Θ is a polar and ϕ is an azimuthal angles, $P_l^m(\cos\Theta)$ is the associated Legendre polynomial; the constant l determines the order of a multipole (that is, 2^l-th pole) and $-l \leq m \leq l$. This close analogy between electrostatic and elastic colloidal multipoles may help devising approaches for self-assembly of colloidal mesostructured composite materials.

To conclude, we describe hexadecapolar NLC colloids with unusual field configurations, highly anisotropic elastic interactions, and versatile forms of self-assembly. Our findings pose a challenge of realizing pure NLC colloidal octupoles, lower order multipoles that have not been observed so far, and provide the means of probing the role of hexadecapolar moments in inter-molecular interactions using colloidal model systems. Self-assembly of hexadecapolar NLC colloids is expected to yield a diverse family of two- and three-dimensional low-symmetry crystal lattices. Beyond the rich experimental platform for fundamental studies, our colloids have potential uses in designing and realizing reconfigurable 'soft' photonic crystals[36] and other NLC-based mesostructured composites[37] with properties that can be pre-engineered through controlling tilted boundary conditions on nanoparticle surfaces[38-40], particle shape, composition and topology,[2,6,8,41] as well as the NLC host properties, particle active behaviour[42] and the use of external stimuli[12,17].

Methods

Preparation of nematic colloidal dispersions. Colloidal PSM particles of radii $r_0 \approx 2.05$, 2.65 and 4 μm and with conically degenerate surface boundary conditions were prepared using one of two different methods as follows[43,44]. In the first method, asymmetric colloidal polystyrene dimers[43,45] were synthesized using a modified seeded polymerization technique[44-46] (Supplementary Fig. 1a). Resulting dimers in ethanol were sonicated for tens of minutes using an ultrasonic bath 8891 (Cole-Parmer) to break apart large and small spherical lobes forming the asymmetric colloidal dimer (Supplementary Fig. 1b,c). Colloidal spheres originating from these dimers were then re-dispersed in pentyl-cyano biphenyl (5CB) obtained from Frinton Laboratories, Inc. The larger (diameter $D = 2r_0 \approx 4.1$ μm) colloidal spheres, which were used as PSMs in our experiments, intrinsically provide conic degenerate surface boundary conditions for the NLC director. In the second preparation method, we used polystyrene divinylbenzene spherical particles PC06N (Bang Laboratories, Inc.) of $D \approx 8$ μm and DC-05 (Thermo Scientific) of $D \approx 5.3$ μm dispersed in 5CB. As dispersed, these particles provide tangential alignment for 5CB molecules and the director field **n**(**r**). However, after keeping these colloidal dispersions in 5CB at the elevated temperature of ≈100 °C for about 12 h, the alignment at the surface of these polymer particles changed to a conical degenerate (Supplementary Fig. 1d–f), similar to that reported in ref. 18 for flat surfaces coated by polystyrene. The tilt angle of the easy axis ψ_e was varying approximately within 20°-75°, depending on details of preparation. The comparison of the surface free-energy density $F_s = W_c(\cos^2\psi - \cos^2\psi_e)^2/2$ costs (ref. 18) for deviating **n**(**r**) to a polar angle ψ away from the easy axis at ψ_e relative to the bulk elastic energy density is characterized by the so-called conical anchoring extrapolation length $K/W_c \sim$ (50-500) nm $\ll r_0$, where[18] $W_c \sim 10^{-5}$ J m^{-2}. Since $K/W_c \ll r_0$, one can assume that the tilted conically degenerate boundary conditions are rigid with respect to changing ψ. To prepare samples with NLC dispersions in a uniformly aligned host, dry PSMs were added to 5CB. The dilute NLC colloidal dispersions were then filled into the cells comprised of two glass plates separated by glass spacers defining the cell gap in the range 15-30 μm, which were sealed at all edges using fast-setting epoxy glue after the filling process. The inner surfaces of cell substrates were treated with polyimide PI2555 (obtained from HD Microsystems) for in-plane homogeneous alignment of the far-field director $\mathbf{n}_0$ defined by unidirectional rubbing.

Imaging techniques and data acquisition. POM (Fig. 1 and Supplementary Figs 1,2), nonlinear optical imaging (Supplementary Fig. 3) and laser manipulations of PSM particles were performed using a single integrated setup capable of simultaneous conventional optical and nonlinear three-photon excitation fluorescence polarizing microscopy (3PEF-PM)[20] and HOT optical control operating at 1,064 nm (ref. 19). The setup was built around an inverted microscope IX81 (Olympus). For 3PEF-PM imaging[20], we have employed a tunable (680-1,080 nm) Ti-Sapphire oscillator (Chameleon Ultra II, Coherent) emitting 140 fs pulses at a repetition rate of 80 MHz. The laser wavelength was tuned to 870 nm for the three-photon excitation of 5CB molecules and the 3PEF-PM signals were collected in epi-detection mode with a photomultiplier tube (H5784-20, Hamamatsu). An Olympus × 100 oil-immersion objective with a high numerical aperture of 1.4 was used for both optical imaging and laser trapping. This experimental setup is described in details elsewhere[19,20]. Motion of colloidal particles was recorded with a charge-coupled camera (Flea, PointGrey) at a rate of 15 fps and their lateral positions versus time were determined from captured sequences of images using motion tracking plugins of ImageJ software (obtained from National Institutes of Health) with the accuracy[8] of 7-10 nm.

Measuring anisotropic diffusion of elastic NLC hexadecapoles. PSMs with conic anchoring distort the homogeneous $\mathbf{n}(\mathbf{r})$ and form elastic hexadecapoles in the aligned NLC (Fig. 1a–e and Supplementary Figs 1b,d–f, 2,3b). These hexadecapoles stay suspended in the bulk of NLC (Supplementary Fig. 3b), which is facilitated by elastic repulsion of PSMs from the confining substrates and their Brownian motion because of thermal fluctuations. Using bright field microscopy and video tracking, one can determine a position of the particle within each frame and then analyse its translational displacements at regular time steps corresponding to the video frame rates. Following the well-established methodology[29] and using the histograms of displacements shown in Supplementary Fig. 4, it is possible to experimentally determine two independent diffusion coefficients $D_{\parallel} = 1.63 \times 10^{-3}\,\mu m^2\,s^{-1}$ and $D_{\perp} = 1.05 \times 10^{-3}\,\mu m^2\,s^{-1}$, which characterize, respectively, the diffusion of elastic hexadecapoles along and perpendicular to the far-field director $\mathbf{n}_0$. As $D_{\parallel} > D_{\perp}$, it is easier for the NLC hexadecapoles to move along $\mathbf{n}_0$ than in other directions. The ratio $D_{\parallel}/D_{\perp} = 1.55$ is not much different from values determined for other elastic multipoles studied previously[29,30] (Fig. 4). These coefficients are also used to determine friction coefficients from the Stokes–Einstein relation, which are then subsequently used in the calculation of anisotropic interaction forces[27] between hexadecapoles.

References

1. Chaikin, P. M. & Lubensky, T. C. *Principles of Condensed Matter Physics* (Cambridge University Press, 2000).
2. Manoharan, V. N. Colloidal matter: packing, geometry, and entropy. *Science* **349,** 1253751 (2015).
3. Anderson, V. J. & Lekkerkerker, H. N. W. Insights into phase transition kinetics from colloid science. *Nature* **416,** 811–815 (2002).
4. Sacanna, S., Irvine, W. T. M., Chaikin, P. M. & Pine, D. J. Lock and key colloids. *Nature* **464,** 575–578 (2010).
5. Jones, M. R. *et al.* DNA-nanoparticle superlattices formed from anisotropic building blocks. *Nat. Mater.* **9,** 913–917 (2010).
6. Senyuk, B. *et al.* Topological colloids. *Nature* **493,** 200–205 (2013).
7. Poulin, P., Holger, S., Lubensky, T. C. & Weitz, D. A. Novel colloidal interactions in anisotropic fluids. *Science* **275,** 1770–1773 (1997).
8. Lapointe, C. P., Mason, T. G. & Smalyukh, I. I. Shape-controlled colloidal interactions in nematic liquid crystals. *Science* **326,** 1083–1086 (2009).
9. Ruhwandl, R. W. & Terentjev, E. M. Long-range forces and aggregation of colloid particles in a nematic liquid crystal. *Phys. Rev. E* **55,** 2958–2961 (1997).
10. Gu, Y. & Abbott, N. L. Observation of Saturn-ring defects around solid microspheres in nematic liquid crystals. *Phys. Rev. Lett.* **85,** 4719–4722 (2000).
11. Ramaswamy, S., Nityananda, R., Raghunathan, V. A. & Prost, J. Power-law forces between particles in a nematic. *Mol. Cryst. Liq. Cryst. Sci. Technol. Sect. A* **288,** 175–180 (1996).
12. Loudet, J. C. & Poulin, P. Application of an electric field to colloidal particles suspended in a liquid-crystal solvent. *Phys. Rev. Lett.* **87,** 165503 (2001).
13. Wood, T. A., Lintuvuori, J. S., Schofield, A. B., Marenduzzo, D. & Poon, W. C. K. A self-quenched defect glass in a colloid-nematic liquid crystal composite. *Science* **334,** 79–83 (2011).
14. Nelson, D. R. Toward a tetravalent chemistry of colloids. *Nano. Lett.* **2,** 1125–1129 (2002).
15. Loudet, J. C., Barois, P. & Poulin, P. Colloidal ordering from phase separation in a liquid- crystalline continuous phase. *Nature* **407,** 611–613 (2000).
16. Ognysta, U. M. *et al.* Square colloidal lattices and pair interaction in a binary system of quadrupolar nematic colloids. *Phys. Rev. E Stat. Nonlin. Soft Matter Phys.* **83,** 041709 (2011).
17. Liu, Q., Yuan, Y. & Smalyukh, I. I. Electrically and optically tunable plasmonic guest-host liquid crystals with long-range ordered nanoparticles. *Nano Lett.* **14,** 4071–4077 (2014).
18. Ramdane, O. O. *et al.* Memory-free conic anchoring of liquid crystals on a solid substrate. *Phys. Rev. Lett.* **84,** 3871–3874 (2000).
19. Trivedi, R. P., Engström, D. & Smalyukh, I. I. Optical manipulation of colloids and defect structures in anisotropic liquid crystal fluids. *J. Optics* **13,** 044001 (2011).
20. Lee, T., Trivedi, R. P. & Smalyukh, I. I. Multimodal nonlinear optical polarizing microscopy of long-range molecular order in liquid crystals. *Opt. Lett.* **35,** 3447–3449 (2010).
21. Smalyukh, I. I., Kuzmin, A., Kachynski, A., Prasad, P. N. & Lavrentovich, O. D. Optical trapping of colloidal particles and measurement of the defect line tension and colloidal forces in a thermotropic nematic liquid crystal. *Appl. Phys. Lett.* **86,** 021913 (2005).
22. Stark, H. Physics of colloidal dispersions in nematic liquid crystals. *Phys. Rep.* **351,** 387–474 (2001).
23. Lubensky, T. C., Pettey, D., Currier, N. & Stark, H. Topological defects and interactions in nematic emulsions. *Phys. Rev. E* **57,** 610–625 (1998).
24. Chernyshuk, S. B. High-order elastic terms, boojums and general paradigm of the elastic interaction between colloidal particles in the nematic liquid crystals. *Eur. Phys. J. E* **37,** 6 (2014).
25. Lev, B. I., Chernyshuk, S. B., Tomchuk, P. M. & Yokoyama, H. Symmetry breaking and interaction of colloidal particles in nematic liquid crystals. *Phys. Rev. E* **65,** 021709 (2002).
26. Tovkach, O. M., Chernyshuk, S. B. & Lev, B. I. Theory of elastic interaction between arbitrary colloidal particles in confined nematic liquid crystals. *Phys. Rev. E* **86,** 061703 (2012).
27. Muševič, I. Optical manipulation and self-assembly of nematic colloids: colloidal crystals and superstructures. *Liq. Cryst. Today* **9,** 2–12 (2010).
28. Poulin, P. & Weitz, D. A. Inverted and multiple nematic emulsions. *Phys. Rev. E* **57,** 626–637 (1998).
29. Loudet, J. C., Hanusse, P. & Poulin, P. Stokes drag on a sphere in a nematic liquid crystal. *Science* **306,** 1525 (2004).
30. Stark, H. & Ventzki, D. Stokes drag of spherical particles in a nematic environment at low Ericksen numbers. *Phys. Rev. E* **64,** 031711 (2001).
31. Ruhwandl, R. W. & Terentjev, E. M. Friction drag on a particle moving in a nematic liquid crystal. *Phys. Rev. E* **54,** 5204–5210 (1996).
32. Ruhwandl, R. W. & Terentjev, E. M. Monte Carlo simulation of topological defects in the nematic liquid crystal matrix around a spherical colloid particle. *Phys. Rev. E* **56,** 5561–5565 (1997).
33. Chernyshuk, S. B., Tovkach, O. M. & Lev, B. I. Elastic octopoles and colloidal structures in nematic liquid crystals. *Phys. Rev. E* **89,** 032505 (2014).
34. Jackson, J. D. *Classical Electrodynamics* (John Wiley & Sons, Inc., 1962).
35. Kaplan, I. G. *Intermolecular Interactions: Physical Picture, Computational Methods and Model Potentials* (John Wiley & Sons, Ltd., 2006).
36. Ravnik, M., Alexander, G. P., Yeomans, J. M. & Žumer, S. Three-dimensional colloidal crystals in liquid crystalline blue phases. *Proc. Natl Acad. Sci. USA* **108,** 5188–5192 (2011).
37. Araki, T. & Tanaka, H. Colloidal aggregation in a nematic liquid crystal: topological arrest of particles by a single-stroke disclination line. *Phys. Rev. Lett.* **97,** 127801 (2006).
38. Giomi, L., Bowick, M. J., Ma, X. & Majumdar, A. Molecular tilt on monolayer-protected nanoparticles. *Europhys. Lett.* **97,** 36005 (2012).
39. Love, J. C., Estroff, L. A., Kriebel, J. K., Nuzzo, R. G. & Whitesides, G. M. Self-assembled monolayers of thiolates on metals as a form of nanotechnology. *Chem. Rev.* **105,** 1103–1169 (2005).
40. Liu, Q. *et al.* Plasmonic complex fluids of nematic-like and helicoidal self-assemblies of gold nanorods with negative order parameter. *Phys. Rev. Lett.* **109,** 088301 (2012).
41. Lapointe, C. P., Hopkins, S., Mason, T. G. & Smalyukh, I. I. Electrically-driven multi-axis rotational dynamics of colloidal platelets in nematic liquid crystals. *Phys. Rev. Lett.* **105,** 178301 (2010).
42. Smalyukh, I. I., Butler, J., Shrout, J. D., Parsek, M. R. & Wong, G. C. L. Elasticity-mediated nematic-like bacterial organization in model extracellular DNA matrix. *Phys. Rev. E Stat. Nonlin. Soft Matter Phys.* **78,** 030701 (2008).
43. Senyuk, B. *et al.* Gourd-shaped colloidal particles in cholesteric liquid crystal. *Soft Matter* **10,** 6014–6023 (2014).
44. Kim, J. W., Larsen, R. J. & Weitz, D. A. Synthesis of nonspherical colloidal particles with anisotropic properties. *J. Am. Chem. Soc.* **128,** 14374–14377 (2006).
45. Ma, F., Wang, S., Smith, L. & Wu, N. Two-dimensional assembly of symmetric colloidal dimers under electric fields. *Adv. Funct. Mater.* **22,** 4334–4343 (2012).
46. Wang, S., Ma, F., Zhao, H. & Wu, N. Bulk synthesis of metal-organic hybrid dimers and their propulsion under electric fields. *ACS Appl. Mater. Interfaces* **6,** 4560–4569 (2014).

Acknowledgements

We thank T. Lubensky, S. Wang and N. Wu for discussions and acknowledge support of the National Science Foundation Grant DMR-1410735.

Author contributions

B.S. and O.P. performed the experimental work. B.S., O.M.T., S.B.C. and I.I.S. analysed the data. I.I.S. and B.S. wrote the manuscript with an input from all co-authors. I.I.S. conceived and directed the project.

Additional information

Competing financial interests: The authors declare no competing financial interests.

18

Control of antiferromagnetic domain distribution via polarization-dependent optical annealing

Takuya Higuchi[1] & Makoto Kuwata-Gonokami[1,2]

The absence of net magnetization inside antiferromagnetic domains has made the control of their spatial distribution quite challenging. Here we experimentally demonstrate an optical method for controlling antiferromagnetic domain distributions in MnF_2. Reduced crystalline symmetry can couple an order parameter with non-conjugate external stimuli. In the case of MnF_2, time-reversal symmetry is macroscopically broken reflecting the different orientations of the two magnetic sublattices. Thus, it exhibits different absorption coefficients between two orthogonal linear polarizations below its antiferromagnetic transition temperature under an external magnetic field. Illumination with linearly polarized laser light under this condition selectively destructs the formation of a particular antiferromagnetic order via heating. As a result, the other antiferromagnetic order is favoured inside the laser spot, achieving spatially localized selection of an antiferromagnetic order. Applications to control of interface states at antiferromagnetic domain boundaries, exchange bias and control of spin currents are expected.

[1] Department of Physics, The University of Tokyo 7-3-1 Hongo, Bunkyo-ku, Tokyo 113-0033, Japan. [2] Photon Science Center, The University of Tokyo 7-3-1 Hongo, Bunkyo-ku, Tokyo 113-0086, Japan. Correspondence and requests for materials should be addressed to T.H.
(email: higuchi@gono.phys.s.u-tokyo.ac.jp).

In antiferromagnetic materials, the magnetic moments of atoms or molecules, most likely produced due to electron spin, are aligned in a staggered manner. In the simple case of a two-sublattice antiferromagnet, there are two possible states: one sublattice is occupied by up spins and the other by down spins—or vice versa. This bistable magnetism is a manifestation of spontaneous symmetry breaking. As a result, macroscopic domains are formed depending on which antiferromagnetic state is realised, as is schematically depicted in Fig. 1a. The control of the spatial distribution of these antiferromagnetic domains is of particular importance[1–4] because they determine the functionalities of antiferromagnetic materials with existing and potential applications, such as exchange coupling between adjacent antiferromagnetic and ferromagnetic orders[5], dense non-volatile memory[6,7] and conductivity by topologically protected metallic states confined to the antiferromagnetic domain boundaries[8,9].

However, to date, such spatial control of antiferromagnetic domains remains challenging. A staggered magnetic field is a force correspondingly conjugate to antiferromagnetism, but generation of such a staggered field requires a flip in the sign of magnetic field on atomic length scales and is extremely difficult. Therefore, one needs to couple other macroscopic stimuli with the antiferromagnetic order for its control. This is not possible when the sublattices overlap each other by a translation of half a unit cell because a macroscopic stimulus cannot distinguish this microscopic translation that changes the sign of antiferromagnetic order parameter. Therefore, it is essential to break this symmetry for control of antiferromagnetic order other than the staggered field. For example, the linear magnetoelectric effect can couple electric field and antiferromagnetic order parameters, but its symmetrical requirement restricts its applications to particular materials[3]. Other than the linear magnetoelectric effect, control of the antiferromagnetic domains has been limited to methods employing indirect parameters, such as stress[10,11] and magnetic field via higher-order magnetic susceptibilities[12]. Selection of an antiferromagnetic state over a whole crystal has been achieved by these methods, but the difficulty of local control of these macroscopic parameters has hindered spatial control of antiferromagnetic domain

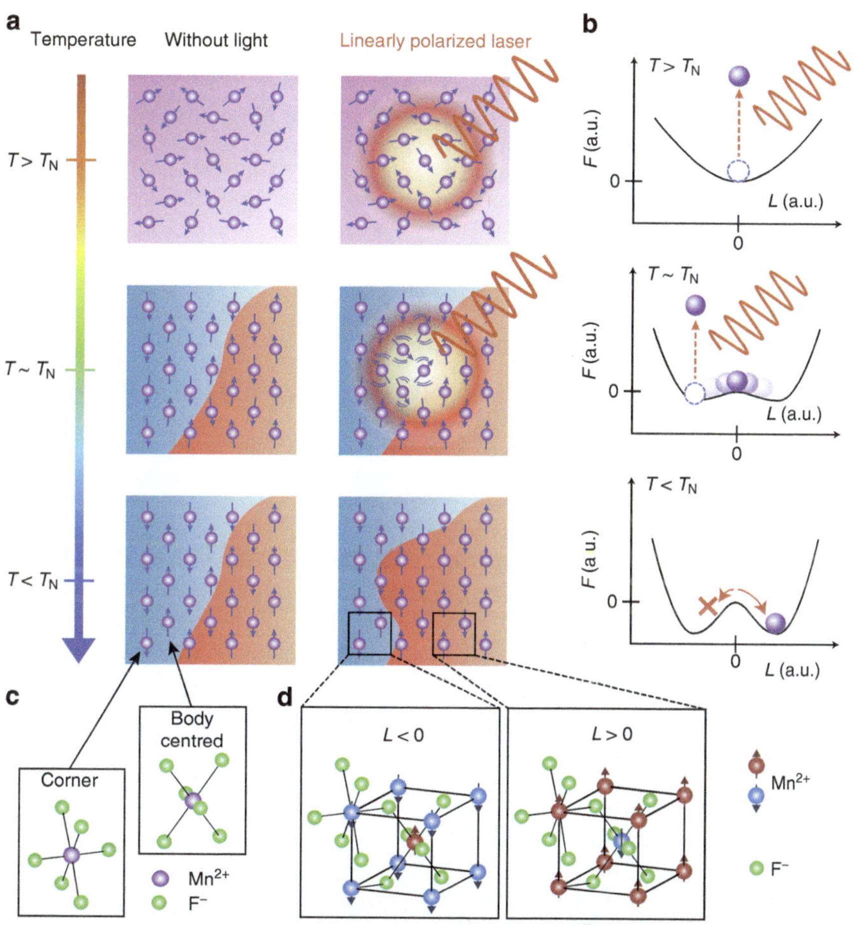

Figure 1 | Schematics of the proposed method for selecting a particular antiferromagnetic state. **(a)** Schematic of the selective heating of undesired precursor antiferromagnetic order. The spins align antiferromagnetically below T_N, and the domain distribution (red and blue areas) is determined by the residual stress and the spatial inhomogeneity of the magnetic field. When linearly polarized light illuminates the boundary in domain distribution during cooling, its absorption depends on the antiferromagnetic order through MLD, and thus it selectively destructs a particular antiferromagnetic order. As a result, the boundary of domain distribution is formed at a different position from that without illumination. **(b)** Schematic plots of how the Ginzburg-Landau energy F evolves as one cools down the temperature of the system. F has two degenerate minima below its Neel temperature, resulting in bistable orders. The purple ball shows the magnetic state, which prefers a low-energy state. The selective heating of particular antiferromagnetic order occurs below $T \sim T_N$, which prevents the system from entering into the antiferromagnetic state that has a larger optical absorption (that is, the left valley in this case), and the other order (the right valley) is chosen. **(c)** Schematics of two Mn^{2+} sites with different orientations of F^- ions surrounding the Mn^{2+}. **(d)** Crystal structure of MnF_2 with two possible antiferromagnetic orders having negative and positive L.

structures, and an external stimulus with a better control for local manipulation has been awaited. In this study, we employ the coupling between antiferromagnetic order and asymmetric optical absorption in an antiferromagnetic material without time-reversal symmetry.

Optical methods have provided unique opportunities in spatial control of (ferro-) magnetic domain structures. For example, demagnetization through optical absorption[13] is a widely employed mechanism in magneto-optic recording for data storage. More recently, methods to control magnetic order through polarization-dependent light–magnetism interactions have been proposed and demonstrated[14]. Ferromagnetic domains are controlled via illumination with a circularly polarized light. The mechanism behind this phenomenon is considered to be combinations of various effects, such as the inverse Faraday effect[14,15], the destruction of magnetic ordering through selective absorption due to magnetic circular dichroism (MCD)[16,17] and optical spin pumping[18]. The timescales of these mechanisms vary, but one common feature in these mechanisms is that there are clear selection rules for the light–matter interaction, on the basis of the optical polarization state and the magnetic ordering parameter. Therefore, the magnetic ordering can be chosen by the polarization state of the pumping light, with the other conditions kept the same.

If a similar process employing optical polarization degrees of freedom directly coupled with antiferromagnetic order parameter is available, this will open the way for controlling the spatial distribution of antiferromagnetic domains, which is otherwise difficult as mentioned heretofore. However, to date, optical control of magnetic domains has been mainly achieved on the basis of selectivity of magnetic order through relations between the helicity of circularly polarized light and net magnetization; thus, it is not directly applicable to antiferromagnetic systems. Note that in hexagonal antiferromagnetic $ScMnO_3$, linearly polarized light was found to induce photomagnetic instability of the antiferromagnetic orders having different spin orientations (a few degrees)[19], but this is not applicable to select one of the two antiferromagnetic orders having opposite spin directions of two-sublattice antiferromagnet.

In this study, we propose a method for controlling anti-ferromagnetic order through optical annealing employing linearly polarized light. We demonstrate that illumination with linearly polarized light during field cooling of an antiferromagnetic material, MnF_2, across its Neel temperature can determine the spatial distribution of antiferromagnetic states. This is possible when an antiferromagnetic material exhibits magnetic linear dichroism (MLD), the difference in optical absorption coefficients under a magnetic field between two cross-linear polarizations[20–22]. Such a coupling of an order parameter with non-conjugate external stimuli requires certain reduction of symmetry. For the MLD to exist, breaking macroscopic time-reversal symmetry is required, as discussed more in detail later. A noteworthy feature of the MLD is that it is odd with respect to the antiferromagnetic order parameter **L** (that is, the difference between sublattice magnetizations, $\mathbf{M}_1 - \mathbf{M}_2$). Hence, as illustrated in Fig. 1a, one can selectively heat the domains of a particular magnetic order by illumination with linearly polarized light. When the crystal is cooled under this selective heating, the spins prefer the other antiferromagnetic state that is less exposed to the heat. We experimentally demonstrate that this scheme actually achieves selection of an antiferromagnetic state in MnF_2, an antiferromagnetic insulator. By rotating the polarization azimuth angle by 90°, we can also choose the other antiferromagnetic state. Therefore, the desired antiferromagnetic state can be locally chosen by tuning the polarization azimuth of the illumination.

Results

Design of experiments for domain control with the MLD. To find evidence that light can affect antiferromagnetic ordering, we focus on MnF_2, an exemplary antiferromagnet. This crystal has two different Mn^{2+} sites at the corner and body centre in a unit cell, as schematically shown in Fig. 1b (ref. 23). Below its Neel temperature, these sites are occupied either by up and down spins or by down and up, resulting in two possible antiferromagnetic states. The spins in antiferromagnetic MnF_2 align along the [001] axis; thus, **L** can be described as $L\mathbf{e}_z$, where L is its amplitude and $\mathbf{e}_z$ is the unit vector along the [001] axis. Any spatial translation cannot overlap one order state to the other because the corner and body-centred sites have different fluorine environments: an additional rotation by 90° is required for this overlap. Therefore, the time-reversal symmetry is macroscopically broken[24]. This is important for applying optical methods to the control of antiferromagnetic order, because optical fields are macroscopic stimuli for the magnetic ordering; that is, the light cannot distinguish a half-unit-cell translation.

The time-reversal symmetry breaking in antiferromagnetic MnF_2 manifests itself in the terms of the optical susceptibility tensor χ_{ij} that are odd with respect to $\mathbf{L} = L\mathbf{e}_z$, leading to MLD. Note that if time-reversal symmetry is maintained (that is, one order can overlap the other with a half-unit-cell translation), such odd terms cannot survive; **L** switches sign under this translation, but the macroscopic susceptibility should not. Of interest here within these terms that are odd to **L** are the ones that are also odd with respect to an external magnetic field, $\mathbf{B} = B\mathbf{e}_z$. Onsager's reciprocal relation[25] provides $\chi_{ij}(L,B) = \chi_{ji}(-L, -B)$. Therefore, the terms that are odd with respect to both L and B are symmetric with respect to the interexchange of subscripts[21]. These symmetric terms induce difference in the optical absorption coefficients of the two cross-linear polarizations (parallel to the [110] and [−110] axes in MnF_2) under a magnetic field. This results in MLD. This is in contrast to the more commonly observed MCD, which originates from asymmetric off-diagonal terms that are odd with respect to only **B** (no contribution of **L**). The symmetric restriction for an antiferromagnetic material exhibiting MLD is the same as that of the piezomagnetic effect, which is allowed by 66 magnetic crystal classes[22,26]. Various other antiferromagnetic materials that break time-reversal symmetry also exhibit MLD and associated birefringence, such as CoF_2 (ref. 20), $Dy_3Al_5O_{12}$ (ref. 27), $MnCO_3$, $CoCO_3$ and $CsMnF_3$ (ref. 28).

Characterization of MLD. We first characterise the magneto-optical properties of MnF_2 around its Neel temperature, $T_N = 67.7$ K. Figure 2a shows the temperature dependence of the absorption coefficient of MnF_2 without an external field. The main peak corresponds to the magnon–exciton pair creation process[29–31]. Below T_N, a prominent MLD is observed around this peak (Fig. 2b). The MLD changes its sign between the two antiferromagnetic order states prepared by a field-cooling method employing nonlinear magnetic susceptibility (see Methods for details). On the other hand, the standard MCD does not depend on the antiferromagnetic order. Figure 2c shows the temperature dependence of the MLD. The peak amplitude, $\alpha_{MLD}(T)$, of the MLD shows a critical behaviour as a function of temperature T: $\alpha_{MLD}(T) = (T_N - T)^{\gamma}$, $\gamma = 0.30 \pm 0.03$. This critical exponent coincides with that of the antiferromagnetic order parameter[32]; this agreement strongly indicates that the MLD is proportional to L.

Characterization of domain distribution. On the basis of this well-characterized magneto-optical property of MnF_2, we design an experiment to vary distributions of antiferromagnetic

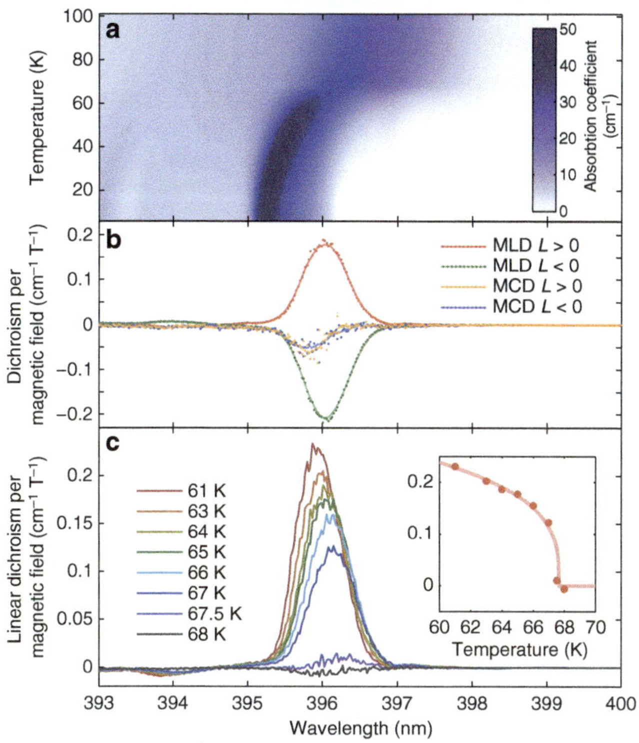

Figure 2 | Optical and magneto-optical properties of MnF_2 around its Neel temperature. (**a**) Temperature dependence of the absorption coefficient at $B=0$. (**b**) MLD and MCD spectra of MnF_2 at 64 K, just below its Neel temperature. MLD is $\alpha_{[-110]}(B)-\alpha_{[110]}(B)$ and MCD is $\alpha_L(B)-\alpha_R(B)$, where α is the absorption coefficient and the subscripts denote the [−110]-linear, [110]-linear, left-circular and right-circular polarizations, respectively. They are both linear to B for a range of $-5\,T<B<5\,T$, and the slopes of the dichroisms per magnetic field strength are plotted. Dots are raw data, and smooth curves are moving averages over seven data points. (**c**) Temperature dependence of the MLD of the MnF_2 crystal with $L>0$. Inset shows the peak amplitude of MLD derived by the fitting of the MLD spectra with Gaussian functions.

domains, as schematically illustrated in Fig. 3a. A continuous-wave laser is separated into pump and probe beams. By scanning the probe beam in one direction and measuring MLD (Fig. 3b), we obtain the one-dimensional spatial distribution of L. We compare the resultant distribution with and without optical pumping to extract the influence of light illumination. In this material, the size of the antiferromagnetic domains is typically between hundreds of micrometres and a few millimetres, as was previously observed by neutron topography measurements[33,34], and is consistent with our results. To date, various other methods have been reported to observe spatial distribution of antiferromagnetic orders, such as polarized neutron topography[33,34], second harmonic generation[35], nonreciprocal reflection[36,37] and X-ray diffraction[38,39]. Optical techniques such as the MLD have an advantage among them providing *in situ* information of spatial domain distribution. Even though the linear optical susceptibility of a bulk purely antiferromagnetic material alone does not contain terms that depends on the antiferromagnetic order parameter, higher-order susceptibility [35] or non-locality (change in wave number) in reflection[36,37] introduces difference between them. In the case of the MLD, the order parameter sensitivity is introduced by an external magnetic field that perturbs the ideal, pure antiferromagnetic state.

Without the laser, the distributions of the antiferromagnetic order parameters are described using the nonlocal Ginzburg–Landau equation[40] as follows:

$$\frac{\partial L(\mathbf{r},t)}{\partial t}=-a\left(\frac{T(\mathbf{r},t)-T_N}{T_N}\right)L(\mathbf{r},t)-bL^3(\mathbf{r},t)+c\Delta L(\mathbf{r},t)+H_{\text{eff}}(\mathbf{r})L(\mathbf{r},t). \quad (1)$$

Here the first two terms are the Ginzburg-Landau energy according to the mean-field approximation, where $T(\mathbf{r},t)$ and $L(\mathbf{r},t)$ are local temperature and antiferromagnetic order parameter, respectively. a and b are parameters for describing the Ginzburg–Landau free energy. The third term describes the nonlocal interactions of the order parameter, and c is a parameter describing this non-locality. The last term includes the effective staggered field, H_{eff}, mainly via magneto-elastic effects[10,11,41] and higher-order magnetic susceptibilities[12]. Previous studies of choosing an antiferromagnetic state over a whole crystal of MnF_2 were achieved by controlling this last term.

The distribution of H_{eff} is given by the residual stress, local impurities and weak transverse components of the magnetic field. Therefore, when we fix these parameters, the antiferromagnetic domain distributions are unchanged. Of these parameters, the external magnetic field is the only variable. We confirm that the domain distributions are robust for a fixed magnetic field. In particular, we cool down the sample three times under a magnetic field of 0.5 T (this condition is used for the following experiments) to see this reproducibility. The s.d.'s among these three experimental sets are shown as error bars in Fig. 3b. On the other hand, the magnetic field applied during the cooling changes the spatial distribution of the domains, as is shown in Fig. 3b. Note that the magnetic field is applied parallel to the ordered sublattice magnetization, and the magnitude of the field is 0.5 T and is much smaller than the spin-flop field of 11.5 T at the temperature (63 K) at which we characterized the domain distribution. The magnetization changes only around the vicinity of the spin-flop magnetic field (around ±2 T), and thus the induced magnetization at the field strength of 0.5 T can still be treated as a weak perturbation[42,43]. Indeed, the MLD showed linear dependence to the external magnetic field at around this magnetic field, which suggests that the external magnetic field is still sufficiently weak to be taken as a perturbation. On top, the induced magnetization alone does not break the degeneracy between the two possible antiferromagnetic orders because the magnetic field is parallel to the spin orientation axis. Even though the induced magnetization also causes ordinary MCD, it cannot distinguish antiferromagnetic order parameters and thus is irrelevant for this study.

Control of domain distribution with the MLD. Next, we cool down the sample across T_N under spatially localized optical illumination. A magnetic field of 0.5 T is applied. The optical absorption coefficient depends on the antiferromagnetic order parameter and the polarization state of light due to the MLD. This can be described as a source of L-dependent heating in the thermal diffusion equation:

$$C_V\frac{\partial T(\mathbf{r},t)}{\partial t}=\lambda\Delta T(\mathbf{r},t)+I(\mathbf{r})(A_0-A_{\text{MLD}}(B)L(\mathbf{r},t)\cos(2\theta)), \quad (2)$$

where C_V is the heat capacity, λ is the thermal conductivity and $I(\mathbf{r})$ is the optical intensity. The second term describes the heat induced by the optical absorption. A_0 is the absorption ratio of the light that does not depend on polarization. The polarization-dependent absorption ratio is $-A_{\text{MLD}}(B)L(\mathbf{r},t)\cos(2\theta)$, where θ is the polarization azimuthal angle measured from the [110] axis of

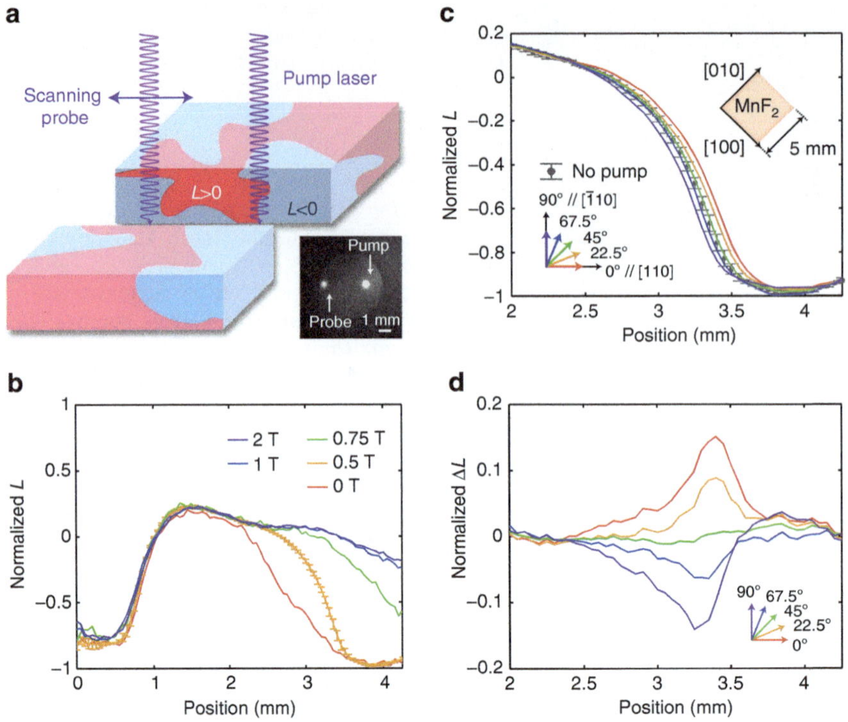

Figure 3 | Experimental demonstration of optical control over the spatial distribution of antiferromagnetic states. (**a**) Schematic of the experiment, where the spatial distribution of antiferromagnetic states depends on the pump azimuth. Fluorescent image of the pump and probe beams are also shown. (**b**) Spatial distribution of L after cooling under various magnetic field values. The L values are normalized by that of a single antiferromagnetic domain. For the measurement with $B = 0.5$ T, the L values are measured after three independent cooling processes and the mean values are plotted. Error bars show the s.d.'s among these three data sets. (**c**) Spatial distribution of L under various pump azimuthal angles. Dotted curve shows the baseline distribution without optical annealing and $B = 0.5$ T. Error bars on the no-pump curve are same as in **b**. (**d**) Optically induced change in L from the baseline distribution.

MnF_2. For linearly polarized light with $\theta = 0°$, heat by the polarization-dependent absorption in the domains with positive L is smaller than that in domains with negative L. Increases in temperature make the antiferromagnetic order unstable according to equation (1). Therefore, the domains with positive L are preferred because they are less heated. We experimentally observe this change in the distribution of antiferromagnetic domains, as shown in Fig. 3d. The region with positive L has a larger volume than that without pump light. On the other hand, light with the cross-linear polarization, that is, with $\theta = 90°$, is expected to affect the domain distributions oppositely, because the sign of the selective heating changes. As expected, the preference of the sign of L reverses when we rotate the polarization azimuth angle of the pump beam by 90°. We also observe the domain distribution for intermediate polarization azimuthal angles and obtain a systematic change. The systematic change is highlighted by subtracting the baseline domain distribution (that is, cooling under $B = 0.5$ T without light) from the domain distributions after cooling under light illumination (Fig. 3d). The change in domain distribution is limited to the region where the light was illuminated.

Numerical simulation of domain distribution. To confirm the scenario that the domain distribution is determined via selective heating by light, we simulate the spatial distribution of antiferromagnetic order parameters according to equations (1) and (2). In this simulation, the sample is cooled down from its edges, which determines the boundary condition for the thermal diffusion equation (Fig. 4a). An effective staggered field, $H_{\mathrm{eff}}(\mathbf{r})$, having a uniform gradient along the x axis, $H_{\mathrm{eff}}(x) = xH_{\mathrm{eff},0}$, is assumed to emulate the creation of an antiferromagnetic domain boundary, which is achieved by considering a magnetic field slightly canted from the z axis. When the crystal is cooled down in the absence of light, the domain boundary is formed at $x = 0$.

When a linearly polarized light is incident on the crystal, the resultant antiferromagnetic domain distribution after cooling differs from that without light. Figure 4a shows the temperature dependence of the domain distribution along the x axis, where optical polarization $\theta = 0°$ is assumed. The domain with positive L has larger volume than the case without light (boundary at $x = 0$). We simulate the polarization dependence of the domain distributions (Fig. 4b), as well as its difference from the baseline distribution (Fig. 4c). These simulated results show qualitative agreement with the experimental observations. The polarization dependence is well reproduced, and the spatial asymmetry in ΔL observed in experiment is also present in the simulation. Note that the spatial distribution of the domains appears blunter in the experiments. This is probably because the domain boundary was not formed parallel to the z axis and the boundary is likely to be formed with mosaic-like microdomain structures, and thus the optical probe integrated over the thickness of crystal provides a blurred distribution.

Discussion

A further progress on the basis of this proof-of-concept study is envisioned. In the current study, the annealing process was carried out with control of the temperature of the whole crystal. It is attracting to replace this temperature control by local and fast heating, for example, via heating with pulsed lasers. Also, recent studies on lifetime of magnons in by means of neutron scattering

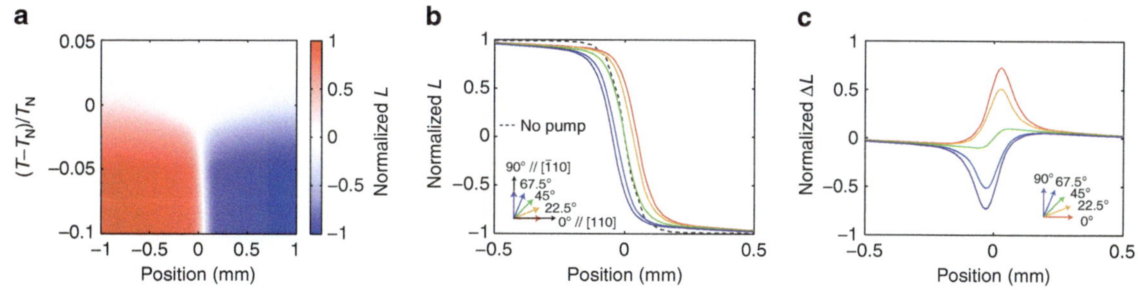

Figure 4 | Simulated distribution of antiferromagnetic order parameters. (**a**) Temperature dependence of L along the x axis. Linearly polarized light with an azimuthal angle of 0 (that is, parallel to the x axis) illuminates the sample. (**b**) Spatial distribution of L at $(T-T_N)/T_N = -0.1$ after polarization-dependent optical annealing. The baseline distribution without optical pumping is also plotted. (**c**) Optically induced change in L from its baseline distribution without optical annealing.

suggested that there are mosaic-like smaller submicrometre domain structures MnF_2 (ref. 44), which should exist at the boundary we observed. With our optical resolution (200 μm), these substructures are averaged out, and one can only see the vertically averaged distributions at larger scales, which is supported not only by the magnetic ordering but also by the other macroscopic effect such as the associated lattice distortion[45]. As the symmetry requirement for the selective heating does not rely on the domain size, but rather on the magnetic crystalline symmetry, it should be applicable to these micro domains as well. Higher optical resolution together with a sample thinner than a single domain is promising to address control of the microdomain structures.

In summary, we showed that illumination with linearly polarized light under an external magnetic field acts as a staggered stimulus to spins in antiferromagnetic MnF_2 via selective heating. The linearly polarized light distinguished antiferromagnetic states through MLD, and its absorption heats the undesired state, thereby realising the desired state. With this method, we were able to select a particular antiferromagnetic state. The scope of our method is solely restricted by the symmetry of crystals that allow MLD (which is the same as the restriction of the piezomagnetic effect), which is found in various antiferromagnetic materials[22,26]. Therefore, numerous possible extensions are expected. For example, all-in/all-out-type antiferromagnetic materials, known as hosts of Weyl fermions[8], satisfy this condition[39,46], and possible control of their ordering provides unique opportunities in the growing field of topologically protected surface states.

Methods

Measurement of dichroism. In this study, we employed two experimental setups: one to measure the dichroism of a single antiferromagnetic domain and one to control the spatial distribution of antiferromagnetic orders. In both experiments, a single crystal of MnF_2 with a [001] face, a thickness of 1 mm, and an area of $5 \times 5\,mm^2$ was placed in a sample holder in a cryostat (Microstat MO; Oxford Instruments). The cryostat had a superconducting magnet that generated a magnetic field from -5 to $+5$ T.

For measuring dichroism under a magnetic field, a light-emitting diode with a central wavelength of 397 nm and a bandwidth of 10 nm was employed as the light source. The light transmitted through the specimen was introduced into a monochromator (SpectraPro-300i; Acton Research) with a grating with a groove density of 2,400 G mm^{-1} and a blaze wavelength of 240 nm. Spectra of the transmitted light were recorded by a charge-coupled device-based optical multichannel analyser (OMA, LN/CCD-1,100 PG/UV; Princeton Instruments).

The difference in the absorption coefficients between the two orthogonal polarizations was measured as a function of the external magnetic field along the [001] axis. MLD is given by $\alpha_{[-110]}(B) - \alpha_{[110]}(B)$ and MCD is given by $\alpha_L(B) - \alpha_R(B)$, where α is the absorption coefficient and the subscripts denote the [−110]-linear, [110]-linear, left-circular and right-circular polarizations, respectively. These dichroisms showed linear dependence to the external magnetic field; thus, their slopes as functions of the field are plotted in Fig. 2b,c.

To compare the dichroism spectra of two possible antiferromagnetic orders, we needed to prepare one particular magnetic order over the entire crystal. This was achieved by cooling the crystal under a magnetic field slightly canted from the [001] axis (~2°). We considered nonlinear terms in the free energy of the form $\chi_{ijkl} L_i B_j B_k B_l$; here the terms $\chi_{zxxz} = \chi_{zyyz}$ are nonzero with respect to the magnetic point group of antiferromagnetic MnF_2, 4'/mmm'. These free-energy terms are odd with respect to the antiferromagnetic vector **L**. Therefore, canting the magnetic field from the [001] axis in the [110] direction resulted in additional differences in the forming energies of the two antiferromagnetic orders. This determined the antiferromagnetic order. This method to control antiferromagnetic order was employed by Kharchenko *et al.*[12], and we employed the same technique.

Observation and manipulation of antiferromagnetic domains. To observe and manipulate antiferromagnetic domains in MnF_2, we employed a continuous-wave diode laser (Toptica) lasing at a wavelength of 396.25 nm selected by an external cavity with a grating. The laser beam was separated into probe and pump beams by a 10:90 separator. The weaker and stronger beams were the probe and pump beams, respectively. Both beams were focused onto the sample with spot sizes of ~200 μm. The pump power was 13 mW and the light was linearly polarized. A half-wave plate was placed so as to control its polarization azimuth angle.

To increase the measurement sensitivity, the modulation method was employed for probing. The probe beam was sampled by a chopper with a frequency, f_{ch}, of 200 Hz and modulated by a photoelastic modulator (PEM). In the PEM, a crystal oscillated at a frequency, f_{PEM}, of 50 kHz, which changed the polarization state of the light with this frequency. The transmitted light was detected by a silicon photodiode, and the signal was analysed by three lock-in amplifiers having lock-in frequencies of f_{ch}, f_{PEM} and $2f_{PEM}$, respectively. The signal for f_{ch} indicated the total transmitted intensity, and the $2f_{PEM}$ signal divided by f_{ch} was proportional to the MLD signal. The f_{PEM} signal contained information on both the MCD and the linear dichroism in the diagonal direction (in between [100] and [010]). This (non-magnetic) linear dichroism appeared when the probe beam was canted from a right angle to the sample surface that was useful for aligning the beam angle.

The positions of the pump and probe beams were observed by fluorescence from MnF_2. When excitons were created by the laser, orange fluorescence was observed around a wavelength of 600 nm. To observe this fluorescence, a charge-coupled device camera was placed behind a coloured glass that filtered the laser. This fluorescence was useful for imaging because the emitted light was incoherent, and thus free from speckles.

If the magnetic field was canted from the [001] axis, its transverse component determined the antiferromagnetic order over the entire crystal, as was done in the first experiment. To make the crystal inhomogeneously contain both antiferromagnetic orders, we placed the crystal in the normal direction to reduce the effect of transverse magnetic fields. By matching the reflected beams from the sample surface and the window of the cryostat, this alignment was achieved within 0.1°.

Numerical simulation. Solutions of the thermal diffusion equation and the non-local Ginzburg–Landau equation are simultaneously obtained via numerical integration using the Runge–Kutta method. A square crystal of MnF_2 with an edge of 5 mm is discretized into a two-dimensional spatial mesh with a mesh size of 0.05 mm. The laser beam spot size is 0.1 mm in radius. Boundary conditions for the thermal diffusion equation are given by the temperature at the crystal edges, which decreases as a function of time to simulate the cooling of the crystal from its edges. The cooling rate is chosen to be slow enough so that the system undergoes an equilibrium state at the edge temperature during cooling. Values for the thermal diffusion constant and heat capacity of MnF_2 are taken from refs 47,48, respectively. The parameters a and b in the Ginzburg–Landau equation are taken from ref. 32. The non-locality parameter c of the magnetic order parameter is not known, and a best parameter is chosen to reproduce the experimental results.

References

1. Catalan, G., Seidel, J., Ramesh, R. & Scott, J. F. Domain wall nanoelectronics. *Rev. Mod. Phys.* **84,** 119 (2012).
2. Bode, M. *et al.* Atomic spin structure of antiferromagnetic domain walls. *Nat. Mater.* **5,** 477–481 (2006).
3. Zhao, T. *et al.* Electrical control of antiferromagnetic domains in multiferroic $BiFeO_3$ films at room temperature. *Nat. Mater.* **5,** 823–829 (2006).
4. Goltsev, A. V., Pisarev, R. V., Lottermoser, T. h. & Fiebig, M. Structure and interaction of antiferromagnetic domain walls in hexagonal $YMnO_3$. *Phys. Rev. Lett.* **90,** 177204 (2003).
5. Nogués, J. & Schuller, I. K. Exchange bias. *J. Magn. Magn. Mater.* **192,** 203–232 (1999).
6. Loth, S., Baumann, S., Lutz, C. P., Eigler, D. M. & Heinrich, A. J. Bistability in Atomic-scale antiferromagnets. *Science* **335,** 196–199 (2012).
7. Marti, X. *et al.* Room-temperature antiferromagnetic memory resistor. *Nat. Mater.* **13,** 367–374 (2014).
8. Witczak-Krempa, W., Chen, G., Kim, Y. B. & Balents, L. Correlated quantum phenomena in the strong spin-orbit regime. *Ann. Rev. Cond. Mat. Phys.* **5,** 57 (2014).
9. Yamaji, Y. & Imada, M. Metallic interface emerging at magnetic domain wall of antiferromagnetic insulator—fate of extinct Weyl electrons. *Phys. Rev. X* **4,** 021035 (2014).
10. Wolf, W. P. & Huan, C. H. A. Magnetoelastic effects on antiferromagnetic phase transitions. *J. Appl. Phys.* **63,** 3904 (1988).
11. Clark, G. F. & Tanner, B. K. Antiferromagnetic domain wall motion under external stress. *Phys. Status Solidi (a)* **59,** 241–249 (1980).
12. Kharchenko, N. F., Miloslavskaya, O. V. & Milner, A. A. Odd magnetic dichroism of linearly polarized light in the antiferromagnet MnF_2. *Low Temp. Phys.* **31,** 825–830 (2005).
13. Beaurepaire, E., Merle, J.-C., Daunois, A. & Bigot, J.-Y. Ultrafast Spin Dynamics in Ferromagnetic Nickel. *Phys. Rev. Lett.* **76,** 4250–4253 (1996).
14. Stanciu, C. D. *et al.* All-optical magnetic recording with circularly polarized light. *Phys. Rev. Lett.* **99,** 047601 (2007).
15. Kirilyuk, A., Kimel, A. V. & Rasing, T. Ultrafast optical manipulation of magnetic order. *Rev. Mod. Phys.* **82,** 2731–2784 (2010).
16. Genkin, G. M., Nozdrin, Y. u. N., Okomel'kov, A. V. & Tokman, I. D. Magnetization of a garnet film through a change in its multidomain structure under circularly polarized light. *Phys. Rev. B* **86,** 024405 (2012).
17. Khorsand, A. R. *et al.* Role of magnetic circular dichroism in all-optical magnetic recording. *Phys. Rev. Lett.* **108,** 127205 (2012).
18. Gridnev, V. N. Optical spin pumping and magnetization switching in ferromagnets. *Phys. Rev. B* **88,** 014405 (2013).
19. Fiebig, M., Fröhlich, D., Lottermoser, T. h. & Pisarev, R. V. Photoinduced instability of the magnetic structure of hexagonal $ScMnO_3$. *Phys. Rev. B* **65,** 224421 (2002).
20. Eremenko, V. V., Kharchenko, N. F., Beliy, L. I. & Tutakina, O. P. Birefringence of the antiferromagnetic crystals linear in a magnetic field. *J. Magn. Magn. Mater.* **15-18,** 791–792 (1980).
21. Higuchi, T. & Kuwata-Gonokami, M. Microscopic origin of magnetic linear dichroism in the antiferromagnetic insulator MnF_2. *Phys. Rev. B* **87,** 224405 (2013).
22. Erementko, V. V., Kharchenko, N. F., Litvinenko, Y. u. G. & Naumenko, V. M. *Magneto-Optics and Spectroscopy of Antiferromagnets* (Springer-Verlag, 1992).
23. Strong, S. L. Crystal structure distortions in MnO_2 and MnF_2 at their antiferromagnetic curie temperature. *J. Phys. Chem. Solids* **19,** 51–53 (1961).
24. Cracknell, A. P. Group theory and magnetic phenomena in solid. *Rep. Prog. Phys.* **32,** 633–707 (1969).
25. Onsager, L. Reciprocal relations in irreversible processes. *I. Phys. Rev.* **37,** 405–426 (1931).
26. Birss, P. R. *Symmetry and Magnetism, in Selected Topics in Solid State Physics.* (ed. Wohlfarth, E. P.) Vol. III, 125–145 (North-Holland, 1966).
27. Dillon, J. J. F., Talley, L. D. & Chen, E. Y. Optical birefringence in $Dy_3Al_5O_{12}$(DAG) in fields along [111]. *AIP Conf. Proc.* **34,** 388–390 (1976).
28. Borovik-Romanov, A. S., Kreines, I. M., Pankov, A. A. & Talalaev, M. A. Magnetic birefringence of light in the antiferromagnetic $MnCO_3$ $CoCO_3$, and $CsMnF_3$. *J. Exp. Theor. Phys.* **39,** 378–382 (1974).
29. Sell, D. D., Greene, R. L. & White, R. M. Optical exciton-magnon absorption in MnF_2. *Phys. Rev.* **158,** 489–510 (1967).
30. Meltzer, R. S., Lowe, M. & McClure, D. S. Magnon sidebands in the optical absorption spectrum of MnF_2. *Phys. Rev.* **180,** 561–578 (1969).
31. Tanabe, Y., Moriya, T. & Sugano, S. Magnon-induced electric dipole transition moment. *Phys. Rev. Lett.* **15,** 1023–1025 (1965).
32. De Jongh, L. J. & Miedema, A. R. Experiments on simple magnetic model systems. *Adv. Phys.* **23,** 1–260 (1974).
33. Alperin, H. A., Brown, P. J., Nathans, R. & Pickart, S. J. Polarized neutron study of antiferromagnetic domains in MnF_2. *Phys. Rev. Lett.* **8,** 237 (1962).
34. Baruchel, J., Schlenker, M. & Barbara, B. 180° Antiferromagnetic domains in MnF_2 by neutron topography. *J. Mag. Mag. Mater.* **15-18,** 1510–1512 (1980).
35. Fiebig, M., Fröhlich, D., Leute, S. t. & Pisarev, R. V. Topography of antiferromagnetic domains using second harmonic generation with an external reference. *Appl. Phys. B* **66,** 265–270 (1998).
36. Krichevtsov, B. B., Pavlov, V. V., Pisarev, R. V. & Gridnev, V. N. Spontaneous non-reciprocal reflection of light from antiferromagnetic Cr_2O_3. *J. Phys. Condens. Matter* **5,** 8233–8244 (1993).
37. Pisarev, R. V. Crystal optics of magnetoelectrics. *Ferroelectrics* **162,** 191–209 (1994).
38. Schlenker, M. & Baruchel, J. Possibilities of X-ray and neutron topography for domain and phase coexistence observations. *Ferroelectrics* **162,** 299–306 (1994).
39. Tardif, S. *et al.* All-in–all-out magnetic domains: X-ray diffraction imaging and magnetic field control. *Phys. Rev. Lett.* **114,** 147205 (2015).
40. Fujisaka, H., Tutu, H. & Rikvold, P. A. Dynamic phase transition in a time-dependent Ginzburg-Landau model in an oscillating field. *Phys. Rev. E* **63,** 036109 (2001).
41. Scott, R. A. M. & Anderson, J. C. Indirect observation of antiferromagnetic domains by linear magnetorestriction. *J. Appl. Phys.* **37,** 234–237 (1966).
42. Felcher, G. P. & Kleb, R. Antiferromagnetic domains and the spin-flop transition of MnF_2. *Europhys. Lett.* **36,** 455–460 (1996).
43. Shapira, Y. & Foner, S. Magnetic phase diagram of MnF_2 from ultrasonic and differential magnetization measurements. *Phys. Rev. B* **1,** 3083 (1970).
44. Bayrakci, S. P. *et al.* Lifetimes of antiferromagnetic magnons in two and three dimensions: experiment, theory, and numerics. *Phys. Rev. Lett.* **111,** 017204 (2013).
45. Baruchel, J. *et al.* Piezomagnetism and domains in MnF_2. *J. Phys. Colloq.* **49,** C8-1895–C8-1896 (1988).
46. Arima, T. Time-reversal symmetry breaking and consequent physical responses induced by all-in-all-out type magnetic order on the Pyrochlore lattice. *J. Phys. Soc. Jpn* **82,** 013705 (2013).
47. Slack, G. A. Thermal conductivity of CaF_2, MnF_2, CoF_2, and ZnF_2 crystals. *Phys. Rev.* **122,** 1451 (1961).
48. Boo, W. O. J. & Stout, J. W. Heat capacity and entropy of MnF_2 from 10 to 300°K. Evaluation of the contributions associated with magnetic ordering. *J. Chem. Phys.* **65,** 3929 (1976).

Acknowledgements

We thank Y. Tanabe for his fruitful discussion, H. Tamaru for his experimental support and discussion, and K. Konishi and K. Yoshioka for their technical support. We also thank Y. Yamaji, T Kurumaji and T. Oka for helpful discussions. The crystal employed for this study was provided by N. Nagasawa, and backup crystals were provided by Tokuyama Co. This work was supported partly by a Grant-in-Aid for Scientific Research (KAKENHI) on Innovative Areas, 'Optical Science of Dynamically Correlated Electrons (grant no. 2003)' (20104002), from the Ministry of Education, Culture, Sports, Science and Technology of Japan, and by Japan Society for the Promotion of Science (JSPS) through its FIRST Program. T.H. acknowledges support from a JSPS Research Fellowship.

Author contributions

T.H. conceived and conducted the experiments, analysed the data and provided the microscopic theory and the simulation. M.K.-G. supervised the project. All authors discussed the results and implications at all stages and contributed to improvement of the manuscript text.

Additional information

Competing financial interests: The authors declare no competing financial interests.

Efficient linear phase contrast in scanning transmission electron microscopy with matched illumination and detector interferometry

Colin Ophus[1], Jim Ciston[1], Jordan Pierce[2], Tyler R. Harvey[2], Jordan Chess[2], Benjamin J. McMorran[2], Cory Czarnik[3], Harald H. Rose[4] & Peter Ercius[1]

The ability to image light elements in soft matter at atomic resolution enables unprecedented insight into the structure and properties of molecular heterostructures and beam-sensitive nanomaterials. In this study, we introduce a scanning transmission electron microscopy technique combining a pre-specimen phase plate designed to produce a probe with structured phase with a high-speed direct electron detector to generate nearly linear contrast images with high efficiency. We demonstrate this method by using both experiment and simulation to simultaneously image the atomic-scale structure of weakly scattering amorphous carbon and strongly scattering gold nanoparticles. Our method demonstrates strong contrast for both materials, making it a promising candidate for structural determination of heterogeneous soft/hard matter samples even at low electron doses comparable to traditional phase-contrast transmission electron microscopy. Simulated images demonstrate the extension of this technique to the challenging problem of structural determination of biological material at the surface of inorganic crystals.

[1] National Center for Electron Microscopy, Molecular Foundry, Lawrence Berkeley National Laboratory, 1 Cyclotron Road, Berkeley, California 94720, USA. [2] Department of Physics, University of Oregon, 1585 E 13th Avenue, Eugene, Oregon 97403, USA. [3] Gatan Inc., 5794 W Las Positas Boulevard, Pleasanton, California 94588, USA. [4] Department of Physics, Center for Electron Microscopy, Ulm University, Albert-Einstein-Allee 11, 89069 Ulm, Germany. Correspondence and requests for materials should be addressed to C.O. (email: cophus@gmail.com) or to P.E. (email: percius@lbl.gov).

Structural analysis at atomic resolution is commonly used to provide deep insight into the functionality of structures in both biological and physical sciences. Recent examples include the enormous progress in dynamic structural biology[1], direct imaging of screw dislocations via optical sectioning[2], measurements of TiO_6 octahedra in perovskite superlattices[3] and many others. Transmission electron microscopy (TEM) and scanning TEM (STEM) are ubiquitous techniques for high-resolution analysis of both hard and soft matter structures due to the focusing capabilities of electron optics. The theoretical resolution limit for traditional TEM and STEM is 1–2 Å, which has further been extended to 0.5 Å with aberration correction[4–6]. This has significantly improved quantitative structural analysis in both TEM and STEM to sub-Å resolution and single-picometre precision in materials science where many materials can tolerate very high electron doses. Compared with STEM, TEM phase-contrast imaging is overwhelmingly preferred by the biological community because it provides an efficient means of imaging weak-phase objects at doses at or below $10\,e^{-}\,Å^{-2}$. Resolution for biological materials is limited by the achievable signal-to-noise ratio (SNR) before the target structure is damaged or destroyed, rather than TEM information transfer[7–11]. The primary method currently used to solve the structure of dose-sensitive samples is single-particle reconstruction, from cryo-electron microscopy (cryo-EM). This method is very effective, but typically requires many thousands of identical particles isolated from each other so that the defocused signals from adjacent particles do not interfere. Further, information transfer depends on the defocus used, which has a large nonlinear effect on the contrast in the final image[12–15]. This effect is especially prevalent for phase-contrast high-resolution TEM (HRTEM) and requires careful inspection, numerical aberration correction and/or computer simulation for direct structural interpretation.

Image interpretation in STEM is typically simpler. The contrast and efficiency of STEM is controlled by the geometry of the post specimen, monolithic detectors which simply integrate over specific scattering angles in reciprocal space. The two most common STEM imaging techniques are annular dark field (ADF) that can produce incoherent image contrast roughly proportional to the projected mass thickness of the sample and bright field (BF), which can produce coherent image contrast similar to traditional TEM[15,16]. An alternative method called annular bright field is a new technique designed to directly image weakly scattering materials such as lithium and oxygen, but this imaging technique is difficult to optimize since the detector inner/outer angle ratio is set by the physical detector size[17,18]. ADF-STEM with incoherent image contrast is commonly used to produce interpretable images at atomic resolution based on high-angle scattering, but this process is relatively inefficient per incident electron, commonly requiring significantly more dose per unit area compared with TEM. This is especially limiting for light elements such as carbon, which scatter electrons very weakly. Thus, STEM is more commonly used for materials science and infrequently in biological sciences[19,20].

Coherent, phase-contrast imaging in STEM is also possible using a differential measurement of the BF centre disk as initially proposed by Rose[21]. Dekkers and De Lang, Rose and Haider *et al.*[22–24] also proposed using a STEM probe aberration corrector to form a probe containing a reference wave, which can be directly interpreted using a segmented detector. Ptychography is another dose-efficient phase-contrast method that utilizes full images of the transmitted electron diffraction pattern to reconstruct both the complex (real and imaginary) probe image and complex sample potential. Ptychography was demonstrated at atomic resolution for the first time by Nellist *et al.*[25], with recent improvements in both computer algorithms and detectors[26–28]. One recent example of how phase-contrast imaging can be achieved in STEM is given by the work of Pennycook *et al.*[29,30]. Their method implements an elegant ptychographic reconstruction algorithm that uses subregions recorded on a pixel array detector to form efficient phase-contrast images.

STEM experiments can also be expanded using methods other than advanced detector geometries and computational algorithms. One example is the recent use of structured phase in electron microscopy, typically performed by placing a phase or amplitude plate in the probe-forming aperture to produce an electron probe with the desired properties. Diffraction gratings have been used in STEM to create vortex beams with orbital angular momentum[31–33] and Bessel beams for very long depth-of-field imaging[34].

Although all of these methods are important steps to improving STEM imaging beyond the simple mass-thickness contrast of ADF-STEM, it is highly desirable to develop a high-resolution imaging technique with directly interpretable contrast that can also operate with high efficiency to reduce beam damage. This could expand the use of STEM to solve important questions in the biological field, as well as hybrid hard/soft materials[35,36].

In this manuscript, we present simulations and a proof-of-principle experiment for a new kind of phase-contrast electron microscopy called matched illumination and detector interferometry (MIDI)–STEM. MIDI–STEM combines the concepts of phase gratings, aberration correction, high-speed pixelated direct electron detectors and phase reconstruction using an interference pattern (such as in electron holography). The MIDI–STEM method produces almost ideal linear phase-contrast images over a wide range of spatial frequencies with very high efficiency and could potentially be used to image soft matter and beam-sensitive samples at atomic resolution.

Results

Description of MIDI–STEM experimental set-up. A simplified diagram comparing a conventional STEM (Fig. 1a) and MIDI–STEM (Fig. 1b) experimental set-up is shown in Fig. 1. In conventional STEM experiments, the probe is formed by a plane wave incident on a circular condenser aperture. Lens elements are used to create a circular electron beam with (approximately) constant phase, which converges to an atomic-scale probe at the sample plane. Electromagnetic deflectors scan the probe over the sample surface in a 2D grid pattern. As shown in Fig. 1a, post specimen, monolithic detectors integrate over regions of the scattered (dark field) or unscattered (bright field) electron diffraction pattern. Two common detector configurations are shown, an ADF detector and a BF detector.

In a MIDI–STEM experiment, diagrammed in Fig. 1b, a patterned phase plate is placed at the probe-forming aperture position. The phase plate consists of alternating concentric trenches with equal area where a thin SiN film has been patterned by a focused ion beam. Each alternating ring applies either a 0 or $\pi/2$-phase shift due to the local SiN thickness. This phase plate generates a probe with a built-in reference wave, which is then scanned across the sample as in traditional STEM imaging. In this case, the high-angle scattering signal is recorded by a traditional ADF detector, and a pixelated direct electron detector is used to record an image of the transmitted centre beam at each scanned position. The ADF detector produces a signal that is very similar to a conventional ADF-STEM experiment. The images of the transmitted centre beam are processed by fitting a virtual detector to match the

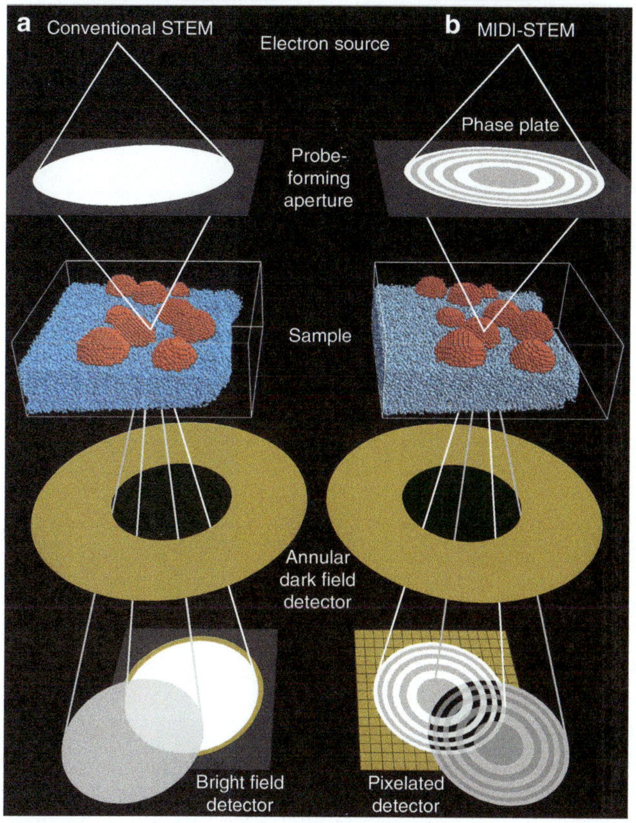

Figure 1 | Experimental set-up for STEM experiments. (**a**) Conventional set-up, with a round probe-forming aperture and monolithic, single-pixel ADF and BF detectors below the sample. (**b**) MIDI-STEM set-up, with a patterned phase plate placed in the probe-forming aperture and a pixelated detector below the sample.

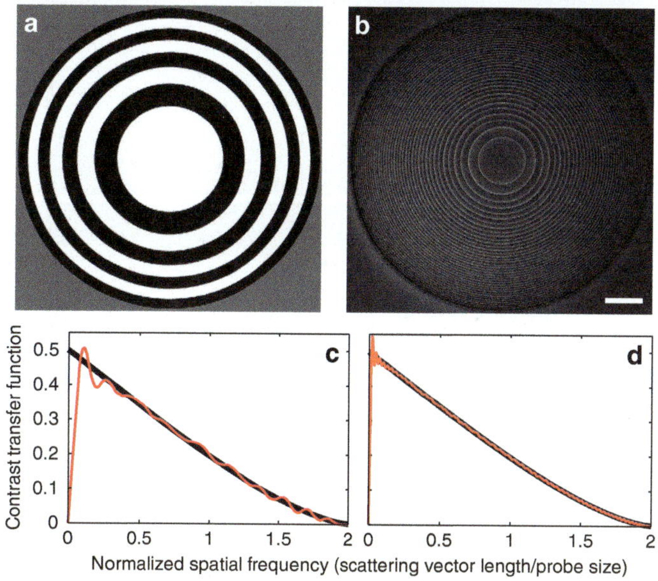

Figure 2 | MIDI-STEM phase plates and resulting CTFs. (**a**) Schematic of a phase plate with 4 ring pairs, and (**b**) scanning electron microscopy image of the patterned phase plate with 20 ring pairs used in this study. Scale bar, 5 μm. The calculated CTFs for (**a**,**b**) are plotted in (**c**,**d**), respectively. Black diagonal lines show the CTF for an ideal phase-contrast STEM experiment.

geometry of the phase plate producing an approximately linear phase signal. The virtual detector consists of the even and odd-numbered annular rings formed by the phase plate, where the phase signal is given by the difference between the sum of all odd ring intensities minus the sum of all even ring intensities. Precise alignment of the virtual detector rings can be achieved using an image of the centre beam in vacuum or by averaging all diffraction pattern images and fitting ellipses to the ring edges[37]. The ability to match the virtual detector to the phase-plate geometry using post processing makes MIDI-STEM highly flexible to compensate for any errors in the phase plate itself or in the scanning electronics. It is also capable of utilizing almost any pre-specimen phase-plate design. Further technical details of the MIDI-STEM model are given in Supplementary Note 1 and Supplementary Fig. 1.

The contrast transfer function (CTF) of a microscopy technique describes the measured contrast as a function of the scattering angle or spatial frequency[15]. A monotonically decreasing CTF that passes both low and high spatial frequencies is desirable for easy image interpretation. The CTF of MIDI-STEM can be calculated from the overlap region of the 0 and $\pi/2$ regions of the probe for the unscattered centre disk and a scattered disk. An example of a MIDI-STEM CTF is plotted in Fig. 2c, where the phase-plate geometry is shown in Fig. 2a. The scanning electron microscopy image in Fig. 2b shows the exact phase plate used in this study, which produces the CTF plotted in Fig. 2d. The geometric construction used to calculate MIDI-STEM CTFs is shown in more detail in Supplementary Figs 2, 3 and 4 and Supplementary Note 2.

A MIDI-STEM experiment. We performed an MIDI-STEM experiment to image a highly heterogeneous sample consisting of randomly oriented cuboctahedral gold nanoparticles (NPs) supported on a thin amorphous carbon film to demonstrate the linear imaging capabilities of this technique. The average image of the centre beam from all probe positions is plotted in Fig. 3a, with the fitted edges of the virtual detector overlaid on the right half of the image as red lines. Note that the contrast has been scaled up to make the phase-plate rings visible. The ADF detector is not visible at this contrast level, but was positioned such that the inner detector angle was just beyond the edge of the outer-most phase-plate ring at 20 mrad.

In Fig. 3b,c, we show the MIDI-STEM image reconstructed using the matched virtual detector and the simultaneously recorded ADF detector image, respectively. The ADF image shows strong contrast for the gold NPs, and the atomic planes are visible in several NPs typically with the (111) plane spacing. The carbon support is very faintly visible in the ADF image, and though it can be distinguished from vacuum, no structural information can be obtained.

The MIDI-STEM image shown in Fig. 3b also shows good contrast for the NPs, with a similar SNR for the atomic planes as the ADF image in Fig. 3c. Furthermore, the MIDI-STEM image also shows very strong contrast for the carbon support, especially at the vacuum edge. The ADF image was used to color the area occupied by the gold NPs in Fig. 3d to emphasize the surrounding carbon structure. Inside the carbon film, we observe regions of correlated intensity between adjacent pixels in both the fast (horizontal) and slow (vertical) directions. Because each image pixel is a separate probe position representing a completely independent measurement, we ascribe these features to the atomic clustering characteristics of filament-like structures known to exist in amorphous carbon[38]. The gold NPs also have significant additional contrast that we interpret as amorphous carbon clustering around the particles. The source of this carbon could be from the sample fabrication process, from the surrounding substrate, or contamination from previous STEM

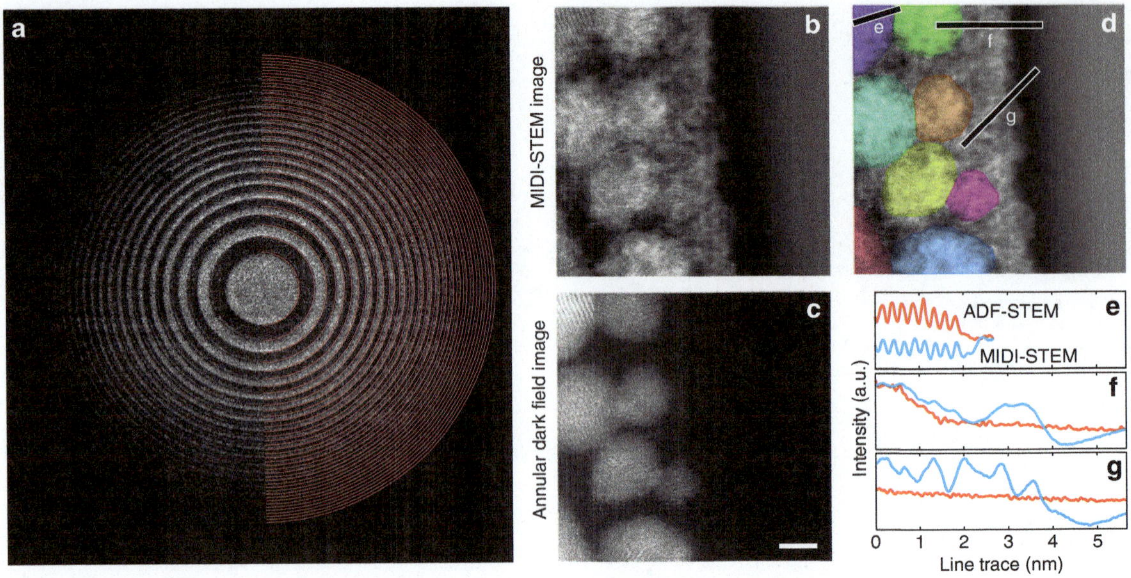

Figure 3 | MIDI-STEM experiment of a heterogeneous sample. (**a**) The image of the centre beam averaged from all diffraction patterns. Simultaneously, recorded images of gold NPs on a thin carbon support (**b**) using a virtual detector (edges outlined by red lines) shown in (**a,c**) using a conventional ADF-STEM detector. Scale bar, 2 nm. (**d**) Same as (**b**), with gold NPs shaded with random colours to emphasize the surrounding carbon. (**e–g**) Line traces with positions shown in (**d**) for images (**b,c**).

scans used for focusing. This additional contrast is not visible in the ADF images (except perhaps as some weak 'fuzziness') demonstrating that ADF imaging suppresses weakly scattering atoms such as carbon, which makes samples appear to be cleaner than they really are. ADF-STEM is essentially biased towards highly scattering materials, and this experiment demonstrates the capabilities of MIDI-STEM to simultaneously image low- and high-scattering materials.

Three line traces taken from the MIDI-STEM and ADF-STEM images of Fig. 3b,c are plotted in Fig. 3e–g. Figure 3e shows that both MIDI- and ADF-STEM are sensitive to the atomic lattice planes of the NPs, with approximately the same SNR. The left side of Fig. 3f shows that both methods show thickness contrast for an off-zone-axis NP, but the carbon substrate edge at the vacuum is essentially invisible in the ADF-STEM image, while strongly visible in the MIDI-STEM image. Finally, the trace in Fig. 3g along only amorphous carbon shows strong structural fluctuations in MIDI-STEM, and again no contrast in ADF-STEM. The MIDI-STEM image shows slowly varying intensity in the vacuum because it is a differential phase technique (similar to a high-pass filter) that cannot retrieve the d.c. component or very low spatial frequencies. This is evident in the CTF curves plotted in Fig. 2.

Contrast transfer of MIDI-STEM images from simulation. To validate our experimental results, we have simulated a MIDI-STEM experiment of a similar sample using the multislice method. The projected potential of the sample and atomistic model are shown in Fig. 4a,b. The simulated sample consists of randomly oriented cuboctahedral gold NPs attached to a wedge-shaped substrate of amorphous carbon. The realistic amorphous carbon atomic coordinates from[38] were tiled into a wedge with a maximum thickness of 5 nm on the left side and a minimum thickness at the substrate/vacuum edge of 3 nm. No additional carbon was added on top of the gold NPs as seen in the experimental results. The simulated STEM scan of 300×300 probe positions with a probe position spacing of 0.5 Å was confined to the green box overlaid on the projected potential.

Three detector configurations are considered in Fig. 4c–k; a BF-STEM detector constructed by summing over the central probe disk from 0 to 17 mrad, a (low angle) ADF-STEM detector from 20 to 95 mrad and a MIDI-STEM virtual detector consisting of the electrons recorded on odd rings minus those recorded on even rings. Detector geometries were chosen to match the experimental set-up presented earlier although only experimental ADF- and MIDI-STEM signals were experimentally available. All simulated images are plotted using infinite dose (no noise) and with an intermediate dose of 500 e^{-} Å^{-2}. In addition, the images are quantitatively evaluated by plotting the measured signal intensity at infinite dose as a fraction of the total incident electrons versus the projected potential of the sample at each probe position. The pixels are separated into two groups corresponding to probe positions at only amorphous carbon (blue) or probe positions including both carbon and gold (red) in projection. A polynomial trend line (black) was fitted to all points as a guide for the eye. Note that at an accelerating voltage of 300 kV, a projected potential of 1,500 V Å roughly corresponds to a π-phase shift of the incident electron wave, and therefore this specimen does not obey the weak-phase approximation.

The BF- and ADF-STEM simulations are essentially complementary, as expected when using such a low inner angle for the ADF detector. Both show strong contrast for the gold NPs and weak contrast for the carbon substrate at infinite electron dose. However, when using a dose of only 500 e^{-} Å^{-2}, the BF-STEM signal is overwhelmed by noise; only faint outlines of the NPs are visible and the substrate is almost invisible. The ADF-STEM image produces relatively better contrast at lower dose, as the NPs show high contrast both for atomic columns and atomic planes. In this ADF-STEM image, the substrate can be differentiated from the vacuum, but no structural information can be obtained. A more typical high-angle ADF-STEM image with a large detector inner angle ($>$50 mrad) of this sample produces a slightly better image of the gold, but significantly less contrast for the carbon substrate.

The simulated MIDI-STEM image by comparison shows very strong contrast for both the NPs and the amorphous substrate. Even at an electron dose much lower than typical STEM

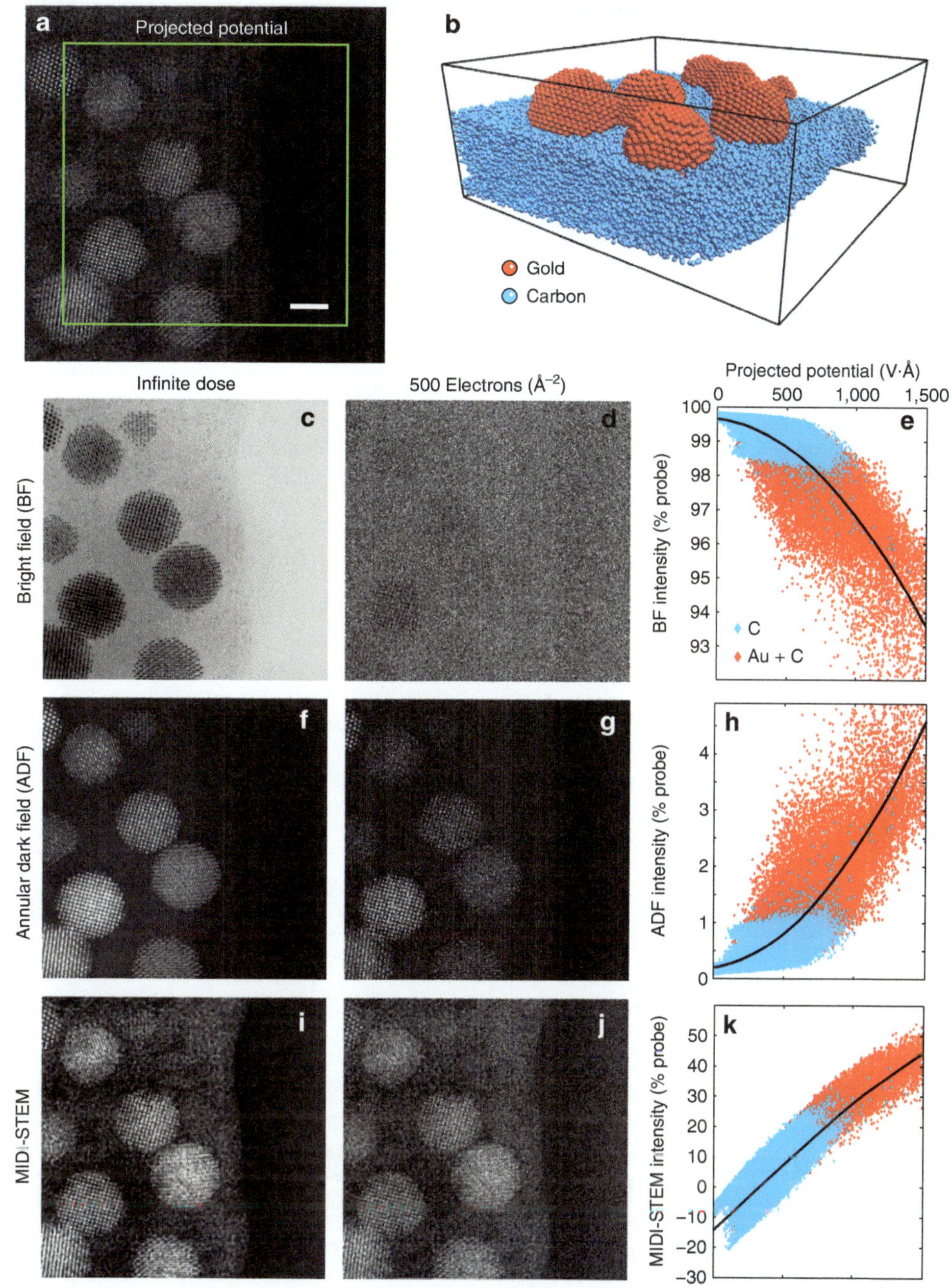

Figure 4 | MIDI-STEM simulations of a heterogeneous sample. (**a**) The projected potential and (**b**) corresponding atomistic three-dimensional model. Scale bar, 2 nm. Images were generated from scanning over the area enclosed in the green box using two common STEM imaging modes, (**c–e**) BF-STEM and (**f–h**) ADF-STEM, and (**i–k**) MIDI-STEM. Simulations are shown for (**c,f,i**) infinite and (**d,g,j**) moderate dose. (**e,h,k**) Quantitative comparison between the projected potential and the measured infinite dose signal shows that MIDI-STEM is a significantly more linear measurement than traditional methods.

experiments, atomic positions and planes are visible in the NPs, with contrast roughly equivalent to the ADF-STEM images. However, the contrast of the amorphous substrate has been significantly increased, and the atomic-scale details of the projected potential are visible in the infinite dose images. The finite dose MIDI-STEM image also shows many of the same structural details when comparing with the fine structure of the projected potential. Qualitatively, the efficiency of MIDI-STEM is explained by its ability to measure small scattering events due to the alternating rings of the phase plate. The plot of projected potential versus measured MIDI-STEM signal shows it is far more linear than BF- or ADF-STEM and has a much tighter distribution of measurements. Importantly, both the carbon and gold plus carbon signals fall on the same roughly linear curve. The primary sources of non-linearity in the MIDI-STEM measurement are the high-pass filtering effect of MIDI-STEM (a minor effect on the scale of a 20 nm field of view) and the decreased number of electrons available to scatter for thicker regions of the sample.

Discussion

It is important to emphasize that the key strengths of this method are its applicability for beam-sensitive soft materials and the advantages of linear contrast transfer towards the study of hard/soft interfaces in materials science. We have reported both the experimental and theoretical validity of MIDI–STEM for a sample of gold NPs on an amorphous carbon support as a very general case of a highly heterogeneous sample as a proof of principle. To explore the limits of the MIDI–STEM method, we performed a multislice simulation of a DNA snippet connecting two gold NPs on a single layer of graphene, plotted in Fig. 5. DNA was chosen for its well-known structure with weak scattering and moderate dose sensitivity. The same microscope parameters and detectors as Fig. 4 were used, and electron doses of infinity, 500 and 100 e^{-} Å^{-2} were simulated. However, unlike in Fig. 4c–h, in Fig. 5c–h the BF- and ADF-STEM simulations were performed using a conventional STEM probe without a phase plate. As above, neither the BF- nor ADF-STEM images show any appreciable contrast in the DNA section at a non-infinite electron dose. Conversely, the MIDI–STEM images (Fig. 5i–k) show linear contrast even at fairly low electron doses. Even at such low doses, MIDI–STEM produces enough contrast to identify not only the presence or absence of a bio-molecule, but also the shape envelope and orientation, while being in focus. MIDI–STEM is therefore a promising technique for imaging relatively radiation hard bio-molecules and heterostructures such as hard/soft interfaces. For the most dose-sensitive bio-molecules, highly defocused cryo-EM is a more efficient imaging method, but many samples cannot meet the cryo-EM requirements of many well-separated identical structures, with no strongly scattering components. The same atomic coordinates are used for a comparison with phase contrast, defocused HRTEM imaging in Supplementary Note 3 and Supplementary Fig. 5, using the deconvolution methods described in ref. 14. These simulations show that more information can be recovered from HRTEM imaging, but this requires large defocus values that can produce delocalization artifacts.

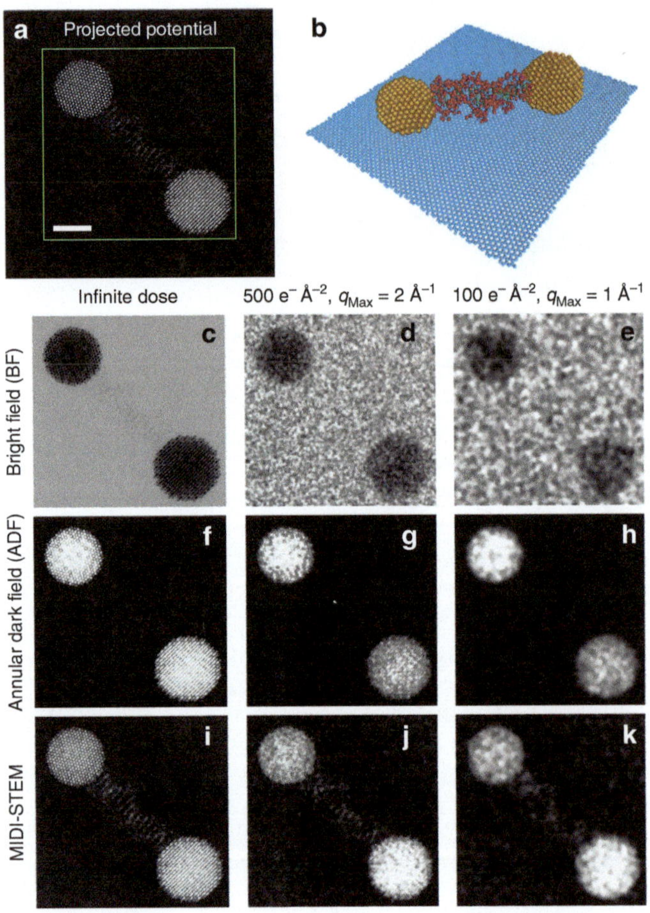

Figure 5 | MIDI-STEM simulations of a biological-inorganic interface. (**a**) Projected potential of a DNA snippet connecting two gold NPs on a single layer of graphene substrate, and (**b**) corresponding atomistic three-dimensional model. Scale bar, 2 nm. Images were generated from scanning over the area enclosed in the green box using (**c**-**e**) BF-STEM, (**f**-**h**) ADF-STEM and (**i**-**k**) MIDI-STEM at infinite and two low electron doses. Conventional STEM probes were used for BF- and ADF-STEM, while 20 ring pairs were used for MIDI-STEM.

In summary, we have experimentally demonstrated the MIDI–STEM imaging method with great promise for improving the contrast in STEM images for weakly scattering materials. We also performed multislice simulations of a sample realistically modelled after our experiment to confirm our interpretation of the experimental results. In this experiment, we imaged gold NPs on an amorphous carbon support, using a pixelated direct electron detector to construct the virtual detectors necessary for MIDI–STEM while simultaneously recording an ADF-STEM image. The MIDI–STEM image simultaneously showed atomic-plane contrast for highly scattering gold NPs and the amorphous structure of carbon regions. Structural features on the near-atomic scale were clearly visible in the amorphous carbon film, showing that MIDI–STEM is a promising candidate to directly image samples consisting of both hard and soft matter at atomic or near-atomic resolution using relatively low electron doses. The primary advantages of MIDI–STEM are high signal efficiency, good transfer of low spatial frequency information and the ability to image while in focus to minimize signal delocalization. MIDI–STEM should also allow the possibility of post-acquisition software aberration correction using pychographic methods.

Methods

Experimental. All experimental results presented in this paper were recorded on TEAM I, an aberration-corrected FEI Titan 80–300 operated in STEM mode at 300 kV with a convergence semi-angle of 17.2 mrad. The phase-plate geometry used to form the MIDI–STEM probes were equal-area Fresnel zone plates with 20 ring pairs, fabricated using focused ion beam milling of a SiN membrane[33]. The transmitted electron diffraction pattern at each probe position was recorded using a Gatan K2 IS direct electron detector with 3,840 × 3,712 pixels, operated at 400 frames per second and binned by 2. The camera acquisition and probe scanning were synchronized using a Gatan Digiscan. The probe was scanned over the 14.5 nm field of view with 256 × 256 probe positions to create a 256 × 256 × 1,920 × 1,792 four-dimensional STEM data set consisting of 420 GB of raw images.

Analysis and Simulation. Post processing to fit the virtual detector was done using custom scripts in MATLAB. All multislice simulations were performed using custom MATLAB codes that follows the methods of Kirkland[15], using the same microscope parameters as in the experiment and eight frozen phonon configurations. STEM probes were spaced by 0.5 Å, and the simulation pixel size was 0.2 Å. Experimental and simulated microscope parameters were optimized to give the highest contrast. A model detailing the geometric CTF calculations is given in the Supplementary Notes 1 and 2.

References

1. van den Bedem, H. & Eraser, J. S. Integrative, dynamic structural biology at atomic resolution – it's about time. *Nat. Methods* **12,** 307–318 (2015).
2. Yang, H. *et al.* Imaging screw dislocations at atomic resolution by aberration-corrected electron optical sectioning. *Nat. Commun.* **6,** 7266 (2015).
3. Zhu, Y., Withers, R. L., Bourgeois, L., Dwyer, C. & Etheridge, J. Direct mapping of Li-enabled octahedral tilt ordering and associated strain in nanostructured per-ovskites. *Nat. Mat.* **14,** 1142–1149 (2015).
4. Batson, P. E., Dellby, N. & Krivanek, O. L. Sub-angstrom resolution using aberration corrected electron optics. *Nature* **418,** 617–620 (2002).
5. Rose, H. Prospects for aberration-free electron microscopy. *Ultramicroscopy* **103,** 1-6 (2005).
6. Dahmen, U. *et al.* Background, status and future of the transmission electron aberration-corrected microscope project. *Philos. Trans. R. Soc. Lond. A* **367,** 3795–3808 (2009).

7. Glaeser, R. M. Limitations to significant information in biological electron microscopy as a result of radiation damage. *J. Ultrastruct. Res.* **36,** 466–482 (1971).
8. Isaacson, M., Johnson, D. & Crewe, A. V. Electron beam excitation and damage of biological molecules; its implications for specimen damage in electron microscopy. *Rad. Res.* **55,** 205–224 (1973).
9. Henderson, R. The potential and limitations ofneutrons, electrons and x-rays for atomic resolution microscopy of unstained biological molecules. *Quart. Rev. Biophys.* **28,** 171–193 (1995).
10. Egerton, R. E., Li, P. & Malac, M. Radiation damage in the tem and sem. *Micron* **35,** 399–409 (2004).
11. Egerton, R. E. Mechanisms of radiation damage in beam-sensitive specimens, for tem accelerating voltages between 10 and 300 kv. *Micro. Res. Tech.* **75,** 1550–1556 (2012).
12. Erank, J. Single-particle imaging of macromolecules by cryo-electron microscopy. *Ann. Rev. Biophys. Biomol. Struct.* **31,** 303–319 (2002).
13. Frank, J. *Three-dimensional electron microscopy of macromolecular assemblies: visualization of biological molecules in their native state* (Oxford Univ. Press, 2006).
14. Downing, K. H. & Glaeser, R. M. Restoration of weak phase-contrast images recorded with a high degree of defocus: the twin image problem associated with CTE correction. *Ultramicroscopy* **108,** 921–928 (2008).
15. Kirkland, E. J. *Advanced computing in electron microscopy* (Springer Science & Business Media, 2010).
16. Pennycook, S. J. & Nellist, P. D. *Scanning transmission electron microscopy: imaging and analysis* (Springer Science & Business Media, 2011).
17. Eindlay, S. D. *et al.* Dynamics of annular bright field imaging in scanning transmission electron microscopy. *Ultramicroscopy* **110,** 903–923 (2010).
18. Hovden, R. & Muller, D. A. Efficient elastic imaging of single atoms on ultrathin supports in a scanning transmission electron microscope. *Ultramicroscopy* **123,** 59–65 (2012).
19. Muller, D. A. Structure and bonding at the atomic scale by scanning transmission electron microscopy. *Nat. Mat.* **8,** 263–270 (2009).
20. Williams, D. B. & Carter, C. B. *Transmission electron microscopy: a textbook for materials science* Vol. 3 (Springer Science & Business Media, 2009).
21. Rose, H. Phase contrast in scanning transmission electron microscopy. *Optik* **39,** 416–436 (1974).
22. Dekkers, N. H. & De Lang, H. Differential phase contrast in a STEM. *Optik* **41,** 452–456 (1974).
23. Rose, H. Nonstandard imaging methods in electron microscopy. *Ultramicroscopy* **2,** 251–267 (1977).
24. Haider, M., Epstein, A., Jarron, P. & Boulin, C. A versatile, software configurable multichannel STEM detector for angle-resolved imaging. *Ultramicroscopy* **54,** 41–59 (1994).
25. Nellist, P. D., McCallum, B. C. & Rodenburg, J. M. Resolution beyond the'information limit'in transmission electron microscopy. *Nature* **374,** 630–632 (1995).
26. Hue, F., Rodenburg, J. M., Maiden, A. M., Sweeney, F. & Midgley, P. A. Wave-front phase retrieval in transmission electron microscopy via ptychography. *Phys. Rev. B* **82,** 121415 (2010).
27. Humphry, M. J., Kraus, B., Hurst, A. C., Maiden, A. M. & Rodenburg, J. M. Ptychographic electron microscopy using high-angle dark-field scattering for sub-nanometre resolution imaging. *Nat. Commun.* **3,** 730 (2012).
28. Ou, X., Horstmeyer, R., Yang, C. & Zheng, G. Quantitative phase imaging via fourier ptychographic microscopy. *Opt. Lett.* **38,** 4845–4848 (2013).
29. Pennycook, T. J. *et al.* Efficient phase contrast imaging in STEM using a pixelated detector. part 1: experimental demonstration at atomic resolution. *Ultramicroscopy* **151,** 160–167 (2015).
30. Yang, H., Pennycook, T. J. & Nellist, P. D. Efficient phase contrast imaging in STEM using a pixelated detector. part ii: optimisation of imaging conditions. *Ultramicroscopy* **151,** 232–239 (2015).
31. Verbeeck, J., Tian, H. & Schattschneider, P. Production and application of electron vortex beams. *Nature* **467,** 301–304 (2010).
32. McMorran, B. J. *et al.* Electron vortex beams with high quanta of orbital angular momentum. *Science* **331,** 192–195 (2011).
33. Harvey, T. R. *et al.* Efficient diffractive phase optics for electrons. *New J. Phys.* **16,** 093039 (2014).
34. Grillo, V. *et al.* Generation of nondiffracting electron bessel beams. *Phys. Rev. X* **4,** 011013 (2014).
35. Peplow, M. Materials science: the hole story. *Nature* **520,** 148–150 (2015).
36. Hallinan, Jr D. T. & Balsara, N. P. Polymer electrolytes. *Ann. Rev. Mat. Res.* **43,** 503–525 (2013).
37. Fitzgibbon, A., Pilu, M. & Fisher, R. B. Direct least square fitting of ellipses. *IEEE Trans. Pat. Anal. Mach. Int.* **21,** 476–480 (1999).
38. Ricolleau, C., Le Bouar, Y., Amara, H., Landon-Cardinal, O. & Alloyeau, D. Random vs realistic amorphous carbon models for high resolution microscopy and electron diffraction. *J. Appl. Phys.* **114,** 213504 (2013).

Acknowledgements

Work at the Molecular Foundry was supported by the Office of Science, Office of Basic Energy Sciences, of the US Department of Energy under Contract No. DE-AC02-05CH11231. Work at University of Oregon was supported by the US Department of Energy, Office of Science, Basic Energy Sciences under Award No. DE-SC0010466.

Author contributions

C.O. developed the MIDI-STEM method to combine a structured phase electron probe with a matching virtual detector. H.H.R. initially proposed the alternating-zone interference imaging mode used in MIDI-STEM. Ercius performed the experiments, using STEM coil synchronization, recording scripts and four-dimensional STEM experimental protocols developed by J.Ci. and C.C. J.P., T.R.H., J.Ch. and B.J.M. fabricated and tested the phase plates. C.O. ran all multislice simulations and performed the analysis of experiments and simulations. C.O., P.E. and J.Ci. wrote the manuscript with input from all other authors. C.O. wrote the Supplementary Notes with input from T.R.H., J.Ch. and B.J.M.

Additional information

Competing financial interests: The authors declare no competing financial interests.

20

Visualizing the orientational dependence of an intermolecular potential

Adam Sweetman[1], Mohammad A. Rashid[1], Samuel P. Jarvis[1], Janette L. Dunn[1], Philipp Rahe[1] & Philip Moriarty[1]

Scanning probe microscopy can now be used to map the properties of single molecules with intramolecular precision by functionalization of the apex of the scanning probe tip with a single atom or molecule. Here we report on the mapping of the three-dimensional potential between fullerene (C_{60}) molecules in different relative orientations, with sub-Angstrom resolution, using dynamic force microscopy (DFM). We introduce a visualization method which is capable of directly imaging the variation in equilibrium binding energy of different molecular orientations. We model the interaction using both a simple approach based around analytical Lennard-Jones potentials, and with dispersion-force-corrected density functional theory (DFT), and show that the positional variation in the binding energy between the molecules is dominated by the onset of repulsive interactions. Our modelling suggests that variations in the dispersion interaction are masked by repulsive interactions even at displacements significantly larger than the equilibrium intermolecular separation.

[1] School of Physics and Astronomy, University of Nottingham, Nottingham NG7 2RD, UK. Correspondence and requests for materials should be addressed to A.S. (email: adam.sweetman@nottingham.ac.uk).

The nature of intermolecular interactions underpins a vast array of physical and chemical phenomena, and is a scientific theme that straddles the disciplines of physics, chemistry and biology. Particular impetus has been given to the study of intermolecular forces at the single-molecule level due to the stunning advances in ultrahigh resolution scanning probe imaging pioneered by Gross *et al.*[1]. Three-dimensional (3D) force maps were acquired over planar organic molecules that bore a striking resemblance to the classic textbook 'ball-and-stick' models. These advances were first realized via the controllable functionalization of the scanning probe tip with a single pre-selected atom or molecule, which provides a unique level of control with which to investigate the atomic and molecular scale properties of matter, and also helps to eliminate the most troublesome aspect of scanning probe experiments, that is, the uncertainty surrounding the tip structure.

Although this tip functionalization strategy is now commonly applied to single CO molecules to allow intramolecular imaging[1,2], the technique has application well beyond imaging, and similar protocols have also been used to study the electronic[3,4] and mechanical[5] properties of single molecules trapped in the tip-sample junction, and to quantitatively measure intermolecular interactions[6–8]. There has also been considerable interest centred around the possibility of using this technique to directly visualize intermolecular interactions[9], although considerable debate surrounds the interpretation of these results[2,10–12].

In this paper, we discuss the results of a series of experiments—and their interpretation on the basis of both simple analytical potentials and DFT—that map the orientational dependence of the 3D potential between two-complex molecules. By measuring the full 3D potential we are able to apply a novel visualization method that directly shows the variation in the equilibrium binding energy for the molecular system for different relative orientations of the molecules. We also discuss the feasibility of detecting the variation in dispersion forces due to molecular rotation via DFM.

Results

Experimental results. Figure 1 shows representative constant height Δf images, taken from a 3D grid, acquired at decreasing tip-sample separation over three surface-adsorbed C_{60} molecules in different orientations, using a C_{60}-terminated tip.

At larger separations a featureless circular attractive interaction is observed (Fig. 1b), but on closer approach intricate intramolecular features are resolved (Fig. 1c,d), followed by their intense 'sharpening' (Fig. 1e,f). This evolution in contrast is similar to the onset of sub-molecular features during imaging of planar molecules with flexible tips[1,12]. However, because in this experiment both molecules have a complex structure, the intramolecular features in these images cannot be easily assigned to the molecular structure of the surface molecule as is the case for images taken with simple (that is, atomic point-like) tip terminations.

Converting the acquired Δf grid into a map of potential allows us to create similar constant height images of the tip-sample force and potential (Fig. 2a,b and Supplementary Fig. 1). Although constant height slices of force and energy provide the closest visual analogue to how the data are collected, these images necessarily conflate the value of the tip-sample energy and the topographic height of the molecule at a given position. Consequently, topographically higher features dominate the constant height image due to their being effectively shifted in z, even if the range of energies at these positions is identical to other locations over the molecule.

Representative single $U(z)$ curves may be extracted (Fig. 2c) and allow a selection of the energy minimum values at different positions to be observed, but this is an indirect, and not necessarily intuitive, method of analysing the variation in intermolecular potential across the molecules.

In Fig. 2d, we instead show an image constructed by searching each vertical column in the 3D data set (that is, each $U(z)$ curve) for the value of the potential energy minimum, and then projecting this minimum value over the xy plane of the grid, which we hereafter refer to as a 'U_{min}' image. This provides an

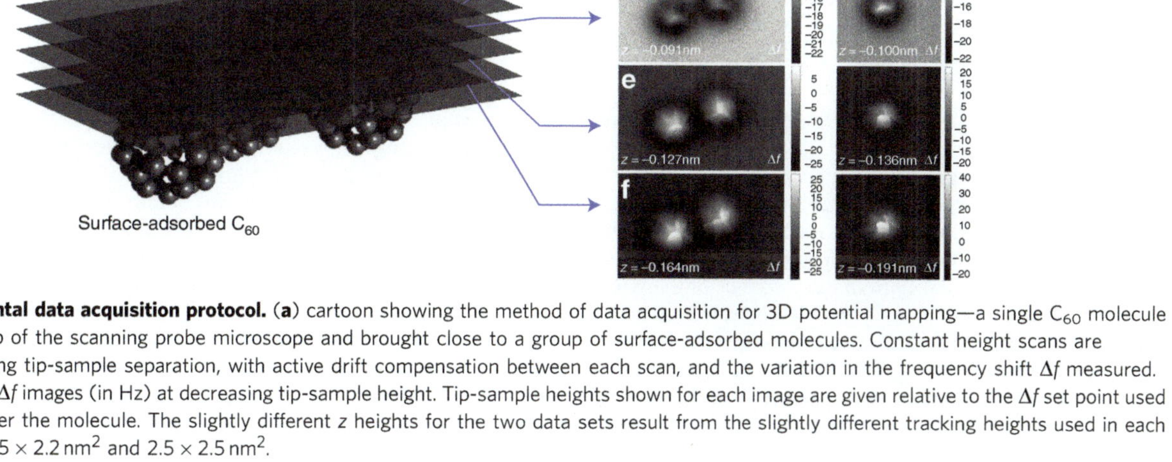

Figure 1 | Experimental data acquisition protocol. (**a**) cartoon showing the method of data acquisition for 3D potential mapping—a single C_{60} molecule is attached to the tip of the scanning probe microscope and brought close to a group of surface-adsorbed molecules. Constant height scans are acquired at decreasing tip-sample separation, with active drift compensation between each scan, and the variation in the frequency shift Δf measured. (**b-f**) representative Δf images (in Hz) at decreasing tip-sample height. Tip-sample heights shown for each image are given relative to the Δf set point used for atom-tracking over the molecule. The slightly different z heights for the two data sets result from the slightly different tracking heights used in each case. Image sizes: $3.5 \times 2.2\,nm^2$ and $2.5 \times 2.5\,nm^2$.

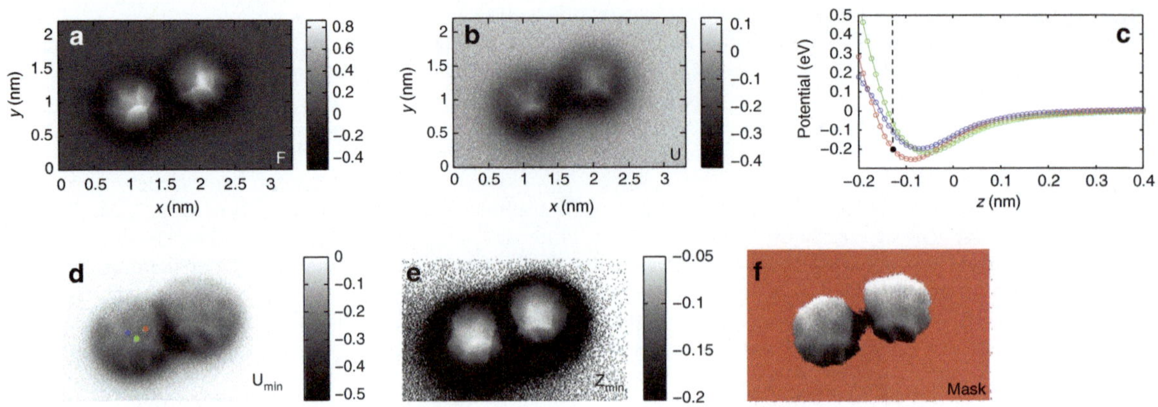

Figure 2 | Experimental measurement of variation in potential between C_{60} molecules in different orientations. Constant height images of (**a**) force (in nN) and (**b**) energy (in eV). (**c**) Representative $U(z)$ curves taken at different positions across the left hand C_{60} molecule, dotted line shows the height of the force and energy slices shown in **a**,**b**. (**d**) Image showing the variation in the value of the energy at the minimum in the $U(z)$ curve (in eV) at each position in the grid. The positions of the curves shown in **c** are marked. (**e**) As for **d** but showing instead the z height at which the minimum occurs, note that the black regions indicate parts of the grid where the minimum is found at the lowest tip-sample separation (that is, no turnaround detected). (**f**) Variation in energy minimum masked using the minimum in z position. Red shading indicates locations where the minimum in the intermolecular potential is not present in the $U(z)$ curve.

immediate and intuitive way of visualizing the strength of the equilibrium interaction as the relative position of the tip- and sample-adsorbed molecules is varied. We note that, because of the near-unique high-rotational symmetry of the C_{60} molecules, displacements in the xy plane should be equivalent to changes in rotational orientation.

We highlight here that some care must be taken in the interpretation of these images, as the value of the minimum in the potential energy curve only has a directly interpretable physical meaning when the actual minimum of the potential is present in a given $U(z)$ curve (that is, the turn-around in the $U(z)$ curve is present in the data set). If the minimum is not reached, then the closest point of approach will usually be identified as the minimum value. We therefore also map the height of the potential energy minimum in terms of z, which yields a complementary map of z_{min}. By masking the U_{min} map with the z_{min} map we can exclude those curves which do not contain the $U(z)$ turn-around, and visualize only the region of the image, which can be interpreted directly as representing the intermolecular interaction minimum (Fig. 2f). Application of this visualization technique also reveals a gradient in the value of the minimum in the potential across the molecule, most likely related to an asymmetric mounting of the molecule on the tip. Since this gradient directly affects the spread in the energy values we therefore only discuss the variation observed in the region located over the centre of the molecule, where the variation due to the gradient is small compared to the variation produced by the changes in molecular orientation (Supplementary Figs 16–18).

The same technique may also be applied to the 3D force field and raw 3D Δf measurements (see Supplementary Fig. 3). Although these maps do not have such a direct physical interpretation as for the minimum in the potential, they still provide an extremely powerful technique for visualizing the relative interaction over the molecule. Interestingly, we note a strong qualitative similarity in the appearance of these images and recent data acquired using a profile-corrected constant height technique by Moreno *et al.*[13]. We also note that Mohn *et al.*[14] recovered a pssuedo-topographic Δf image from a 3D data set, and experimentally it has been shown how to operate in the $\Delta f = 0$ regime[15], which might, in principle, produce similar imaging if applied to intermolecular measurements. Critically, however, none of these earlier works directly measured and visualised the physical quantity of interest here: the variation in the value of the minimum in the intermolecular potential.

Our data demonstrate that as the relative orientations of the tip and surface C_{60} molecules are varied the potential minimum between the two molecules varies of the order 60 meV. A key question is therefore—what is the origin of this variation? A common approach to evaluating the C_{60}–C_{60} intermolecular interaction is to model molecular energy variation using the Girifalco potential[7,16], but this simplified model assumes a uniform spherical interaction, and does not give any information about sub-molecular variation in the potential. In particular, given the extended 3D nature of the molecule, it is not immediately clear how the attractive and repulsive components of the intermolecular potential contribute to the variation in the magnitude, and position, of the energy minimum. Following recent studies investigating the variation in dispersion force as a function of molecular size[6], there is also an open question as to whether the difference in the dispersion interaction can be observed for changes in the orientation of extended molecules. C_{60}, with its near-spherical symmetry represents a particularly important test bed for this hypothesis.

Computational results. To interpret our results, we modelled our experimental system with two different approaches (as described in the Methods section). First, we used a simple Lennard–Jones (L–J) potential for two C_{60} molecules, coupled with a modified version of the flexible tip model introduced by Hapala *et al.*[12], to simulate the C_{60} interaction. The simple nature of the model means that it is computationally inexpensive and thus can be exploited to generate high-resolution 3D data sets of comparable data density to those we obtain experimentally (Fig. 3d–g). Second, to test the validity of our empirical model, we compare the results of the L–J calculations to simulations of the same C_{60}–C_{60} interaction performed using the *ab initio* CP2K DFT code. The significant computational cost of the *ab initio* simulations precludes the calculation of a full 3D grid as for the L–J simulations, and we therefore instead compare 2D *xz* slices taken across the centre of the molecule–molecule interaction (Fig. 3b,c). In this comparison, we modelled a prototypical high symmetry orientation (hexagon face on hexagon face, hereafter referred to as

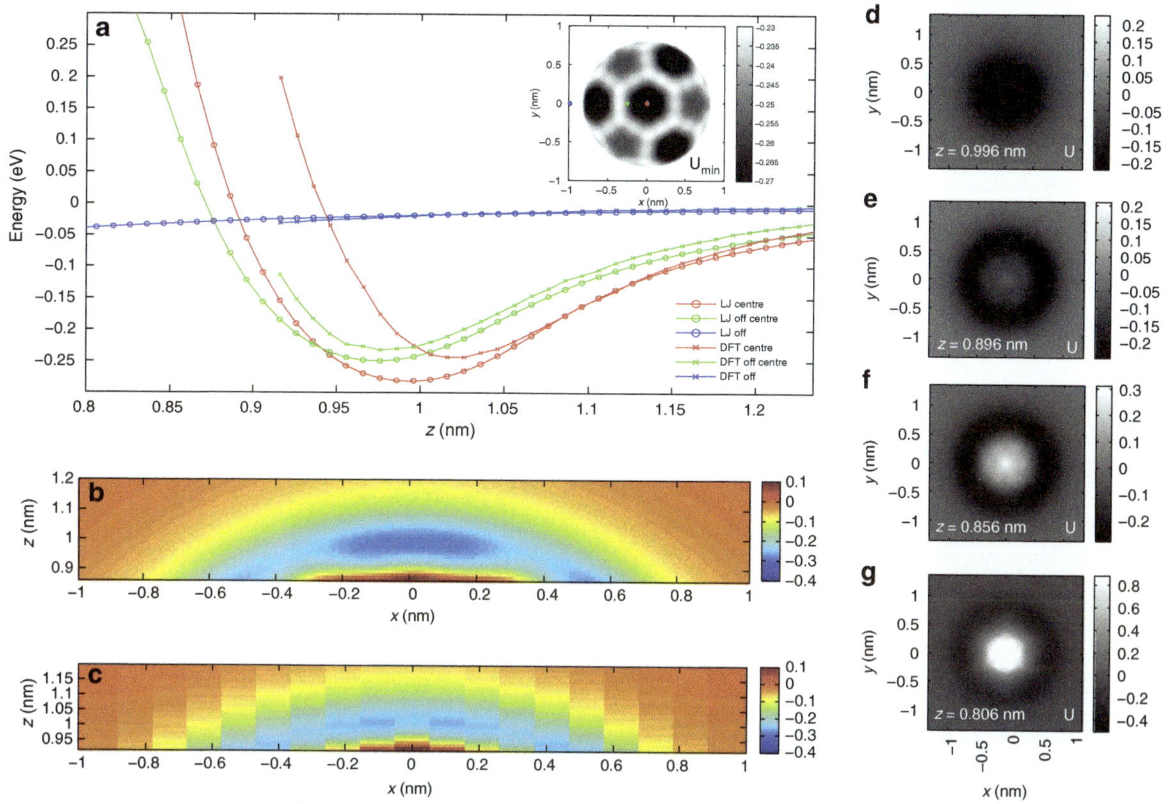

Figure 3 | Modelling the C_{60}-C_{60} interaction. (**a**) Comparison of potentials calculated by L-J modelling and DFT. *z* heights are defined as the vertical separation of the two molecules measured from the molecular centres. Curves are shown for the same initial *xy* coordinates in both simulations. Inset: Complete *xy* U_{min} (in eV) image for L-J simulation. (**b**,**c**) *xz* plot of calculated energies for L-J and DFT simulations, respectively. (**d**–**g**) Constant height *xy* energy images (in eV) from L-J simulations, relative heights labelled on each image.

Hex–Hex) for the two molecules using both simulations methods. In general, we find good agreement between the two techniques, noting in particular that the potential gradients in both the attractive and repulsive branches of the potential curve are very similar. We do, however, observe some quantitative differences between the empirical L-J simulations and *ab initio* DFT. Specifically, for the L-J parameters chosen ($\epsilon_a = 2.5$ meV and $r_a = 1.966$ Å (ref. 16)), the maximum depth and width of the well is slightly larger that for the DFT simulations, as is the variation in the range of U_{min} (ΔU_{min} for L-J *xz* plot ~ 50 meV compared with ~ 20 meV for the DFT *xz* plot over the same range (Supplementary Figs 20–22)). We note, however, that variation in the positions of the minimum is almost identical (ΔZ_{min} for L-J *xz* plot ~ 0.09 nm, compared with $\Delta Z_{min} \sim 0.11$ nm for the DFT *xz* plot). Furthermore, it is clear that tuning the choice of L-J parameters based on the DFT results could improve the quantitative agreement between the two simulation methods, but here we prefer to use those L-J parameters derived from previous experimental work and which are also consistent with our earlier publications, rather than arbitrarily adjusting the L-J parameters. These results imply that while the L-J model is a simplification of the complex intermolecular interaction, it nonetheless appears to be sufficient to model much of the essential physics underpinning the variation in intermolecular potential.

Although we stress that the high-symmetry Hex–Hex configuration used in the simulations is not the configuration of the C_{60} molecules in the experimental data set shown in Fig. 2, we nonetheless observe a number of qualitatively similar features in both the simulations and experiment. In particular, the simulations reproduce the 'sharpening' of the features observed in the constant-height experimental images, in line with the sharpening reported for CO-terminated tips. In addition, the appearance of the simulated U_{min} image is qualitatively similar to that acquired in the experiment, which reveals the complex variation in potential minimum as the molecular positions are varied. Interestingly, the L-J simulations overestimate the depth of the potential relative to the DFT calculations, but better reproduce the variation in U_{min} observed experimentally, with a variation of ~ 50 to 60 meV in the U_{min} image depending on molecular orientation (Supplementary Figs 16–18). We also note that simulations performed with other tip-sample molecular configurations, such as those found for C_{60} adsorption on the Si(111)-7 × 7 substrate, produce much more complex patterns in the constant height, and U_{min}, images (see Supplementary Figs 7–13), qualitatively similar to those observed experimentally.

Discussion

Because of the simple additive, and analytical, nature of the L-J model, it is possible to decompose the interaction into its attractive and repulsive components, and ascertain if we might in principle be able to observe rotational variation in the dispersion interaction between the C_{60} molecules. To assess the relative influence of the repulsive and attractive elements of the potential on the value of the potential minimum we investigated the change in the potential for several orientations of the tip and sample C_{60} (Fig. 4a). We then plot the modulus (i.e. the absolute value) of the differences in the total energies, and the separate energies from the r^6 and r^{12} terms, between these orientations and the high symmetry 'Hex-Hex' configuration (Fig. 4b), and then extract the differences in these ΔU_{r^6} and $\Delta U_{r^{12}}$ terms (Fig. 4c). Specifically,

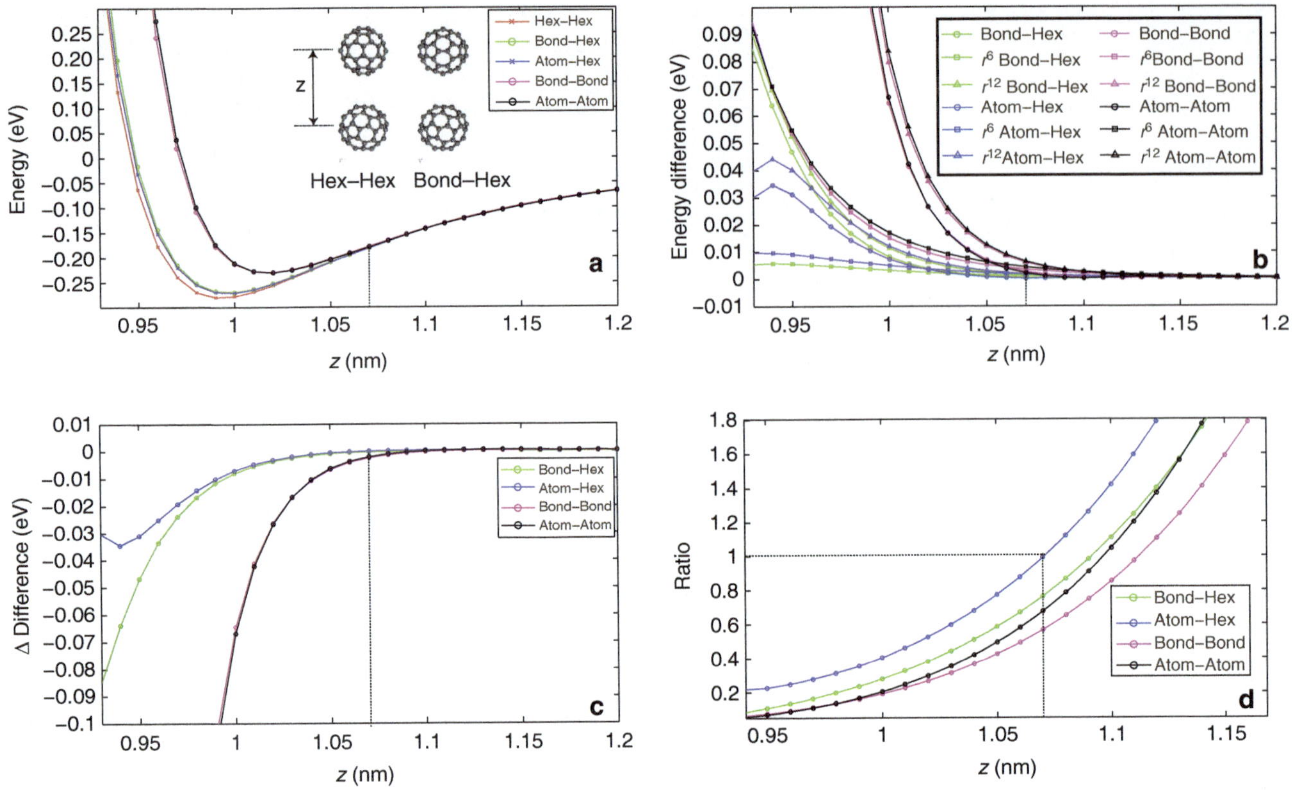

Figure 4 | Investigating the origin of the variation in potential energy minimum. (**a**) Variation in potential energy between two C_{60} molecules due to variation in rotational orientation, the terms in the legend refer to the facing part of the molecule. Inset, representative ball-and-stick models showing two of the simulated orientations. (**b**) Modulus of the difference in U_{total}, U_{r^6}, and $U_{r^{12}}$, between stated orientation and the Hex-Hex orientation. (**c**) Difference of variation in U_{r^6} and $U_{r^{12}}$ terms (that is, $|\Delta U_{r^6}| - |\Delta U_{r^{12}}|$). Negative values indicate the $U_{r^{12}}$ term dominates the difference in energy between the two configurations. (**d**) Ratio of variation in U_{r^6} and $U_{r^{12}}$ terms. Values above 1 indicate the U_{r^6} term has a larger contribution. Dotted lines indicate the separation at which the U_{r^6} term first becomes dominant.

we define $|\Delta U_{r^6}| - |\Delta U_{r^{12}}|$ as the modulus of the difference of the ΔU_{r^6} term (between the stated orientation and the Hex–Hex orientation), minus the modulus of the difference in the $\Delta U_{r^{12}}$ terms (between the stated orientation and the Hex–Hex orientation).

This quantity gives the relative influence of the two terms in defining the difference in the energy curves between the two orientations. If the difference in r^6 terms at a given separation is greater (that is, dispersion forces vary significantly between different orientations), then this quantity will be positive. If the difference in r^{12} terms is large (that is, Pauli forces vary significantly between different orientations), then it will be negative. If both quantities contribute equally to the difference in energy, then the term will be approximately zero. Surprisingly, we observe that the difference is negative and, consequently, the differences observed in the total energies, even in the part of the well where the potential gradient is positive, are dominated by repulsive interactions. Here we wish to make it explicitly clear that, for intermolecular separations greater than the equilibrium value (that is, before the energy 'turn-around'), the magnitude of the r^6 term is indeed larger in all cases, and dominates the r^{12} term, but that contribution of the r^6 term is very similar for all the orientations.

It must be noted, however, that the interplay between the two terms is somewhat subtle. If we examine the ratio of the differences (Fig. 4d), then it is clear that the r^6 term does begin to dominate the difference in the energies at around 1.06–1.1 nm separation. However, by reference to (Fig. 4c) it becomes clear that at this separation the difference in the potentials is <5 meV, that is, below our experimental sensitivity. Therefore, our modelling suggests that at the point at which the potential curves for different orientations become experimentally distinguishable, the difference between them is dominated by repulsive, rather than dispersive, interactions. As such it seems likely that although the magnitude of the variation in energy due to the variation in dispersion interaction under rotation of the molecule might in principle be within the noise limit of current DFM techniques, its direct measurement will always be hindered by the intrinsic convolution of the variation in energy due to repulsive forces, even at intermolecular separations significantly greater than the equilibrium value, where the gradient in the potential is positive.

We have presented 3D mapping of the variation in intermolecular interaction under changes in rotational orientation of a complex molecule with sub-Angstrom resolution via the functionalization of a scanning probe tip. Using a novel visualization method we can directly observe the variation in the value and position of the minimum of the potential energy as the orientation of the molecules is varied. By comparison of our results to both simple analytical and *ab initio* simulations, we are able to show that the variation in binding energy across the molecule is dominated by the onset of repulsive interactions between the front-most parts of the molecules. Surprisingly, we also find that variation in the net attractive part of the potential due to rotation of the molecules is still dominated by the repulsive forces, and the majority of the molecule only adds a uniform background to the potential. We anticipate that similar experimental techniques to those described here could be utilized to intuitively visualise the reactivity across complex interatomic and intermolecular potentials, including molecules with polarized or hydrogen bonding end groups.

Methods

Experimental methods. Clean Si(111)-7 × 7 surfaces were prepared by flash annealing a silicon wafer to 1,200 °C, rapid cooling to 900 °C and then slow cooling to room temperature. A low coverage of C_{60} was prepared by depositing the molecules from a home made tantalum pocket deposition source onto the room temperature substrate. Post-deposition, the sample was transferred into the scan head of an Omicron Nanotechnology LT DFM operating in UHV at cryogenic temperatures, and left to cool to 5 K before imaging.

Commercial qPlus sensors (Omicron Nanotechnology GmbH) with electrochemically etched tungsten wire tips were introduced into the scan head without any further preparation. The sensors were first prepared on clean Si(111)-7 × 7 surfaces by standard STM techniques until good STM/DFM resolution was achieved. Single C_{60} molecules were transferred to the tip by close approach to surface-adsorbed molecules, and the functionalization of the tip was checked by inverse imaging of the tip adsorbed molecule on the surface adatoms (Supplementary Figs 2,4 and 5)[7]. In all experiments an oscillation amplitude (A_0) of 110 pm was used, and the tip-sample bias was set to 0 V. Three-dimensional Δf volumes over the molecules were collected via the 'slice' method[17] and site specific (short-range) Δf values were extracted using the 'on–off' method[18,19] then converted to potentials using the Sader–Jarvis algorithm[20]. Due to the long acquisition times required, residual thermal drift and piezoelectric creep were corrected using a custom atom-tracking and scripting setup[21,22]. Further details on the experimental setup, data processing steps and additional experimental data sets may be found in the Supplementary Methods.

Flexible tip model and simulated spectroscopy procedure. To simulate DFM images, we adapted the method proposed by Hapala *et al.*[12] to model the interaction between a sample and a CO-functionalized DFM tip. In our simulation the functionalized tip is assumed to consist of a tip base (outermost atom of the tip) and a probe. The probe is the flexible end of the model tip, and is allowed to move around the tip base. In our simulation, the probe is a C_{60} molecule consisting of 60 carbon atoms acting as a single effective probe particle attached to the tip base (Supplementary Fig. 6). Each atom in the probe experiences three forces; (i) a Lennard-Jones (L-J) force due to the tip base, (ii) a sum of all pairwise L-J forces due to interactions with atoms in the sample and (iii) a lateral harmonic force from the tip base. The net force on the probe is calculated by summing up all the forces experienced by each atom on the probe. The L-J interactions between atoms α and β are written as

$$\mathbf{F}_{\alpha\beta}(\mathbf{R}) = 12\epsilon_{\alpha\beta}\mathbf{R}\left(\frac{r^{12}_{\alpha\beta}}{r^{14}} - \frac{r^{6}_{\alpha\beta}}{r^{8}}\right) \quad (1)$$

$$U_{\alpha\beta}(r) = \epsilon_{\alpha\beta}\left(\frac{r^{12}_{\alpha\beta}}{r^{12}} - \frac{2r^{6}_{\alpha\beta}}{r^{6}}\right), \quad (2)$$

where $r = |\mathbf{R}|$ is the distance between atoms α and β, $\epsilon_{\alpha\beta} = \sqrt{\epsilon_\alpha \epsilon_\beta}$ is the pair-binding energy and $r_{\alpha\beta} = r_\alpha + r_\beta$ is the equilibrium separation of the two atoms with ϵ_α and r_α being the atomic parameters. In our calculations the L-J parameters for the carbon atoms were set to ϵ_α = 2.5 meV and r_α = 1.966 Å to ensure consistency with the work of Girifalco *et al.*[16] and our own earlier work[7]. For the tip base a value of r_α = 5.0 Å was chosen, in order to take into account the larger size of the C_{60} molecule. The probe lateral stiffness and apex L-J parameter were set to $k_{xy} = 0.5\ \mathrm{N\,m^{-1}}$ and ϵ_α = 1, 000 meV, respectively (Supplementary Figs 14–15). We acquired the simulation data by scanning the sample laterally with a step of Δx, Δy = 0.1 Å. At each lateral position we placed the tip base at an initial separation z_0 = 22 Å from the surface molecule and approached the sample (in our simulations another C_{60} molecule) in steps of Δz = 0.1 Å until z = 17.5 Å allowing the probe position to be relaxed at each step due to the combined force of the sample and tip base. Note, however, that for ease of comparison to the DFT simulations all molecular separations discussed in the paper are given relative to the initial vertical core-core separation of the probe C_{60} from the surface C_{60}.

Density functional theory. DFT calculations were performed using the same initial high symmetry geometry (as described in main text) as the L-J simulations using the open source CP2K/Quickstep code[23,24] utilising a hybrid Gaussian and plane-wave method[25]. Goedecker, Teter and Hutter pseudopotentials[26] and the Perdew Burke Ernzerhof generalized gradient approximation method[27] were used with a 300 Ry plane-wave energy cutoff. To account for dispersion interactions we employed the Grimme DFT-D3 method[28], which well reproduced the C_{60}–C_{60} pair potential (Supplementary Fig. 19) A double-zeta Gaussian basis set plus polarization (DZVP-MOLOPT)[29] was used with a force convergence criterion for geometry relaxation of 0.05 eV Å^{-1}. Geometry relaxation was carried out by allowing all atoms to relax other than the hexagonal faces of each molecule furthest apart from one another (to simulate attachment to the surface/tip).

References

1. Gross, L., Mohn, F., Moll, N., Liljeroth, P. & Meyer, G. The chemical structure of a molecule resolved by atomic force microscopy. *Science* **325,** 1110–1114 (2009).
2. Jarvis, S. P. Resolving intra-and inter-molecular structure with non-contact atomic force microscopy. *Int. J. Mol. Sci.* **16,** 19936 (2015).
3. Schull, G., Frederiksen, T., Arnau, A., Sanchez-Portal, D. & Berndt, R. Atomic-scale engineering of electrodes for single-molecule contacts. *Nat. Chem.* **6,** 23–27 (2010).
4. Schull, G., Frederiksen, T., Brandbyge, M. & Berndt, R. Passing current through touching molecules. *Phys. Rev. Lett.* **103,** 11–14 (2009; 0910.1281.
5. Hauptmann, N. *et al.* Force and conductance during contact formation to a C_{60} molecule. *New. J. Phys.* **14,** 073032 (2012).
6. Wagner, C. *et al.* Non-additivity of molecule-surface van der Waals potentials from force measurements. *Nat. Commun.* **5,** 5568 (2014).
7. Chiutu, C. *et al.* Precise orientation of a single C_{60} molecule on the tip of a scanning probe microscope. *Phys. Rev. Lett.* **108,** 268302 (2012).
8. Sun, Z., Boneschanscher, M. P., Swart, I., Vanmaekelbergh, D. & Liljeroth, P. Quantitative atomic force microscopy with carbon monoxide terminated tips. *Phys. Rev. Lett.* **106,** 46104 (2011).
9. Zhang, J. *et al.* Real-space identification of intermolecular bonding with atomic force microscopy. *Science* **342,** 611–614 (2013).
10. Sweetman, A. M. *et al.* Mapping the force field of a hydrogen-bonded assembly. *Nat. Commun.* **5,** 3931 (2014).
11. Hämäläinen, S. K. *et al.* Intermolecular contrast in atomic force microscopy images without intermolecular bonds. *Phys. Rev. Lett.* **113,** 186102 (2014).
12. Hapala, P. *et al.* Mechanism of high-resolution STM/AFM imaging with functionalized tips. *Phys. Rev. B* **90,** 085421 (2014).
13. Moreno, C., Stetsovych, O., Shimizu, T. K. & Custance, O. Imaging three-dimensional surface objects with submolecular resolution by atomic force microscopy. *Nano Lett.* **15,** 2257 (2015).
14. Mohn, F., Schuler, B., Gross, L. & Meyer, G. Different tips for high-resolution atomic force microscopy and scanning tunneling microscopy of single molecules. *Appl. Phys. Lett.* **102,** 073109 (2013).
15. Rode, S., Schreiber, M., Kühnle, A. & Rahe, P. Frequency-modulated atomic force microscopy operation by imaging at the frequency shift minimum: The dip-df mode. *Rev. Sci. Instrum.* **85,** 043707 (2014).
16. Girifalco, L. a. Interaction potential for carbon (C_{60}) molecules. *J. Phys. Chem.* **95,** 5370–5371 (1991).
17. Neu, M. *et al.* Image correction for atomic force microscopy images with functionalized tips. *Phys. Rev. B* **89,** 205407 (2014).
18. Lantz, M. A. *et al.* Quantitative measurement of short-range chemical bonding forces. *Science* **291,** 2580–2583 (2001).
19. Sweetman, A. & Stannard, A. Uncertainties in forces extracted from non-contact atomic force microscopy measurements by fitting of long-range background forces. *Beilstein J. Nanotechnol.* **5,** 386–393 (2014).
20. Sader, J. E. & Jarvis, S. P. Accurate formulas for interaction force and energy in frequency modulation force spectroscopy. *Appl. Phys. Lett.* **84,** 1801 (2004).
21. Abe, M. *et al.* Room-temperature reproducible spatial force spectroscopy using atom-tracking technique. *Appl. Phys. Lett.* **87,** 173503 (2005).
22. Rahe, P. *et al.* Flexible drift-compensation system for precise 3D force mapping in severe drift environments. *Rev. Sci. Instrum.* **82,** 063704 (2011).
23. Hutter, J, Iannuzzi, M., Schiffmann, F. & VandeVondele, J. CP2K: atomistic simulations of condensed matter systems. *WIREs Comput. Mol. Sci.* **4,** 1525 (2013).
24. VandeVondele, J. *et al.* Quickstep: Fast and accurate density functional calculations using a mixed gaussian and plane waves approach. *Comput. Phys. Commun.* **167,** 103–128 (2005).
25. Lippert, G., Jurg, H. & Parinello, M. A hybrid gaussian and plane wave density functional scheme. *Mol. Phys.* **92,** 477–488 (1997).
26. Goedecker, S., Teter, M. & Hutter, J. Separable dual-space gaussian pseudopotentials. *Phys. Rev. B* **54,** 1703–1710 (1996).
27. Perdew, J. P., Burke, K. & Ernzerhof, M. Generalized gradient approximation made simple. *Phys. Rev. Lett.* **77,** 3865–3868 (1996).
28. Grimme, S., Antony, J., Ehrlich, S. & Krieg, H. A consistent and accurate ab initio parametrization of density functional dispersion correction (dft-d) for the 94 elements h-pu. *J. Chem. Phys.* **132,** 154104 (2010).
29. VandeVondele, J. & Hutter, J. Gaussian basis sets for accurate calculations on molecular systems in gas and condensed phases. *J. Chem. Phys.* **127,** 114105 (2007).

Acknowledgements

We gratefully acknowledge valuable discussions with P. Hapla and S. Hämäläinen regarding implementation of the L-J model, and important comments on the manuscript by J. Leaf. A.S. gratefully acknowledges the support of the Leverhulme Trust via fellowship ECF-2013-525. S.P.J. thanks the Engineering and Physical Sciences Research Council (EPSRC) for the award of fellowship EP/J500483/1, and The Leverhulme Trust for the award of fellowship ECF-2015-005, respectively. P.J.M. thanks EPSRC and the Leverhulme Trust, respectively, for Grants No. EP/G007837/1 and F00/114 BI. M.R.A. acknowledges funding from the University of Nottingham via the Vice-Chancellor's Scholarship for Research. P.R. received funding from the People Programme (Marie Curie Actions) of the European Union's Seventh Framework Programme (FP7/2007–2013) under REA grant agreement no. [628439]. We also acknowledge the support of the University of Nottingham High Performance Computing Facility (in particular, Dr Colin Bannister).

Author contributions

A.S. and P.M. conceived and designed the experiments. A.S. carried out the experiments and analysed the experimental data. A.S., M.A.R., J.L.D and S.P.J. designed and ran the L-J calculations. S.P.J. performed and analysed the DFT calculations. P.R. designed the atom tracking equipment and scripting. A.S. and P.M. wrote the paper. All authors read and commented on the final manuscript.

Additional information

Competing financial interests: The authors declare no competing financial interests.

21

A dilation-driven vortex flow in sheared granular materials explains a rheometric anomaly

K.P. Krishnaraj[1] & Prabhu R. Nott[1]

Granular flows occur widely in nature and industry, yet a continuum description that captures their important features is yet not at hand. Recent experiments on granular materials sheared in a cylindrical Couette device revealed a puzzling anomaly, wherein all components of the stress rise nearly exponentially with depth. Here we show, using particle dynamics simulations and imaging experiments, that the stress anomaly arises from a remarkable vortex flow. For the entire range of fill heights explored, we observe a single toroidal vortex that spans the entire Couette cell and whose sense is opposite to the uppermost Taylor vortex in a fluid. We show that the vortex is driven by a combination of shear-induced dilation, a phenomenon that has no analogue in fluids, and gravity flow. Dilatancy is an important feature of granular mechanics, but not adequately incorporated in existing models.

[1]Department of Chemical Engineering, Indian Institute of Science, Bangalore 560012, India. Correspondence and requests for materials should be addressed to P.R.N. (email: prnott@chemeng.iisc.ernet.in).

Flowing granular materials exhibit superficial similarities with fluids, but the details of their mechanical response are significantly different, as the micromechanics of grain interactions is quite distinct from that of fluid molecules[1,2]. The design of industrial processes and understanding of natural phenomena involving granular flows call for the development of reliable continuum models. Although several advances have been made in this direction[2–9], most models are confined to narrow-flow regimes, or leave out key features. For critical evaluation and refinement of available models, experimental measurement of the rheology is indispensable, for which the protocols of fluid rheometry may be usefully employed.

The rheological properties of fluids are most conveniently measured using devices that generate a class of simple shear flows[10]. The cylindrical Couette cell is such a device, but when it is used for granular materials, unexpected behaviour emerges. Recent studies that used this device for granular rheometry found a striking anomaly in the stress[11,12]: the vertical shear stress changes sign on shearing, and the magnitudes of all components of the stress increase roughly exponentially with depth. This behaviour is contrary to previous experiments[13,14] and the predictions of plasticity theories[8,13], which yield a fluid-like stress. After considering several plausible mechanisms, it was speculated[11,12] that an anisotropic microstructure[15,16] is the likely cause of the anomalous stress.

Here we show the cause to be one of the least-expected mechanisms—a single vortex spanning the entire granular column. Vortices arise in fluids too when the rotation rate of the inner cylinder exceeds a critical value, due to the centrifugal Taylor–Couette instability[17]; this instability has also been observed in a fluidized granular bed sheared at high rates[18]. We show that the vortex in a slowly sheared granular material is fundamentally different in its origin and manifestation.

Results

DEM simulations and experiments. To understand the cause of the anomalous stress, we conducted simulations using the discrete element method (DEM)[19,20] (see Methods). The simulations were complemented by video imaging the free surface of an experimental Couette device. In the simulations, the annular gap of a cylindrical Couette cell (Fig. 1) with smooth or rough walls is filled with spherical grains of density ρ_p and mean diameter d_p to a fill height H by two methods, raining and dense packing. The inner cylinder is then rotated at a constant angular speed Ω until a state of steady shear is reached. We varied Ω over a range that spans the quasistatic and lower end of the intermediate regimes, which are delineated by the Savage number Sa, defined as the ratio of the stress due to impulsive grain collisions to the total stress (Supplementary Note 1). Unless stated otherwise, the DEM results shown in the paper are for smooth walls, filling by raining and $Sa = 2 \times 10^{-6}$; however, the qualitative features of the kinematics and stress are independent of the wall roughness, the Savage number, the parameters in the grain contact model and the method of filling. In the experiments, the steady-state velocity profile at the free surface is determined by video imaging and particle image velocimetry (see Methods).

Validation of the DEM simulations. The results of our DEM simulations are first validated by comparison with experimental data. The kinematics of the azimuthal flow is compared with the data obtained from video imaging (Fig. 2a). The simulations reproduce the exponential decay of v_θ with distance from the inner cylinder, a feature that is characteristic of slow granular flows[21–23]. The mechanics is validated by comparing the stress at the outer cylinder with the data of Gutam *et al.*[12] (Fig. 2b,c). The simulations reproduce qualitatively all the features of the

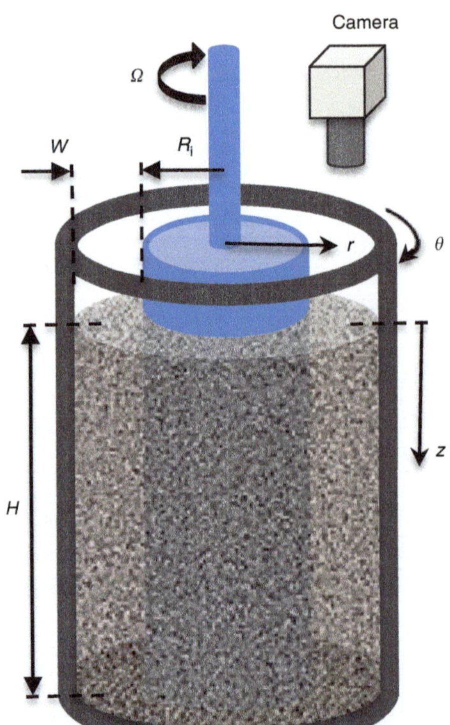

Figure 1 | Schematic diagram of the cylindrical Couette cell. The granular material is placed in the annular gap between the two coaxial cylinders of radii R_i and $R_i + W$, and sheared by rotating the inner cylinder at constant angular speed Ω. The DEM simulations are for a Couette cell of dimensions $R_i = 37\ d_p$, $W = 16\ d_p$, and $H = 30$, 60 and 90 d_p. The experiments use glass beads of mean diameter $d_p = 0.83$ mm and a Couette cell of dimensions $R_i = 72\ d_p$, $W = 18\ d_p$ and $H = 300\ d_p$.

anomalous stress observed in the experiments: in a static column, the shear stress σ_{rz} is positive (that is, the vertical component on the traction on the wall is downwards) and all components of the stress saturate asymptotically with the depth z, in accord with the classical Janssen solution[24,25] (Supplementary Note 2); on shearing, σ_{rz} changes sign and the normal stress σ_{rr} acquires a positive curvature. The quantitative differences between the simulation and experimental data may be partly due to the Couette cell in the simulations being much smaller, for computational tractability; moreover, we do not attempt to tune the parameters of the contact model to achieve agreement, as our primary aim is to gain a qualitative understanding of the cause of the anomalous stress. Nevertheless, our qualitative validation is useful because the features we verify are significant: the exponential decay of $v_\theta(r)$, a characteristic feature of slow granular flows, and the reversal in the sign of σ_{rz} on shearing, a key finding of our previous experimental studies.

Starting from an initial state generated by raining, shearing results in overall compaction, as seen from the fall of the free surface (Fig. 3a,b) and the profiles of the solids fraction (Fig. 3d), because raining yields a loosely packed initial state. The low-density layers adjacent to the two cylinders are due to the wall-imposed constraints on packing. Nevertheless, significant dilation in the shear layer is evident from the broadening of the low-density layer near the inner cylinder, in accord with numerous experimental observations[2,26]. The volume-fraction distributions at steady state for the two filling methods are almost identical (Fig. 3b–d), leading us to the conclusion that the initial preparation of the granular bed has little bearing on the properties at steady state.

The secondary vortex. Further examination of the kinematics reveals a secondary flow in the r-z plane, superimposed over the azimuthal flow. Although its velocity scale is small compared to that of the azimuthal flow, the radial and vertical velocities v_r and v_z are easily measurable, and show a systematic trend (Fig. 4a,b). The shapes of the v_r profiles at the free surface determined from experiment and DEM simulations are in good agreement. When the velocities at all the locations are combined to construct the streamlines, a single vortex that extends over the entire width and height of the Couette gap emerges (Fig. 4c; Supplementary Movie 1). Two aspects of the secondary flow are notable: The first is that a thin layer adjacent to the free surface flows radially inwards (Supplementary Movie 2). In contrast, the Taylor vortex in a fluid causes an outward radial flow near the free surface, if the fluid column is shallow enough to have only one vortex[27]. Second, there is always a single vortex for the range of fill heights H we have studied (Fig. 4c–e), whereas the size of Taylor vortices scales with W, resulting in a vertical train of counter-rotating vortices. The qualitative features of the vortex are independent of the Savage number and the wall roughness (Fig. 4e; Supplementary Fig. 1a–c).

We note that the free surface slopes downward from the outer to inner cylinders (Fig. 3b,c), a feature that is also observed in the experiments; while the time taken for the free surface slope and the secondary flow to reach steady state roughly coincide, our evidence is insufficient to infer whether the slope is caused by the secondary flow, or another aspect of the mechanics (such as normal stress differences[10]).

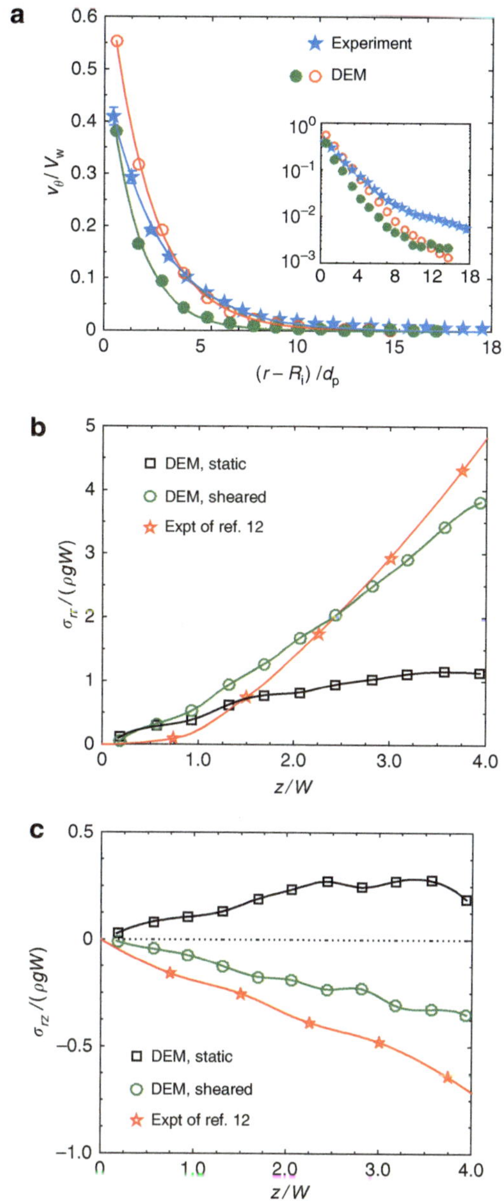

Figure 2 | Validation of the DEM simulations. (**a**) The azimuthal velocity v_θ (scaled by $V_w \equiv R_i\Omega$) at the free surface as a function of radial distance from the inner cylinder; the inset shows the same plot in linear-log scale. The green filled circles are for rough walls, and the red open circles are for smooth walls. The lines are exponential fits $v_\theta = v_0 \exp[-(r-R_i)/\beta]$ of the data, with $\beta = 2.5\ d_p$ (experiment), $2\ d_p$ (DEM, rough walls) and $1.5\ d_p$ (DEM, smooth walls). The error bars, indicating one standard deviation about the mean, are obtained from eight independent measurements; error bars for the DEM simulations are smaller than the size of the symbols. (**b**,**c**) Variation with depth of the normal stress and the vertical shear stress on the outer cylinder for static and sheared columns with rough walls; here ρ is the average bulk density in the Couette cell. The lines are cubic spline fits of the data, and the experimental data are from ref. 12. In all the panels, the DEM results are for $H = 90\ d_p$.

The vortex explains the anomalous stress. To show that the vortex flow explains the anomalous stress, we consider the simplest plasticity model for the stress tensor $\boldsymbol{\sigma}$[28,29],

$$\boldsymbol{\sigma} = p_c(\phi)(1-\mu_b\nabla\cdot\mathbf{v}/\dot{\gamma})^m\boldsymbol{\delta} - 2\mu_s p_c(\phi)\mathbf{D}/\dot{\gamma} \quad (1)$$

where $\mathbf{v}$ is the velocity vector, $\mathbf{D}$ the deviatoric part of the deformation rate tensor (with scalar norm $\dot{\gamma} \equiv [2\mathbf{D}:\mathbf{D}]^{1/2}$), $\boldsymbol{\delta}$ the identity tensor, and $p_c(\phi)$ is the pressure at the critical state[2,5]. The parameters in the model are the bulk and shear plastic moduli, μ_b and μ_s, and the exponent m. The above is a model for rate-independent plasticity if μ_b and μ_s are

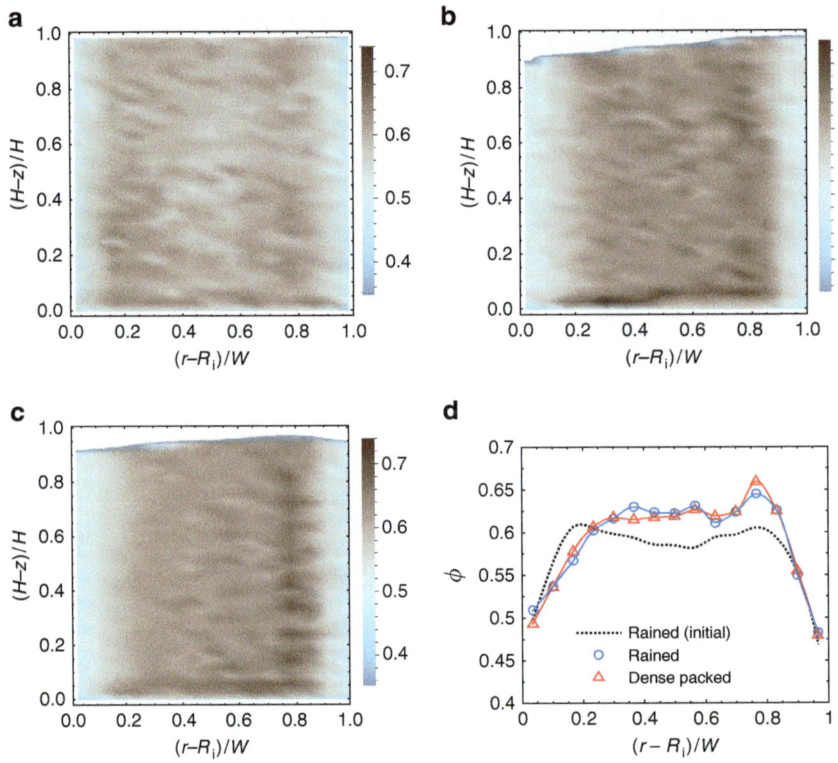

Figure 3 | Density distribution and dilation. (**a**-**c**) Distribution of the solids fraction ϕ before commencement of shear after filling by raining (**a**), at the sheared steady state after filling by raining (**b**), and at the sheared steady state after filling by dense packing (**c**). The colours indicate the value of ϕ. (**d**) The radial profile of ϕ at $z = 15\ d_p$. The results in all the panels are for fill height $H = 30\ d_p$ and rough walls.

independent of $\dot{\gamma}$, and rate-dependent plasticity otherwise[9,30]. Applying the model to the problem at hand, at the two cylinders impenetrability requires $v_r = 0$, whence $D_{rz} = \partial v_z/\partial r$. From the radial variation of v_z, we see that $\partial v_z/\partial r > 0$ at the outer cylinder (inset of Fig. 4b). Equation (1) then implies that $\sigma_{rz} < 0$, which is in accord with experimental observations[11,12] (Fig. 2c). Thus, the vortex explains the reversal in the sign of σ_{rz} when the granular column is sheared. This result, in conjunction with Coulomb friction at the boundaries, explains the sharp rise of all components of the stress with z[11,12] (Supplementary Note 2; Supplementary Fig. 2), and thereby all aspects of the anomalous stress. Although we have used a simple plasticity model to illustrate the link between the vortex and the anomalous stress, more elaborate models[2,29] lead to the same conclusion.

Mechanism for the secondary vortex. We now address the cause of the secondary flow. The sense of the vortex indicates that it is not centrifugal in origin, but for a definitive confirmation, we conduct a simulation of shear between two plane vertical walls (Fig. 5a). Figure 5b shows the secondary flow that results—two counter-rotating vortices placed symmetrically about the mid-plane of the Couette gap (see also Supplementary Movie 3). Symmetry arises here because the two walls are indistinguishable, unlike in cylindrical Couette flow. Thus, the vortices closely resembling those in Fig. 4c–e, arise in the complete absence of the centrifugal force.

What then drives the vortex? This question is answered by considering the transient evolution of the secondary flow soon after initiation of shear. The streamlines in Fig. 5c–e show that the flow is initially radially outwards and upwards, and concurrently, there is downward flow close to the inner cylinder. These two flows combine to curl the streamlines towards the inner cylinder at later time, eventually establishing a steady vortex (Fig. 4c) within a 90° rotation of the inner cylinder. When gravity is turned off and a frictionless wall used to confine the material from the top, we find no secondary vortex. It is now clear that the vortex is driven by the combination of shear-induced dilation and gravity flow—dilation causes the material to flow away from the shear layer and ultimately to the free surface, and the downward flow of grains near the inner cylinder sustains the vortex. The sense of the vortex is determined by the asymmetry in dilation between the inner and outer cylinders (Fig. 3), as a result of the shear rate of the primary flow decaying exponentially with radial distance from the inner cylinder (Fig. 2a).

We have used the transient evolution of the vortex to elucidate the mechanism, but it applies equally well at steady state. In the absence of gravity flow, dilation would cease after the initial transient, and a steady state would be reached wherein the flow is purely azimuthal and there is a radial density gradient. While gravity flow is essential to sustain the vortex, we surmise that it is initiated by dilation, as we cannot think of another physical mechanism that would cause a radial flow.

Importantly, we find the secondary vortex to be present even when the upper surface is not traction-free. Supplementary Figure 3 shows the streamlines at steady state for plane Couette flow when the material is confined at the top by a horizontal rigid plate of fixed weight, which is allowed to move vertically. The presence of two counter-rotating vortices is evident, as in the case of a traction-free surface (Fig. 5b), though their strength and symmetry decrease as the weight of the confining plate increases. We find the vortex to persist in cylindrical Couette flow too when confined at the top, and its basic structure is largely independent of the confinement condition and plate roughness (K.P.K., P. V. Dsouza, T. Murthy and P.R.N., manuscript in preparation).

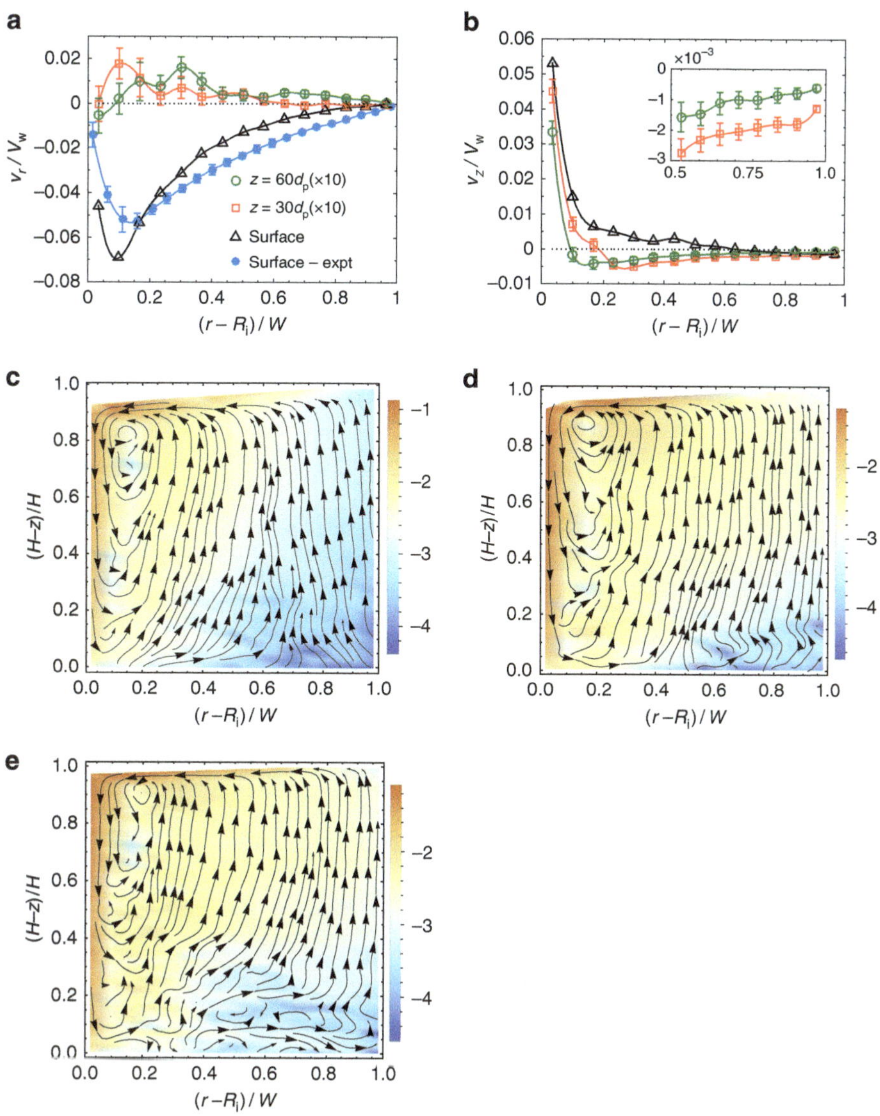

Figure 4 | The form of the secondary flow. (**a**,**b**) The velocities in the radial and vertical directions as a function of r at different z. The open symbols are DEM results for $H=90\ d_p$ and the filled circles are experimental data. The error bars are obtained in the manner described in the caption of Fig. 2. (**c**–**e**) Streamlines of the secondary flow at steady state for fill heights $H=30\ d_p$ (**c**), 60 d_p (**d**) and 90 d_p (**e**). The background colour indicates the value of $\log_{10} s$, where $s \equiv \left(v_r^2+v_z^2\right)^{1/2}/V_w$ is the magnitude of the secondary flow.

Thus, a traction-free surface is not essential for the formation of the vortex—the combined effects of dilation and gravity break the up–down symmetry, and determine the sense of the vortex.

Discussion

The dilation-driven secondary flow that we report is a novel phenomenon that has no analogue in fluids. Although the scale of the secondary flow is smaller than the primary azimuthal flow, its rheological signature is large, owing to the dependence of the shear stresses on the pressure (equation (1)). Evidence of a secondary flow was provided by two previous studies[31,32], the former hypothesizing dilatancy as a possible cause and the latter finding convection to vanish under microgravity conditions; however, they could not discern the detailed form of the secondary flow and its origin. Our results establish that there is a single vortex (for the range of fill heights we have explored), resulting from the combined effects of dilatancy and gravity flow.

It has been known for long that dilation accompanies deformation in dense granular materials, and, conversely, compaction when the material is loose[2,33]. This key feature distinguishes plastic deformation of granular materials from that of metals, and of course fluids. Existing rheological models for granular flows are inadequate for capturing compressibility; while density change along a streamline is predicted (such as in flow through a hopper), to our knowledge no model captures the shear-induced density gradient across streamlines[8]. It is usually assumed that the effect of dilatancy is confined to the (typically) narrow shear layer, but our study establishes that it can act as a driving force for a large-scale flow. Thus, despite some advances, much remains to be done towards formulating a robust and effective rheological model for granular materials—we expect that our study will provide a useful impetus in this direction.

The Taylor vortex in fluids arises from an instability of the base state of azimuthal flow, due to an imbalance between the centrifugal force and the radial pressure gradient[17]. It is quite possible that the granular vortex arises from a similar

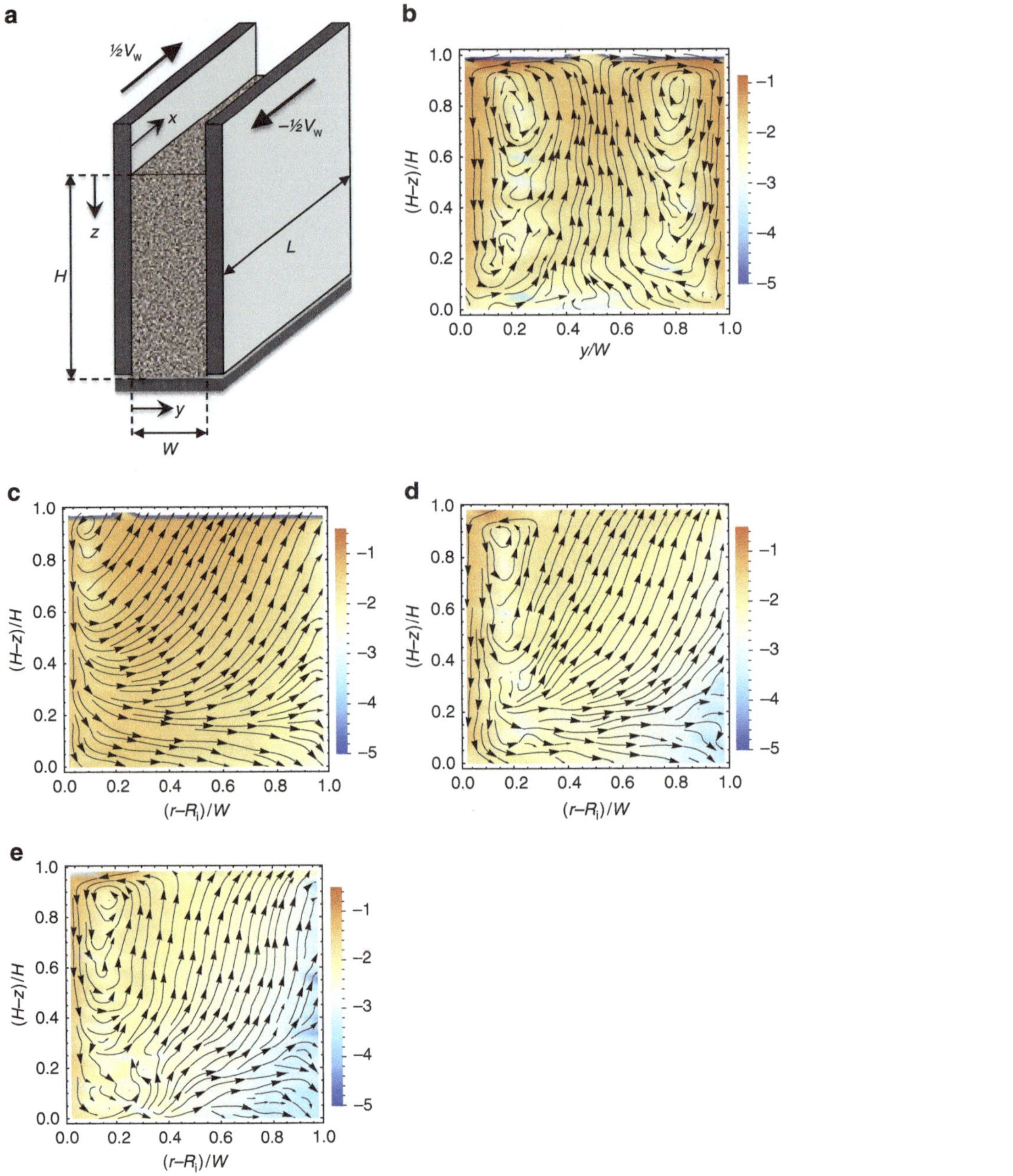

Figure 5 | The cause of the secondary vortex. (**a**) Schematic diagram of plane shear between vertical walls. (**b**) Streamlines of the secondary flow in plane shear. The unit cell of dimensions $W = 16\ d_p$, $H = 30\ d_p$ and $L = 50\ d_p$ is replicated periodically in the x direction. (**c–e**) Transient evolution of the secondary flow in a cylindrical Couette cell after initiation of shear: the streamlines after an inner cylinder rotation of 3.4° (**c**), 22° (**d**) and 45° (**e**). The background colour indicates the value of $\log_{10} s$, where $s \equiv (v_r^2 + v_z^2)^{1/2}/V_w$ is the magnitude of the secondary flow. In **c–e** the Couette cell is filled by the dense packing method to a fill height $H = 30\ d_p$ and the walls are rough; the velocity fields are averaged over the preceding 3.4° in (**c**) and 8° in (**d**,**e**).

hydrodynamic instability, with dilation as the driving force. However, a stability analysis must await the formulation of a rheological model that captures cross-streamline dilation. We finally note that the anomalous stress reported by our group[11,12] was then ascribed to microstructural anisotropy; while we now offer a more compelling explanation, there is sufficient evidence in the literature to motivate the incorporation of an anisotropic fabric in a comprehensive rheological model.

Methods

DEM simulations. DEM is a widely used computational tool for granular mechanics, where the positions and interactions of all the particles are tracked in time using simple models for grain interactions. We used the soft-particle contact model[19,20], wherein the grains are treated as deformable spheres, but rather than calculate their deformation in detail, they are allowed to overlap. The interaction forces are written in terms of the overlap and its time rate of change. The normal and tangential forces have elastic and viscous components, and the tangential force incorporates an additional Coulomb slider to capture rate-independent friction (Supplementary Fig. 4a). Considering the contact of grains i and j centred at position vectors $\mathbf{r}_i$ and $\mathbf{r}_j$, the normal and tangential forces on particle i are[34]

$$\mathbf{F}_n^{(i)} = k_n \delta \mathbf{n} - m_{\text{eff}} \gamma_n \mathbf{v}_n, \tag{2}$$

$$\mathbf{F}_t^{(i)} = \begin{cases} -k_t \Delta \mathbf{s} - m_{\text{eff}} \gamma_t \mathbf{v}_t & \text{if } \left| \mathbf{F}_t^{(i)} / \mathbf{F}_n^{(i)} \right| < \mu \\ -\mu \left| \mathbf{F}_n^{(i)} \right| \mathbf{v}_t / |\mathbf{v}_t| & \text{otherwise} \end{cases} \tag{3}$$

respectively. Here $\delta \equiv R_i + R_j - |\mathbf{r}_i - \mathbf{r}_j|$ is the overlap, $\mathbf{n}$ is the unit normal at the point of contact pointing towards the centre of i, $\mathbf{v}_n$ and $\mathbf{v}_t$ are the relative velocities at the point of contact in the normal and tangential directions, respectively, $\Delta\mathbf{s}$ is the tangential displacement and $m_{\text{eff}} \equiv (1/m_i + 1/m_j)^{-1}$ is the effective mass that determines the damping force. The spring stiffness constants k_n, k_t, the damping constants γ_n, γ_t and the Coulomb friction coefficient μ are the parameters that define the interaction. The motion of each particle is determined by integrating Newton's second law, assuming pairwise additivity of the interaction forces. The computationally intensive simulations, involving the tracking of $\sim 5 \times 10^5$ particles for durations corresponding to several rotations of the inner cylinder, were conducted using the LAMMPS package[35].

To avoid crystalline order, a mixture of grains of diameter 0.9 d_p, d_p and 1.1 d_p (of mass fractions 0.3, 0.4 and 0.3, respectively) was used. Simulations were conducted for two types of boundaries (cylinders and bottom wall): the surfaces were either smooth, or coated with a rigid, close-packed triangular lattice of grains of diameter 0.9 d_p; the two cases are referred to as smooth and rough walls, respectively. For both types of walls, the grain–wall interactions were computed by treating the walls as bodies of infinite mass, but with the same stiffness, damping and friction constants as for grain–grain interactions.

For values of the spring constants k_n and k_t that are appropriate for hard particles like sand and glass beads, accurately resolving each contact requires so small a time step of integration that the computation time becomes prohibitively high. It is therefore standard practice to optimize their values such that they are low enough for the computations to be tractable, but high enough that the macroscopic behaviour mimics that of hard particles. The values chosen for the simulations are $k_n = 10^6 m_p g/d_p$, $k_t = 2/7 k_n$, as per previous studies that have modelled hard grains[16,34]. Here m_p is the mass of the particle of diameter d_p and g is the gravitational acceleration. The damping coefficients were set to $\gamma_n = 317\, m_p \sqrt{g/d_p}$, $\gamma_t = 1/2\gamma_n$ and the friction coefficient to $\mu = 0.5$. The chosen value of γ_n yields a normal coefficient of restitution for a collision of 0.7 (ref. 16). To verify that the results do not depend qualitatively on the parameter values[36], simulations were conducted for three values of k_n varying over two decades. The profiles of the normal and vertical shear stress at the outer cylinder are shown in Supplementary Fig. 4b,c. Increasing k_n results in larger magnitudes of the stress components, but the qualitative features that characterize the stress anomaly remain independent of k_n.

The Couette cell was filled by two methods, raining and dense packing. In the former, the mixture of grains (of sizes mentioned above) was poured uniformly over the annular gap from a reservoir at the top of the Couette cell, until the fill height H was reached. In the latter, grains of uniform size 1.1 d_p were placed in a body-centred cubic lattice within the Couette cell up to a fill height H, in the absence of gravity; the grains were then randomly shrunk to achieve the required size distribution, after which gravity was turned on. The two methods gave an initial average volume fraction of $\langle\phi\rangle = 0.576$ and 0.602, respectively.

The continuum variables were obtained by averaging over space and time. The velocity and volume fraction at coordinates (r, z) were obtained by averaging over all the particles in an annular cylinder of thickness $\Delta r = d_p$ (centred at r) and depth $\Delta z = 3\, d_p$ (centred at z); the wall stresses were obtained by averaging over all particle–wall contacts in such a cylinder adjacent to the outer cylinder. The steady-state variables (Figs 2–4 and 5b) were time-averaged for half a period of rotation of the inner cylinder, but the transient velocity fields were averaged over much shorter durations (see caption of Fig. 5).

Experiments. The experiments used spherical beads of soda-lime glass of density $\rho_p = 2{,}500\,\text{kg m}^{-3}$ and a narrow size distribution with mean diameter $d_p = 0.83$ mm. The Couette apparatus (dimensions in caption of Fig. 1) was filled with the glass beads, and the inner cylinder rotated at 2 revolutions per minute (corresponding to a Savage number $Sa = 3.3 \times 10^{-6}$) for a period of about 30 min to reach steady state. Video images of the free surface were then acquired using a digital video camera (Prosilica GE680) at a rate of 200 frames per second and resolution 640 × 480 pixels (scale factor 0.112 mm per pixel). The instantaneous velocity field was determined by cross-correlation of the intensity maps (averaged over bins of 12 × 12 pixels) in successive frames using the PIVlab[37] toolbox in MATLAB. The steady-state velocity profile was determined by time-averaging the instantaneous velocity field for 30 s.

References

1. Jaeger, H. M., Nagel, S. R. & Behringer, R. P. Granular solids, liquids, and gases. *Rev. Mod. Phys.* **68,** 1259–1273 (1996).
2. Rao, K. K. & Nott, P. R. *An Introduction to Granular Flow* (Cambridge University Press, 2008).
3. Haff, P. K. Grain flow as a fluid-mechanical phenomenon. *J. Fluid Mech.* **134,** 401–430 (1983).
4. Lun, C. K. K., Savage, S. B., Jeffrey, D. J. & Chepurniy, N. Kinetic theories for granular flow: inelastic particles in Couette flow and slightly inelastic particles in a general flow field. *J. Fluid Mech.* **140,** 223–256 (1984).
5. Jackson, R. in *Theory of Dispersed Multiphase Flow.* (ed Meyer, R. E.) 291–337 (Academic Press, 1983).
6. Johnson, P. C. & Jackson, R. Frictional-collisional constitutive relations for granular materials, with application to plane shearing. *J. Fluid Mech.* **176,** 67–93 (1987).
7. Savage, S. B. Analyses of slow high-concentration flows of granular materials. *J. Fluid Mech.* **377,** 1–26 (1998).
8. Mohan, L. S., Rao, K. K. & Nott, P. R. A frictional Cosserat model for the slow shearing of granular materials. *J. Fluid Mech.* **457,** 377–409 (2002).
9. Jop, P., Forterre, Y. & Pouliquen, O. A constitutive law for dense granular flows. *Nature* **441,** 727–730 (2006).
10. Bird, R. B., Armstrong, R. C. & Hassager, O. *Dynamics of Polymeric Liquids* Vol. 1 (John Wiley, 1977).
11. Mehandia, V., Gutam, K. J. & Nott, P. R. Anomalous stress profile in a sheared granular column. *Phys. Rev. Lett.* **109,** 128002 (2012).
12. Gutam, K. J., Mehandia, V. & Nott, P. R. Rheometry of granular materials in cylindrical Couette cells: anomalous stress caused by gravity and shear. *Phys. Fluids* **25,** 070602 (2013).
13. Tardos, G., Khan, M. & Schaeffer, D. Forces on a slowly rotating, rough cylinder in a Couette device containing a dry, frictional powder. *Phys. Fluids* **10,** 335–341 (1998).
14. Tardos, G., McNamara, S. & Talu, I. Slow and intermediate flow of a frictional bulk powder in the Couette geometry. *Powder Technol.* **131,** 23–39 (2003).
15. Magnanimo, V. & Luding, S. A local constitutive model with anisotropy for ratcheting under 2d axial-symmetric isobaric deformation. *Granular Matter* **13,** 225–232 (2011).
16. Sun, J. & Sundaresan, S. A constitutive model with microstructure evolution for flow of rate-independent granular materials. *J. Fluid Mech.* **682,** 590–616 (2011).
17. Taylor, G. I. Stability of a viscous liquid contained between two rotating cylinders. *Philos. Trans. R. Soc. London Ser. A* **223,** 289–343 (1923).
18. Conway, S. L., Shinbrot, T. & Glasser, B. J. A Taylor vortex analogy in granular flows. *Nature* **431,** 433–436 (2004).
19. Cundall, P. A. & Strack, O. D. L. A discrete numerical model for granular assemblies. *Géotechnique* **29,** 47–65 (1979).
20. Walton, O. R. & Braun, R. L. Viscosity, granular temperature, and stress calculations for shearing assemblies of inelastic frictional disks. *J. Rheol.* **30,** 949–980 (1986).
21. Losert, W., Bocquet, L., Lubensky, T. C. & Gollub, J. P. Particle dynamics in sheared granular matter. *Phys. Rev. Lett.* **85,** 1428–1431 (2000).
22. Mueth, D. M. *et al.* Signatures of granular microstructure in dense shear flows. *Nature* **406,** 385–389 (2000).
23. Ananda, K. S., Moka, S. & Nott, P. R. Kinematics and statistics of dense, slow granular flow through vertical channels. *J. Fluid Mech.* **610,** 69–97 (2008).
24. Janssen, H. A. Versuche ueber Getreidedruck in Silozellen (experiments on grain pressure in silo). *Z. Ver. Deut. Ing.* **39,** 1045–1049 (1895).
25. Sperl, M. Experiments on corn pressure in silo cells—translation and comment of Janssen's paper from 1895. *Granular Matter* **8,** 59–65 (2006).
26. Kabla, A. J. & Senden, T. J. Dilatancy in slow granular flows. *Phys. Rev. Lett.* **102,** 228301 (2009).
27. Toya, Y., Nakamura, I., Yamashita, S. & Ueki, Y. An experiment on a Taylor vortex flow in a gap with a small aspect ratio bifurcation of flows in an asymmetric system. *Acta Mech.* **102,** 137–148 (1994).
28. Srivastava, A. & Sundaresan, S. Analysis of a frictional-kinetic model for gas-particle flow. *Powder Technol.* **129,** 72–85 (2003).
29. Nott, P. R. Classical and Cosserat plasticity and viscoplasticity models for slow granular flow. *Acta. Mech.* **205,** 151–160 (2009).
30. Elaskar, S. A., Godoy, L. A., Gray, D. D. & Stiles, J. M. A viscoplastic approach to model the flow of granular solids. *Int. J. Solids Struct.* **37,** 2185–2214 (2000).
31. Khosropour, R., Zirinsky, J., Pak, H. K. & Behringer, R. P. Convection and size segregation in a couette flow of granular material. *Phys. Rev. E* **56,** 4467–4473 (1997).
32. Murdoch, N. *et al.* Granular convection in microgravity. *Phys. Rev. Lett.* **110,** 018307 (2013).
33. Reynolds, O. On the dilatancy of media composed of rigid particles in contact. With experimental illustrations. *Phil. Mag.* **20,** 469–481 (1885).
34. Silbert, L. E. *et al.* Granular flow down an inclined plane: Bagnold scaling and rheology. *Phys. Rev. E* **64,** 051302 (2001).
35. Plimpton, S. Fast parallel algorithms for short-range molecular dynamics. *J. Comp. Phys.* **117,** 1–19 (1995).
36. Singh, A., Magnanimo, V., Saitoh, K. & Luding, S. The role of gravity or pressure and contact stiffness in granular rheology. *New J. Phys.* **17,** 043028 (2015).
37. Thielicke, W. & Stamhuis, E. Pivlab—towards user-friendly, affordable and accurate digital particle image velocimetry in matlab. *J. Open Res. Software.* **2** http://dx.doi.org/10.5334/jors.bl (2014).

Acknowledgements

The authors gratefully acknowledge Peter Varun Dsouza for assistance with imaging, and G. Subramnaian, J.D. Goddard and K. Kesava Rao for useful discussions. Funding for this work was provided by the Science and Engineering Research Board and the DST-ANR collaborative research programme. Thanks are also due to the anonymous reviewers whose comments helped in significantly improving the manuscript.

Author contributions

P.R.N. conceived the study, K.P.K. conducted the DEM simulations, P.R.N. and K.P.K. analysed the data and interpreted the results and P.R.N. along with K.P.K. prepared the manuscript.

Additional information

Competing financial interests: The authors declare no competing financial interests.

22

Large elasto-optic effect and reversible electrochromism in multiferroic $BiFeO_3$

D. Sando[1,*,†], Yurong Yang[2,*], E. Bousquet[3], C. Carrétéro[1], V. Garcia[1], S. Fusil[1], D. Dolfi[4], A. Barthélémy[1], Ph. Ghosez[3], L. Bellaiche[2] & M. Bibes[1]

The control of optical fields is usually achieved through the electro-optic or acousto-optic effect in single-crystal ferroelectric or polar compounds such as $LiNbO_3$ or quartz. In recent years, tremendous progress has been made in ferroelectric oxide thin film technology—a field which is now a strong driving force in areas such as electronics, spintronics and photovoltaics. Here, we apply epitaxial strain engineering to tune the optical response of $BiFeO_3$ thin films, and find a very large variation of the optical index with strain, corresponding to an effective elasto-optic coefficient larger than that of quartz. We observe a concomitant strain-driven variation in light absorption—reminiscent of piezochromism—which we show can be manipulated by an electric field. This constitutes an electrochromic effect that is reversible, remanent and not driven by defects. These findings broaden the potential of multiferroics towards photonics and thin film acousto-optic devices, and suggest exciting device opportunities arising from the coupling of ferroic, piezoelectric and optical responses.

[1] Unité Mixte de Physique, CNRS, Thales, Univ. Paris-Sud, Université Paris-Saclay, 91767 Palaiseau, France. [2] Department of Physics and Institute for Nanoscience and Engineering, University of Arkansas, Fayetteville, Arkansas 72701, USA. [3] Theoretical Materials Physics, Université de Liège, B-5, B-4000 Sart-Tilman, Belgium. [4] Thales Research and Technology France, 1 Avenue Augustin Fresnel, 91767 Palaiseau, France. * These authors contributed equally to this work. † Present address: School of Materials Science and Engineering, University of New South Wales, Sydney 2052, Australia. Correspondence and requests for materials should be addressed to D.S. (email: daniel.sando@unsw.edu.au) or to M.B. (email: manuel.bibes@thalesgroup.com).

Bismuth ferrite ($BiFeO_3$—BFO) is multiferroic at room temperature with strong ferroelectric polarisation[1] and G-type antiferromagnetic ordering with a cycloidal modulation of the Fe spins[2]. Most research on this material has been driven by the prospect of electrically controlled spintronic devices[3]. More recently, however, BFO has revealed further remarkable multifunctional properties. Notable discoveries include conductive domain walls[4], a strain-driven morphotropic phase boundary[5] and a specific magnonic response that can be tuned by epitaxial strain[6] or electric field[7]. Moreover, with a bandgap ($\sim$2.7 eV) in the visible[8], large birefringence[9] (0.25–0.3), a strong photovoltaic effect[10] and sizeable linear electro-optic coefficients[11], BFO is garnering interest in photonics and plasmonics[12].

Most of these physical properties are intimately linked to structural parameters, and may thus be tuned in thin films by epitaxial strain. Strain engineering[13] is a powerful tool through which, for instance, ferroelectricity is strongly enhanced in $BaTiO_3$ (ref. 14), or induced in otherwise non-ferroelectric materials such as $SrTiO_3$ (ref. 15). In BFO, two structural instabilities are sensitive to epitaxial strain: the polar distortion—responsible for the ferroelectricity—and antiferrodistortive (FeO_6 octahedra) rotations. In strained BFO films, the competition between both instabilities and their coupling to ferroic order parameters yields rich phase diagrams, revealing new structural, ferroelectric and magnetic phases[6,16], as well as large variations in the ferroelectric Curie temperature[17] and the spin direction[6].

Here, we present a combined experimental and theoretical study demonstrating that strain induces a very large change in the refractive index of BFO, which corresponds to an effective elasto-optic coefficient larger than in any ferroelectric, and larger than that of quartz[18]. This effect is accompanied by a shift of the optical bandgap, reminiscent of pressure-induced changes in light-absorption[19], a phenomenon known as piezochromism in other materials systems[20]. The trends in the optical properties as a function of strain are well reproduced by our first-principles calculations, and we are able to clarify precisely why the optical bandgap of tetragonal-like BFO is larger than that of the rhombohedral-like phase. Finally, we show how an electric field can be used to toggle between two strain states with different light absorption, corresponding to an electrochromic effect that is intrinsic, reversible and non-volatile.

Results

Sample preparation and structural characterization. Fully strained BFO thin films were grown using pulsed laser deposition on (001)-oriented substrates (in pseudocubic notation, which we use throughout this paper) spanning a broad range of lattice mismatch (from -7.0% to $+1.0\%$; Methods section). At low strain—compressive or tensile—the films crystallize in the so-called R-like phase of BFO, derived from the bulk rhombohedral (R3c) phase. At high compressive strain ($\leq 4\%$), the films grow in the T-like phase[16] with a large tetragonality ratio $c/a \approx 1.26$ (cf. Fig. 1b). Reciprocal space maps around the $(113)_{pc}$ or $(223)_{pc}$ reflections (Supplementary Fig. 1) reveal that all our BFO films possess a monoclinic structure (M_A or M_B for R-like, M_C for T-like[12,21]), and further scans (not shown) indicate the presence of two structural domain variants (see the sketch in Fig. 1a). The in-plane and out-of-plane pseudocubic lattice parameters are presented in Fig. 1b. The in-plane parameter shows a monotonic decrease with compressive strain, while the out-of-plane parameter concomitantly increases, albeit with a sharp jump at $\sim -3.5\%$ corresponding to the structural transition between the R-like and T-like phases.

First-principles calculations. To explore the effect of strain on the optical properties of BFO thin films, we performed first-principles calculations, using the Heyd–Scuseria–Ernzerhof (HSE) hybrid functional (see Methods section for details). In the following, we denote the electronic bandgap as the energy difference between the valence band maximum (VBM) and the conduction band minimum (CBM), while the optical bandgap corresponds to the extrapolation of the linear region of the Tauc plot (cf. Fig. 2c); theoretically, the optical bandgap is computed from the complex dielectric function.

Figure 1c,d show the computed electronic density of states for the R-like phase of BFO (at 0, $+2$ and -3% strain) and for the T-like phase (-5 and -7%). The insets of Fig. 1c,d show that the electronic bandgap is lower for the T-like phase than for the R-like phase, particularly for -7% strain, consistent with previous studies[22,23]. In the R-like phase, both compressive and tensile strains yield an increase of the electronic bandgap, similar to the situation in $SrTiO_3$ (ref. 24).

The partial density of states of Fig. 1e-l shows that the VBM mostly consists of O 2p orbitals for any considered misfit strain, and the CBM mainly comprises Fe d_{xy}, d_{xz}, d_{yz} orbitals for both the R-like and T-like phases. In R-like BFO, strain-induced changes in the FeO_6 octahedra rotations and the polar modes conspire[24] to slightly lift the degeneracy of the Fe 3d orbitals, but the nature of the electronic states at the VBM and CBM is globally preserved. In contrast, the pyramidal coordination of the FeO_5 unit in highly elongated T-like BFO yields a large splitting of the 3d states with the d_{xy} orbital sitting 300 meV lower in energy (Fig. 1h). Figure 1l shows that this d_{xy} state is weakly hybridized with O states and the states near the CBM in the T-like phase have very little O 2p character. This suggests that optical transitions from the VBM to those states should be very weak, and that the main optical transitions in the T-like phase should occur from the VBM to d_{xz} and d_{yz} states that lie in energy $\sim$300 meV above the CBM. In other words, the optical bandgap should be higher in T-like BFO than in R-like BFO despite the opposite trend in the electronic bandgap. This presumption is confirmed by the energy dependence of the extinction coefficient derived from our calculations (Fig. 2b): the absorption edge appears at least 200 meV higher in the T-like phase than in the R-like phase.

Optical characterization. Figure 2a presents the experimental energy dependence of the extinction coefficient, extracted from spectroscopic ellipsometry measurements (Supplementary Fig. 2). These data confirm the theoretical prediction of a larger optical bandgap for the T-like phase, and are consistent with previous studies. The agreement between the experimental and calculated extinction coefficient curves (Fig. 2a,b) is very good, particularly for the onset of absorption. The corresponding experimental Tauc plots (cf. Methods section) for these three samples are shown in Fig. 2c, indicating that the optical bandgap for T-like BFO is 3.02 eV, while for R-like BFO, a compressive strain of 2.6% induces an increase in the bandgap from 2.76 to 2.80 eV. Figure 2d summarizes the strain dependence of the experimental and calculated optical bandgap. In the R-like phase, both compressive and tensile strains induce an increase of the optical bandgap, and the T-like polymorph exhibits an optical bandgap $\sim$0.25 eV larger than the R-like phase, consistent with previous reports[25].

Our observation of a strain-induced change in optical bandgap and thus optical absorption is reminiscent of an effect called piezochromism[20], which corresponds to changes in light absorption driven by hydrostatic pressure. Piezochromic effects have been identified in several organic compounds, but for

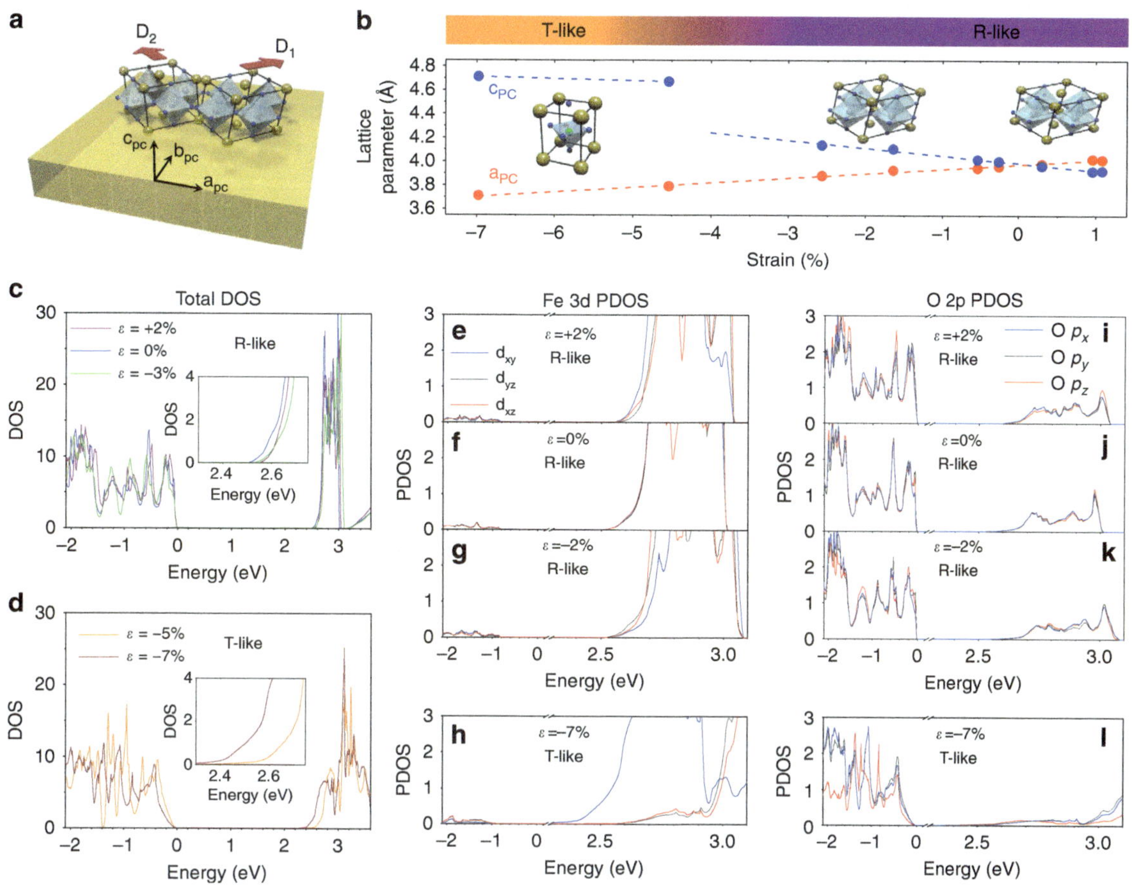

Figure 1 | Electronic structure of strained $BiFeO_3$ thin films. (**a**) Sketch of the two structural variants present in our monoclinic (M_A or M_B) R-like $BiFeO_3$ films. The red arrows indicate the direction of the monoclinic distortion for the two variants D_1 and D_2. (**b**) In-plane and out-of-plane lattice parameters of our strained BFO films. Total density of state (DOS) for R-like (**c**) and T-like phases (**d**). The insets show the DOS near the CBM. Partial density of states (PDOS) of iron 3d (**e-h**) and oxygen (**i-l**) states for R-like and T-like BFO. Note the break between 0.2 and 2.3 eV in the horizontal axes in (**e-l**). For all panels, only the spin-up channel states are shown; the spin-down channel states are the same as the spin up due to the antiferromagnetic order.

inorganics mainly in the $CuMoO_4$ family[26,27]. In this compound, the application of hydrostatic pressure triggers a first order transition between two polytypes having different optical absorption spectra due to changes in the oxygen cage surrounding the Cu ions[26]. Interestingly, in both $CuMoO_4$ and $BiFeO_3$ the absorption is stronger when the transition metal cation is in an octahedral oxygen environment, which suggests a possible trend, and strategies for engineering piezochromic effects in other perovskites.

In bulk BFO, the bandgap is known to decrease with pressure[19], particularly below 3.4 GPa and at the structural phase transition near 9.5 GPa. This corresponds to a piezochromic effect of amplitude 0.058 eV GPa^{-1} at low pressure, and of 0.027 eV GPa^{-1} on average between ambient pressure and 18 GPa (ref. 19). In our films, from the strain values and Young's modulus[28], we estimate the amplitude of the piezochromic effect in $BiFeO_3$ at $\sim$0.12 eV GPa^{-1}. Importantly, working with thin films may be advantageous for several applications[29]. For instance, the thin film geometry allows the application of large electric fields to toggle between two optical polytypes, thereby producing potentially high-speed electrochromic effects.

We have explored this possibility in BFO thin films with coexisting R-like and T-like regions[5]. We applied an electric field to transform the mixed R−T BFO into nominally pure T-like BFO over $10 \times 10\,\mu m^2$ regions (cf. Fig. 3a) and probed the local optical transmission in and out of this area using a conventional optical microscope. Figure 3b shows a transmission image with a dielectric filter (bandwidth 10 nm) centred at 420 nm inserted between the white light source and the sample. Clearly, a $10 \times 10\,\mu m^2$ square with a higher intensity than the background is visible in the image. Remarkably, this effect is reversible: applying a voltage with the opposite polarity restores a mixed R+T state (cf. Fig. 3c), which restores a stronger optical absorption, see Fig. 3d. This contrast is stable for several weeks.

We have acquired similar optical images using various dielectric filters, recorded the transmitted intensity in and out of the $10 \times 10\,\mu m^2$ square and calculated the contrast difference as a function of wavelength. The contrast is maximal between 420 and 450 nm, see Fig. 3e. This dependence agrees very well with the expected contrast difference, calculated from the extinction coefficients of pure R-like $BiFeO_3$ and pure T-like $BiFeO_3$ films of Fig. 2a. This confirms that the contrast in the optical images is indeed due to the intrinsic modulation of the optical bandgap induced by the electrical poling, rather than by defect-mediated processes as in Ca-doped BFO (ref. 30) or WO_3 (ref. 20).

Finally, we focus on the influence of strain on the real part of the complex refractive index n. In Fig. 4a, we highlight representative results of the variation of n with wavelength for R-like BFO that is weakly strained (on $SmScO_3$, SSO), strongly compressively strained (on (La,Sr)(Al,Ta)O_3 (LSAT)), and T-like BFO (on $LaAlO_3$ (LAO)). Below the optical bandgap (that is, for wavelengths longer than $\sim$460 nm), the refractive index systematically decreases with increasing strain. This is also visible in

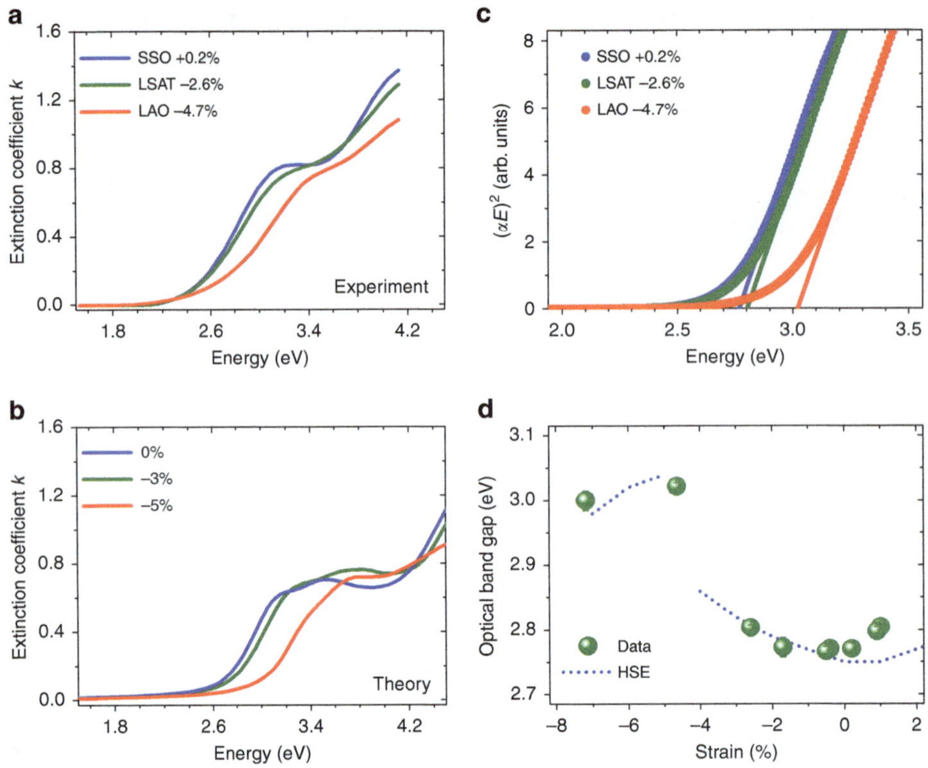

Figure 2 | Optical absorption properties of strained $BiFeO_3$ films. (**a**) Measured extinction coefficient for three representative strain levels. (**b**) Calculated extinction coefficient for strain levels comparable to those displayed in **a**. (**c**) Tauc plots generated from measurements for representative samples. (**d**) Summary of optical bandgap versus strain results, comparing theory and experiment. The error bars were determined by generating the dispersion laws using the upper and lower bounds of the Tauc-Lorentz oscillator parameters (from their uncertainties) and finding the resultant maximum variation in the bandgap.

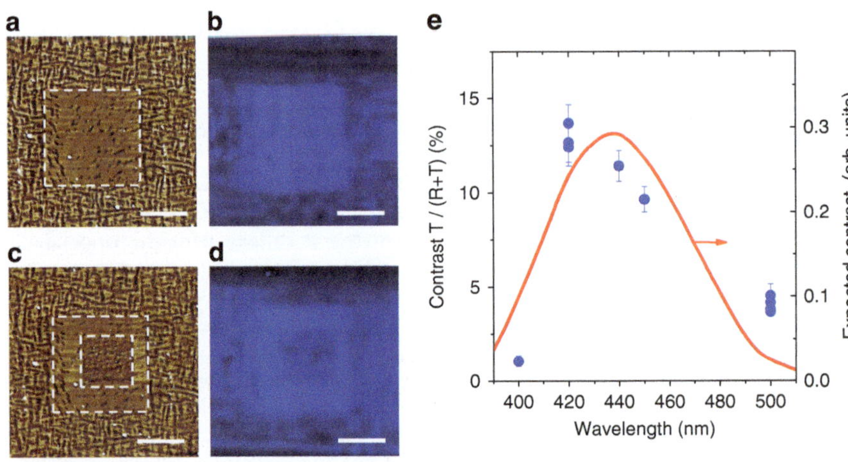

Figure 3 | Electrochromism in $BiFeO_3$ thin films. (**a**) Topography image after poling a $10 \times 10\,\mu m^2$ square, locally transforming R + T BFO into T-like BFO. (**b**) Transmission optical image acquired in the same region with a dielectric filter centred at 420 nm (bandwidth 10 nm). (**c**) Topography image of the same area after poling a $5 \times 5\,\mu m^2$ region with an opposite voltage, restoring the R + T structure. (**d**) Transmission optical image with a 420-nm filter. The horizontal dark features are due to twin boundaries in the $LaAlO_3$ substrate. All white scale bars are 5 μm. (**e**) Blue symbols: normalized difference in transmitted light in (T) and out (R + T) of the square in **a** with dielectric filters centred at different wavelengths. Red line: expected contrast calculated from the transmission of pure R-like and T-like films and the transmission function of the dielectric filters. The error bars in **e** are derived from the s.d. of the image pixel values in zones in and out of the T and (R + T) regions.

Fig. 4b, which displays the strain dependence of n at various wavelengths for all samples. The strain-induced change in refractive index measured at 633 nm is reproduced in Fig. 4c and compared with first-principles calculations. The refractive index is higher in the R-like phase and globally decreases with strain, both compressive and tensile.

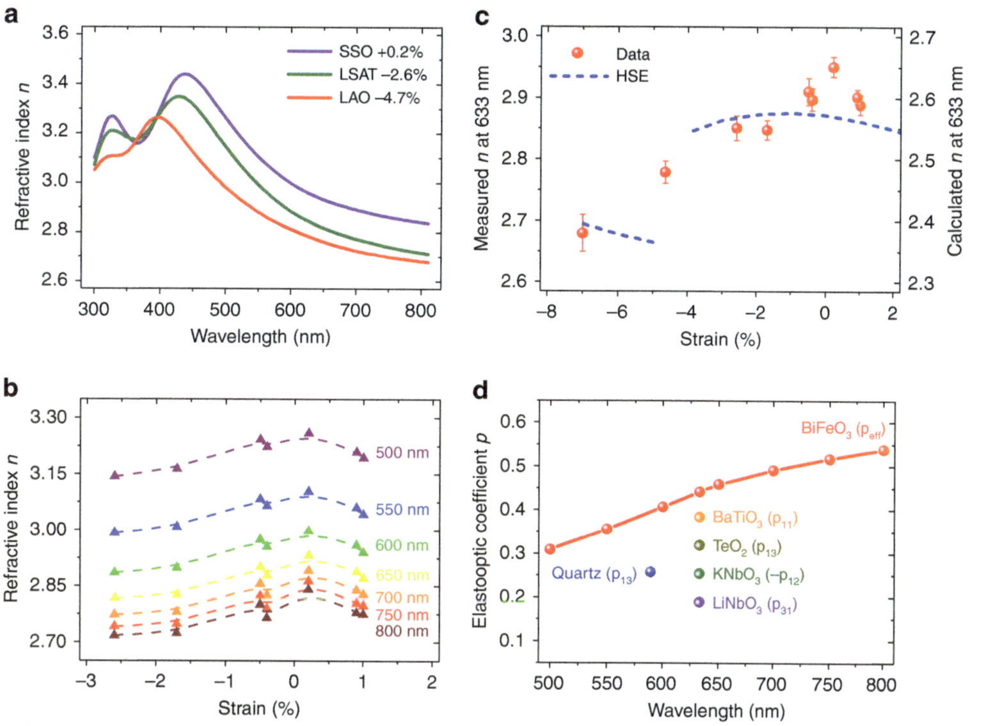

Figure 4 | Optical refractive index and elasto-optic coefficients in strained $BiFeO_3$ films. (**a**) Measured refractive index *n* as a function of wavelength for various strain levels. (**b**) Measured refractive index as a function of strain for various wavelengths, for the R-like phase only. The lines serve as guides to the eye. (**c**) Measured and calculated refractive index at 633 nm as a function of strain. The error bars were determined by generating the dispersion laws using the upper and lower bounds of the Tauc-Lorentz oscillator parameters (from their uncertainties) and finding the resultant maximum variation in the refractive index. (**d**) Effective elasto-optic coefficient of BFO as a function of wavelength. Representative reported largest elasto-optic coefficients of various other crystalline materials are plotted for comparison.

Discussion

The experimental results of Fig. 4 indicate that BFO exhibits a strong elasto-optic effect (change in refractive index on physical strain). Taking the slope of the change in $1/n^2$ with strain (Supplementary Fig. 3) for the weakly compressively strained BFO samples at various wavelengths larger than the BFO bandgap (at ~460 nm), we obtain an effective elasto-optic coefficient for BFO, as shown in Fig. 4d (Supplementary Note 1). In this figure we also plot the elasto-optic tensor element with the largest magnitude for various elasto-optic media[18,31]. The results suggest that BFO should be a robust elasto-optic medium and, more specifically, that BFO has an effective elasto-optic coefficient at least twice as large as $LiNbO_3$.

Combined with its relatively low Young's modulus[28] and sound speed[32], the large elasto-optic coefficient of BFO yields an acousto-optic figure-of-merit[18] $M=\frac{n^6p^2}{Ev_{ac}}$ as large as $M=365\times10^{-15}\,s^3\,kg^{-1}$, a value much larger than for any other material for longitudinal acoustic waves (cf. in TeO_2, $M=23\times10^{-15}\,s^3\,kg^{-1}$ and in $LiNbO_3$ $M=1.8\times10^{-15}\,s^3\,kg^{-1}$, ref. 18). Importantly, as BFO may easily be grown[33] on isostructural perovskites with giant piezoelectric responses (such as PMN-PT or PZN-PT)[34], this huge figure-of-merit opens the way towards thin film acousto-optical components[35] with potential performance orders of magnitude greater than those currently based on single crystals. This would extend the potential of BFO to devices to deflect or modulate optical fields, and towards the emerging field of optomechanics, from back-action and laser cooling, to highly integrated sensors and frequency references[36].

More generally, our work has implications for the design of multifunctional devices exploiting the magnetic, ferroelectric or piezoelectric response of $BiFeO_3$ in conjunction with these unique optical properties. Importantly, the mechanisms that we identify to modulate bandgap and absorption are not specific to $BiFeO_3$ and can be transposed to many perovskites. Strain-induced elasto-optic and piezochromic effects even larger than in $BiFeO_3$, possibly by one order of magnitude or more, could be awaiting discovery in other oxide materials, particularly in Mott insulators[37] in which the bandgap falls between *3d* states and is lower than in $BiFeO_3$. Giant, electric field-controllable optical absorption in the visible range could thus be exploited, opening the way towards devices harvesting both mechanical and solar energies.

Methods

$BiFeO_3$ thin film growth and structural characterization. Single phase R-like or T-like films of BFO were prepared by pulsed laser deposition (using the conditions of ref. 38) on the following single crystal substrates: $YAlO_3$, LAO, LSAT, $SrTiO_3$, $DyScO_3$, $TbScO_3$, $SmScO_3$, $NdScO_3$ and $PrScO_3$. The scandates and $YAlO_3$ were (110)-oriented (orthorhombic notation) while cubic $SrTiO_3$, LSAT and rhombohedral LAO were (001)-oriented. The nominal biaxial strain induced by these substrates ranges from −7.0% (compressive) to +1.0% (tensile). The thickness of the films was 50–70 nm as determined by X-Ray reflectometry, and confirmed by spectroscopic ellipsometry measurements. High-angle X-Ray diffraction 2θ–θ scans, collected with a Panalytical Empyrean diffractometer using $CuK_{\alpha1}$ radiation, indicated that the films were epitaxial and grew in a single phase. Mixed R+T phase BFO (nominal thickness 100 nm) were grown using a KrF excimer laser at 540 °C and 0.36 mbar on (001)-oriented LAO substrates after the growth of a 10-nm-thick $LaNiO_3$ bottom electrode at 640 °C and 0.36 mbar.

Optical characterization. The films were characterized using spectroscopic ellipsometry with a UVISEL spectral-scanning near infrared spectroscopic phase-modulated ellipsometer from HORIBA Jobin-Yvon. The incidence angle was 70° and the wavelength range was 300–840 nm (0.62–4.13 eV). This range was imposed by the ellipsometer (maximum energy ~4.13 eV), while absorption peaks arising from colour centres in the scandate substrates limited the maximum wavelength to 840 nm. These boundaries do not adversely affect the present analysis since the optical bandgap of BFO is well within the explored spectral range. The raw ellipsometry data were fitted to a multilayer model consisting a semi-infinite substrate, BFO layer and roughness layer implemented by the Bruggeman approximation with a void and BFO mixture. The dispersion law of the BFO layer was described by three Tauc–Lorentz oscillators[8], where the central energies correspond to charge transfer transitions. An example of a typical fit (in this case for BFO on $NdScO_3$) is shown in Supplementary Fig. 2b. The complex dispersion law ($\tilde{n} = n + ik$) of BFO was determined for each sample, and for all fits the mean square error, χ^2, was <2 (as indicated in Supplementary Table 1). To extract the bandgap from the dispersion laws, Tauc plots of $(\alpha E)^2$ versus E were constructed, and the linear region was extrapolated to the E axis (Fig. 2c), yielding the gap value. For each sample, ellipsometric data were collected and the dispersion law and bandgap calculated a minimum of four times, and the results averaged. The error bars displayed in Figs 2d and 4c were determined by generating the dispersion laws using the upper and lower bounds of the Tauc–Lorentz oscillator parameters (from their uncertainties) and finding the resultant maximum variation in the bandgap and refractive index.

First-principles calculations. Calculations were performed within density-functional theory, as implemented in the Vienna *ab initio* simulation package[39,40]. An energy cutoff of 550 eV was used, and the set of projector-augmented wave potentials was employed to describe the electron–ion interaction. We considered the following valence electron configuration: $5d^{10}6s^26p^3$ for Bi, $3p^63d^64s^2$ for Fe and $2s^22p^4$ for O. Supercells containing 20 atoms were used, and G-type antiferromagnetism was adopted. Electronic relaxations converged within 10^{-6} eV and ionic relaxation was performed until the residual force was <1 meV Å^{-1}. We used the PBEsol + U functional[41] (selecting $U = 4$ eV for the Fe ions) to relax the structures, and used both this PBEsol + U functional and the HSE hybrid functional[42] to calculate physical properties such as electronic structure and the dielectric function. These two methods yielded very similar results (hence we only report results for HSE in Figs 1, 2 and 4), with the exception that the PBEsol + U functional underestimated the electronic bandgap by 0.4 eV, while HSE overestimated this bandgap by 0.8 eV. The imaginary part of the dielectric tensor was obtained via

$$\varepsilon''_{\alpha\beta}(\omega) = \frac{4\pi^2 e^2}{\Omega} \lim_{q\to 0} \frac{1}{q^2} \sum_{c,v,\mathbf{k}} 2\omega_{\mathbf{k}} \delta(\epsilon_{c\mathbf{k}} - \epsilon_{v\mathbf{k}} - \omega) \times \langle u_{c\mathbf{k}+\mathbf{e}_\alpha q} \mid u_{v\mathbf{k}} \rangle \langle u_{c\mathbf{k}+\mathbf{e}_\beta q} \mid u_{v\mathbf{k}} \rangle^*, \quad (1)$$

where the indices c and v refer to conduction and valence band states, respectively, $u_{c\mathbf{k}}$ is the cell periodic part of the orbitals at the k-point $\mathbf{k}$, and $\mathbf{e}_\alpha$ is a unit vector along the α Cartesian direction[43]. Finally, the real part of the dielectric tensor $\varepsilon'_{\alpha\beta}$ was obtained through the Kramers–Kronig transformation $\varepsilon'_{\alpha\beta}(\omega) = 1 + \frac{2}{\pi} \mathrm{P} \int_0^\infty \frac{\varepsilon''_{\alpha\beta}(\omega')\omega'}{\omega'^2 - \omega^2} d\omega'$, where P denotes the principal value. We then obtained the extinction coefficient k and refractive index n by $\tilde{\varepsilon} = \varepsilon' + i\varepsilon'' = (n + ik)^2$. Note that local field effects were neglected in our calculations. The optical bandgap determined from the calculated dielectric function was seen to overestimate the experiment by 0.8 eV; therefore, in all figures in this manuscript, the conduction band has been systematically shifted by 0.8 eV with respect to the VBM to reflect this scissors correction.

For the refractive index we find a systematic quantitative difference of ~ 0.3 between experiment and theory, which can be understood by the fact that first-principles calculations consider defect-free samples, neglect local field and temperature effects, and only incorporate the average between the different components of the dielectric function tensor (which is, additionally, a quantity rather difficult to simulate precisely by *ab initio* methods).

References

1. Lebeugle, D. *et al.* Room-temperature coexistence of large electric polarization and magnetic order in $BiFeO_3$ single crystals. *Phys. Rev. B* **76,** 024116 (2007).
2. Sosnowska, I., Peterlin-Neumaier, T. & Steichele, E. Spiral magnetic ordering in bismuth ferrite. *J. Phys. C Solid State Phys.* **15,** 4835–4846 (1982).
3. Heron, J. T. *et al.* Deterministic switching of ferromagnetism at room temperature using an electric field. *Nature* **516,** 370–373 (2014).
4. Seidel, J. *et al.* Conduction at domain walls in oxide multiferroics. *Nat. Mater.* **8,** 229–234 (2009).
5. Zeches, R. J. *et al.* A strain-driven morphotropic phase boundary in $BiFeO_3$. *Science* **326,** 977–980 (2009).
6. Sando, D. *et al.* Crafting the magnonic and spintronic response of $BiFeO_3$ films by epitaxial strain. *Nat. Mater.* **12,** 641–646 (2013).
7. Rovillain, P. *et al.* Electric-field control of spin waves at room temperature in multiferroic $BiFeO_3$. *Nat. Mater.* **9,** 975–979 (2010).
8. Allibe, J. *et al.* Optical properties of integrated multiferroic $BiFeO_3$ thin films for microwave applications. *Appl. Phys. Lett.* **96,** 182902 (2010).
9. Rivera, J.-P. & Schmid, H. On the birefringence of magnetoelectric $BiFeO_3$. *Ferroelectrics* **204,** 23–33 (1997).
10. Choi, T., Lee, S., Choi, Y. J., Kiryukhin, V. & Cheong, S.-W. Switchable ferroelectric diode and photovoltaic effect in $BiFeO_3$. *Science* **324,** 63–66 (2009).
11. Sando, D. *et al.* Linear electro-optic effect in multiferroic $BiFeO_3$ thin films. *Phys. Rev. B* **89,** 195106 (2014).
12. Sando, D., Barthélémy, A. & Bibes, M. $BiFeO_3$ epitaxial thin films and devices: past, present and future. *J. Phys. Condens. Matter* **26,** 473201 (2014).
13. Schlom, D. G. *et al.* Elastic strain engineering of ferroic oxides. *MRS Bull.* **39,** 118–130 (2014).
14. Choi, K. J. *et al.* Enhancement of ferroelectricity in strained $BaTiO_3$ thin films. *Science* **306,** 1005–1008 (2004).
15. Haeni, J. H. *et al.* Room-temperature ferroelectricity in strained $SrTiO_3$. *Nature* **430,** 758–761 (2004).
16. Béa, H. *et al.* Evidence for room-temperature multiferroicity in a compound with a giant axial ratio. *Phys. Rev. Lett.* **102,** 217603 (2009).
17. Infante, I. C. *et al.* Bridging multiferroic phase transitions by epitaxial strain in $BiFeO_3$. *Phys. Rev. Lett.* **105,** 057601 (2010).
18. Dieulesaint, E. & Royer, D. *Elastic Waves in Solids II* (Springer-Verlag Berlin Heidelberg, 2000).
19. Gómez-Salces, S. *et al.* Effect of pressure on the band gap and the local FeO_6 environment in $BiFeO_3$. *Phys. Rev. B* **85,** 144109 (2012).
20. Bamfield, P & Hutchings, M. G. *Chromic Phenomena: Technological Applications of Colour Chemistry.* 2nd edn (Royal Society of Chemistry, 2010).
21. Vanderbilt, D. & Cohen, M. H. Monoclinic and triclinic phases in higher-order Devonshire theory. *Phys. Rev. B* **63,** 094108 (2001).
22. Ju, S. & Cai, T. Ab initio study of ferroelectric and nonlinear optical performance in $BiFeO_3$ ultrathin films. *Appl. Phys. Lett.* **95,** 112506 (2009).
23. Dong, H., Liu, H. & Wang, S. Optical anisotropy and blue-shift phenomenon in tetragonal $BiFeO_3$. *J. Phys. D Appl. Phys.* **46,** 135102 (2013).
24. Berger, R. F., Fennie, C. J. & Neaton, J. B. Band gap and edge engineering via ferroic distortion and anisotropic strain: the case of $SrTiO_3$. *Phys. Rev. Lett.* **107,** 146804 (2011).
25. Chen, P. *et al.* Optical properties of quasi-tetragonal $BiFeO_3$ thin films. *Appl. Phys. Lett.* **96,** 131907 (2010).
26. Rodríguez, F., Hernández, D., Garcia-Jaca, J., Ehrenberg, H. & Weitzel, H. Optical study of the piezochromic transition in $CuMoO_4$ by pressure spectroscopy. *Phys. Rev. B* **61,** 16497–16501 (2000).
27. Gaudon, M. *et al.* Unprecedented 'one-finger-push'-induced phase transition with a drastic color change in an inorganic material. *Adv. Mater.* **19,** 3517–3519 (2007).
28. Redfern, S. A. T., Wang, C., Hong, J. W., Catalan, G. & Scott, J. F. Elastic and electrical anomalies at low-temperature phase transitions in $BiFeO_3$. *J. Phys. Condens. Matter* **20,** 452205 (2008).
29. Greenberg, C. B. Optically switchable thin films: a review. *Thin Solid Films* **251,** 81–93 (1994).
30. Seidel, J. *et al.* Prominent electrochromism through vacancy-order melting in a complex oxide. *Nat. Commun.* **3,** 799 (2012).
31. Weber, M. J. *Handbook of Optical Materials* (CRC Press, 2003).
32. Smirnova, E. P. *et al.* Acoustic properties of multiferroic $BiFeO_3$ over the temperature range 4.2–830 K. *Eur. Phys. J. B* **83,** 39–45 (2011).
33. Biegalski, M. D. *et al.* Strong strain dependence of ferroelectric coercivity in a $BiFeO_3$ film. *Appl. Phys. Lett.* **98,** 142902 (2011).
34. Park, S.-E. & Shrout, T. R. Ultrahigh strain and piezoelectric behavior in relaxor based ferroelectric single crystals. *J. Appl. Phys.* **82,** 1804–1811 (1997).
35. Lean, E. G. H., White, J. M. & Wilkinson, C. D. W. Thin-film acoustooptic devices. *Proc. IEEE* **64,** 779–788 (1976).
36. Aspelmeyer, M., Kippenberg, T. J. & Marquardt, F. Cavity optomechanics. *Rev. Mod. Phys.* **86,** 1391–1452 (2014).
37. Imada, M., Fujimori, A. & Tokura, Y. Metal-insulator transitions. *Rev. Mod. Phys.* **70,** 1039–1263 (1998).
38. Béa, H *et al.* Influence of parasitic phases on the properties of $BiFeO_3$ epitaxial thin films. *Appl. Phys. Lett.* **87,** 072508 (2005).
39. Kresse, G. & Furthmüller, J. Efficient iterative schemes for ab initio total-energy calculations using a plane-wave basis set. *Phys. Rev. B* **90,** 11169–11186 (1996).
40. Kresse, G. & Hafner, J. Ab initio molecular dynamics for liquid metals. *Phys. Rev. B* **47,** 558–561 (1993).
41. Perdew, J. *et al.* Restoring the density-gradient expansion for exchange in solids and surfaces. *Phys. Rev. Lett.* **100,** 136406 (2008).
42. Krukau, A. V., Vydrov, O. A., Izmaylov, A. F. & Scuseria, G. E. Influence of the exchange screening parameter on the performance of screened hybrid functionals. *J. Chem. Phys.* **125,** 224106 (2006).
43. Gajdoš, M., Hummer, K., Kresse, G., Furthmüller, J. & Bechstedt, F. Linear optical properties in the projector-augmented wave methodology. *Phys. Rev. B* **73,** 045112 (2006).

Acknowledgements

This work was supported by the French Research Agency (ANR) projects 'Méloïc', 'Nomilops' and 'Multidolls,' the European Research Council Advanced Grant 'FEMMES'

(Contract No. 267579) and the European Research Council Consolidator Grant 'MINT' (Contract No. 615759). Y.Y. and L.B. thank the financial support of ONR Grant No N00014-12-1-1034 and DARPA Grant No. HR0011-15-2-0038 (Matrix program), and the Arkansas High Performance Computer Center for the use of its supercomputers. Ph.G. acknowledges a Research Professorship from the Francqui Foundation, financial support of the ARC project 'AIMED' and F.R.S.-FNRS PDR project ' HiT4FiT' as well as access to Céci-HPC facilities funded by F.R.S.-FNRS (Grant No 2.5020.1) and the Tier-1 supercomputer of the Fédération Wallonie- Bruxelles funded by the Walloon Region (Grant No 1117545). This work was also supported (E.B.) by F.R.S.-FNRS Belgium, and calculations were partly performed within the PRACE projects TheoMoMuLaM and TheDeNoMo. We thank J.-L. Reverchon for assistance with the ellipsometry measurements.

Author contributions

M.B. conceived and supervised the study with the help of A.B. and D.D. D.S. and C.C. grew the samples and characterized them with X-ray diffraction. D.S. performed ellipsometry measurements and analysed the data. V.G., S.F. and M.B. characterized the electrochromic response. Y.Y, E.B., Ph.G. and L.B. performed first-principles calculations. D.S. and M.B. wrote the manuscript with input from all authors.

Additional information

Competing financial interests: The authors declare no competing financial interests.

Magnon spectrum of the helimagnetic insulator Cu_2OSeO_3

P.Y. Portnichenko[1], J. Romhányi[2], Y.A. Onykiienko[1], A. Henschel[3], M. Schmidt[3], A.S. Cameron[1], M.A. Surmach[1], J.A. Lim[1], J.T. Park[4], A. Schneidewind[5], D.L. Abernathy[6], H. Rosner[3], Jeroen van den Brink[7] & D.S. Inosov[1]

Complex low-temperature-ordered states in chiral magnets are typically governed by a competition between multiple magnetic interactions. The chiral-lattice multiferroic Cu_2OSeO_3 became the first insulating helimagnetic material in which a long-range order of topologically stable spin vortices known as skyrmions was established. Here we employ state-of-the-art inelastic neutron scattering to comprehend the full three-dimensional spin-excitation spectrum of Cu_2OSeO_3 over a broad range of energies. Distinct types of high- and low-energy dispersive magnon modes separated by an extensive energy gap are observed in excellent agreement with the previously suggested microscopic theory based on a model of entangled Cu_4 tetrahedra. The comparison of our neutron spectroscopy data with model spin-dynamical calculations based on these theoretical proposals enables an accurate quantitative verification of the fundamental magnetic interactions in Cu_2OSeO_3 that are essential for understanding its abundant low-temperature magnetically ordered phases.

[1] Institut für Festkörperphysik, TU Dresden, Helmholtzstraße 10, D-01069 Dresden, Germany. [2] Max Planck Institute for Solid State Research, Heisenbergstraße 1, D-70569 Stuttgart, Germany. [3] Max Planck Institute for Chemical Physics of Solids, Nöthnitzer Straße 40, D-01187 Dresden, Germany. [4] Heinz Maier-Leibnitz Zentrum (MLZ), TU München, Lichtenbergstraße 1, D-85747 Garching, Germany. [5] Jülich Centre for Neutron Science (JCNS), Forschungszentrum Jülich GmbH, Outstation at Heinz Maier-Leibnitz Zentrum (MLZ), Lichtenbergstraße 1, D-85747 Garching, Germany. [6] Quantum Condensed Matter Division, Oak Ridge National Laboratory (ORNL), Oak Ridge, Tennessee 37831, USA. [7] Leibniz Institute for Solid State and Materials Research, IFW Dresden, Helmholtzstraße 20, D-01069 Dresden, Germany. Correspondence and requests for materials should be addressed to D.S.I. (email: dmytro.inosov@tu-dresden.de).

Chiral magnets form a broad class of magnetic materials in which long-range dipole interactions, magnetic frustration, or relativistic Dzyaloshinskii–Moriya (DM) interactions twist the initially ferromagnetic spin arrangement, thus leading to the formation of noncollinear incommensurate helical structures with a broken space-inversion symmetry[1–4]. Helical magnetic states occur in a very broad range of compounds including metals and alloys, semiconductors and multiferroics. From a fundamental point of view, the significance of such materials is due to the richness of their possible magnetic structures as compared to ordinary (commensurate and collinear) magnets. Perhaps the most prominent example is given by the topologically non-trivial long-range ordered states known as skyrmion lattices[5–7].

The interest in the multiferroic ferrimagnet Cu_2OSeO_3 with a chiral crystal structure has been intensified in recent years after the discovery of a skyrmion-lattice phase in this system[8], thus making it the first known insulator that exhibits a skyrmion order. It was shown soon afterwards that this skyrmion lattice can be manipulated by the application of either magnetic[9] or electric[10] field, and more recently also by chemical doping[11], which indicates a delicate balance of the magnetic exchange and DM interactions with the spin anisotropy effects. Understanding the complex magnetic phase diagram of Cu_2OSeO_3 therefore requires detailed knowledge of the spin Hamiltonian with precise quantitative estimates of all interaction parameters, which can be obtained from measurements of the spin-excitation spectrum. Microscopic theoretical models that were recently proposed for the description of spin arrangements in Cu_2OSeO_3 include five magnetic exchange integrals and five anisotropic DM couplings between neighbouring $S=\frac{1}{2}$ copper spins[12–16], yet their experimental verification was so far limited to thermodynamic data[14], terahertz electron spin resonance (ESR)[15], far-infrared[17] and Raman spectroscopy[18] that can only probe zone-centre excitations in reciprocal space. Here we present the results of inelastic neutron scattering (INS) measurements that reveal the complete picture of magnetic excitations in Cu_2OSeO_3 accessible to modern neutron spectroscopy over the whole Brillouin zone and over a broad range of energies and demonstrate good quantitative agreement with spin-dynamical model calculations both in terms of magnon dispersions and dynamical structure factors.

Results

Crystal structure and Brillouin zone unfolding. The cubic copper(II)-oxoselenite Cu_2OSeO_3 crystallizes in a complex chiral structure with 16 formula units per unit cell (space group $P2_13$, lattice constant $a=8.925$ Å)[19]. This structure has been visualized, for instance, in refs 8,14. For the discussion of magnetic properties relevant for our study, it is useful to consider only the magnetic sublattice of copper ions, which can be approximated by a face-centred cubic (fcc) lattice of identical Cu_4 tetrahedra, as shown in Fig. 1a. This reduces the volume of the primitive unit cell fourfold with respect to the original simple-cubic structure, resulting in only four Cu atoms per primitive cell. In reciprocal space, this simplified lattice possesses a larger 'unfolded' Brillouin zone with a volume of $4(2\pi/a)^3$ as shown in Fig. 1b with blue lines.

Existing theoretical models[13–15] suggested that individual Cu_4 tetrahedra represent essential magnetic building blocks of Cu_2OSeO_3. Because of the structural inequivalence of the copper sites within every tetrahedron, its magnetic ground state orients one of the four Cu^{2+} spins antiparallel to three others that are coupled ferromagnetically, resulting in a state with the total spin $S=1$. The interactions between neighbouring tetrahedral clusters are at least 2.5 times weaker than intra-tetrahedron couplings[13–15] and lead to a ferromagnetic arrangement of their total spins below the Curie temperature, $T_C=57$ K. In addition, weak DM interactions twist the resulting ferrimagnetic state into an incommensurate helical spin structure propagating along the ⟨001⟩ direction with a pitch $\lambda_h=63$ nm, that is 100 times larger than the distance between nearest-neighbour tetrahedra, which corresponds to the propagation vector $k_h=0.010$ Å^{-1} derived from small-angle neutron-scattering data[9].

As the next step, we note that the same structure can be represented as a half-filled fcc lattice of individual Cu atoms. Indeed, if one supplements the lattice with an equal number of imaginary Cu_4 tetrahedra by placing them within the voids of the original structure, as shown in Fig. 1a, a twice smaller fcc unit cell with lattice parameter $a/2$ can be introduced (red dashed lines). Its primitive cell contains on average one half of a copper atom and corresponds to the large 'unfolded' Brillouin zone with a volume of $32(2\pi/a)^3$ that is shown with red lines in Fig. 1b. As will be seen in the following, the dynamical structure factor

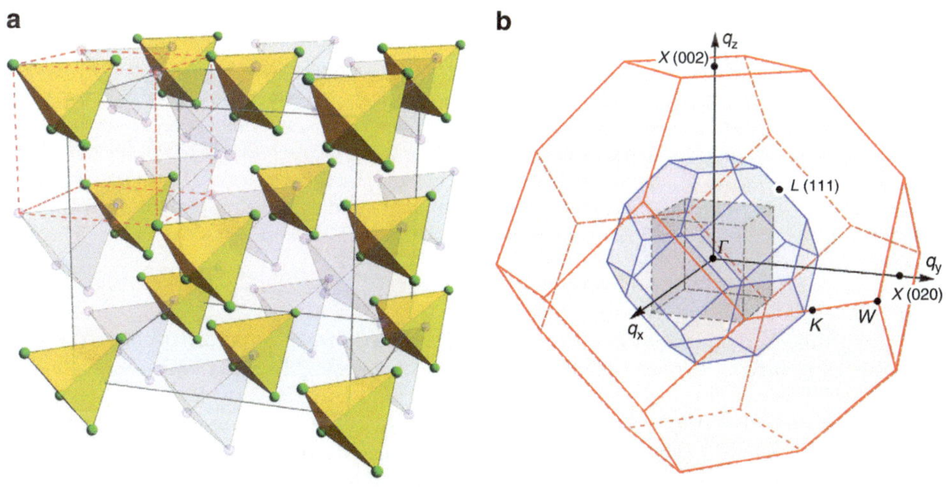

Figure 1 | Simplified structure of the magnetic Cu sublattice in Cu_2OSeO_3 and Brillouin zone unfolding. (**a**) The sublattice of Cu atoms approximated by an fcc arrangement of identical tetrahedral Cu_4 clusters, the corresponding unit cell shown with grey lines. Introduction of imaginary tetrahedra (faded colour) within the voids allows to introduce a twice smaller structural unit cell (dashed red lines) that corresponds to a half-filled fcc lattice of Cu atoms. (**b**) Brillouin zones that correspond to the original crystallographic simple-cubic unit cell with the lattice parameter a (inner grey cube), the fcc unit cell with lattice parameter a (smaller truncated octahedron, blue lines) and the fcc unit cell with lattice parameter $a/2$ (larger truncated octahedron, red lines).

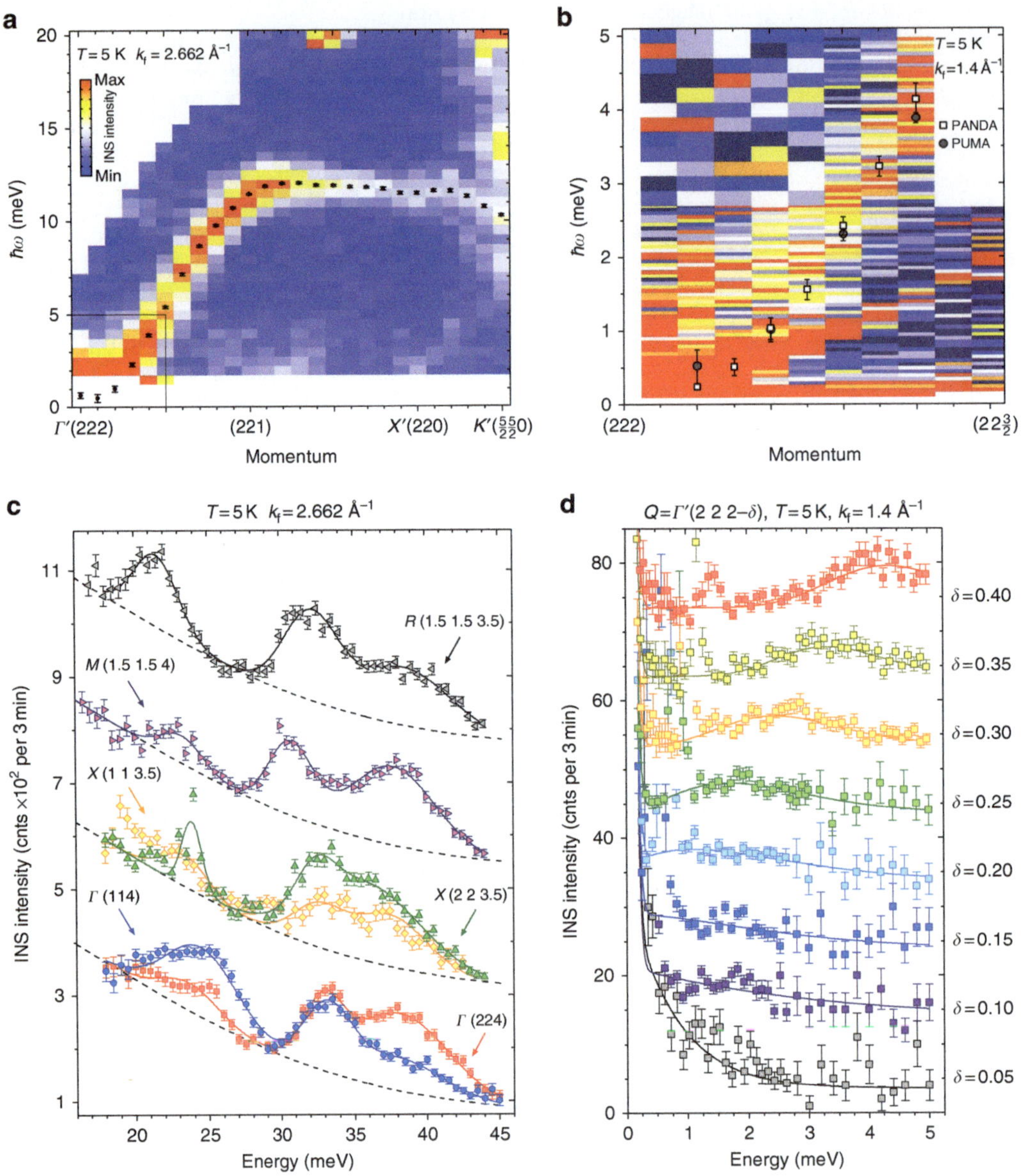

Figure 2 | Magnon excitations in Cu_2OSeO_3 as measured by triple-axis neutron spectroscopy. All the presented data were measured in the low-temperature spin-spiral state ($T=5$ K). (**a**) Dispersion of the low-energy ferromagnon branch along the (00*H*) direction mapped out with the thermal-neutron spectrometer PUMA. Data points represent peak positions that resulted from fitting the energy profiles for every momentum. (**b**) The lower part of the same dispersion within the region shown by a black rectangle in **a** measured with higher resolution at the cold-neutron spectrometer PANDA. The magnon corresponds to the higher-energy broadened peak (squares), the sharp weaker peaks below it originate from spurious Bragg scattering. The peak positions from the thermal-neutron data in **a** are also shown here for comparison (grey circles). (**c**) Raw energy scans across the upper magnon bands measured at several high-symmetry points using the PUMA spectrometer. Here the spectra are grouped and marked according to the small (original crystallographic) cubic Brillouin zone. (**d**) Individual energy cuts from the cold-neutron dataset in **b** for different momenta along (2 2 2-δ). The fits shown with solid lines neglect the spurious peaks at the low-energy side of the spectrum. The error bars in all panels represent 1 s.d. of the count rate.

representing the distribution of magnon intensities in Cu_2OSeO_3 inherits the symmetry of this large Brillouin zone. The described unfolding procedure is therefore useful for reconstructing the irreducible reciprocal-space volume for the presentation of INS data. Throughout this paper, we will denote momentum space coordinates in reciprocal lattice units corresponding to the crystallographic simple-cubic unit cell ($1\ \text{r.l.u.} = 2\pi/a$), whereas high-symmetry points will be marked in accordance to the large unfolded Brillouin zone as explained in Fig. 1b. Equivalent points from higher Brillouin zones will be marked with a prime.

Triple-axis neutron spectroscopy. We start the presentation of our experimental results with the triple-axis spectroscopy (TAS) data shown in Fig. 2. At low energies, the spectrum is dominated by an intense Goldstone mode with a parabolic dispersion, which emanates from the $\Gamma'(222)$ wave vector and has the highest intensity at this point, as can be best seen from the thermal-neutron data in Fig. 2a. Because this point corresponds to the centre of the large unfolded Brillouin zone, we can associate this low-energy mode with a ferromagnetic spin-wave branch anticipated from the theory for the collinear magnetically ordered state[13]. Upon moving away from the zone centre along the (001)

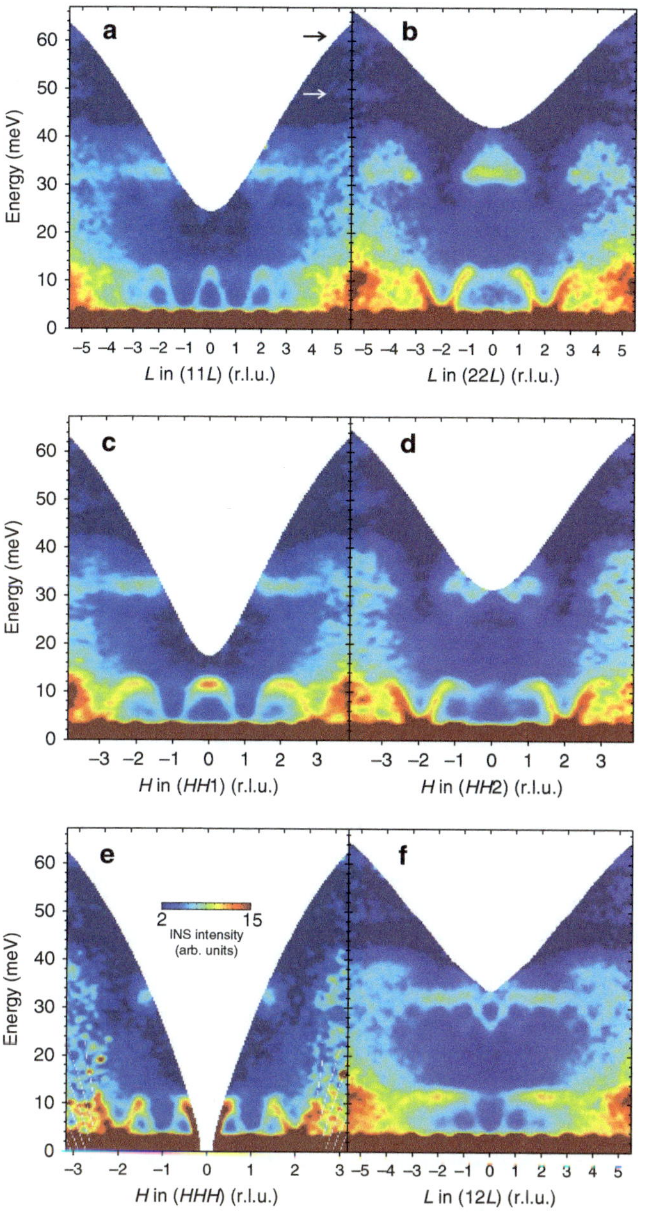

Figure 3 | Energy-momentum cuts through the TOF data. The colour maps show INS intensity distributions along several representative high-symmetry directions: (**a**) (11*L*); (**b**) (22*L*); (**c**) (*HH*1); (**d**) (*HH*2); (**e**) (*HHH*); and (**f**) (12*L*). All the presented data were measured at the base temperature of 4.5 K. Black and white arrows mark two high-energy optical phonon modes.

direction, its dispersion reaches a maximum of ~12 meV, while its intensity is reduced continuously towards the unfolded-zone boundary. Much weaker replicas of the same ferromagnon branch can be also recognized at other points with integer coordinates, like (221) or (220), as they coincide with zone centres of the original crystallographic Brillouin zone but not the unfolded one. We have fitted individual energy scans in Fig. 2a with Gaussian peak profiles to extract the measured magnon dispersion shown as black dots.

For a closer examination of the low-energy part of the ferromagnon branch around the $\Gamma'(222)$ point, we also performed TAS measurements at a cold-neutron spectrometer providing higher energy resolution, as shown in Fig. 2b,d. Here we see a parabolically dispersing ferromagnetic spin-wave branch (squares) in good agreement with the dispersion obtained from the thermal-neutron measurements (grey circles). The data are additionally contaminated with spurious Bragg scattering that appears as a sharp straight line below the magnon peak in Fig. 2b or as a strong low-energy peak in Fig. 2d. While fitting the experimental data to extract the magnon dispersion (solid lines in Fig. 2d), this spurious peak had to be masked out. As a result, the ferromagnon mode appears to be gapless within our experimental resolution, which agrees with the earlier microwave absorption measurements, where the spin gap value of only 3 GHz (12 μeV) has been reported[20]. By fitting the experimental dispersion at low energies with a parabola, $\hbar\omega = Dq^2$, where q is the momentum transfer along (001) measured relative to the (222) structural Bragg reflection, we evaluated the spin-wave stiffness $D = 52.6\,\text{meV}\,\text{Å}^2$, which is comparable to that in the prototypical metallic skyrmion compound MnSi (ref. 21).

The magnetic Goldstone mode is significantly broadened in both energy and momentum and has a large intrinsic width that is considerably exceeding the instrumental resolution of ~0.14 meV. This can be naturally explained as a result of the incommensurability of the spin-spiral structure and its deviation from the collinear ferrimagnetic order. In the helimagnetic state, the low-energy magnetic excitations are expected to split into so-called helimagnons emanating from the first-, second- and higher-order magnetic Bragg reflections[22–24]. The incommensurability parameter k_h is too small in our case to be resolved with a conventional triple-axis spectrometer. The typical helimagnon energy scale in Cu_2OSeO_3 can be estimated as $Dk_h^2 \approx 10.5\,\mu\text{eV}$, that is an order of magnitude lower than in MnSi and also lies well below the energy resolution of the instrument. As a result, multiple unresolved helimagnon branches merge into a single broadened peak as seen in Fig. 2d.

According to the theoretical calculations[13], a higher-energy band of dispersive magnon excitations is expected between ~25 and 40 meV. In Fig. 2c, we show several spectra measured in this energy range at a number of high-symmetry points that reveal at least three distinct peaks. Here we have grouped the spectra according to their location in the original crystallographic Brillouin zone (grey cube in Fig. 1b) to emphasize that the peak intensities are strongly affected by the dynamical structure factors, whereas their positions in energy in different Brillouin zones remain essentially the same for equivalent wave vectors. The theory anticipates that these high-energy modes represent an intricate tangle of multiple magnon branches, so that several of them may contribute to every experimentally observed peak. A direct comparison with the existing spin-dynamical calculations, therefore, appears difficult with the limited TAS data and requires a complete mapping of the momentum–energy space using time-of-flight (TOF) neutron spectroscopy that we present below.

Time-of-flight neutron spectroscopy. The benefit of TOF neutron scattering is that it provides access to the whole four-dimensional energy–momentum space ($\hbar\omega$, **Q**) in a single measurement, which is particularly useful for mapping out complex magnetic dispersions that persist over the whole Brillouin zone. The data can be then analysed to extract any arbitrary one- or two-dimensional cut from this data set as we describe in the Methods. Figure 3 presents several typical energy–momentum cuts taken along different high-symmetry lines parallel to the (001), (110) and (111) crystallographic axes. One clearly sees both the low-energy ferromagnon band that is most intense around $\Gamma'(222)$ and the intertwined higher-energy dispersive magnon modes between 25 and 40 meV whose intensity anticorrelates with that of the low-energy modes: it vanishes near $\Gamma'(222)$ yet is maximized near the $X'(220)$ point

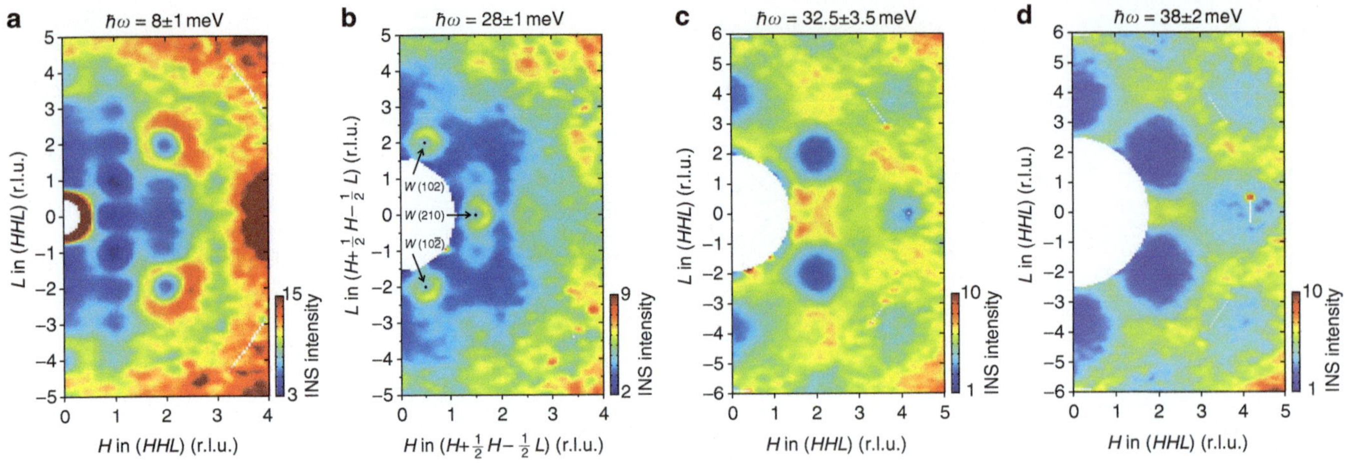

Figure 4 | Constant-energy slices through the TOF data parallel to the (*HHL*) plane. (**a**) Along the (*HHL*) plane at $\hbar\omega = 8 \pm 1$ meV cutting through the low-energy magnon band with intense rings of scattering from the ferromagnon modes around Γ points; (**b**) Along the offset $(H-\frac{1}{2}\ H+\frac{1}{2}\ L)$ plane at $\hbar\omega = 28 \pm 1$ meV through the bottom part of the upper magnon band, showing rings of intensity around *W* points; (**c,d**) Along the (*HHL*) plane through the upper magnon band at $\hbar\omega = 32.5 \pm 3.5$ meV and 38 ± 2 meV, respectively. In all panels, the data were integrated within ± 0.15 r.l.u. from the indicated plane along the $(1\bar{1}0)$ direction that is orthogonal to the plane of the figure. Corresponding energy integration ranges given above each panel are also shown in Fig. 5a with vertical bars labelled A-D.

at the unfolded-zone boundary, where the low-energy modes are weak. The low- and high-energy modes are separated by a broad energy gap in the range $\sim$13–25 meV. In addition, at much higher energies one can see two relatively weak flat modes centred around 49 and 61 meV (marked with arrows in Fig. 3a). Although the theory predicts a flat magnetic mode around 54 meV (ref. 13), the fact that their intensity monotonically increases towards higher $|\mathbf{Q}|$ speaks rather in favour of optical phonons. It is likely that the 54 meV magnetic mode is far too weak compared with the phonon peaks and therefore cannot be clearly seen in our data. As the data in Fig. 3 extend over several equivalent Brillouin zones in momentum space, one can see an increase of the nonmagnetic background and a simultaneous decrease of the magnetic signal towards higher $|\mathbf{Q}|$, resulting in a rapid reduction of the signal-to-background ratio. Therefore, in the following we restrict our analysis mainly to the irreducible part of the reciprocal space corresponding to the shortest possible wave vectors between $\Gamma(000)$ and $\Gamma'(222)$.

It is interesting to note how the hierarchical structure of Brillouin zones introduced in Fig. 1 is directly reflected in the magnon intensities. Along the (*HHH*) direction in Fig. 3e, the strongest spin-wave modes are seen around centres of the large unfolded zones, $\Gamma(000)$ and $\Gamma'(222)$, while a somewhat weaker mode appears at the $L(111)$ point that coincides with the zone centre for the smaller fcc Brillouin zone (shown in blue in Fig. 1b). Other points with integer coordinates that are centres of the smallest simple-cubic zones but of none of the larger ones, such as (110) or (001), contain only much weaker replica bands that are barely seen in the data (for example, Figs 2a and 3a,c). This leads to the appearance of rather unusual features in certain cuts through the spectrum, such as the one at the (001) point in the middle of Fig. 3c, which visually resembles the bottom part of the famous hourglass-like dispersion in some transition-metal oxides[25–27]. However, here its intense top part stems from the nearest unfolded-zone centre $\Gamma(000)$ that lies outside of the image plane, while the weaker downward-dispersing branches originate from the L points at $(11\bar{1})$ and (111), producing such a peculiar shape. The situation is even more complicated at higher energies, where the strongest well-resolved downward-pointing modes with a bottom around 26 meV are located at *W* points of the large unfolded zone, such as (102), (210) or (320), with weaker replicas shifted by the (111) or equivalent vector. Interestingly, without Brillouin zone unfolding such a particularity of the *W* points would not at all be obvious without explicit calculations of the dynamical structure factors, as they are all equivalent to the zone centre in the original crystallographic Brillouin zone notation.

The pattern of INS intensity in momentum space is visualized by the constant-energy slices presented in Fig. 4. All of them are oriented parallel to the (*HHL*) plane, which was horizontal in our experimental geometry. The low-energy cut integrated around 8 ± 1 meV (Fig. 4a) passes through the origin and contains both the (222) and (111) wave vectors, where rings of scattering from the stronger and weaker magnetic Goldstone modes, respectively, can be clearly seen. The next cut at 28 ± 1 meV in Fig. 4b is shifted by the $(\frac{1}{2}\frac{\bar{1}}{2}0)$ vector in such a way that it crosses the previously mentioned downward-pointing high-energy modes, which appear as small circles around *W* points. Finally, the last two panels (Fig. 4c,d) show integrated intensity in the (*HHL*) plane at energies corresponding to the middle (32.5 ± 3.5 meV) and top (38 ± 2 meV) of the upper magnon band, respectively. While individual magnon dispersions are intertwined at these energies and cannot be resolved separately, in both panels we observe deep minima of intensity at the $\Gamma'(222)$, $\Gamma''(400)$ and equivalent wave vectors. They correspond to the suppression of the dynamical structure factor at the centre of the unfolded Brillouin zone, where the low-energy ferromagnon mode, on the contrary, is most intense.

To get a more complete picture of magnetic excitations in Cu_2OSeO_3 and to compare them with the results of spin-dynamical calculations, we have symmetrized the TOF data (see Methods) and assembled energy–momentum profiles along a polygonal path involving all high-symmetry directions in **Q**-space into a single map presented in Fig. 5a. Due to the kinematic constraints, the data in the first Brillouin zone are always limited to lower energies as can be seen in Fig. 3, but have the best signal-to-noise ratio. Therefore, for the best result we also underlayed data from higher Brillouin zones to show the higher-energy part of the spectrum at equivalent wave vectors in the unfolded notation.

Spin-dynamical calculations. Based on previous theoretical and experimental results[13–15], we used a tetrahedron-factorized

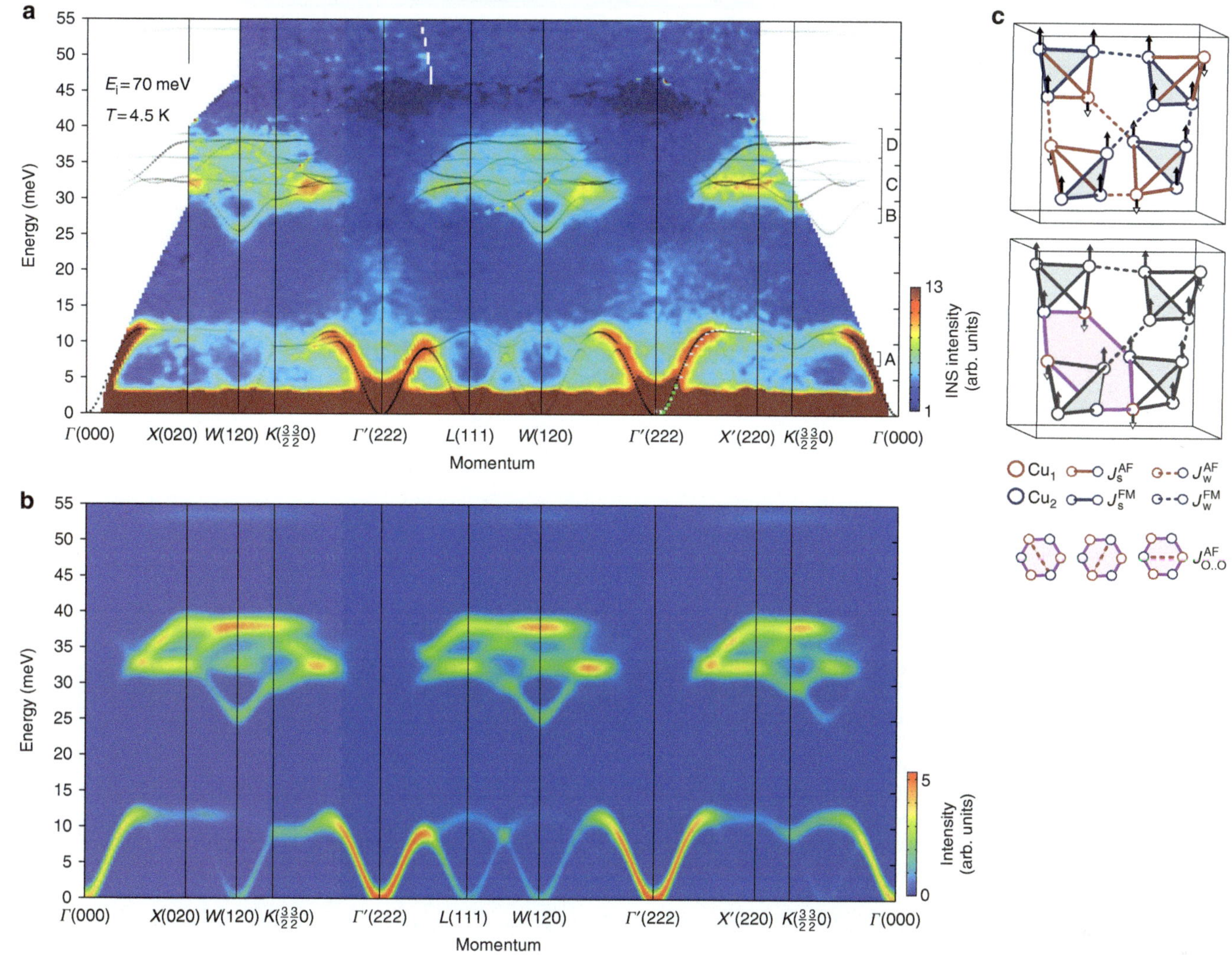

Figure 5 | Symmetrized TOF data along high-symmetry directions. (**a**) Energy-momentum profile of the symmetrized low-temperature TOF data ($T=4.5$ K) along a polygonal path involving most of the high-symmetry directions in **Q**-space. The data within every segment have been averaged with all equivalent cuts in the same Brillouin zone. The peak positions extracted from the fits to the thermal- and cold-neutron TAS data between the $\Gamma'(222)$ and $X'(220)$ points are shown here with white and green data points. The vertical bars marked as A–D on the right-hand side of the panel show energy integration ranges of the corresponding constant-energy cuts in Fig. 4. Dotted lines show the calculated dispersions, their transparency reflecting the dynamical structure factors (calculated intensities) as described in the text. (**b**) Calculated scattering intensity presented as a colour map that was obtained by broadening the results of our spin-dynamical calculations with a Gaussian function meant to model the experimental resolution. (**c**) Exchange paths in Cu_2OSeO_3. Top: First-neighbour ferromagnetic (blue) and antiferromagnetic (red) couplings. Weak bonds are indicated with dashed lines. Bottom: Further-neighbour interaction $J_{O..O}^{AF}$ realized on the diagonals of hexagons formed by alternating Cu_1 and Cu_2 sites.

multi-boson method to calculate the magnon spectrum for the collinear ferrimagnetically ordered state. We consider the Heisenberg-type Hamiltonian:

$$\mathcal{H} = \sum_{i,j} J_{ij} S_i \cdot S_j, \tag{1}$$

where J_{ij} denote five different kinds of exchange couplings between sites i and j that are explained in Fig. 5c. There are two strong couplings, a ferromagnetic (J_s^{FM}) and an antiferromagnetic (J_s^{AF}) one, residing on the Cu_4 tetrahedra that form the elementary units of our tetrahedra-factorized multiboson theory. Furthermore, there are weak ferromagnetic (J_w^{FM}) and antiferromagnetic (J_w^{AF}) first-neighbour interactions connecting the tetrahedra. The last coupling is an antiferromagnetic longer-range interaction $(J_{O..O})$ connecting Cu_1 and Cu_2 across the diagonals of hexagonal loops formed by alternating Cu_1 and Cu_2 sites. These five interactions are treated on a mean-field level, whereas the antisymmetric DM interactions are neglected for simplicity, as they would affect only the lowest frequency range of the spectrum in the vicinity of the zone centre. This same model has been discussed in detail in ref. 13, for details see also Supplementary Notes 1 and 2 with the Supplementary Table 1. Having a complete picture of magnetic excitations in the entire Brillouin zone up to high energies, we can refine the previously established interaction parameters determined from *ab-initio* calculations[14] and ESR experiments[15].

The values for weak interaction parameters were found to be $J_w^{AF}=27$ K, $J_w^{FM}=-50$ K and $J_{O..O}^{AF}=45$ K, as defined earlier by fitting the magnetization data[14]. These parameters, in fact, provide low-lying modes in agreement with the TOF data as shown in Fig. 5a. Strong coupling constants within the tetrahedra, J_s^{AF} and J_s^{FM}, are mainly responsible for the positions of high-energy modes and govern the intra-tetrahedron excitations. We keep $J_s^{AF}=145$ K as it was determined from fitting the ESR spectrum[15]. Modifying J_s^{FM} to -170 K, we can reproduce the high-energy modes in excellent agreement with experiment, as seen in Fig. 5a, where the result of the spin-dynamical calculations is shown with dotted lines (see also Supplementary Note 1 with Supplementary Figs 1 and 2).

To fully test our model, we also calculated the scattering cross section. At zero temperature this corresponds to

$$\frac{d^2\sigma}{d\Omega d\omega} = \sum_{\alpha,\beta} \left(\delta_{\alpha\beta} - \hat{k}_\alpha \hat{k}_\beta\right) S^{\alpha\beta}(\mathbf{k}, \omega), \quad (2)$$

where α, $\beta = \{x, y, z\}$, $\hat{k}_\alpha = k_\alpha/|k|$, and $S^{\alpha\beta}(\mathbf{k}, \omega)$ is the dynamical spin structure factor. Here we used the original crystallographic unit cell and the corresponding cubic Brillouin zone, taking the exact positions of each copper atom within the unit cell into account. The theoretically established scattering intensity plot (artificially broadened to model the experimental resolution) is shown in Fig. 5b, demonstrating strikingly good agreement with the INS data.

Discussion

We have presented the complete overview of spin excitations in Cu_2OSeO_3 throughout the entire Brillouin zone and over a broad energy range. The data reveal low-energy ferromagnetic Goldstone modes and a higher-energy band of multiple intertwined dispersive spin-wave excitations that are separated by an extensive energy gap. All the observed features can be excellently described by the previously developed theoretical model of interacting Cu_4 tetrahedra, given that one of the strong interaction parameters, J_s^{FM} is set to a new value of -170 K. The complete list of the dominant parameters for the magnetic Hamiltonian determined here is now able to describe all the available experimental results (magnetization, ESR and INS data) simultaneously. Our model can serve as a starting point for more elaborate low-energy theories that would be able to explain the complex magnetic phase diagram of Cu_2OSeO_3, including the helimagnetic order and skyrmion-lattice phases, which still represents a challenge of high current interest in solid-state physics.

Methods

Sample preparation. The compound Cu_2OSeO_3 has first been synthesized by a reaction of CuO (Alfa Aesar 99.995%) and SeO_2 (Alfa Aesar 99.999%) at 300 °C (2 days) and 600 °C (7 days) in evacuated fused silica tubes. Starting from this microcrystalline powder, single crystals of Cu_2OSeO_3 were then grown by chemical transport reaction using NH_4Cl as a transport additive, which decomposes in the vapour phase into ammonia and the transport agent HCl. The reaction was performed in a temperature gradient from 575 °C (source) to 460 °C (sink), and a transport additive concentration of 1 mg cm^{-3} NH_4Cl (Alfa Aesar 99.999%)[18]. The resulting single crystals with typical sizes of 30–60 mm^3 were selectively characterized by magnetization, dilatometry, and X-ray diffraction measurements, showing good crystallinity and reproducible behaviour of physical properties. The crystals were then coaligned into a single mosaic sample for neutron-scattering measurements with a total mass of ~2.5 g using a digital X-ray Laue camera.

Triple-axis INS measurements. The measurements presented in Fig. 2 have been performed at the PUMA thermal-neutron spectrometer with a fixed final neutron wave vector $k_f = 2.662$ Å^{-1} and the PANDA cold-neutron spectrometer with $k_f = 1.4$ Å^{-1}, both at the FRM-II research reactor in Garching, Germany. A pyrolytic graphite or cold beryllium filter, respectively, was installed after the sample to suppress higher order scattering contamination from the monochromator. The sample was mounted in the (*HHL*) scattering plane and cooled down with a closed-cycle refrigerator to the base temperature of ~5 K in both experiments.

Time-of-flight measurements. TOF data in Figs 3–5 were collected using the wide angular range chopper spectrometer ARCS[28] at the Spallation Neutron Source, Oak Ridge National Laboratory, with the incident neutron energy set to $E_i = 70$ meV. Here the base temperature of the sample was at 4.5 K. The reciprocal space was mapped by performing a full 360° rotation of the sample about the vertical $(1\bar{1}0)$ axis in steps of 1° counting ~20 min per angle. The experimental energy resolution measured as the full width at half maximum of the elastic line was ~2.5 meV.

Data analysis. The TOF data were normalized to a vanadium reference measurement for detector efficiency correction, combined and transformed into energy–momentum space using the open-source MATLAB-based HORACE analysis software. Because of the high symmetry of the cubic lattice, the TOF data set could be symmetrized in order to improve statistics in the data and thereby improve the signal-to-noise ratio considerably. After the establishment of the fact, that the intensity of the INS signal follows the symmetry of the large unfolded Brillouin zone introduced in Fig. 1b, we have assumed this symmetry for the symmetrization. This means that every energy–momentum cut in Fig. 3 and every segment of Fig. 5a has been averaged with all possible equivalent cuts within the same Brillouin zone (cuts from different Brillouin zones with different |**Q**| were not averaged). For example, the *LW* or *LK* segments forming the radius and apothem of the hexagonal face of the unfolded Brillouin zone boundary, respectively, (Fig. 1b) have 48 equivalent instances along six non-parallel directions in the Brillouin zone that can be averaged. As this kind of data analysis is very time consuming and requires significant computational efforts, it still awaits to become the standard practice in neutron-scattering research.

Software availability. The open-source MATLAB-based HORACE software package is available from http://horace.isis.rl.ac.uk.

References

1. Tokura, Y. & Seki, S. Multiferroics with spiral spin orders. *Adv. Mater.* **22,** 1554–1565 (2010).
2. Braun, H.-B. Topological effects in nanomagnetism: from superparamagnetism to chiral quantum solitons. *Adv. Phys.* **61,** 1–116 (2012).
3. Kimura, T. Magnetoelectric hexaferrites. *Annu. Rev. Condens. Matter Phys.* **3,** 93–110 (2012).
4. Nagaosa, N. & Tokura, Y. Topological properties and dynamics of magnetic skyrmions. *Nat. Nanotechnol.* **8,** 899–911 (2013).
5. Bogdanov, A. N. & Yablonskii, D. A. Thermodynamically stable "vortices" in magnetically ordered crystals. The mixed state of magnets. *Sov. Phys. JETP* **68,** 101–103 (1989).
6. Rößler, U. K., Bogdanov, A. N. & Pfleiderer, C. Spontaneous skyrmion ground states in magnetic metals. *Nature* **442,** 797–801 (2006).
7. Mühlbauer, S. *et al.* Skyrmion lattice in a chiral magnet. *Science* **323,** 915–919 (2009).
8. Seki, S., Yu, X. Z., Ishiwata, S. & Tokura, Y. Observation of skyrmions in a multiferroic material. *Science* **336,** 198–201 (2012).
9. Seki, S. *et al.* Formation and rotation of skyrmion crystal in the chiral-lattice insulator Cu_2OSeO_3. *Phys. Rev. B* **85,** 220406 (R) (2012).
10. White, J. S. *et al.* Electric field control of the skyrmion lattice in Cu_2OSeO_3. *J. Phys. Condens. Matter* **24,** 432201 (2012).
11. Wu, H. C. *et al.* Unexpected observation of splitting of skyrmion phase in Zn doped Cu_2OSeO_3. *Sci. Rep.* **5,** 13579 (2015).
12. Yang, J. H. *et al.* Strong Dzyaloshinskii-Moriya interaction and origin of ferroelectricity in Cu_2OSeO_3. *Phys. Rev. Lett.* **109,** 107203 (2012).
13. Romhányi, J., van den Brink, J. & Rousochatzakis, I. Entangled tetrahedron ground state and excitations of the magnetoelectric skyrmion material Cu_2OSeO_3. *Phys. Rev. B* **90,** 140404 (R) (2014).
14. Janson, O. *et al.* The quantum nature of skyrmions and half-skyrmions in Cu_2OSeO_3. *Nat. Commun.* **5,** 5376 (2014).
15. Ozerov, M. *et al.* Establishing the fundamental magnetic interactions in the chiral skyrmionic mott insulator Cu_2OSeO_3 by terahertz electron spin resonance. *Phys. Rev. Lett.* **113,** 157205 (2014).
16. Chizhikov, V. A. & Dmitrienko, V. E. Microscopic description of twisted magnet Cu_2OSeO_3. *J. Magn. Magn. Mater.* **382,** 142–151 (2015).
17. Miller, K. H. *et al.* Magnetodielectric coupling of infrared phonons in single-crystal Cu_2OSeO_3. *Phys. Rev. B* **82,** 144107 (2010).
18. Gnezdilov, V. P. *et al.* Magnetoelectricity in the ferrimagnetic Cu_2OSeO_3: Symmetry analysis and Raman scattering study. *Low Temp. Phys.* **36,** 550–557 (2010).
19. Effenberger, H. & Pertlik, F. Die Kristallstrukturen der Kupfer(II)-oxo-selenite $Cu_2O(SeO_3)$ (kubisch und monoklin) und $Cu_4O(SeO_3)_3$ (monoklin und triklin). *Monatsh. Chem.* **117,** 887–896 (1986).
20. Kobets, M. I. *et al.* Microwave absorption in the frustrated ferrimagnet Cu_2OSeO_3. *Low Temp. Phys.* **36,** 176–179 (2010).
21. Ishikawa, Y., Shirane, G., Tarvin, J. A. & Kohgi, M. Magnetic excitations in the weak itinerant ferromagnet MnSi. *Phys. Rev. B* **16,** 4956–4970 (1977).
22. Janoschek, M. *et al.* Helimagnon bands as universal excitations of chiral magnets. *Phys. Rev. B* **81,** 214436 (2010).
23. Kugler, M. *et al.* Band structure of helimagnons in MnSi resolved by inelastic neutron scattering. *Phys. Rev. Lett.* **115,** 097203 (2015).
24. Schwarze, T. *et al.* Universal helimagnon and skyrmion excitations in metallic, semiconducting and insulating chiral magnets. *Nat. Mater.* **14,** 478–483 (2015).

25. Hinkov, V. *et al.* Spin dynamics in the pseudogap state of a high-temperature superconductor. *Nat. Phys.* **3,** 780–785 (2007).
26. Boothroyd, A. T., Babkevich, P., Prabhakaran, D. & Freeman, P. G. An hour-glass magnetic spectrum in an insulating, hole-doped antiferromagnet. *Nature* **471,** 341–344 (2011).
27. Drees, Y. *et al.* Hour-glass magnetic excitations induced by nanoscopic phase separation in cobalt oxides. *Nat. Commun.* **5,** 5731 (2014).
28. Abernathy, D. L. *et al.* Design and operation of the wide angular-range chopper spectrometer ARCS at the Spallation Neutron Source. *Rev. Sci. Instrum.* **83,** 015114 (2012).

Acknowledgements

We thank S. Zherlitsyn and Y. Gritsenko for sound velocity measurements that assisted our data interpretation and M. Rotter for helpful discussions at the start of this project. The work at the TU Dresden was financially supported by the German Research Foundation within the collaborative research centre SFB 1143, research training group GRK 1621, and the individual research grant no. IN 209/4-1. J.R. acknowledges partial funding from the Hungarian OTKA Grant K106047. Research at ORNL's Spallation Neutron Source was sponsored by the Scientific User Facilities Division, Office of Basic Energy Sciences, the US Department of Energy.

Author contributions

A.H. and M.S. have grown single crystals for this study; P.Y.P., A.S.C., M.A.S., J.A.L. and D.S.I. conducted the INS experiments; J.T.P., A.S. and D.L.A. provided instrument support at different neutron spectrometers; P.Y.P. and Y.A.O. analysed the neutron data and prepared the figures; J.R. and J.v.d.B. provided theory support and performed the spin-dynamical calculations; P.Y.P., J.R., H.R., J.v.d.B. and D.S.I. interpreted the results; D.S.I. supervised the project and wrote the manuscript; all authors discussed the results and commented on the paper.

Additional information

Competing financial interests: The authors declare no competing financial interests.

Directional charge separation in isolated organic semiconductor crystalline nanowires

J.A. Labastide[1,*,†], H.B. Thompson[1,*], S.R. Marques[1], N.S. Colella[2,†], A.L. Briseno[2] & M.D. Barnes[1,3]

One of the fundamental design paradigms in organic photovoltaic device engineering is based on the idea that charge separation is an extrinsically driven process requiring an interface for exciton fission. This idea has driven an enormous materials science engineering effort focused on construction of domain sizes commensurate with a nominal exciton diffusion length of order 10 nm. Here, we show that polarized optical excitation of isolated pristine crystalline nanowires of a small molecule n-type organic semiconductor, 7,8,15,16-tetra-azaterrylene, generates a significant population of charge-separated polaron pairs along the π-stacking direction. Charge separation was signalled by pronounced power-law photo-luminescence decay polarized along the same axis. In the transverse direction, we observed exponential decay associated with excitons localized on individual monomers. We propose that this effect derives from an intrinsic directional charge-transfer interaction that can ultimately be programmed by molecular packing geometry.

[1] Department of Chemistry, University of Massachusetts, Amherst Massachusetts 01003, USA. [2] Department of Polymer Science and Engineering, University of Massachusetts, Amherst Massachusetts 01003, USA. [3] Department of Physics, University of Massachusetts, Amherst Massachusetts 01003, USA.
* These authors contributed equally to this work. † Present address: Department of Physics, University of Massachusetts, Amherst Massachusetts 01003, USA (J.A.L.); Department of Chemistry and Chemical Biology, Harvard University 12 Oxford Street, Cambridge, Massachusetts 02138, USA (N.S.C.).
Correspondence and requests for materials should be addressed to M.D.B. (email: mdbarnes@umass.edu).

In the conventional view of the mechanistic operation of organic photovoltaic (OPV) devices, optically prepared Frenkel excitons migrate via diffusive transport (hopping) to a heterojunction where the interfacial energy drives exciton dissociation into polaron (electron and hole) pairs[1–3]. However, radiative recombination[3–8], along with other quenching losses, contribute to a significant reduction in charge harvesting efficiency; a problem that has driven an enormous international research effort aimed at controlling donor and acceptor domain sizes to minimize such losses.

Diffusional transport of localized excitons in organic semiconductors is mediated by coupling between neighbouring chromophores, where the coupling mechanism is usually taken to be a Coulombic dipole–dipole (Frenkel exciton, FE) coupling[9]. Both the sign and magnitude of the matrix element, V_{ij}, describing the nearest-neighbour coupling are important in aggregates of conjugated organic materials: the exciton diffusion coefficient scales as $|V_{ij}|$, while the sign of the interaction determines the energy ordering of coupled states (the lowest energy state for $V_{ij} > 0$ contains *N-1* nodes, where *N* is the number of coupled chromophores, whereas the lowest energy state for $V_{ij} < 0$ is nodeless). This gives rise to the familiar H- or J-aggregate spectroscopic behaviour in vibronic structure in organic semiconductors[10]. Recently, Spano and Yamagata proposed that short-range charge-transfer (CT) coupling—augmenting the usual Coulombic dipole–dipole coupling mechanism—gives rise to hybrid HJ aggregate behaviour that includes both dipole–dipole and CT coupling[11,12]. Since CT interactions depend on wavefunction overlap between adjacent chromophores, the sign and magnitude of the interaction is highly sensitive to the precise packing geometry within the unit cell[13,14]. The combination of known crystal structure and high-level calculations of directional Frenkel exciton and CT couplings, along with experimental techniques for isolating individual crystalline nanowires, presents an interesting opportunity for investigation of directional coupling in small-molecule semiconductor assemblies.

A recently discovered n-type semiconductor of the rylene family of dyes, 7-8-15-16, tetraazaterrylene (TAT) readily forms high-quality, defect-free single (monoclinic) crystals by either physical vapour transport (PVT) or solution-phase self-assembly[15,16]. Figure 1 shows photoluminescence images of TAT nanowire crystals formed from self-assembly in toluene solution. Details of the crystal structure have been reported[15], but briefly, the TAT molecules pack in a herringbone type arrangement in the *b*–*c* plane, and π-stacking occurs along the *a* axis. Several features of the TAT crystal structure are important to point out here. The molecular chromophore orientation is nearly orthogonal ($\sim 80°$) to the π-stacking axis. In addition, the chromophore arrangement in the *b*–*c* plane is such that dipole–dipole coupling is nearly zero, and the molecules are sufficiently separated spatially so that CT coupling is also very weak (approximately 5 meV for in-plane coupling, versus 150 meV along the π-stacking direction. Thus both dipole–dipole (Frenkel exciton) and CT coupling occur almost exclusively along the crystallographic *a* direction.

The combined effects of Frenkel exciton and CT coupling were illustrated recently in the quantitative spectral simulations by Yamagata and Spano[13]. Superficially, the absorption and photoluminescence spectra of isolated TAT extended crystals appear like a simple H-type aggregate, with 0–0 peak intensity diminished with respect to that of the 0–1 vibronic transition[10]. However, simulation of the absorption spectrum revealed a competition between a long-range H-type Frenkel exciton component, and a short-range J-type CT component, where interference between the two couplings results in J-like spectral

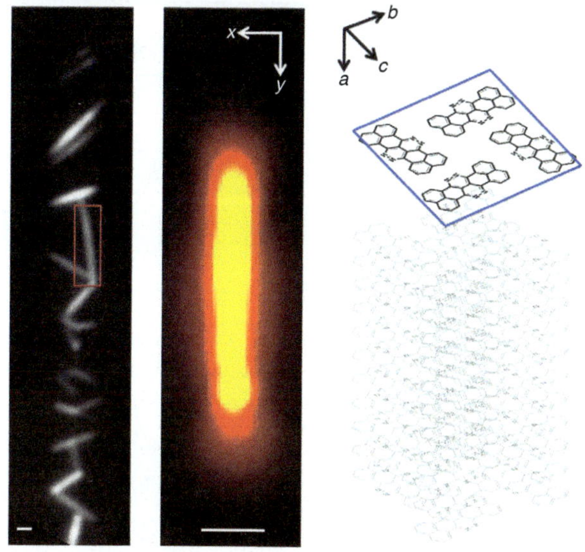

Figure 1 | Orientation and structural schematic of TAT nanowire crystals. (Left) Photoluminescence (PL) of TAT nanowire crystals formed from solution-phase self assembly. (Center) PL image of isolated TAT nanowire with π-stacking direction oriented along laboratory *Y* axis. (Right) Structural schematic of TAT nanowire crystal showing relative orientation of crystallographic axes {*a*, *b*, *c*} with respect to laboratory *X*-*Y* plane. Scale bar, 300 nm.

features at low energies, and H-like spectral features at higher energies, ultimately yielding absorption (and photoluminescence) spectra with hybrid H/J character. Since the chromophore and π-stacking directions are approximately orthogonal (transverse and parallel to the crystal long axis respectively)[15], absorption and emission polarized along the π-stacking axis must be mediated by the CT interaction, as dipole–dipole coupling preserves (transverse) polarization. We observed similar effects in isolated P3HT nanofibers where the interplay between CT and Frenkel exciton coupling interconverts inter- and intra-chain excitons at different rates depending on the degree of π-stacking registration between lamellar stacks.[17]

Here we demonstrate directional charge separation within pristine isolated single crystals of a small molecule n-type semiconductor TAT via manipulation of excitation polarization. Enabled by a strong charge-transfer coupling along specific orientations in the crystal[11,13], pulsed illumination polarized along the π-stack direction (crystallographic *a* axis) forms short-lived inter-chromophore excitations that decompose into polaron pairs. This charge separation is signalled by power-law decay in the photoluminescence whose amplitude and decay exponent depend on the aspect ratio of the crystal. This study demonstrates the potential for both enhancement and directional control of photogenerated charge-separation efficiencies in small-molecule semiconductor assemblies without the need for an interface.

Results

Photoluminecence imaging of isolated nanowire crystals. Figure 2 shows PL images along with corresponding emission spectra of different TAT assemblies formed from chloroform solutions cast on glass substrates, illustrating the spectral evolution of combined H/J coupling with increasing crystal size and presumed π-stacking order. Comparison of PL spectra from sampled solid-state assemblies with those from solution phase indicated that essentially all the aggregation is heavily seeded in solution, and promotes larger scale assembly during solvent

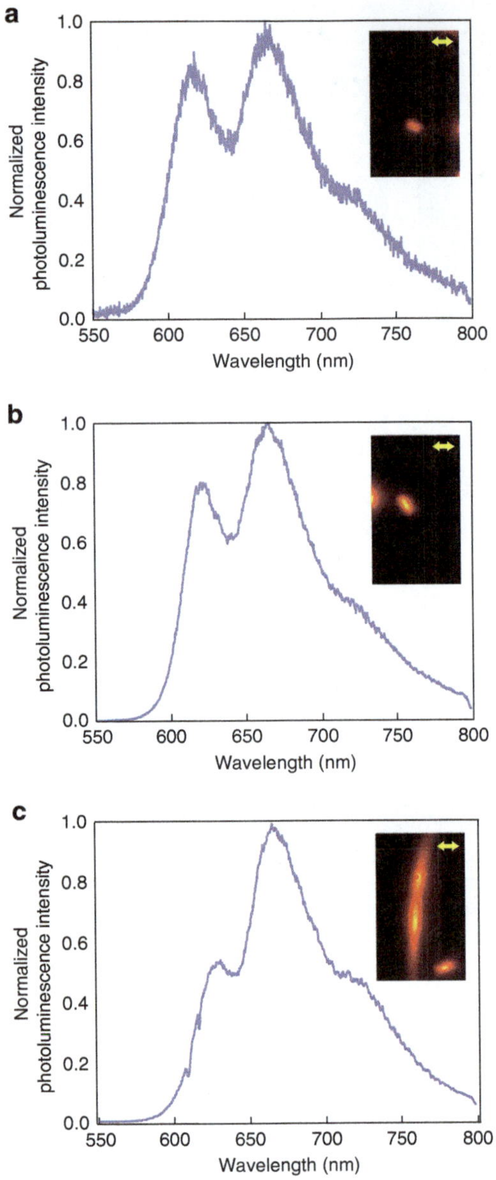

Figure 2 | PL spectral evolution with increased crystal aspect ratio. PL spectra of (**a**) a nanocrystallite with an AR approximately equal to 1, (**b**) a crystal with an AR >1 and <3 and (**c**) a nanowire with an AR ≥3 with corresponding PL images, Scale bar, 300 nm. Note the qualitative changes in PL spectra indicating the structural envelope.

evaporation. The different intensities in vibronic features in PL indicate that the dominant coupling depends on aggregation state; in the small (diffraction limited) nanocrystallite shown in Fig. 2a, the emission spectrum is characterized by an intensity ratio of vibronic components close to unity, while the corresponding intensity ratio for the extended crystal PL spectrum (Fig. 2c) is ~0.5, and is essentially indistinguishable from those of physical vapour transport (PVT) grown crystals[16]. The origin of this spectral evolution will be discussed in detail elsewhere, but the spectral differences suggest different contributions of short-range charge-transfer and longer range Coulomb interactions, depending on the spatial extent of the crystal.

Figure 3 shows measured excitation polarization anisotropies (total PL intensity versus excitation polarization orientation) for crystals of varying aspect ratio (AR). Interestingly, the smallest nanocrystallites ($AR \approx 1$ as determined from the PL image) show the highest degree of polarization anisotropy ($M \approx 0.6$), while elongated crystals show only a modest polarization contrast ($M \approx 0.2$) in excitation. It is important to note that, in the absence of any interaction along the π-stacking direction (nominally orthogonal to the chromophore axis), the photoluminescence response to polarized excitation along this direction should approach zero. This also supports the picture of crystal size dependence of the relative CT and Frenkel exciton contributions to crystal absorption and PL, where the Frenkel exciton contribution increases with increasing crystal size and π-stacking order which increases the absorption strength along the *a* axis. The inset in Fig. 2 shows the variation of PL intensity with rotating excitation polarization, referenced to the crystal *a* axis (grey trace). The PL maximum occurs at $\phi = 78 \pm 3°$, in good agreement with the chromophore orientation determined from the crystal structure. It is important to point out that while smaller excitation anisotropy parameters are usually associated with disordered systems, the PL spectra of the large aspect ratio crystals ($AR > 3$), are identical to those of pristine PVT-grown crystals, and therefore not associated with disorder.

Time- and polarization-resolved photoluminescence imaging. Figure 4 shows the time evolution of the polarization contrast in photoluminescence from a TAT crystal for excitation polarized parallel (Y-polarized, blue triangles) and perpendicular (X-polarized, red squares) to the crystal *a* axis. Here, we defined time-dependent polarization contrast as $M(t) = (N_X(t) - N_Y(t))/(N_X(t) + N_Y(t))$, where *t* is the photon arrival time relative to the excitation pulse. In order to keep the uncertainty in $M(t)$ approximately constant, we used a nonlinear binning such that the total number of counts was the same for each selected time step. Note that at early times (Fig. 4d, ~80% of the

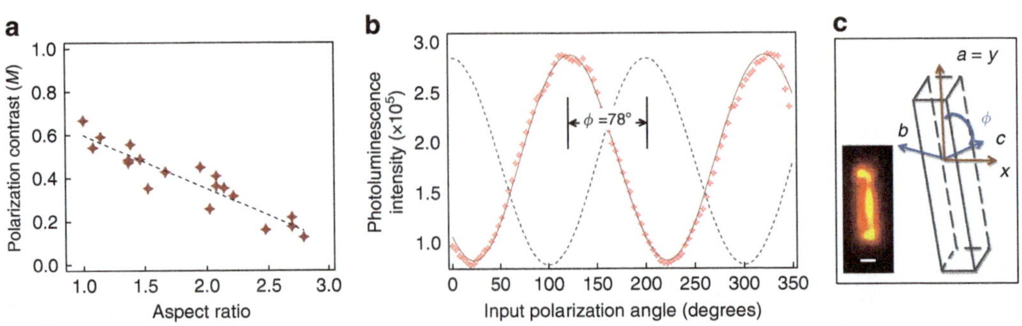

Figure 3 | Polarization contrast in excitation with varying crystal aspect ratio. (**a**) Polarization contrast ($M = (I_{max} - I_{min})/(I_{max} + I_{min})$) for TAT crystallites of varying aspect ratios fit with a linear regression (dashed line). (**b**) The intensity trajectory over $\sim 2\pi$ rotation of excitation polarization, referenced to the crystal long axis (dashed line). (**c**) Chromophore orientation relative to the crystallographic *a* axis (long axis) as determined from the phase of the intensity modulation. Scale bar , 300 nm.

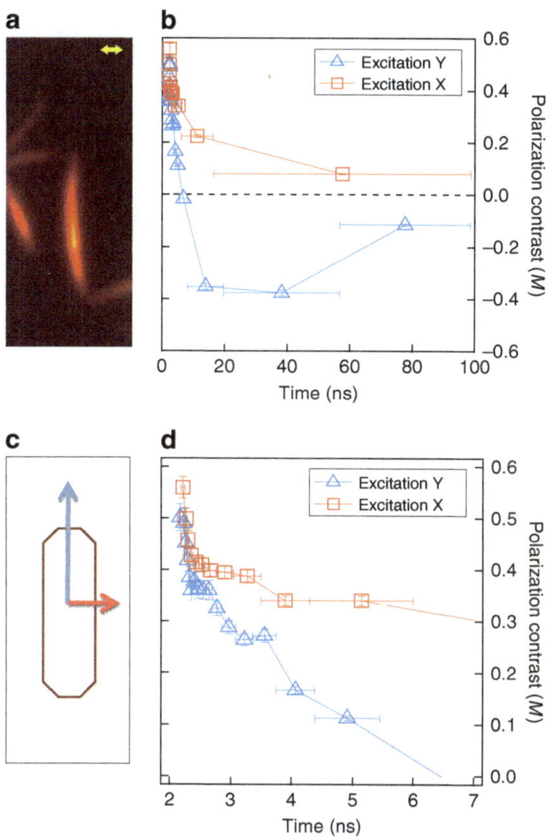

Figure 4 | Time-resolved polarization contrast in photoluminescence for X- and Y-polarized light. (**a**) PL image of a TAT nanowire, Scale bar, 300 nm. (**b**) Time evolution of polarization contrast in PL for X-polarized (red squares) and Y-polarized (blue triangles) excitation. (**c**) Polarization detection scheme. (**d**) Short time behaviour of the polarization contrast for the different excitation polarizations. Vertical error bars for (**b**,**d**) represent 1 s.d. in polarization contrast for each point containing the same total number of counts, and horizontal error bars were determined by the averaged time in each step; thus nonlinear binning in time increases the horizontal uncertainty with increasing time.

photoluminescence is X-polarized (parallel to the chromophore axis) irrespective of excitation polarization, with only a small difference of $M(t=0)$ of 0.58 for X-polarized excitation versus 0.5 for Y-polarized excitation. The polarization contrast in PL near $t \approx 0$ suggests that initially delocalized excitations relax on an ultrafast timescale to create Frenkel excitons, localized primarily individual molecules, and a relatively small fraction ($\sim$25%) on adjacent π-stacked pairs. In both cases, the polarization contrast decreases rapidly (see inset detail) but depends sensitively on excitation polarization. For X-polarized excitation, the PL becomes almost completely depolarized after about 40 ns. For Y-polarized excitation, the effect is far more marked, showing an inversion of the sign of $M(t)$ after $\sim$5 ns, indicating the formation of longer-lived species with transition moments polarized approximately along the π-stack axis.

Discussion

The mechanistic origin of this behaviour was revealed by looking at the PL decay in each polarization channel in detail. Figure 5 shows the PL images from a nanowire (Fig. 5a), crystal (Fig. 5d) and nanocrystallite (Fig. 5g), along with PL decay decomposed into X- and Y-polarization components with excitation polarized along the Y- direction (parallel to the π-stack axis). At short times, the PL follows an exponential decay in both channels characteristic of excitonic emission. Interestingly, at long times, the PL decay dynamics of the nanowire are determined by the excitation polarization; the Y-polarized luminescence (Fig. 5b) follows a power-law decay ($I(t) \approx A \cdot t^{-(1+\mu)}$, where A is an amplitude, and μ is the power-law exponent) with $\mu = 1.3$, while the X-polarized PL decay (Fig. 5c) is exponential. In contrast, the PL decay from the crystal, being thicker in the transverse dimension and having a smaller aspect ratio than the nanowire, follows a power-law in both X and Y polarization channels with $\mu_X = 1.7$ and $\mu_Y = 1.6$. For the diffraction limited nanocrystallite, the PL decay follows a power-law in both X and Y polarization channels, but with much larger characteristic decay constants ($\mu_X \approx \mu_Y = 3.5$).

As demonstrated by Silva and coworkers[3], power-law decay in photoluminescence from assemblies of organic semiconductors is the signature of two-body polaron pair recombination, with amplitudes and exponents reporting on the efficiency of polaron generation and characteristic timescales for their recombination. In semicrystalline P3HT films, Silva and coworkers showed that exciton regeneration occurs via tunnelling of geminate charge-pairs at amorphous-crystalline interfaces, thus concluding that the mechanism for charge separation in P3HT was extrinsic, requiring an interface between amorphous and crystalline domains to drive exciton dissociation. This was corroborated by recent work by Silva and Stingelin[18], which highlighted the role of mixed domains, as well as work on isolated crystalline P3HT nanofibres[17] (which show only very small power-law components) and semicrystalline P3HT nanoparticles[19], which show very robust power-law photoluminescence. The question here is whether the delayed photoluminescence in TAT crystals is derived from exciton regeneration by tunnelling (similar to P3HT) or through interaction of mobile carriers within the crystal.

Polaron-pairs can be formed either by unimolecular exciton fission or via bi-molecular triplet–triplet interactions. In pentacene for example, the very low-lying triplet energy (0.86 eV) enables efficient conversion of singlet excitons into two triplets, which then undergo annihilation to form charge pairs[20–22]. Like P3HT films, the delayed PL in TAT crystals scales linearly with pump intensity corresponding to exciton formation via charge-pair recombination with distributed kinetics, thus ruling out bi-molecular triplet–triplet annihilation, which would be expected to show a nonlinear dependence on pump intensity. The question remains whether the exciton regeneration mechanism in TAT crystals occurs via tunnelling between more-or-less immobile charge-pairs similar to the mechanism in P3HT, or two-body encounters between mobile carriers. Estimates of carrier diffusion distance based on computed electron/hole hopping energies in TAT crystals (50 and 4.5 meV for electron and hole, respectively) suggest that anisotropic carrier mobility on the experimental time scale should significantly contribute the polaron pair recombination kinetics.[23]

To address this issue, we looked at the temperature dependence of the delayed photoluminescence. Figure 6 shows the temperature dependence of both amplitude and power-law exponent in Y-polarized luminescence (with Y-polarized excitation) obtained from the same TAT nanocrystal. In contrast with the results by Silva and coworkers[3], we find that both amplitude and exponent scale approximately like $T^{1/2}$ clearly indicating that exciton regeneration is not associated with tunnelling, but rather diffusion of carriers along the a axis. This suggests a physical picture in which exciton regeneration occurs via encounter of diffusing carriers within a certain capture radius. While more work is required to quantitatively identify the temperature

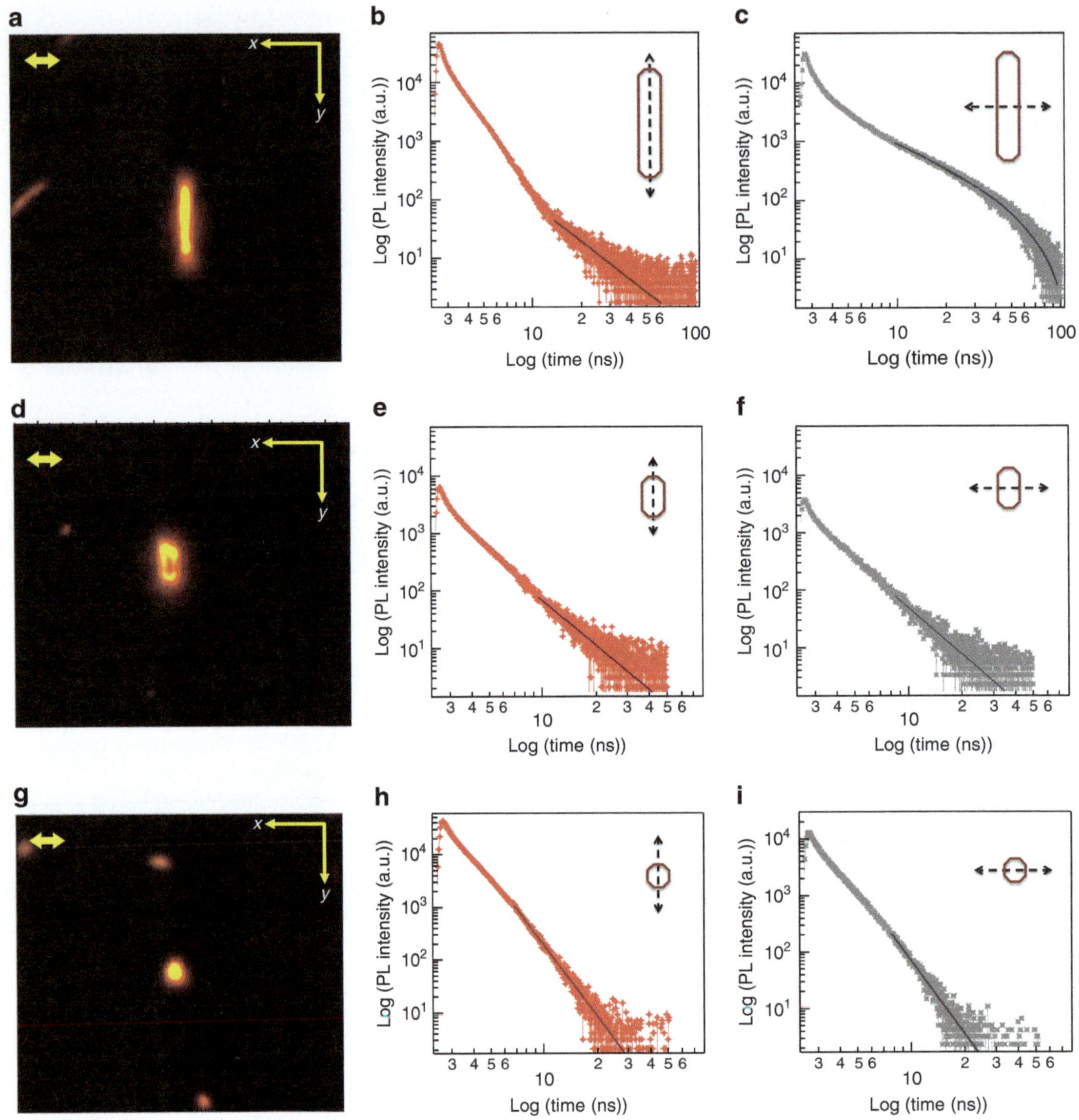

Figure 5 | Time- and polarization-resolved PL decay dynamics with varying crystal aspect ratio. (**a**,**d**,**g**) PL images of TAT nanowire, crystal and nanocrystallite, respectively, Scale bar, 300 nm. (**b**,**c**,**e**,**f**,**h**,**i**) X- and Y-polarized PL decays from nanowire, crystal, and nanocrystallite for excitation along the *y* axis. For the nanowire (**a**), the Y-polarized PL (**b**) follows a power-law decay ($\mu = 2.3$), while the X-polarized PL decay (**c**) is bi-exponential ($\gamma 1 = 5.1$ ns and $\gamma 2 = 20.2$ ns). For the crystal (**d**), both polarization channels (**e**,**f**) show power-law decay with $\mu_X = 2.7$ and $\mu_Y = 2.6$. For the nanocrystallite (**g**), luminescence also follows a power-law decay in both polarization channels (**h**,**i**) with $\mu_X \approx \mu_Y = 4.5$.

dependence, the physical picture appears to be analogous to bi-molecular reactive kinetics, with an increase of regenerated excitons per unit time, which appears in our measurement as increased amplitude of the power-law signal. Note that if the initial charge-separation process was thermally activated, we should expect to see nonlinear (exponential) increase in amplitude with increasing temperature. Within the same analogy, we can interpret the temperature dependence in the power-law exponent as a decrease in the mean radius resulting in faster exciton regeneration and photoluminescence decay.

In the small, diffraction limited nanocrystallites, we observed large, polarization–isotropic power-law exponents and relatively early power-law onset ($t < 7$ ns). Here the short-range charge-transfer interaction dominates inter-chromophore coupling, and promotes the geminate formation and efficient dissociation of both inter and intra-chromophore excitations. The larger power-law exponents are indicative of smaller average electron-hole radius, likely due to lower mobilities associated with π-stacking disorder in the smaller assemblies. The increase in π-stacking order in the extended crystals shifts the balance of interactions towards Coulomb coupling, reducing the number of geminate polarons formed, and the rate of Frenkel exciton dissociation, but also increases the diffusive mobility along the *a* axis. The combination of these factors leads to later power-law onsets, smaller amplitudes, and factor of 2 smaller exponents seen in the power-law decay from extended crystals. In these crystals, the mean electron-hole distances calculated from the combination of experimentally determined power-law exponents and calculated zero field mobilities are too large for recombination via tunnelling. We propose that exciton dissociation and diffusion-mediated polaron-pair recombination, rather than geminate polaron pairs that recombine via tunnelling, is responsible for the power-law signal polarized in the *b*–*c* plane seen from the large crystals.

A fascinating directionality of the PL decay mechanism in isolated TAT crystals was revealed using time- and polarization-

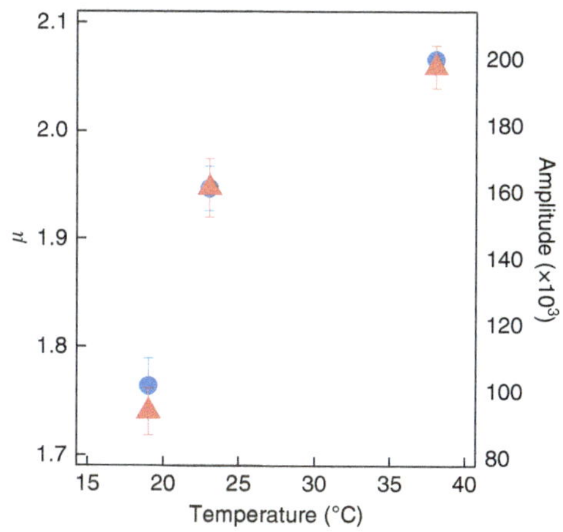

Figure 6 | Temperature dependence of power-law exponent and amplitude. Power-law exponent μ (blue circles) and amplitude (red triangles) recovered from fits to delayed photoluminescence from the same TAT crystal for Y-polarized excitation and emission as a function of temperature, with s.d. error bars. Both follow an approximate $T^{1/2}$ dependence.

resolved photoluminescence spectroscopy. For the crystalline nanowires studied, excitation polarized along the π-stacking axis results in PL polarized along the same axis that decays according to a power-law with exponent whose magnitude depends on crystal aspect ratio, while the PL polarized transverse to the crystallographic a axis (approximately aligned along the chromophore axis) follows an exponential decay characteristic of Frenkel excitons localized on individual molecular sites. For thicker crystals with smaller aspect ratios, power-law decays characteristic of polaron pair recombination was seen for PL polarized in all directions, perhaps indicating a minimum spatial extent required for non-geminate polaron pair generation in the lateral dimension. In contrast with P3HT films, for example, the mechanism for charge-pair formation in TAT crystals appears to be intrinsic, deriving from a strong charge-transfer coupling along the π-stacking axis. These results illustrate the potential of harnessing directional charge-transfer interactions in crystalline molecular assemblies for enhanced charge-separation without the need for diffusion and interfacial dissociation. While interfacial engineering efforts within the field have generated invaluable insights to mitigate associated efficiency losses, not having to rely on an interface to generate charge-separated states is of considerable advantage. We have shown that by simple manipulation of the crystal aspect ratio and the excitation polarization, it is possible to directly access charge-transfer interactions that could offer a significant advantage to improving power conversion efficiencies in organic photovoltaic devices. Further, the ability to make macroscopically directionally ordered films of TAT single crystals has been well demonstrated, and represents a platform with which we can exert directional control over energy and charge flow in small-molecule semiconductor assemblies. Finally, it is important to point out that the observations reported here are not unique to TAT; the design principles that dictate direction and magnitude of Coulomb and charge-transfer interactions apply to a broad class of small molecule and oligomeric semiconductors materials whose packing geometry can be tuned by molecular architecture.

Methods

Experimental methodology. We employed a variant of time-tagged time-resolved (T^3R) photon counting[24,25] called time- and polarization-resolved photoluminescence[17,19] to probe directional coupling in isolated TAT crystals. Photons emitted from the sample are sorted into two orthogonal polarization components detected by avalanche photodiodes whose outputs are processed through a high-speed router, thus identifying the arrival time and polarization state of each individual detected photon[17,19], enabling direct measurement of the time-evolution of polarization contrast with resolution ultimately limited by the time-to-digital converter (4 ps in our implementation). This technique is particularly well suited for probing polarization evolution mechanisms in crystalline assemblies because the orientation of the crystal long axis in the lab frame is easily discernable from the photoluminescence image, thus defining an orientation with respect to the excitation polarization. Because the molecular packing structure is known, the photoluminescence in different polarization channels can be referenced to either the molecular chromophore orientation or π-stacking direction.

Sample preparation. Isolated TAT nanowires and nanocrystallites were prepared by sonication of a 0.5 μM TAT in chloroform solution for roughly 1 h. A small amount of solution was then drop-cast onto a plasma cleaned glass coverslip resulting in well-dispersed nanowires and nanocrystallites with a density of ~10 crystals per 100 μm². The solution was left for 1–2 days for subsequent aggregation, allowing for the formation of decreased aspect ratio crystals. After 1–2 days, a small amount of aggregated solution was drop-cast onto a plasma cleaned glass coverslip with ~5 crystals per 100 μm².

Optical measurements. The time- and polarization- resolved photoluminescence measurements were done using a home built microscope described by ref. 17. Isolated nanostructures were excited with a pulsed diode laser (485 nm) with an average input intensity of approximately 16 W cm^{-2}, a pulse width (FWHM) of 90 ps, and a repetition rate of 10 MHz. The PL spectra were taken on a separate microscope set-up with a continuous-wave excitation using an air-cooled argon ion laser (488 nm).

References

1. Scholes, G. D. & Rumbles, G. Excitons in nanoscale systems. *Nat. Mater.* **5,** 683–696 (2006).
2. Scharber, M. C. *et al.* Design rules for donors in bulk-heterojunction solar cells—towards 10% energy-conversion efficiency. *Adv. Mater.* **18,** 789–794 (2006).
3. Paquin, F. *et al.* Charge separation in semicrystalline polymeric semiconductors by photoexcitation: is the mechanism intrinsic or extrinsic? *Phys. Rev. Lett.* **106,** 197401 (2011).
4. Cowan, S. R., Banerji, N., Leong, W. L. & Heeger, A. J. Charge formation, recombination, and sweep-out dynamics in organic solar cells. *Adv. Func. Mater.* **22,** 1116–1128 (2012).
5. Barnes, M. D. & Baghar, M. Optical probes of chain packing structure and exciton dynamics in polythiophene films, composites, and nanostructures. *J. Polym. Sci. Pol. Phys.* **50,** 1121–1129 (2012).
6. Singh, S. & Vardeny, Z. V. Ultrafast transient spectroscopy of polymer/fullerene blends for organic photovoltaic applications. *Materials* **6,** 897–910 (2013).
7. Guo, J. M., Ohkita, H., Benten, H. & Ito, S. Near-IR Femtosecond Transient Absorption Spectroscopy of Ultrafast Polaron and Triplet Exciton Formation in Polythiophene Films with Different Regioregularities. *J. Am. Chem. Soc.* **131,** 16869–16880 (2009).
8. Provencher, F. *et al.* Direct observation of ultrafast long-range charge separation at polymer-fullerene heterojunctions. *Nat. Commun.* **5,** 4288 (2014).
9. Yamagata, H., Pochas, C. M. & Spano, F. C. Designing J- and H-aggregates through wave function overlap engineering: applications to poly(3-hexylthiophene). *J. Phys. Chem. B* **116,** 14494–14503 (2012).
10. Yamagata, H. & Spano, F. C. Interplay between intrachain and interchain interactions in semiconducting polymer assemblies: The HJ-aggregate model. *J. Chem. Phys.* **136,** 184901 (2012).
11. Spano, F. C. Excitons in conjugated oligomer aggregates, films, and crystals. *Annu. Rev. Phys. Chem.* **57,** 217–243 (2006).
12. Yamagata, H. *et al.* HJ-Aggregate Behavior of Crystalline 7,8,15,16-Tetraazaterrylene: Introducing a New Design Paradigm for Organic Materials. *J. Phys. Chem. C* **118,** 28842–28854 (2014).
13. Spano, F. C. The spectral signatures of Frenkel polarons in H- and J-aggregates. *Accounts Chem. Res.* **43,** 429–439 (2010).
14. Hestand, N. J., Tempelaar, R., Knoester, J., Jansen, T. L. C. & Spano, F. C. Exciton mobility control through sub-Å packing modifications in molecular crystals. *Phys. Rev. B* **91,** 195315 (2015).
15. Fan, J., Zhang, L., Briseno, A. L. & Wudl, F. Synthesis and characterization of 7,8,15,16-tetraazaterrylene. *Org. Lett.* **14,** 1024–1026 (2012).
16. Wise, A. J. *et al.* Spectroscopy of discrete vertically oriented single-crystals of n-type tetraazaterrylene: understanding the role of defects in molecular semiconductor photovoltaics. *Phys. Chem. Chem. Phys.* **16,** 15825–15830 (2014).
17. Labastide, J. A., Baghgar, M., McKenna, A. & Barnes, M. D. Time- and polarization-resolved photoluminescence decay from isolated polythiophene (P3HT) nanofibers. *J. Phys. Chem. C* **116,** 23803–23811 (2012).
18. Dou, F. *et al.* The effect of phase morphology on the nature of long-lived charges in semiconductor polymer:fullerene systems. *J. Mater. Chem. C* **3,** 3722–3729 (2015).

19. Labastide, J. A. *et al.* Time- and Polarization-Resolved Photoluminescence of Individual Semicrystalline Polythiophene (P3HT) Nanoparticles. *J. Phys. Chem. Lett.* **2,** 2089–2093 (2011).
20. Jundt, C. *et al.* Exciton dynamics in pentacene thin-films studied by pump-probe spectroscopy. *Chem. Phys. Lett.* **241,** 84–88 (1995).
21. Beljonne, D., Yamagata, H., Bredas, J. L., Spano, F. C. & Olivier, Y. Charge-transfer excitations steer the Davydov splitting and mediate singlet exciton fission in pentacene. *Phys. Rev. Lett.* **110,** 226402 (2013).
22. Walker, B. J., Musser, A. J., Beljonne, D. & Friend, R. H. Singlet exciton fission in solution. *Nat. Chem.* **5,** 1019–1024 (2013).
23. Zhao, C. B., Wang, W. L., Yin, S. W. & Ma, Y. Theoretical investigation on electronic, optical, and charge transport properties of 7,8,15, 16-tetraazaterrylene and its derivatives with electron-attracting substituents. *N. J. Chem.* **37,** 2925–2934 (2013).
24. Yang, H. Change-point localization and wavelet spectral analysis of single-molecule time series. *Single Mol. Biophys.* **146,** 219–243 (2012).
25. Yang, H. & Xie, X. S. Statistical approaches for probing single-molecule dynamics photon-by-photon. *Chem. Phys.* **284,** 423–437 (2002).

Acknowledgements

Support from the U.S. Department of Energy Grant # DE-FG02-05ER15695 (Program Manager: Larry Rahn) is gratefully acknowledged. N.S.C. and A.L.B. thank the NSF Center for Hierarchical Manufacturing (CMMI-0531171).

Author contributions

J.A.L. and H.B.T. contributed equally to this work. M.D.B. conceived the concept. J.A.L., H.B.T. and S.R.M. ran experiments and performed data analysis. N.S.C. and A.L.B. provided the sample materials. J.A.L. and M.D.B. wrote the manuscript, and all authors commented on the manuscript at all stages of development.

Additional information

Competing financial interests: The authors declare no competing financial interests.

25

Aharonov-Bohm oscillations in Dirac semimetal Cd_3As_2 nanowires

Li-Xian Wang[1,*], Cai-Zhen Li[1,*], Da-Peng Yu[1,2,3] & Zhi-Min Liao[1,2]

Three-dimensional Dirac semimetals, three-dimensional analogues of graphene, are unusual quantum materials with massless Dirac fermions, which can be further converted to Weyl fermions by breaking time reversal or inversion symmetry. Topological surface states with Fermi arcs are predicted on the surface and have been observed by angle-resolved photoemission spectroscopy experiments. Although the exotic transport properties of the bulk Dirac cones have been demonstrated, it is still a challenge to reveal the surface states via transport measurements due to the highly conductive bulk states. Here, we show Aharonov-Bohm oscillations in individual single-crystal Cd_3As_2 nanowires with low carrier concentration and large surface-to-volume ratio, providing transport evidence of the surface state in three-dimensional Dirac semimetals. Moreover, the quantum transport can be modulated by tuning the Fermi level using a gate voltage, enabling a deeper understanding of the rich physics residing in Dirac semimetals.

[1]State Key Laboratory for Mesoscopic Physics, Department of Physics, Peking University, Beijing 100871, China. [2]Collaborative Innovation Center of Quantum Matter, Beijing, China. [3]Institute of Physics and Electronic Information, Yunnan Normal University, Kunming 650500, China. *These authors contributed equally to this work. Correspondence and requests for materials should be addressed to Z.-M.L. (email: liaozm@pku.edu.cn).

The Dirac fermions in graphene and topological insulators are confined in a two-dimensional plane, while in a three-dimensional (3D) Dirac semimetal the Dirac fermions are extended to the 3D space[1–17]. Recently, Cd_3As_2 and Na_3Bi have been predicted to be Dirac semimetals[5,6], which were further identified by angle-resolved photoemission spectroscopy experiments[7–10]. The Dirac point in 3D Dirac semimetals is constituted by two degenerated Weyl nodes and this degeneracy can be lifted by breaking time reversal or inversion symmetry[5,6]. Owing to its unique bulk Dirac cones, the 3D Dirac semimetal has become an excellent platform to investigate exotic transport properties, such as giant positive magnetoresistance (MR)[11], quantum linear MR[5,12], splitting of Landau levels[13–16] and chiral anomaly induced negative MR[17–21]. On the other hand, non-trivial surface states with arc-like Fermi surface are theoretically predicted and observed by angle-resolved photoemission spectroscopy experiments[5,6,22,23]. Nevertheless, the non-trivial surface states are not accessible by transport measurements in bulk Dirac semimetals because of the bulk-dominant conduction.

The Aharonov–Bohm (A–B) effect of individual nanowires with large surface-to-volume ratio provides an effective method to explore the transport properties of the non-trivial surface states. As the mean free path of the carriers is comparable with the perimeter of the nanowire, the electron wave is confined in the finite boundary condition along the nanowire perimeter. Consequently, the surface energy bands are enforced into discrete sub-bands because of the quantum confinement[24,25]. As sweeping the magnetic field parallel to the nanowire direction, the density of states at the Fermi level is periodically altered as each sub-band crosses the Fermi energy[24,25], and the oscillating period is described by the Φ/Φ_0, where Φ is the magnetic flux enclosed by the path of the surface carriers and $\Phi_0 = h/e$. These quantum oscillations have been observed in topological insulator nanoribbons and nanowires, which are still called as A–B oscillations in the recent literatures[26–28]. Particularly, the helical surface states in topological insulator accumulate an additional π Berry phase after travelling a closed trajectory, resulting in the conductance peaks are situated at the magnetic flux of $\left(n+\frac{1}{2}\right)h/e$ (refs 26–28), which is very distinct from ordinary h/e oscillations in normal metal rings[29].

Here we report the A–B oscillations in individual Cd_3As_2 nanowires to demonstrate the transport properties of the surface states in Dirac semimetals. A π A–B effect that the conductance oscillations peak at odd integers of $h/2e$ with a period of h/e is observed, providing transport evidence of the topological surface states of Cd_3As_2 nanowires. The A–B oscillations are tunable by altering the Fermi level of the nanowire via gate voltage, giving deeper insight of the 3D Dirac semimetal with Dirac fermion bulk states and Fermi arc surface states.

Results

Characterizations. The Cd_3As_2 nanowires were synthesized via chemical vapour deposition method, which are with high aspect ratio and right stoichiometry (Supplementary Fig. 1). The Cd_3As_2 nanowires are of single crystalline nature and grown along the [112] direction, as shown by the high-resolution transmission electron microscopy image in Fig. 1a. The surface of individual nanowires is usually covered by a thin amorphous layer ~5 nm forming a core-cell structure (Supplementary Fig. 2), which may help to maintain the clean Cd_3As_2 surface immunized from the environment[30].

The schematic representation of the device and measurement configuration is shown in Fig. 1b. The temperature-dependent resistance manifests semiconducting-like behaviours (Supplementary Fig. 3), similar to other 3D Dirac semimetal materials, such as Na_3Bi, and is attributed to the low carrier density[17,20]. Notable Shubnikov-de Haas oscillations were also observed under perpendicular magnetic field (Supplementary Fig. 4). The linear extrapolation of the Landau index plot yields the intercept ~0.1, which is close to the intercept of 1/8 in 3D Dirac systems with perfect π Berry phase, giving an evidence for the Dirac fermion transport in Cd_3As_2 nanowires[31].

Aharonov–Bohm oscillations. The A–B oscillations were measured as the magnetic field was applied parallel with the longitudinal axis of the nanowire. Figure 1c shows the resistance of a nanowire with diameter ~115 nm (Device 1) as a function of magnetic field **B** at different temperatures. In low magnetic field regime ($|\mathbf{B}| < 0.3$ T), the MR exhibits an obtuse dip because of the weak anti-localization, which is in agreement with the presence of strong spin–orbit interaction in Cd_3As_2 (ref. 5). At higher fields, the MR starts to oscillate periodically on a smooth negative MR background at 1.5 K. The negative MR has already been observed in Dirac semimetals and Weyl semimetals, which was believed to originate from the chiral anomaly induced charge pumping effect[17–21]. The magnitude of the negative MR is increased from −19% at 1.5 K to −34% at 20 K (Supplementary Fig. 5), which is consistent with previous report[20]. It should be noted that the (quasi-) ballistic transport along the perimeter of the nanowire is necessary to observe the A–B oscillations[28]. Therefore, the A–B oscillations are much more pronounced in the fine nanowire (~115 nm in diameter) in this work than the thick nanowire in ref. 20. As the temperature rises up to 20 K, the oscillations are nearly smeared out, indicative of its quantum nature. To focus on the quantum oscillations, the negative MR background is removed and the oscillating term $\Delta G(\mathbf{B})$ is shown in Fig. 1d. More than 10 oscillation periods (from 0 to ±5 T) can be clearly distinguished that are distributed symmetrically around the zero-field. The fast-Fourier transform (FFT) on $\Delta G(\mathbf{B})$ shown in Fig. 1e indicates the main peak at $\sim 2.55\,T^{-1}$, corresponding to an individual 0.38 T period, giving a remarkable signature of flux-related quantum oscillations. If the quantum interference occurs on the surface, the h/e A–B oscillations yield a cross-section area of the nanowire $\sim 0.11\,\mu m^2$. The cross-section area of the Device 1 was exactly estimated to be $0.11 \pm 0.005\,\mu m^2$ by direct measurements on its cut (Supplementary Fig. 6), which identifies the observed quantum oscillations are from the A–B effect.

The amplitude of oscillating term $\Delta G \sim 0.005 e^2/h$ is much smaller than the quantum conductance e^2/h. Previously, the observed amplitudes of A–B oscillations in topological insulator nanoribbons are also smaller than e^2/h (refs 26,28). The electrons circle along the nanowire perimeter in a quasi-ballistic manner, but drift along the longitudinal direction of the nanowire in a diffusive manner (mean free path $<L$). As probing the longitudinal conductance, the A–B oscillation amplitude of conductance may be reduced due to the diffusive transport in longitudinal direction. The amplitude of the A–B oscillation increases with decreasing temperature, as shown in Fig. 1f. It is found that the FFT peak amplitude obeys to T^{-1} law in the temperature range of 5–20 K, then deviates the T^{-1} dependence below ~5 K and increases more slowly as further decreasing the temperature. The T^{-1} temperature dependence of oscillation amplitude indicates the nature of ballistic transport in circumferential direction[28]. The deviation from T^{-1} dependence below 5 K is due to the low influence of thermal broadening on the electron wave packets[32]. The critical temperature for the deviation of T^{-1} dependence can be estimated by considering the critical thermal

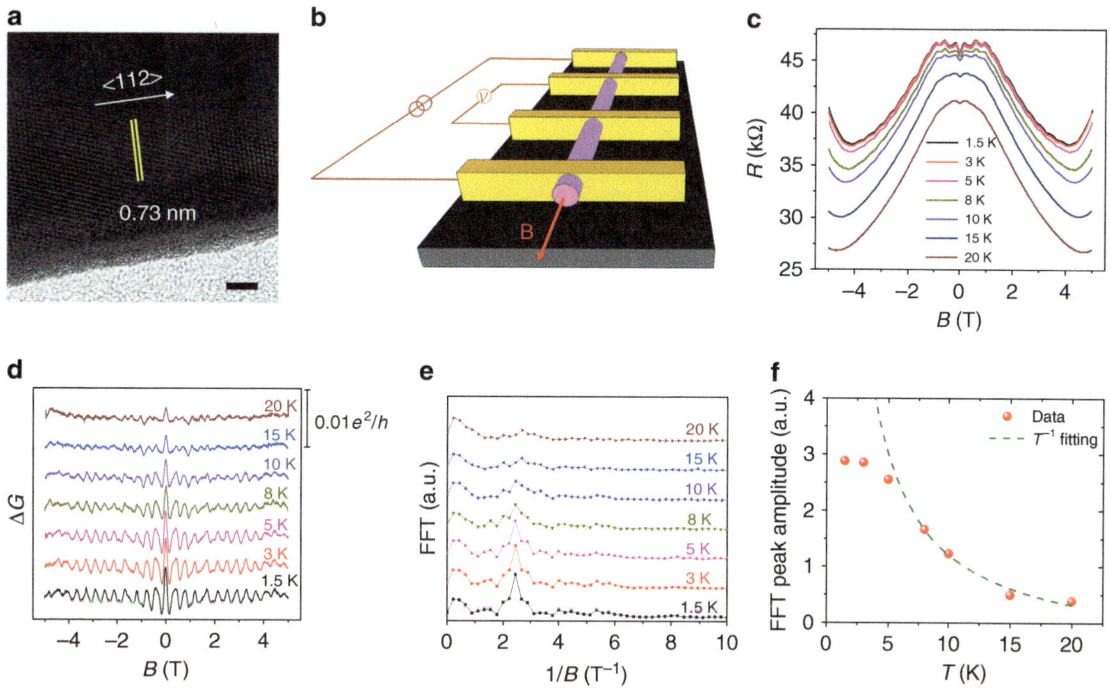

Figure 1 | Magnetotransport of a nanowire with diameter ~115 nm (Device 1). (**a**) High-resolution transmission electron microscopy image of a typical nanowire indicates <112> growth direction with interplanar space ~0.73 nm. Scale bar, 5 nm. (**b**) Schematic diagram of the four-terminal device with applied magnetic field aligned with the length. (**c**) Resistance as a function of magnetic field at different temperatures from 1.5 to 20 K. (**d**) The oscillations in conductance as a function of magnetic field after subtracting background. (**e**) FFT spectrums of the conductance oscillations. (**f**) Plot of the temperature dependence of the FFT peak amplitude. The dashed line indicates the fitting by the T^{-1} dependence.

energy $E_c = \frac{h}{\pi e^2 R}\Delta E$, where R is the nanowire resistance, and ΔE is the sub-band spacing due to geometric confinement[32]. According to nanowire boundary conditions, the sub-band spacing has a form of $\Delta E = \frac{h v_F}{W} = h v_F \sqrt{\frac{1}{4\pi S}}$, where W is the perimeter of nanowire, S is cross-sectional area and v_F is Fermi velocity[26]. Hence, the critical temperature is calculated to be ~7.2 K, which is close to the experimental observation of ~5 K.

The elastic scattering length $L_0 \sim 10$ nm of the longitudinal transport along the nanowire is estimated by the amplitude of oscillating term $\Delta G(\mathbf{B}) \sim 0.005e^2/h$ via the Landauer–Büttiker formula of $G = \frac{e^2/h}{1+L/L_0}$ (ref. 33), where $L \sim 2\,\mu$m is the distance between the two voltage probes. Although the electrons circle along the nanowire perimeter in a quasi-ballistic manner, the carrier drift along the longitudinal direction of the nanowire is in a diffusive manner (mean free path $<L$). Therefore, the obtained L_0 can be underestimated due to the reduced amplitude ΔG. Nevertheless, the ballistic or quasi-ballistic transport along the nanowire perimeter must be satisfied to observe the A–B oscillations[26–28,30,34]. Therefore, it is naturally inferred that the non-trivial surface states in the Cd_3As_2 nanowire should have high carrier mobility with suppressed scatterings by topological protection. Our results are also consistent with the observations in topological insulator nanowires that are not ballistic along the nanowire length but ballistic along the nanowire perimeter[34].

The oscillations in ΔG as a function of magnetic flux Φ in unit of h/e are presented in Fig. 2a. In the two initial periods, the conductance peaks are at the integer quantum flux ($\Phi = h/e$, $2h/e$). Then, a phase-shift takes place in the range from 1 to 2 T, wherein the conductance peaks are at random flux $\Phi = 3.1h/e$ and $4.3h/e$. After the phase-shift region, the conductance peaks are stable at the half-integer quantum flux ($\Phi = (5+1/2)h/e$, $(6+1/2)h/e, \ldots$). The conductance oscillation peaks at half-integer of Φ_0 with a period of Φ_0 are very similar with that observed in topological insulators, which was ascribed to the spin-related π Berry phase of the topological surface states[24,25]. At $\Phi/\Phi_0 = 0$, the inter-spacing between sub-bands due to the finite boundary condition is estimated by $\Delta E = h v_F/W$ (refs 26–28). At $\Phi/\Phi_0 = n$, the surface state is fully gapped at $E = 0$. While at $\Phi/\Phi_0 = n + 1/2$, a gapless surface band appears due to the additional π Berry phase[26]. Intuitively, to observe this anomalous A–B oscillation, it requires that the Fermi-level E_F is close to $E = 0$ or the quantum confinement induced gap is large enough. Because of the high crystal quality with fewer defects, the Cd_3As_2 nanowires are of low carrier density with the order of $\sim 10^{17}$ cm^{-3}, which has been measured by field-effect gating method and found to be the crucial factor for the observation of the negative MR effect in our previous work[20]. The low carrier density also suppresses the bulk conductance with the resistance $R > 25.8$ kΩ at low temperatures, promoting the detection of the surface states of the nanowire with large surface-to-volume ratio.

To understand the phase-shift of the A–B effect in Cd_3As_2 nanowires, we would like to discuss the origin of the additional π Berry phase by considering the evolution of Dirac point under magnetic fields. In Cd_3As_2, the Dirac points are distributed equally away from the Γ point along the [001] direction at $\pm k_D$ (ref. 5). As schematically shown in Fig. 2b, without magnetic field, the Fermi arc surface states connect the two bulk Dirac nodes and each Dirac node contains two degenerate Weyl nodes. Applying an external magnetic field can break the time reversal symmetry and split the two Dirac points into two pairs of Weyl nodes along the magnetic field direction[35]. As schematically shown in Fig. 2c, under magnetic field, each surface Fermi arc connects the Weyl nodes with right-handed chirality and left-handed chirality, respectively. The chirality-non-trivial surface states are non-degenerate. The lifting of electronic state degeneracy results in an additional Berry phase π for cyclotron electrons

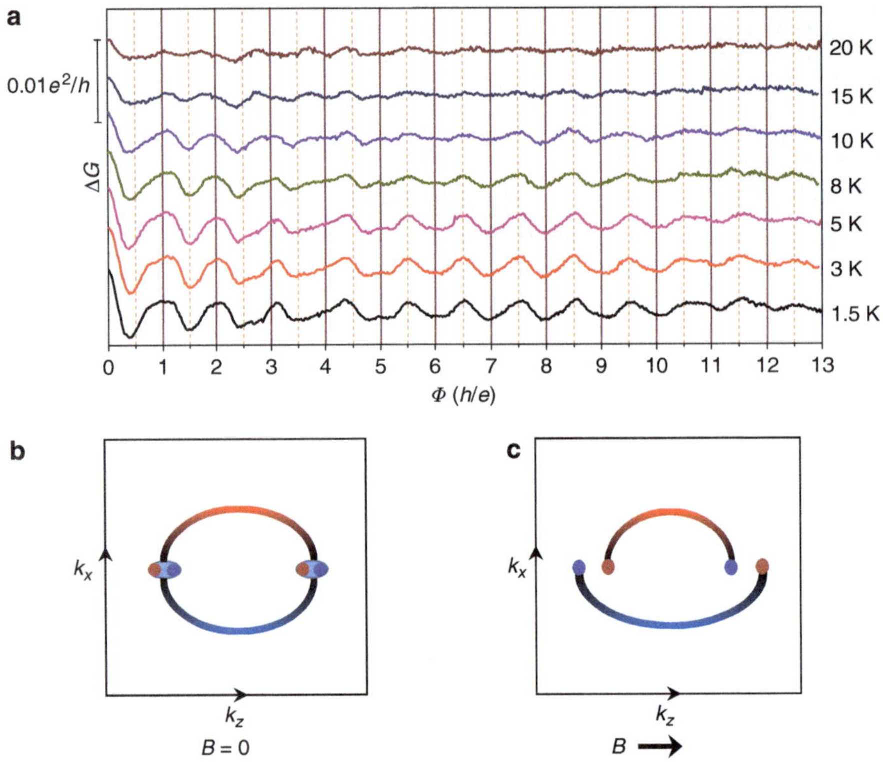

Figure 2 | Aharonov-Bohm oscillations. (**a**) Oscillating conductance as a function of magnetic flux in unit of h/e at variable temperatures. (**b,c**) Schematic diagram of the Weyl nodes projected onto the two-dimensional $k_x - k_z$ surface plane (**b**) without and (**c**) with an external magnetic field. The right-handed chirality and left-handed chirality are denoted by the red ball and blue ball, respectively. The Fermi arcs connect the Weyl nodes with opposite chirality. Without magnetic field, each Dirac node consists of two degenerate Weyl nodes. A magnetic field can separate the two degenerate Weyl nodes in the momentum space.

after cycling around one of the Weyl nodes in the low-energy band. Moreover, the analysis of Shubnikov-de Haas oscillations also suggests the presence of non-trivial π Berry phase (Supplementary Fig. 4)[16]. It is very interesting to compare our results with the A–B oscillations in topological insulator nanoribbons. The surface sub-bands have the 1D band dispersion as $E(n,k,\Phi) = \pm h v_F [\frac{k^2}{4\pi^2} + \left(n+0.5-\frac{\Phi}{\Phi_0}\right)^2 / W^2]^{1/2}$, where the A–B oscillation in conductance is anticipated to peak at odd integers of $\Phi_0/2$ (refs 26–28). Similarly, the A–B oscillation with peaks at odd integers of $\Phi_0/2$ was also found in our Cd_3As_2 nanowires under a magnetic field larger than a critical value. This phase-shift is believed to be in close relationship to the extra π Berry phase released by the magnetic field-induced lifting of degeneracy.

Angular dependence of A–B oscillations. To verify the influence of magnetic flux on the A–B effect, the nanowire direction was tilted, forming an angle θ with the direction of the magnetic field (Fig. 3a). As shown in Fig. 3b, the magnetotransport results illustrate a negative to positive MR transition from $\theta = 0°$ (parallel case) to $\theta = 40°$. The oscillations in ΔG obtained at different tilted angles are presented in Fig. 3c. The weak anti-localization induced conductance peak at $\mathbf{B} = 0$ is not sensitive to the tilt angle, indicating its bulk origin. At low magnetic field, the periodic oscillations at high tilt angles are still discernible. By scaling the magnetic field with Bcosθ, we replot the oscillations at different tilt angles in Fig. 3d. It is found that the oscillation peaks are line up in Fig. 3d, suggesting the dominant magnetic flux along the nanowire longitudinal direction. Meanwhile, the oscillation peaks split at high tilt angles, which may be due to the deflection of the electron trajectory under the Lorentz force from the perpendicular component $B_\perp$ of the magnetic field. It is also found that the periodic oscillations vanish gradually as increasing θ under high magnetic fields. The vanishing of oscillations under high magnetic field may be due to the $B_\perp$-induced breaking of the time reversal symmetry of the surface states and a bandgap opening, which is consistent with that observed in topological insulators[26].

Gate-voltage dependence of A–B oscillations. The periodic conductance oscillation is caused by the rearrangement of the surface sub-bands in energy as varying the magnetic flux[26,28]. Therefore, the occupation of subbands can affect the pattern of oscillation in aspect of magnitude and phase. By employing a back-gate with 285 nm SiO_2, we are able to modulate the sub-band occupation by applying gate voltage (V_g), and explore how the Fermi level affects the A–B oscillations. To achieve this goal, we performed systematic gate-modulation experiments on a very thin Cd_3As_2 nanowire with diameter ~65 nm (Device 2), in which a large sub-band spacing $\Delta E = \frac{h v_F}{W} \sim 21\text{meV}$ can be obtained. As shown in Fig. 4a, the transfer curve indicates the ambipolar field effect of the device. The resistance saturates to a certain value at high gate voltages (>5 V), indicating that the screening effect may hinder further gate-modulating as the Fermi-level is far away from the Dirac point. Nevertheless, as shown in Fig. 4a, the resistance can be largely tuned at small gate voltages V_g, demonstrating the effective gate-modulation as long as the screening effect is not strong enough.

We have measured the magnetotransport properties of the nanowire in the gate voltage range without obvious screening

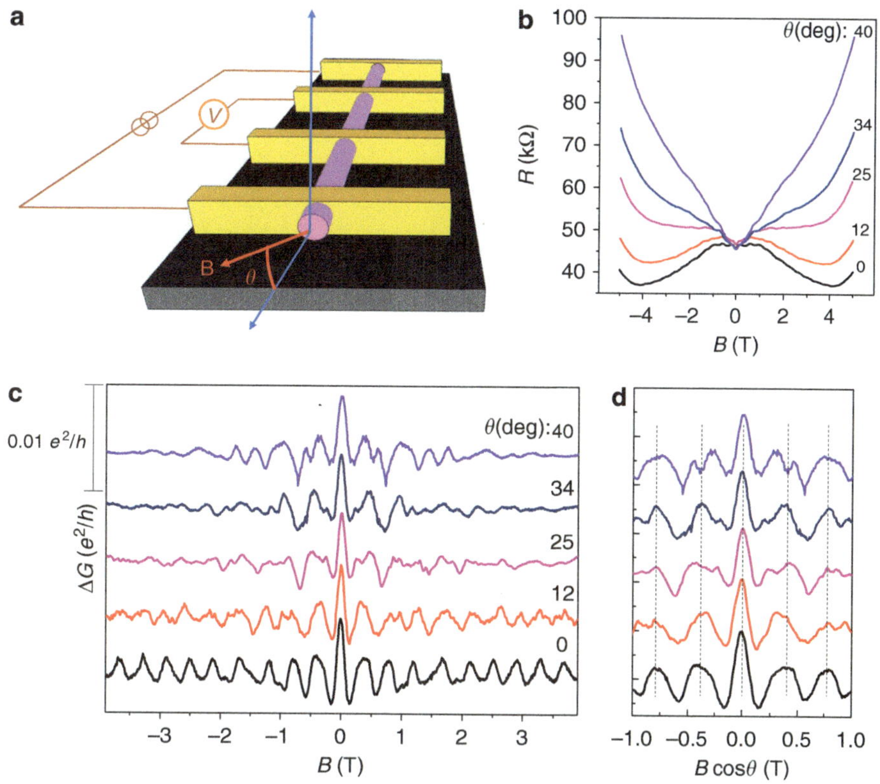

Figure 3 | Aharonov-Bohm oscillations at tilted magnetic field. (**a**) Schematic representation of the measurement configuration. The θ is the angle between the orientation of the magnetic field and the longitudinal direction of the nanowire. (**b**) Magnetoresistance at different θ angles and (**c**) the corresponding conductance oscillations. (**d**) The conductance oscillations as a function of $B\cos\theta$.

effect. As shown in Fig. 4b, the A–B oscillations in ΔG were plotted as a function of magnetic flux at different V_g of -7.5, -3, 0, 3 and 5 V. It is worth noting that recent studies on Cd_3As_2 thin films ($\sim$50 nm in thickness) demonstrate that a considerable bandgap >24.9 meV may be opened at the Dirac point as a result of quantum confinement effect[36]. Because this nanowire is with diameter of $\sim$65 nm, it is very likely that a bandgap is also opened at the Dirac point in the nanowire. Therefore, at $V_g = 0$ V (near the Dirac point as seen in Fig. 4a), the Fermi level is in the bandgap. As the Fermi level is inside the bandgap, the A–B oscillation thus has a normal origin, which may be responsible for the observation of the maxima of ΔG at even integers of $\Phi_0/2$ at $V_g = 0$ V, as shown in Fig. 4b. As tuning the Fermi level into the valence band, there is a phase shift of the A–B oscillations at $V_g = -3$ V. As the Fermi level further enters into the valence band with linear dispersion, the π-A–B effect is observed at $V_g = -7.5$ V, as shown in Fig. 4b, where the maxima of ΔG are at odd integers of $\Phi_0/2$.

As tuning the Fermi level into the conduction band, the A–B oscillations are still clearly observed but with more complicated features at $V_g = 3$ and 5 V, as shown in Fig. 4b. Previous studies indicate that the electron has much higher mobility than that of hole in Cd_3As_2 because of the different Fermi velocity and scattering rate with defects[9,14,37]. The magnetic field can force the bulk electrons to the surface under the classic limit of $R < L_B = \sqrt{\hbar/eB}$, that is, the nanowire radius R is smaller than the magnetic length L_B (refs 38,39). The electrons with high carrier mobility deflected from the bulk channel contribute to the surface conductance significantly, resulting in the irregular A–B oscillations. Furthermore, we have performed more systematic gate-dependent A–B oscillations on another sample with similar size (radius $\sim$37 nm, Device 3), as shown in Supplementary Fig. 7. The oscillation patterns are complicated at relatively weak magnetic fields, implying the co-contribution to the quantum oscillations from both the bulk and surface electrons. Fortunately, further increasing B may give rise to less scattering on the surface from bulk electrons as $L_B < R$, and there is a clear π-A–B effect under high magnetic field (Supplementary Fig. 7b).

In Dirac semimetals, the two bulk Dirac points are separated by $2k_D$ in momentum space. The relative magnitude of k_D and the Fermi wave vector k_F is thus crucial to describe the electronic structure in Dirac semimetals. Further increasing k_F larger than k_D, two Fermi pockets nested at $\pm k_D$ merge into one entire Fermi surface centred at Γ, called the Lifshitz transition[14]. As the system is back to degenerate system, the π Berry phase should disappear. To verify this scenario, we measured a nanowire (Device 4) with back-gate voltage up to 60 V. At $V_g = 60$ V, the oscillation peaks at relatively random positions under low magnetic fields (Supplementary Fig. 8), which may be due to the large magnetic length of the electron orbit and high carrier density. In this situation, the bulk carrier can be scattered to the surface, which makes the A–B oscillations complicated. As increasing magnetic field, the magnetic length of the bulk electron cyclotron motion decreases and the oscillation starts to peak exactly at the integer quantum flux with $\Phi = 5, 6, 7, 8$ h/e (Supplementary Fig. 8). Since the Fermi level is far from the Dirac points, the Fermi surface may enter above the Lifshitz transition, thus leading to the normal A–B oscillations.

Altshuler–Aronov–Spivak effect. For the thick nanowire device, the electron transport along the nanowire perimeter may be in the diffusive regime. Figure 5a shows a thick nanowire with diameter $\sim$200 nm (Device 5). Even in a relatively narrow field range from -1 to 1 T, prominent oscillations superposed on negative MR background can be easily discerned, as shown in Fig. 5b.

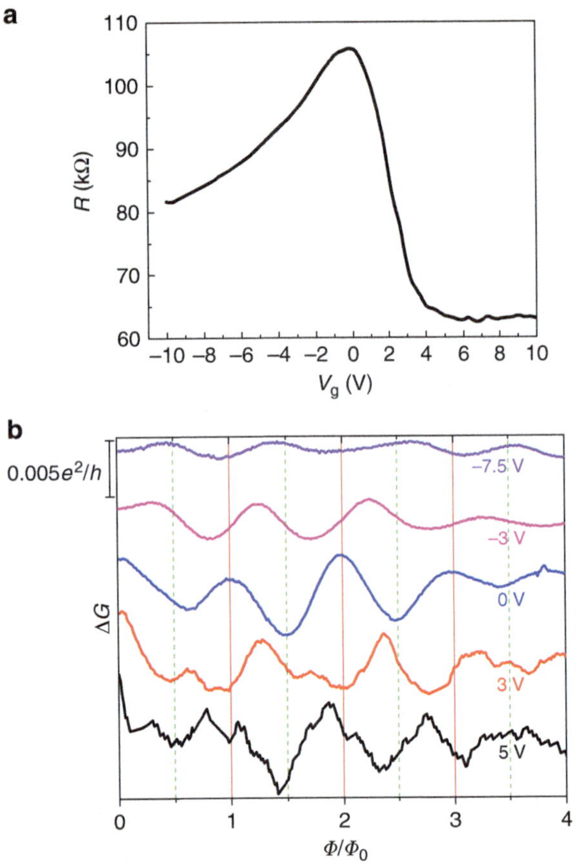

Figure 4 | Gate tuning of Aharonov-Bohm oscillations. (**a**) Gate voltage dependent resistance of a nanowire with diameter ~65 nm (Device 2) at 1.5 K. (**b**) A-B oscillations at gate voltages as denoted and at 1.5 K. The positive and negative gate voltages pull the Fermi level into the conduction band and valence band, respectively.

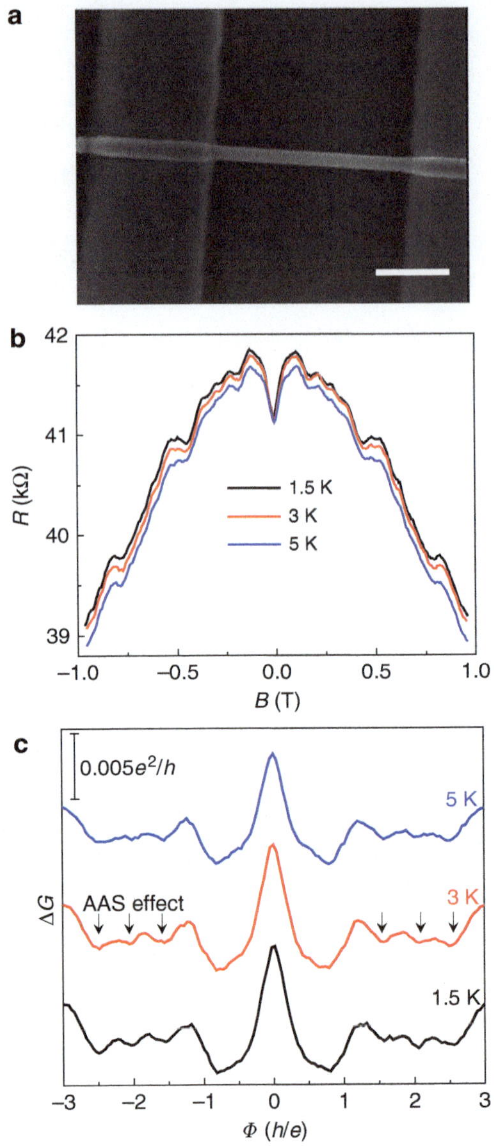

Figure 5 | Altshuler-Aronov-Spivak oscillations. (**a**) Scanning electron microscopic image of a thick nanowire with diameter about 200 nm (Device 5). Scale bar, 1 µm. (**b**) Magnetoresistance at different temperatures. (**c**) The conductance oscillations as a function of magnetic flux, showing the AAS effect.

According to the nanowire cross-sectional area (Supplementary Fig. 6), it is found that there are oscillations with $h/2e$ flux period, as noted by arrows in Fig. 5c. These $h/2e$ oscillations may be attributed to the Altshuler–Aronov–Spivak (AAS) effect in diffusive regime. Notably, the $h/2e$ oscillation peaks at $(n+\frac{1}{2})h/2e$ ($n=2, 3, 4$), indicating a phase shift for the $h/2e$ oscillations that needs further investigations.

Discussion

In summary, we have studied the A–B effect in individual Dirac semimetal Cd_3As_2 nanowires as the magnetic flux penetrates through the cross-section of the nanowire. As increasing magnetic field, a phase shift in A–B oscillations was observed, which is ascribed to the splitting of the Dirac node into Weyl nodes due to the magnetic field-induced time-reversal symmetry breaking. With this lifting of the Dirac node degeneracy, the conductance oscillation peaks at odd integers of $h/2e$ with period of h/e, which provides transport evidence of the topological surface states of Cd_3As_2 nanowires. As tuning the Fermi level by gate voltage, a transition between the 0 A–B effect and the π A–B effect was observed, which is self-consistent with the presence of linear dispersion relation in 1D-like quantum confined system. In thick nanowires, the A–B oscillations were largely reduced and the AAS oscillations presented due to the transition from quasi-ballistic to diffusive transport. The A–B oscillations observed in the Dirac semimetal nanowires show abundant features, offering a route to explore the exotic surface states.

Methods

Growth of the nanowires. The Cd_3As_2 nanowires were synthesized by chemical vapour deposition method in a horizontal furnace. For safety, the whole system was placed in a ventilation closet and the pumper was equipped with a cold trap to collect the dust in the exhaust. The Cd_3As_2 powders (purity >99.99%, Alfa Aesar) placed at the centre of the tube furnace were heated from room temperature to 650 °C in 20 min after pumping and flushing the tube several times to get rid of oxygen. The Si substrates covered by a 5-nm-thick gold film were placed at the downstream away from the source ~15 cm to collect the products. The system was maintained at 650 °C for 10 min and at about 1 atm pressure with ~20 s.c.c.m Argon flow during the growth process. Then, the system cooled down to room temperature naturally.

Device fabrication. The Cd_3As_2 nanowire devices were fabricated on a Si substrate with 285 nm SiO_2 layer. To make the nanowire direction can be parallel with the direction of the magnetic field, microstrips were first prefabricated on a SiO_2/Si square substrate as marks, which are parallel to one side of the square substrate. Then, the nanowire was transferred from the as-grown substrate to the SiO_2/Si square substrate, and the nanowire direction was aligned parallel to the prefabricated marks by a micromanipulator. The individual nanowire devices were fabricated using standard electron beam lithography techniques. To form ohmic contacts between Cd_3As_2 nanowires and electrodes, the amorphous layer of the nanowire was removed *in situ* by Ar^+ etching treatment in the metal deposition chamber. Then, ~150-nm-thick Au electrodes were deposited. The linear current–voltage curve (Supplementary Fig. 9) indicates the Ohmic contacts between the nanowire and the electrodes.

Transport measurements. The SiO_2/Si square substrate with the Cd_3As_2 nanowire devices was mounted onto the sample holder in an Oxford commercial variable temperature insert to make the nanowire parallel to the direction of the magnetic field. The sample holder can be rotated continuously by an electric motor to change the angle between **E** and **B**. The electrical signals were measured using a four-terminal method by Stanford SR830 lock-in amplifiers at frequency of 17.7 Hz.

References

1. Novoselov, K. *et al.* Two-dimensional gas of massless Dirac fermions in graphene. *Nature* **438**, 197–200 (2005).
2. Zhang, Y. B., Tan, Y. W., Stormer, H. L. & Kim, P. Experimental observation of the quantum Hall effect and Berry's phase in graphene. *Nature* **438**, 201–204 (2005).
3. Xu, Y. *et al.* Observation of topological surface state quantum Hall effect in an intrinsic three-dimensional topological insulator. *Nat. Phys.* **10**, 956–963 (2014).
4. Young, S. M. *et al.* Dirac semimetal in three dimensions. *Phys. Rev. Lett.* **108**, 140405 (2012).
5. Wang, Z., Weng, H., Wu, Q., Dai, X. & Fang, Z. Three-dimensional Dirac semimetal and quantum transport in Cd_3As_2. *Phys. Rev. B* **88**, 125427 (2013).
6. Wang, Z. *et al.* Dirac semimetal and topological phase transitions in A(3)Bi (A = Na, K, Rb). *Phys. Rev. B* **85**, 195320 (2012).
7. Liu, Z. K. *et al.* Discovery of a three-dimensional topological Dirac semimetal, Na_3Bi. *Science* **343**, 864–867 (2014).
8. Liu, Z. K. *et al.* A stable three-dimensional topological Dirac semimetal Cd_3As_2. *Nat. Mater.* **13**, 677–681 (2014).
9. Neupane, M. *et al.* Observation of a three-dimensional topological Dirac semimetal phase in high-mobility Cd_3As_2. *Nat. Commun.* **5**, 3786 (2014).
10. Borisenko, S. *et al.* Experimental realization of a three-dimensional Dirac semimetal. *Phys. Rev. Lett.* **113**, 027603 (2014).
11. Liang, T. *et al.* Ultrahigh mobility and giant magnetoresistance in the Dirac semimetal Cd_3As_2. *Nat. Mater.* **14**, 280–284 (2014).
12. He, L. *et al.* Quantum transport evidence for the three-dimensional Dirac semimetal phase in Cd_3As_2. *Phys. Rev. Lett.* **113**, 246402 (2014).
13. Jeon, S. *et al.* Landau quantization and quasiparticle interference in the three-dimensional Dirac semimetal Cd_3As_2. *Nat. Mater.* **13**, 851–856 (2014).
14. Cao, J. *et al.* Landau level splitting in Cd_3As_2 under high magnetic fields. *Nat. Commun.* **6**, 6779 (2015).
15. Xiang, Z. *et al.* Angular-dependent phase factor of Shubnikov-de Haas oscillations in the Dirac semimetal Cd_3As_2. *Phys. Rev. Lett.* **115**, 226401 (2015).
16. Zhao, Y. *et al.* Anisotropic Fermi surface and quantum limit transport in high mobility three-dimensional Dirac semimetal Cd_3As_2. *Phys. Rev. X* **5**, 031037 (2015).
17. Xiong, J. *et al.* Evidence for the chiral anomaly in the Dirac semimetal Na_3Bi. *Science* **350**, 413 (2015).
18. Li, H. *et al.* Negative magnetoresistance in Dirac semimetal Cd_3As_2. *Nat. Commun.* **7**, 10301 (2016).
19. Parameswaran, S., Grover, T., Abanin, D., Pesin, D. & Vishwanath, A. Probing the chiral anomaly with nonlocal transport in three-dimensional topological semimetals. *Phys. Rev. X* **4**, 031035 (2014).
20. Li, C.-Z. *et al.* Giant negative magnetoresistance induced by the chiral anomaly in individual Cd_3As_2 nanowires. *Nat. Commun.* **6**, 10137 (2015).
21. Zhang, C. *et al. Detection of chiral anomaly and valley transport in Dirac semimetals.* Preprint at http://arxiv.org/abs/1504.07698 (2015).
22. Xu, S.-Y. *et al.* Observation of Fermi arc surface states in a topological metal. *Science* **347**, 294–298 (2015).
23. Yi, H. *et al.* Evidence of topological surface state in three-dimensional Dirac semimetal Cd_3As_2. *Sci. Rep.* **4**, 6106 (2014).
24. Zhang, Y. & Vishwanath, A. Anomalous Aharonov-Bohm conductance oscillations from topological insulator surface states. *Phys. Rev. Lett.* **105**, 206601 (2010).
25. Bardarson, J. H., Brouwer, P. & Moore, J. Aharonov-Bohm oscillations in disordered topological insulator nanowires. *Phys. Rev. Lett.* **105**, 156803 (2010).
26. Hong, S. S., Zhang, Y., Cha, J. J., Qi, X.-L. & Cui, Y. One-dimensional helical transport in topological insulator nanowire interferometers. *Nano Lett.* **14**, 2815–2821 (2014).
27. Cho, S. *et al.* Aharonov-Bohm oscillations in a quasi-ballistic three-dimensional topological insulator nanowire. *Nat. Commun.* **6**, 7634 (2015).
28. Jauregui, L. A., Pettes, M. T., Rokhinson, L. P., Shi, L. & Chen, Y. P. Magnetic field-induced helical mode and topological transitions in a topological insulator nanoribbon. *Nat. Nanotech.* Doi:10.1038/nnano.2015.293 (2016).
29. Stone, A. D. Magnetoresistance fluctuations in mesoscopic wires and rings. *Phys. Rev. Lett.* **54**, 2692 (1985).
30. Peng, H. *et al.* Aharonov-Bohm interference in topological insulator nanoribbons. *Nat. Mater.* **9**, 225–229 (2010).
31. Murakawa, H. *et al.* Detection of Berry's phase in a bulk Rashba semiconductor. *Science* **342**, 1490 (2013).
32. Washburn, S., Umbach, C., Laibowitz, R. & Webb, R. A. Temperature dependence of the normal-metal Aharonov-Bohm effect. *Phys. Rev. B* **32**, 4789 (1985).
33. Fagas, G. & Greer, J. C. Ballistic conductance in oxidized Si nanowires. *Nano Lett.* **9**, 1856–1860 (2009).
34. Xiu, F. *et al.* Manipulating surface states in topological insulator nanoribbons. *Nat. Nanotechnol.* **6**, 216–221 (2011).
35. Gorbar, E., Miransky, V. & Shovkovy, I. Engineering Weyl nodes in Dirac semimetals by a magnetic field. *Phys. Rev. B* **88**, 165105 (2013).
36. Liu, Y. *et al.* Gate-tunable quantum oscillations in ambipolar Cd_3As_2 thin films. *NPG Asia Mater.* **7**, e221 (2015).
37. Zhang, E. *et al.* Magnetotransport properties of Cd_3As_2 nanostructures. *ACS Nano* **9**, 8843 (2015).
38. Heremans, J. *et al.* Bismuth nanowire arrays: synthesis and galvanomagnetic properties. *Phys. Rev. B* **61**, 2921 (2000).
39. Nikolaeva, A. A. *et al.* Negative magnetoresistance in transverse and longitudinal magnetic fields in Bi nanowires. *J. Phys. Conf. Ser.* **150**, 022065 (2009).

Acknowledgements

We are grateful to Professors Xincheng Xie, Ji Feng, Kai Chang and Dr Haiwen Liu for inspired discussions. This work was supported by MOST (nos 2013CB934600 and 2013CB932602) and NSFC (nos 11274014, 11234001 and 11327902).

Author contributions

Z.-M.L. conceived and designed the experiments. D.-P.Y. gave scientific advice. L.-X.W. and C.-Z.L. performed the measurements. Z.-M.L. and L.-X.W. wrote the manuscript.

Additional information

Competing financial interests: The authors declare no competing financial interests.

26

Spin-momentum coupled Bose-Einstein condensates with lattice band pseudospins

M.A. Khamehchi[1], Chunlei Qu[2], M.E. Mossman[1], Chuanwei Zhang[2] & P. Engels[1]

The quantum emulation of spin-momentum coupling, a crucial ingredient for the emergence of topological phases, is currently drawing considerable interest. In previous quantum gas experiments, typically two atomic hyperfine states were chosen as pseudospins. Here, we report the observation of a spin-momentum coupling achieved by loading a Bose-Einstein condensate into periodically driven optical lattices. The *s* and *p* bands of a static lattice, which act as pseudospins, are coupled through an additional moving lattice that induces a momentum-dependent coupling between the two pseudospins, resulting in *s*-*p* hybrid Floquet-Bloch bands. We investigate the band structures by measuring the quasimomentum of the Bose-Einstein condensate for different velocities and strengths of the moving lattice, and compare our measurements to theoretical predictions. The realization of spin-momentum coupling with lattice bands as pseudospins paves the way for engineering novel quantum matter using hybrid orbital bands.

[1] Department of Physics and Astronomy, Washington State University, Pullman, Washington 99164, USA. [2] Department of Physics, The University of Texas at Dallas, Richardson, Texas 75080, USA. Correspondence and requests for materials should be addressed to C.Z. (email: chuanwei.zhang@utdallas.edu) or to P.E. (email: engels@wsu.edu).

Spin-momentum coupling (SMC), commonly called spin–orbit coupling, is a crucial ingredient for many important condensed matter phenomena such as topological insulator physics, topological superconductivity, spin Hall effects and so on[1–3]. In this context, the recent experimental realization of SMC in ultracold atomic gases provides a powerful platform for engineering many interesting and novel quantum phases[4–9]. In typical experiments, two atomic hyperfine states act as two pseudospins that are coupled to the momentum of the atoms through stimulated Raman transitions[10,11]. However, ultracold atoms in optical lattice potentials possess other types of degrees of freedom, which can also be used to define pseudospins[12,13]. A natural and important question is whether such new types of pseudospins can be employed to generate SMC.

In optical lattices filled with ultracold atoms, *s*- and *p*-orbital bands are separated by a large energy gap and can be defined as two pseudospin states. One significant difference between hyperfine state pseudospins and lattice band pseudospins lies in the energy dispersion of spin-up and spin-down orientations: the dispersion relations are the same for hyperfine state pseudospins, while they are inverted for lattice band pseudospins. It is well known from topological insulators and superconductor physics that inverted band dispersions, together with SMC, play a central role for topological properties of materials[14–16]. Therefore, it is natural to expect that the inverted band pseudospins, when coupled with the lattice momentum, may lead to interesting topological phenomena in cold atomic optical lattices. Recent experiments with shaken optical lattices (that is, lattices in which the lattice sites are periodically in time shifted back and forth[17]) have realized a simple coupling ($\Omega\sigma_x$ coupling, where Ω is the coupling strength and σ_x a Pauli matrix) between *s*- and *p*-band pseudospins, analogous to Rabi coupling between two regular spins[18]. However, for the exploration of exotic phenomena in optical lattice systems, such as Fulde–Ferrell–Larkin–Ovchinnikov (FFLO) phases[19,20] and Majorana fermions[16], SMC with *s*- and *p*-band pseudospins is highly desirable[21–24].

In our experiments we realize such *s*–*p* band SMC for a Bose-Einstein condensate (BEC) using a weak moving lattice to generate Raman coupling between *s*- and *p*-band pseudospins of a static lattice[25]. The moving lattice acts as a periodic driving field[26–31] and has previously been used to generate an effective magnetic field in the lowest *s* band of a tilted optical lattice[32,33]. In our experiment, the driving frequency of the moving lattice is chosen close to the energy gap between *s* and *p* bands at zero quasimomentum, leading to a series of hybrid *s*-*p* Floquet-Bloch (FB) band structures. FB band structures in optical lattices give rise to interesting and important phenomena in cold atoms and solids[34,35], as is evidenced by the recent experimental realization of a topological Haldane model in a shaken honeycomb optical lattice[36], and the observation of FB states on the surface of a topological insulator[37]. We show that the moving lattice generates two types of coupling between *s*- and *p*-band pseudospins: a momentum-independent Rabi coupling ($\Omega\sigma_x$) and SMC ($\alpha\sigma_x \sin(q_x d)$, where q_x is the quasimomentum and d the lattice period), with strengths of the same order. The coexistence of these two types of coupling leads to asymmetric FB band dispersions[38]. We investigate the FB band structures by measuring the quasimomentum of the BEC. The initial phase of the moving lattice plays a significant role in the Floquet dynamics[29], the effects of which are explored through a quantum quench-induced dynamical coupling of the FB bands. Results are compared with the theoretical predictions from a simple two-band model and from numerical simulations of the Gross–Pitaevskii (GP) equation.

Results

Experimental set up. To generate the *s*–*p* band SMC and FB band structures, we begin with a ^{87}Rb BEC composed of $\sim 5 \times 10^4$ atoms confined in a crossed dipole trap. A static lattice is generated by two perpendicular laser beams with wavelength $\lambda \approx 810$ nm intersecting at the position of the BEC, as schematically shown in Fig. 1a. The harmonic trap frequencies due to the envelope of the static lattice beams and the crossed dipole trap are $(\omega_x, \omega_y, \omega_z) = 2\pi \times (41, 159, 115)$ Hz, where $\mathbf{e}_x$ points along the lattice, $\mathbf{e}_y$ is the horizontal transverse direction and $\mathbf{e}_z$ is the vertical direction. A weak moving lattice with the same lattice period as the static lattice, $d = \pi/k_L$ where $k_L = \sqrt{2}\pi/\lambda$, is then overlaid with the static lattice (Fig. 1b). The moving lattice beams are ~ 180 MHz detuned from the static lattice. A small frequency difference Δ_ω between the two moving lattice beams determines the velocity of the lattice according to $v_{\text{lattice}} = \Delta_\omega / 2k_L$. To induce *s*–*p* orbital band coupling, $|\Delta_\omega|$ is chosen close to the energy gap E_{sp} between the *s* and *p* bands of the static lattice at quasimomentum $q_x = 0$.

One outstanding feature of the coupling scheme employed in these experiments is the asymmetry of the effective *s*–*p* FB bands, which exhibit a local minimum located at a finite quasimomentum $q_x \neq 0$. The direction in which the minimum is shifted away from $q_x = 0$ is determined by the sign of Δ_ω (which determines the direction of motion of the moving lattice) and $|\Delta_\omega| - E_{sp}$ (that is, the detuning of the drive from the bandgap at $q_x = 0$). Before describing experimental results and a formal derivation of

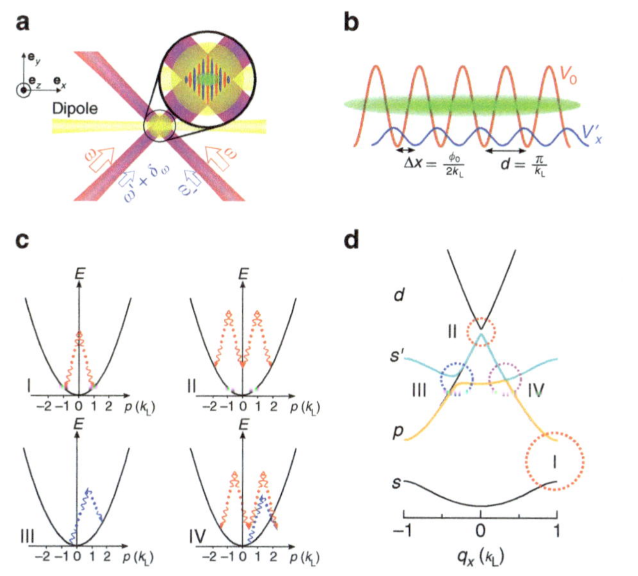

Figure 1 | Experimental set up and schematic lattice illustration. (**a**) Experimental arrangement. The crossed dipole trap beams propagate in the $\mathbf{e}_x$ and $\mathbf{e}_z$ directions. The static and moving lattices have overlapping beams propagating along $\mathbf{e}_x + \mathbf{e}_y$ and $-\mathbf{e}_x + \mathbf{e}_y$. The static lattice is generated using the red beams with laser frequency ω, and moving lattice is generated using the blue beams with laser frequencies ω' and $\omega' + \delta\omega$. (**b**) Lattice potentials along the $\mathbf{e}_x$ direction. The lattice period d is identical for the static lattice V_0 and the moving lattice V'_x. The initial offset between lattice sites of the static and moving lattice, Δx, is dependent on the initial phase ϕ_0 between the two lattices. (**c**,**d**) Illustration of the multi-photon processes for the driven lattice system and the corresponding FB band structure in the first Brillouin zone. The static lattice induces a large energy gap (I) through a two-photon process and a small energy gap (II) through a four-photon process. The moving lattice induces an energy gap when the *s* band and the *p* band are coupled through (III). A smaller energy gap is produced by a combination of the static and moving lattice (IV).

the band structure using Floquet theory[29,30], we lay the groundwork by presenting a multi-photon resonance picture that provides intuitive insights (Fig. 1c,d). In this picture, one starts with the parabolic dispersion of a free atom in the absence of any external potentials. An optical lattice then induces $2n$-photon couplings (with n being an integer number) between points of the dispersion relation due to the absorption and stimulated emission processes. The couplings are centred around pairs of points that fulfil conservation of energy and momentum. At these points, bandgaps open due to avoided crossings. Examples for possible couplings due to the static lattice (red arrows in Fig. 1c) and the moving lattice (blue arrows in Fig. 1c), and the associated bandgaps in the first Brillouin zone are shown in Fig. 1. Different coupling strengths lead to different sizes of bandgaps, which result in an asymmetric band structure.

In another pictorial way, the Floquet band structure for the time-periodic system can be constructed by creating multiple copies of the Bloch band structure of the static lattice that are offset in energy by $|\Delta_\omega|$. The moving lattice couples the p band and the shifted s band (labelled by s' in Fig. 1d) at points where the shifted s band intersects the unshifted p band. The gaps opened by the coupling can formally be calculated using Floquet theory.

Experimental measurements. Adiabatic loading of the BEC into an s-p FB band is achieved by first ramping on the intensity of the static lattice, followed by adiabatically ramping on the moving lattice intensity. In this procedure, the initial relative phase between the two lattices, ϕ_0 (Fig. 1b), becomes irrelevant and can effectively be set to zero. As we shall show in this paper, if the moving lattice is suddenly jumped on instead of adiabatically ramped on, this initial relative phase may manifest itself by markedly changing the dynamics of the system[29].

Figure 2a shows the measurement of the band minimum, $q_{\min}$, for different driving frequencies, Δ_ω, after adiabatically loading a BEC into a FB band. The driving frequencies are chosen such that $\hbar\Delta_\omega$ lies in the gap at $q_x=0$ between the p band (4.64 E_R, where $E_R=\hbar^2k_L^2/2m=h\times1749.5$ Hz) and the d band (5.44 E_R). After adiabatically loading a BEC into a FB band, all lasers are switched off and the BEC is imaged after 14 ms time-of-flight (TOF). In the experimental images clouds with three different kinetic momenta are seen: $q_{\min}$ and $q_{\min}\pm2k_L$. The kinetic momentum of the middle component, $q_{\min}$, is equal to the quasimomentum of the BEC in the hybrid s-p band. Therefore the quasimomentum can be obtained by measuring the position of the three components in the TOF images[39]. Each data point in Fig. 2a is an average over five iterations of the measurement. A shift of the quasimomentum is detected that decreases with increasing driving frequency (Fig. 2a) as the coupling between the p band and shifted s band becomes weaker. The observed shift indicates a shift of the minimum of the upper hybrid band (Fig. 2b) into which the BEC is adiabatically loaded. The solid line in Fig. 2a shows $q_{\min}$ calculated from a simple two-band model (see below) and is in reasonable agreement with the data. The symbols are the results from solution of the Schrödinger equation (squares) and the GP equation (stars), with finite nonlinear interaction strength. The periodically driven dynamics are simulated using the time-dependent GP equation for a two-dimensional system with the same geometry as the experiment. The width of the BEC is $\sim10\,\mu$m. We see that the interaction could modify the single-particle results. Since the atom loss is usually large in the experiment, the numerical results are not intended for direct comparison with the measurements. See Methods for some samples of the GP simulations.

Figure 3 presents a complementary data set for which the driving frequency is set to a constant value with $|\Delta_\omega|<E_{sp}$ (Fig. 3a) or $|\Delta_\omega|>E_{sp}$ (Fig. 3c) and the quasimomentum is determined for various depths of the moving lattice. The sign of Δ_ω determines the direction of motion of the moving lattice. For $|\Delta_\omega|<E_{sp}$ the BEC resides in the lower hybrid s-p FB band (Fig. 3b), while for $|\Delta_\omega|>E_{sp}$ it is in the upper hybrid band (Fig. 3d). This leads to a shift of the $q_{\min}$ into opposite directions for the two cases. For a given driving frequency, the coupling of the two bands is stronger for larger driving field strength (that is, larger depth of the moving lattice) so that the BEC is shifted to a larger absolute value of quasimomentum.

Floquet systems such as the one in our experiment are described by quasienergy bands. They do not have a thermodynamic ground state, and in the presence of many-body interactions their stability can be affected by a variety of factors[40–42]. Resonance-induced collective excitations and modulational instabilities can lead to losses[43]. A prominent feature of band inversion is the fact that hybrid bands can transition between stable and modulationally unstable structures. As shown in Fig. 4a, when the driving frequency approaches the dip α from the left, the lower hybrid band evolves from a globally stable structure to a locally stable structure and eventually enters the globally unstable region at the quasimomentum where BEC mainly resides (Fig. 4b). The instability causes excitations and heating of the BEC. Experimentally, we study the stability of the system by determining the number of condensed atoms

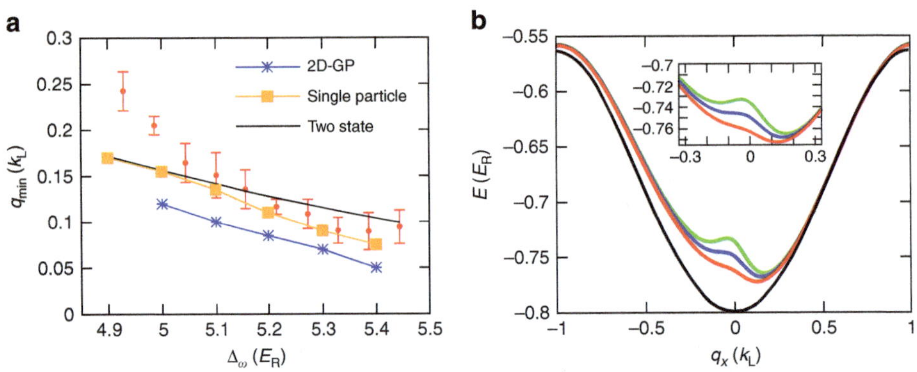

Figure 2 | Effects of the driving frequency. (**a**) Band minimum $q_{\min}$ for the upper hybrid band versus driving frequency Δ_ω. The depth of the moving lattice is 1 E_R. The filled circles are experimental measurements averaged over 5 data points with s.d. errorbars. The black line shows the theoretical prediction of a two-band model. The squares and stars are the results of numerical simulations of the Schrödinger equation and the GP equation, respectively. (**b**) Calculated upper hybrid s-p FB band structure for different driving frequencies $\Delta_\omega=4.99$ E_R, 5.1 E_R and 5.22 E_R from top to bottom. The lowest (black) curve is the s-orbital band without the presence of the driving field.

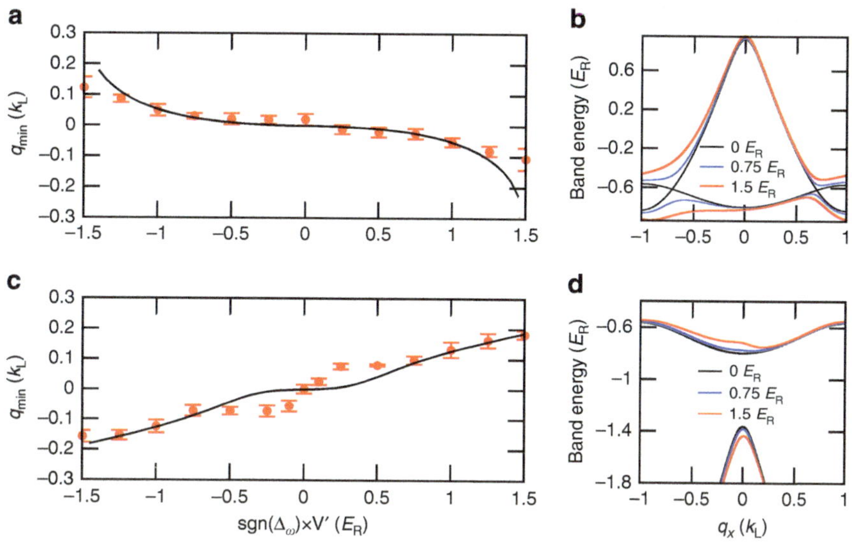

Figure 3 | Effects of the driving strength. Band minimum q_{min} versus driving field strength V' for different driving frequencies of (**a**) $|\Delta_\omega| = 2.92\ E_R$ and (**c**) $|\Delta_\omega| = 5.21\ E_R$. The red points are experimental data averaged over 5 data points with s.d. errorbars, and the solid lines are the the theoretical predictions from a two-band model. $sgn(\Delta_\omega)$ determines the direction of motion of the moving lattice. (**b**,**d**) Corresponding hybrid band structures for different driving field strengths $V'_x = 1.5\ E_R$, 0.75 E_R and 0 E_R (red, blue and black curves, respectively).

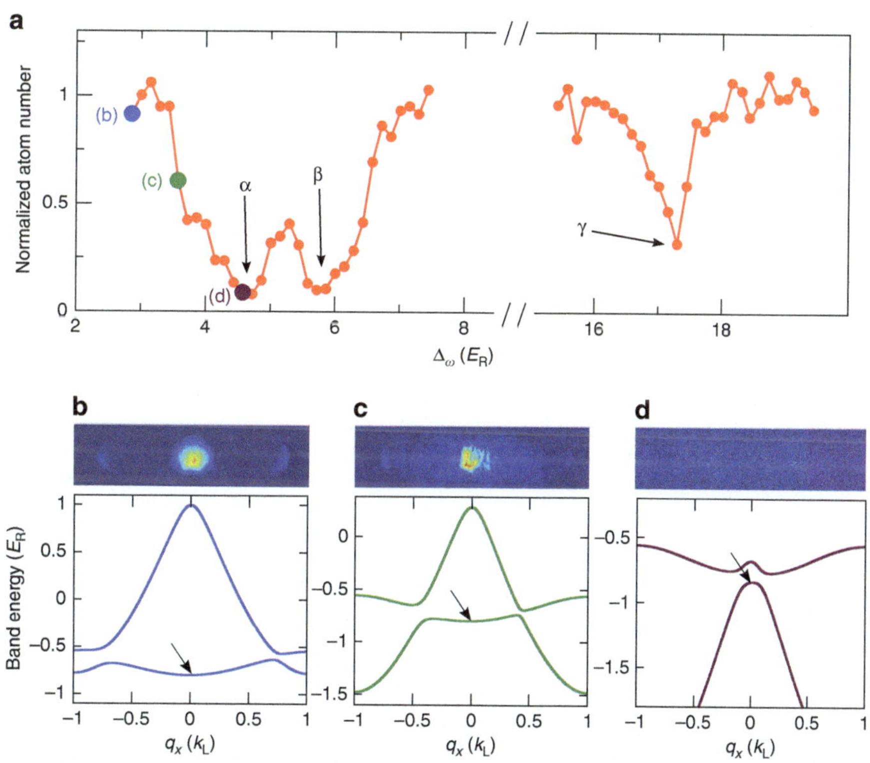

Figure 4 | Stability of the Floquet system. (**a**) Number of atoms remaining in trap after adiabatically loading a BEC into the FB band, normalized to initial atom number determined from independent experimental runs. The static lattice is ramped on to 5.47 E_R in 200 ms. Then the moving lattice is ramped on to a depth of $V' = 0.5\ E_R$ in 60 ms. The dips α, β and γ occur close to the Bloch bands p, d and f. (**b**–**d**) The TOF images and the effective band structures for data points labelled in **a**. The arrows show where the BEC is situated after the moving lattice is adiabatically ramped on.

left in the trap after the static and the moving lattices are successively and adiabatically ramped on. As the instabilities cause heating, the heated atoms are evaporated out of the trap. Therefore absorption imaging reveals an atom loss from the BEC as shown in Fig. 4. The dips α, β and γ in Fig. 4a occur when the driving frequency is chosen such that it leads to a coupling close to the Bloch bands p, d and f of the static lattice at $q_x = 0$, respectively.

Minimal two-band model. The dynamics of the BEC are governed by the full time-dependent GP equation,

$i\hbar\frac{\partial}{\partial t}\psi(\mathbf{r},t)=[H_0(t)+V_{\rm trap}+V_{\rm int}]\psi(\mathbf{r},t)$ where $V_{\rm trap}$ and $V_{\rm int}$ are the external trapping potential and the mean-field interaction, respectively. $H_0(t)$ is the single-particle Hamiltonian,

$$H_0(t)=\frac{p^2}{2m}+V_0\cos^2(k_{\rm L}x)+V'\cos^2\left(k_{\rm L}x+\frac{\phi_0}{2}-\frac{\Delta_\omega t}{2}\right), \quad (1)$$

where the second and the third terms describe the static and moving optical lattices, respectively, and ϕ_0 is the initial relative phase between the two sets of lattices.

When the static lattice depth V_0 is large and when $|\Delta_\omega|$ is close to the energy gap E_{sp}, higher orbital bands are not significantly populated in the driven process and the system is well described by a simple two-band tight-binding model[38]. Following the standard procedure in Floquet theory, we obtain the effective single-particle Hamiltonian (see Methods)

$$H_0^{\rm eff}=\begin{pmatrix}\epsilon_s(q_x) & \Delta_{sp}\\ \Delta_{sp}^* & \epsilon_p(q_x)-|\Delta_\omega|\end{pmatrix}, \quad (2)$$

where

$$\Delta_{sp}=-i[\Omega-\alpha\sin(q_xd)+\beta\cos(q_xd)]e^{-i\phi_0} \quad (3)$$

is the coupling between s- and p-orbital bands that is induced by the moving lattice potential for $\Delta_\omega>0$ (see Methods for more details), and ϵ_s and ϵ_p are the energy dispersions for the uncoupled orbital bands. The three coupling coefficients Ω, α and β are given by $\Omega=\frac{V'}{4}\langle s_i|\sin(2k_{\rm L}x)|p_i\rangle$, $\alpha=\frac{V'}{2}\langle s_i|\cos(2k_Lx)|p_{i+1}\rangle$ and $\beta=\frac{V'}{2}\langle s_i|\sin(2k_{\rm L}x)|p_{i+1}\rangle$, where $|s_i\rangle$ and $|p_i\rangle$ are the maximally localized Wannier orbital states in the ith site. Ω is the coupling between s- and p-orbital states in the same lattice site, while α and β are the couplings between s- and p-orbital states of nearest neighbouring sites. SMC between s–p band psuedospins is represented by $\alpha\sin(q_xd)\sigma_x$ after a spin rotation.

This derivation shows that the inversion symmetry of the FB band structure is broken due to the coexistence of couplings of different parities. When the moving lattice intensity is adiabatically ramped on, the quasimomentum of the BEC gradually shifts away from $q_x=0$ in a definite direction following the hybrid band minimum. This is quite different from previous shaken lattice experiments[17], where the inversion symmetry of the band was preserved and the BEC could spontaneously choose either side of $q_x=0$ as its ground state. In that case, the BEC needed to be accelerated to break the inversion symmetry. In our scheme, the position of the true minimum is uniquely determined by the moving velocity direction, moving lattice depth and driving frequency.

This minimal two-band model captures the essential physics of the driven lattices, as we have seen through the comparison of experimental measurements and theoretical values (Figs 2 and 3), demonstrating the observation of SMC between s–p band pseudospins. However, this model may deviate from the experiment when the modulated dynamics involve additional orbital bands or when the nonlinear interaction is strong such that the single-particle band structure will be renormalized by the interaction term.

Quench dynamics. Since a Floquet system is generated by a time-periodic Hamiltonian, an important question concerns the role of the initial phase of the driving field[29]. For the system considered in this work, this phase determines the relative positions between the moving and static lattice sites. Though the relative phase does not change the effective band structure (equation (2)), and thus the time-averaged dynamics, it can play a crucial role in the micromotion of the BEC. To demonstrate the effect of the initial relative phase, we study the oscillations in the population of the momentum components $k_x=0$, $\pm2k_{\rm L}$ after a quantum quench. Figure 5 presents such quench dynamics after adiabatically ramping on the static lattice to 5.47 $E_{\rm R}$ followed by a sudden jump on of the moving lattice to $V_x'=1\,E_{\rm R}$ with an on-resonant driving frequency $|\Delta_\omega|=E_{sp}$. Figure 5a,b shows the effective band structure before (Fig. 5a) and after (Fig. 5b) the moving lattice is jumped on. The jump projects the BEC on to the new band structure and the BEC acquires components in the new lower and upper hybrid band. Subsequently, these components beat against each other. We focus on the evolution during the first 3 ms, during which the BEC mainly stays at $q_x=0$ without significant dipole motion in the hybrid bands. The TOF images consist of three momentum populations as shown in the two examples in Fig. 5c, which are both taken after 0.5 ms of evolution. In our experiments, the initial phase ϕ_0 between the static and the moving lattice is uncontrolled, and is different in each repetition of the experiment. This leads to different population dynamics in each experimental iteration and explains the differences between the two images in Fig. 5c. The population dynamics measured in 10 subsequent experimental iterations are represented by the rectangles in Fig. 5d,e. For each measurement time, the height of the rectangle indicates the spread in the data. The shaded areas represent the result of numerical GP simulations for a homogeneous spread of relative phases. We find a strong correlation between the experimental spread and the spread predicted by the numerics (Fig. 5d,e). Numerical results calculated for the particular value $\phi_0=0$ are shown by the solid line in Fig. 5d,e. They reveal that for a fixed initial phase there are oscillations on two different time scales. The fast oscillations (with period $T\approx0.12$ ms) corresponds to the micromotion of particles under the high-frequency periodic driving, whereas the slow oscillation (with $T\approx1.75$ ms) corresponds to the time-averaged effective Rabi oscillations between the two hybrid FB bands. For longer holding times, the periodicity is slightly broken due to a small dipole motion. Figure 5f shows the correlation between numerical and experimental spread for the data in Fig. 5d,e. The straight line is a fit showing the correlation trend. A linear correlation of $\rho_{X,Y}=0.64$ is achieved, where $\rho_{X,Y}$ is the Pearson product-moment correlation coefficient defined by $\rho_{X,Y}=\frac{E[(X-E[X])(Y-E[Y])]}{\sigma_X\sigma_Y}$. Here E is the expectation value and σ is the s.d.

Discussion

We have realized and characterized a SMC with lattice bands as pseudospins. This not only provides a powerful tool to control orbital states with a driving field, but also enriches the study of novel quantum matter using hybrid orbital bands. There are many directions that can be taken along this route, for example, the engineering of similar SMC in higher dimensional systems, involving different orbital bands, and quantitative analysis and measurements of the effects of strong interactions on the effective bands. The realization of similar SMC for fermionic atoms such as ^{6}Li and ^{40}K with tunable interactions may open the door for exploring exotic quantum matter. For example, for a spin-balanced Fermi gas loaded into such tilted s–p dressed bands, the Cooper pairs may acquire a finite centre-of-mass momentum due to the broken inversion symmetry that provides a route to search for the long-sought Fulde–Ferrell–Larkin–Ovchinnikov states[38]. Furthermore, the realized spin-orbit coupling with lattice bands as pseudospins may provide a new platform for the study of topological insulators and Majorana fermions[14–16].

Methods

GP simulations. In Fig. 6, we present the results of GP simulations for different moving lattice depths. The modification of the band structure minimum can clearly

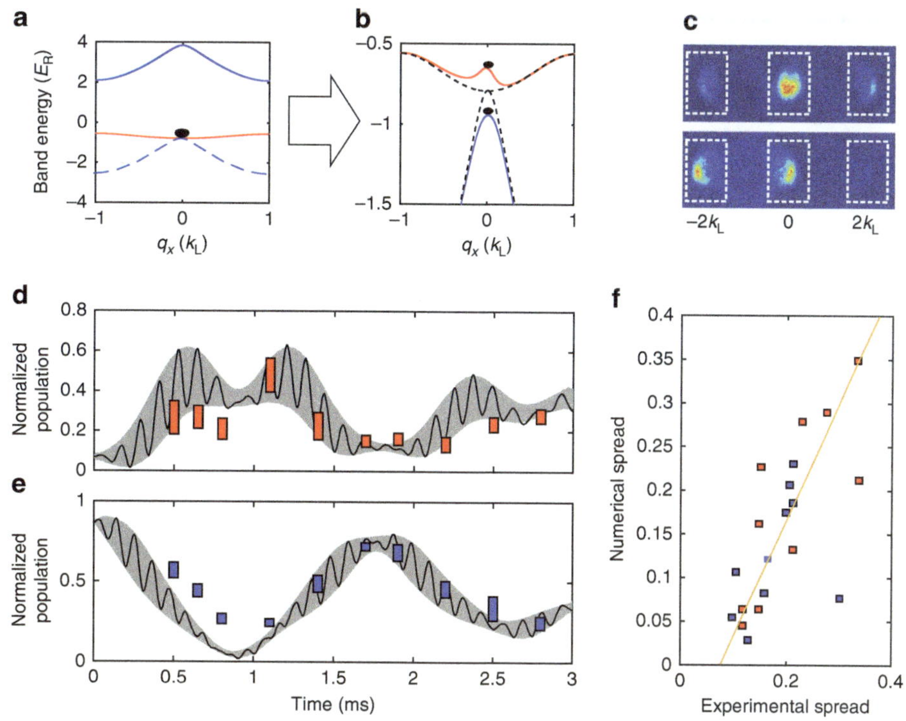

Figure 5 | Quench dynamics after suddenly jumping on the coupling between the *s* and *p* band. Band structure before (**a**) and after (**b**) jumping on the coupling between the *s* and *p* band. (**c**) Two examples for experimental images taken 0.5 ms after the quench. The difference between the images is due to the relative phase between the static and the moving lattice. The experimental data for **d** and **e** are extracted from such images by counting the atom number in the dashed boxes. They indicate the $-2\hbar k$ (left rectangle), 0 (middle rectangle) and $+2\hbar k$ component (right rectangle). (**d**,**e**) The normalized population of the momentum component $-2\hbar k$ and 0 $\hbar k$, respectively. The spread of the experimental data, indicated by the height of the rectangles, is the s.d. of 10 measurements. The shaded areas are the results of numerical GP simulations calculated for a uniform spread of phases ϕ_0. The black curves show the numerical result for $\phi_0 = 0$. (**f**) The correlation between numerical and experimental spread for the data from **d** and **e**. The solid line is a least squares fit to the scatter.

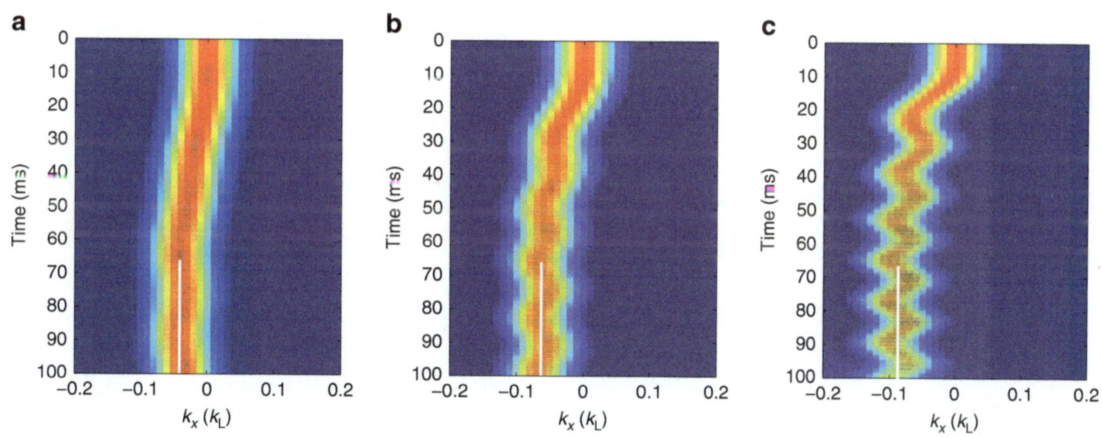

Figure 6 | Plots of the evolution of the BEC momentum space density distribution as a function of k_x for different evolution times. The moving lattice depth is ramped to (**a**) $V' = 0.25\ E_R$, (**b**) $V' = 0.5\ E_R$ and (**c**) $V' = 1.0\ E_R$ in the first 60 ms and then keep this value for the following 40 ms. The white lines indicate the centre of the small dipole osicllation from which the band minimum is determined. The driving frequency is $|\Delta_\omega| = 5.21\ E_R$.

be seen from the shift of the quasimomentum of the BEC during the driving process.

Tight-binding model. We consider a BEC in a combination of a one-dimensional static optical lattice and a moving lattice,

$$V_{\text{static}}(x) = V_0 \cos^2(k_L x), \tag{4}$$

$$V(x,t) = V' \cos^2\left(k_L x + \frac{\phi_0}{2} - \frac{\Delta_\omega}{2} t\right). \tag{5}$$

The single-particle Hamiltonian is given by

$$H_0 = \int dx \Psi^\dagger(x) \left(\frac{p^2}{2m} + V_{\text{static}}(x) + V(x,t)\right) \Psi(x).$$

By expanding the wave function in terms of the two lowest orbital states of the static optical lattice and assuming that all other higher orbital bands are not relevant, we obtain

$$\Psi(x) = \sum_{j,\alpha} b_{j,\alpha} w_\alpha(x - x_j), \tag{6}$$

where $\alpha = s, p$ is the orbital index, j is the site index and $w_\alpha(x - x_j)$ is the Wannier function for orbital state α localized at site x_j with the corresponding annihilation operator $b_{j,\alpha}$. Substituting this wave function into the Hamiltonian, we obtain the tight-binding model

$$H = \sum_{\langle j,k\rangle}\left(-t_s b_{js}^\dagger b_{ks} + t_p b_{jp}^\dagger b_{kp}\right) + \sum_{j\alpha,k\beta} A_{j\alpha,k\beta}(t) b_{j\alpha}^\dagger b_{k\beta} + \sum_j\left(\epsilon_s b_{js}^\dagger b_{js} + \epsilon_p b_{jp}^\dagger b_{jp}\right) \quad (7)$$

$$= \sum_{j,k}\begin{pmatrix} b_{js}^\dagger \\ b_{jp}^\dagger\end{pmatrix}^T\left[\begin{pmatrix} -t_s\delta_{\langle j,k\rangle} + \epsilon_s\delta_{j,k} & 0 \\ 0 & t_p\delta_{\langle j,k\rangle} + \epsilon_p\delta_{j,k}\end{pmatrix} + \begin{pmatrix} A_{js,ks}(t) & A_{js,kp}(t) \\ A_{jp,ks}(t) & A_{jp,kp}(t)\end{pmatrix}\right]\begin{pmatrix} b_{ks} \\ b_{kp}\end{pmatrix} \quad (8)$$

where

$$t_s = -\int dx w_s^*(x - x_j)\left(\frac{p^2}{2m} + V_{\text{static}}(x)\right) w_s(x - x_k),$$
$$t_p = \int dx w_p^*(x - x_j)\left(\frac{p^2}{2m} + V_{\text{static}}(x)\right) w_p(x - x_k),$$
$$\epsilon_s = \int dx w_s^*(x - x_j)\left(\frac{p^2}{2m} + V_{\text{static}}(x)\right) w_s(x - x_j),$$
$$\epsilon_p = \int dx w_p^*(x - x_j)\left(\frac{p^2}{2m} + V_{\text{static}}(x)\right) w_p(x - x_j),$$

are the bare tunnelling elements between the same orbital states on the adjacent sites $|j - k| = 1$ (that is, $\delta_{\langle j,k\rangle} = 1$) and the on-site energies from the static optical lattice. The $\langle\rangle$ in the summations indicates summing over the nearest neighbour sites. The moving lattice induces assisted or dressed tunnelling elements between adjacent sites and also modifies the on-site energies:

$$A_{js,ks}(t) = \int dx w_s^*(x - x_j) V(x,t) w_s(x - x_k),$$
$$A_{jp,kp}(t) = \int dx w_p^*(x - x_j) V(x,t) w_p(x - x_k),$$
$$A_{js,kp}(t) = \int dx w_s^*(x - x_j) V(x,t) w_p(x - x_k),$$
$$A_{jp,ks}(t) = \int dx w_p^*(x - x_j) V(x,t) w_s(x - x_k).$$

Time-independent effective Hamiltonian. To obtain a time-independent effective Hamiltonian, we first eliminate the diagonal-dressed terms by a unitary transformation

$$U(t) = \exp\left[\frac{i}{\hbar}\int_0^t dt' \sum_{j,k}\sum_\alpha A_{j\alpha,k\alpha}(t) b_{j\alpha}^\dagger b_{k\alpha}\right]. \quad (9)$$

The rotated Hamiltonian is

$$H'(t) = U(t)H(t)U^{-1}(t) - i\hbar U(t)\frac{\partial}{\partial t}U^{-1}(t). \quad (10)$$

We make the approximation that the driving does not change the on-site energy and bare tunnelling of the same orbital states:

$$U(t) b_{j\alpha,k\beta}^\dagger b_{j\alpha,k\beta} U^{-1}(t) \approx b_{j\alpha,k\beta}^\dagger b_{j\alpha,k\beta}. \quad (11)$$

We obtain

$$H = \sum_{j,k}\begin{pmatrix} b_{js}^\dagger \\ b_{jp}^\dagger\end{pmatrix}^T\begin{pmatrix} E_s & A_{js,kp}(t) \\ A_{jp,ks}(t) & E_p\end{pmatrix}\begin{pmatrix} b_{ks} \\ b_{kp}\end{pmatrix},$$

where $E_s = -t_s\delta_{\langle j,k\rangle} + \epsilon_s\delta_{j,k}$ and $E_p = t_p\delta_{\langle j,k\rangle} + \epsilon_p\delta_{j,k}$. The s–p_x couplings induced by the moving lattice are

$$\begin{aligned} A_{js,kp}(t) &= \int dx w_s^*(x - x_j) V(x,t) w_p(x - x_k) \\ &= \langle s^j | V'^2\left(k_L x + \tfrac{\phi_0}{2} - \tfrac{\Delta_\omega}{2}t\right) | p^k\rangle \\ &= \tfrac{V'}{2}\langle s^j | [\cos(2k_L x)\cos(\phi_0 - \Delta_\omega t) \\ &\quad - \sin(2k_L x)\sin(\phi_0 - \Delta_\omega t)] | p^k\rangle. \end{aligned} \quad (12)$$

The moving lattice has the same wave vector as the static lattice so that

$$z_0 = \langle s^j|\cos(2k_L x)|p^j\rangle = 0,$$
$$z_1 = \langle s^j|\sin(2k_L x)|p^j\rangle \neq 0,$$
$$z_2 = \langle s^j|\cos(2k_L x)|p^{j+1}\rangle = -\langle s^j|\cos(2k_L x)|p^{j-1}\rangle \neq 0,$$
$$z_3 = \langle s^j|\sin(2k_L x)|p^{j+1}\rangle = +\langle s^j|\sin(2k_L x)|p^{j-1}\rangle \neq 0.$$

After the Fourier transformation

$$b_{js} = \frac{1}{\sqrt{N}}\sum_{k_x} b_{k,s} e^{ik_x x_j}, \quad (13)$$

$$b_{jp} = \frac{1}{\sqrt{N}}\sum_{k_x} b_{k,p} e^{ik_x x_j}, \quad (14)$$

we obtain

$$H(k_x, t) = \begin{bmatrix} \epsilon_s - 2t_s\cos(k_x d) & \prod_{sp}(t) \\ \prod_{sp}^*(t) & \epsilon_p + 2t_p\cos(k_x d)\end{bmatrix},$$

where the s–p coupling is given by

$$\begin{aligned}\prod_{sp}(t) &= ie^{i\phi_0}[\Omega + \alpha\sin(k_x d) + \beta\cos(k_x d)]e^{-i\Delta_\omega t} \\ &\quad + ie^{-i\phi_0}[-\Omega + \alpha\sin(k_x d) - \beta\cos(k_x d)]e^{i\Delta_\omega t}\end{aligned}$$

and the coupling coefficients are defined as

$$\Omega = \frac{V'}{4}z_1 = \frac{V'}{4}\langle s^j|\sin(2k_L x)|p^j\rangle, \quad (15)$$

$$\alpha = \frac{V'}{2}z_2 = \frac{V'}{2}\langle s^j|\cos(2k_L x)|p^{j+1}\rangle, \quad (16)$$

$$\beta = \frac{V'}{2}z_3 = \frac{V'}{2}\langle s^j|\sin(2k_L x)|p^{j+1}\rangle. \quad (17)$$

For $\Delta_\omega > 0$, we perform a unitary transformation followed by the rotating wave approximation,

$$U_2(t) = \begin{bmatrix} 1 & 0 \\ 0 & e^{i\Delta_\omega t}\end{bmatrix}. \quad (18)$$

This leads to the time-independent Hamiltonian

$$\begin{aligned} H_{\text{eff}} &= U_2(t)H(k_x,t)U_2^{-1}(t) - i\hbar U_2(t)\tfrac{\partial}{\partial t}U_2^{-1}(t) \\ &= \begin{bmatrix} \epsilon_s - 2t_s\cos(k_x d) & \prod_{sp}(t)e^{-i\Delta_\omega t} \\ \prod_{sp}^*(t)e^{i\Delta_\omega t} & \epsilon_p + 2t_p\cos(k_x d) - \hbar\Delta_\omega\end{bmatrix},\end{aligned}$$

which returns to our effective Hamiltonian equation (2), where we have $\Delta_{sp} = -i[\Omega - \alpha\sin(q_x d) + \beta\cos(q_x d)]e^{-i\phi_0}$. For $\Delta_\omega < 0$, similar procedures result $\Delta_{sp} = i[\Omega + \alpha\sin(q_x d) + \beta\cos(q_x d)]e^{i\phi_0}$. Note that β is usually much smaller than Ω and α; however, we keep it for completeness.

References

1. Žutić, I., Fabian, J. & Das Sarma, S. Spintronics: fundamentals and applications. *Rev. Mod. Phys.* **76,** 323–410 (2004).
2. Hasan, M. Z. & Kane, C. L. Colloquium: topological insulators. *Rev. Mod. Phys.* **82,** 3045–3067 (2010).
3. Qi, X.-L. & Zhang, S.-C. Topological insulators and superconductors. *Rev. Mod. Phys.* **83,** 1057–1110 (2011).
4. Lin, Y.-J., Jiménez-Garca, K. & Spielman, I. B. Spin-orbit-coupled Bose-Einstein condensates. *Nature* **471,** 83–86 (2011).
5. Zhang, J.-Y. *et al.* Collective dipole oscillations of a spin-orbit coupled Bose-Einstein condensate. *Phys. Rev. Lett.* **109,** 115301 (2012).
6. Wang, P. *et al.* Spin-orbit coupled degenerate Fermi gases. *Phys. Rev. Lett.* **109,** 095301 (2012).
7. Cheuk, L. W. *et al.* Spin-injection spectroscopy of a spin-orbit coupled Fermi gas. *Phys. Rev. Lett.* **109,** 095302 (2012).
8. Hamner, C. *et al.* Dicke-type phase transition in a spin-orbit-coupled Bose-Einstein condensate. *Nat. Commun.* **5,** 4023 (2014).
9. Olson, A. J. *et al.* Tunable Landau-Zener transitions in a spin-orbit-coupled Bose-Einstein condensate. *Phys. Rev. A* **90,** 013616 (2014).
10. Higbie, J. & Stamper-Kurn, D. M. Periodically dressed Bose-Einstein condensate: A superfluid with an anisotropic and variable critical velocity. *Phys. Rev. Lett.* **88,** 090401 (2002).
11. Spielman, I. B. Raman processes and effective gauge potentials. *Phys. Rev. A* **79,** 063613 (2009).
12. Jaksch, D., Bruder, C., Cirac, J. I., Gardiner, C. W. & Zoller, P. Cold bosonic atoms in optical lattices. *Phys. Rev. Lett.* **81,** 3108 (1998).
13. Bloch, I. Ultracold quantum gases in optical lattices. *Nat. Phys.* **1,** 23–30 (2005).
14. Bernevig, B. A., Hughes, T. L. & Zhang, S.-C. Quantum spin Hall effect and topological phase transition in HgTe quantum wells. *Science* **314,** 1757–1761 (2006).
15. König, M. *et al.* Quantum spin Hall insulator state in HgTe quantum wells. *Science* **318,** 766–770 (2007).
16. Fu, L. & Kane, C. L. Superconducting proximity effect and Majorana fermions at the surface of a topological insulator. *Phys. Rev. Lett.* **100,** 096407 (2008).
17. Parker, C. V., Ha, L.-C. & Chin, C. Direct observation of effective ferromagnetic domains of cold atoms in a shaken optical lattice. *Nat. Phys.* **9,** 769–774 (2013).

18. Zheng, W., Liu, B., Miao, J., Chin, C. & Zhai, H. Strong interaction effects and criticality of bosons in shaken optical lattices. *Phys. Rev. Lett.* **113,** 155303 (2014).
19. Fulde, P. & Ferrell, R. A. Superconductivity in a strong spin-exchange field. *Phys. Rev* **135,** 550 (1964).
20. Larkin, A. I. & Ovchinnikov, Y. N. Nonuniform state of superconductors. *Zh. Eksp. Teor. Fiz.* **47,** 1136 (1964).
21. Zheng, Z., Gong, M., Zou, X., Zhang, C. & Guo, G. Route to observable Fulde-Ferrell-Larkin-Ovchinnikov phases in three-dimensional spin-orbit-coupled degenerate Fermi gases. *Phys. Rev. A* **87,** 031602(R) (2013).
22. Wu, F., Guo, G., Zhang, W. & Yi, W. Unconventional superfluid in a two-dimensional Fermi gas with anisotropic spin-orbit coupling and Zeeman fields. *Phys. Rev. Lett.* **110,** 110401 (2013).
23. Zhang, C., Tewari, S., Lutchyn, R. M. & Das Sarma, S. $p_x + ip_y$ superfluid from s-wave interactions of fermionic cold atoms. *Phys. Rev. Lett.* **101,** 160401 (2008).
24. Sato, M., Takahashi, Y. & Fujimoto, S. Non-abelian topological order in s-wave superfluids of ultracold fermionic atoms. *Phys. Rev. Lett.* **103,** 020401 (2009).
25. Müller, T., Fölling, S., Widera, A. & Bloch, I. State preparation and dynamics of ultracold atoms in higher lattice orbitals. *Phys. Rev. Lett.* **99,** 200405 (2007).
26. Lignier, H. *et al.* Dynamical control of matter-wave tunneling in periodic potentials. *Phys. Rev. Lett.* **99,** 220403 (2007).
27. Eckardt, A. *et al.* Frustrated quantum antiferromagnetism with ultracold bosons in a triangular lattice. *EPL* **89,** 10010 (2010).
28. Hauke, P. *et al.* Non-abelian gauge fields and topological insulators in shaken optical lattices. *Phys. Rev. Lett.* **109,** 145301 (2012).
29. Goldman, N. & Dalibard, J. Periodically driven quantum systems: effective Hamiltonians and engineered gauge fields. *Phys. Rev. X* **4,** 031027 (2014).
30. Goldman, N., Dalibard, J., Aidelsburger, M. & Cooper, N. R. Periodically driven quantum matter: The case of resonant modulations. *Phys. Rev. A* **91,** 033632 (2015).
31. D'Alessio, L. & Rigol, M. Long-time behavior of isolated periodically driven interacting lattice systems. *Phys. Rev. X* **4,** 041048 (2014).
32. Aidelsburger, M. *et al.* Realization of the Hofstadter Hamiltonian with ultracold atoms in optical lattices. *Phys. Rev. Lett.* **111,** 185301 (2013).
33. Miyake, H. *et al.* Realizing the Harper Hamiltonian with laser-assisted tunneling in optical lattices. *Phys. Rev. Lett.* **111,** 185302 (2013).
34. Struck, J. *et al.* Quantum simulation of frustrated classical magnetism in triangular optical lattices. *Science* **333,** 996–999 (2011).
35. Struck, J. *et al.* Engineering Ising-XY spin-models in a triangular lattice using tunable artificial gauge fields. *Nat. Phys.* **9,** 738–743 (2013).
36. Jotzu, G. *et al.* Experimental realization of the topological Haldane model with ultracold fermions. *Nature* **515,** 237–240 (2014).
37. Wang, Y. H., Steinberg, H., Jarillo-Herrero, P. & Gedik, N. Observation of Floquet-Bloch states on the surface of a topological insulator. *Science* **342,** 453–457 (2013).
38. Zheng, Z., Qu, C., Zou, X. & Zhang, C. Fulde-Ferrell superfluids without spin-imbalance in three-dimensional driven spinful fermionic optical lattices. Preprint at http://arxiv.org/abs/1501.00448 (2015).
39. Wirth, G., Ölschläger, M. & Hemmerich, A. Evidence for orbital superfluidity in the P-band of a bipartite optical square lattice. *Nat. Phys.* **7,** 147–153 (2011).
40. Choudhury, S. & Mueller, E. J. Transverse collisional instabilities of a Bose-Einstein condensate in a driven one-dimensional lattice. *Phys. Rev. A* **91,** 023624 (2014).
41. Choudhury, S. & Mueller, E. J. Stability of a Floquet Bose-Einstein condensate in a one-dimensional optical lattice. *Phys. Rev. A* **90,** 013621 (2014).
42. Bilitewski, T. & Cooper, N. R. Scattering theory for Floquet-Bloch staes. *Phys. Rev. A* **91,** 033601 (2015).
43. Jimenéz-Garca, K. *et al.* Tunable spin-orbit coupling via strong driving in ultracold-atom systems. *Phys. Rev. Lett.* **114,** 125301 (2015).

Acknowledgements

M.A.K., M.E.M., P.E. are supported by the National Science Foundation (NSF) through Grant No. PHY-1306662. C.Q. and C.Z. are supported by ARO (W911NF-12-1-0334) and AFOSR (FA9550-13-1-0045).

Author contributions

M.A.K., C.Q., C.Z. and P.E. conceived the experiment and theoretical modelling; M.A.K., M.E.M. and P.E. performed the experiments; C.Q., C.Z. performed the theoretical calculations; and C.Z. and P.E. supervised the project.

Additional information

Competing financial interests: The authors declare no competing financial interests.

27

In situ stress observation in oxide films and how tensile stress influences oxygen ion conduction

Aline Fluri[1], Daniele Pergolesi[1], Vladimir Roddatis[2], Alexander Wokaun[1] & Thomas Lippert[1,3]

Many properties of materials can be changed by varying the interatomic distances in the crystal lattice by applying stress. Ideal model systems for investigations are heteroepitaxial thin films where lattice distortions can be induced by the crystallographic mismatch with the substrate. Here we describe an *in situ* simultaneous diagnostic of growth mode and stress during pulsed laser deposition of oxide thin films. The stress state and evolution up to the relaxation onset are monitored during the growth of oxygen ion conducting $Ce_{0.85}Sm_{0.15}O_{2-\delta}$ thin films *via* optical wafer curvature measurements. Increasing tensile stress lowers the activation energy for charge transport and a thorough characterization of stress and morphology allows quantifying this effect using samples with the conductive properties of single crystals. The combined *in situ* application of optical deflectometry and electron diffraction provides an invaluable tool for strain engineering in Materials Science to fabricate novel devices with intriguing functionalities.

[1] Department for Energy and Environment, Paul Scherrer Institut, 5232 Villigen-PSI, Switzerland. [2] Institut für Materialphysik, Universität Göttingen, Friedrich-Hund-Platz 1, Göttingen 37077, Germany. [3] Department of Chemistry and Applied Biosciences, Laboratory of Inorganic Chemistry, Vladimir-Prelog-Weg 1-5/10, ETH Zürich, Zürich CH-8093, Switzerland. Correspondence and requests for materials should be addressed to D.P. (email: daniele.pergolesi@psi.ch) or to T.L. (email: thomas.lippert@psi.ch).

Stress-induced lattice distortions, that is, a tensile or compressive strain of the crystal structure, can significantly influence the physicochemical characteristics of materials by enhancing/inhibiting specific properties or by enabling new functionalities not allowed in the unperturbed structure. Strain engineering offers indeed a new route to tune the characteristics of a material for example by modifying the electronic bandgap[1], multiferroic[2] or catalytic[3] properties, thermal conductivity[4] and charge transport[5–8].

To investigate the effects of strain, epitaxial thin films are typically used where the strain is induced by a lattice mismatch at the interface with the substrate. For highly ordered epitaxial films, established models exist, which assume elastic distortions and coherent growth, that is, a 1:1 matching between all lattice planes[9–11]. The upper limit of the strain is in this case the lattice mismatch between film and substrate. In general, however, we often deal with semi-coherent heterointerfaces where it is energetically favourable to reduce the elastic energy by introducing dislocations. Already a mismatch exceeding $\sim 1\%$ is typically not expected to be fully accommodated[9–11]. The growth mode (layer-by-layer or island-like) also influences the effective strain through the surface morphology and relaxation behaviour[12–14].

Monitoring the stress and the growth mode *in situ* (during the growth) allows investigating fundamental mechanisms of stress generation and evolution from the initial nucleation to the onset of relaxation. Literature concerning the *in situ* diagnostic of stress/strain generation and evolution during the epitaxial growth of oxide films is scarce. Here, we address this topic measuring simultaneously *in situ* the stress with a multi-beam optical stress sensor (MOSS) and the growth mode by reflection high-energy electron diffraction (RHEED) during pulsed laser deposition (PLD).

When a stressed film grows, the substrate bends minimizing the elastic energy. Stoney's equation[10] correlates the substrate curvature $1/\rho$ with the film stress σ as

$$\frac{1}{\rho} = 6\frac{(1-\nu)}{Y\tau_s^2}\tau\sigma \qquad (1)$$

where ρ being the ray of curvature, τ and τ_s the film and substrate thicknesses, ν and Y the Poisson ratio and the Young's modulus of the substrate and σ the average film stress. As sketched in Fig. 1, the MOSS uses the deflection of laser beams to measure the change in curvature, and thus to monitor *in situ* and in-plane the stress of the growing film[10]. In contrast, conventional X-ray diffraction analysis yields the strain, *ex situ* and out-of-plane.

The in- and out-of-plane strain of a film, ε_{xx} and ε_{zz}, are related through the Poisson ratio[9] ν as

$$\varepsilon_{zz} = \frac{2\nu}{\nu-1}\varepsilon_{xx} \qquad (2)$$

For most materials $0<\nu<0.5$, implying that an in-plane tensile (compressive) strain of the unit cell corresponds to an out-of-plane compressive (tensile) strain.

Different *in situ* strain diagnostics have been applied for thin films. The strain evolution of $Ba_{0.5}Sr_{0.5}TiO_3$ thin films grown on MgO was investigated using real-time X-ray diffraction[15]. This is undoubtedly a very powerful approach but it requires a synchrotron light source[16] precluding its application as a routine measurement technique. RHEED was also employed to monitor the in-plane strain evolution, for example, during the growth of $BaTiO_3$ on $SrTiO_3$ (refs 9,14). This specific application of RHEED is very rare, even considering the widespread use of high-pressure RHEED. In fact, to minimize electron scattering, it can only be attempted at low pressure, a growth condition unsuitable for many oxides. A wafer curvature measurement similar to MOSS

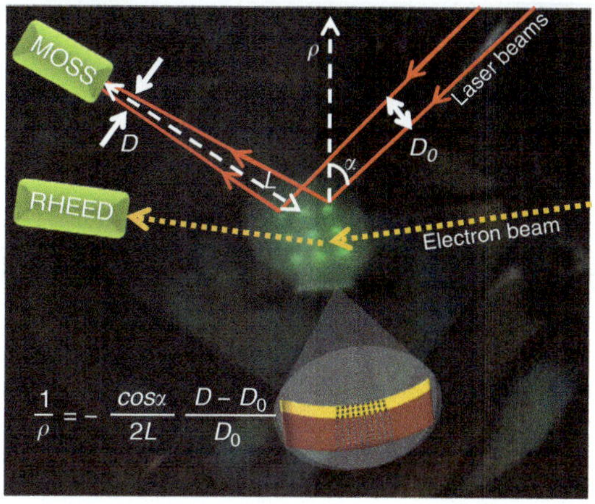

Figure 1 | Working principle of the multi-beam optical stress sensor (MOSS). $10 \times 10\,mm^2$ MgO substrate on the sample holder of the PLD system equipped with MOSS and RHEED. A 3×3 array of parallel laser beams (visible as bright spots on the substrate surface) is reflected by the substrate towards a CCD camera that records the relative distance between the laser spots. The paths of two laser beams of the MOSS and of the electron beam of the RHEED are sketched. The growth of a strained layer induces a change of curvature ($1/\rho$) of the substrate and a change of the direction of the reflected laser beams. The effect of a stress-induced curvature of the substrate is illustrated in cross-section for the case of an in-plane tensile strained film ($\rho>0$). The relative curvature change is obtained by measuring the change in the relative distance $(D-D_0)/D_0$ between the laser beams; D_0 being the distance at the beginning of the growth, L the optical path length and α the incident angle. From the CCD image, the MOSS software calculates the average of $(D-D_0)/D_0$ using all nine beams.

was used during PLD of $BaTiO_3$ and $SrTiO_3$ ultra-thin films on a Pt cantilever[17]. The main advantages of MOSS over similar deflection techniques are that one can use any substrate, no position sensitive detector is needed and that the noise is reduced through multiple beams[10].

Several examples of MOSS (or similar techniques) applications for studying the stress evolution during the growth of semiconductor or metal thin films exist[18–24], whereas reports on MOSS diagnostic with oxide materials are very rare[25]. To our knowledge, this technique has never been applied simultaneously with RHEED during PLD, the most widespread method for the growth of oxide films. In particular, by coupling the MOSS with RHEED, the stress state of the growing film can be correlated with the growth mode from the initial nucleation to stress relaxation, thus providing a unique diagnostic tool.

15 at.% Sm-doped CeO_2 (SDC) is the material selected for this fundamental study. SDC is an oxygen-ion conductor used as electrolyte in solid oxide fuel cells (SOFCs), environmentally sustainable electrochemical energy converters[26]. The oxygen ion conductivity σ_{ion} is described with the Arrhenius equation

$$\sigma_{ion} = \frac{\sigma_0}{T}e^{-\frac{E_A}{k_B T}} \qquad (3)$$

T being the temperature, σ_0 the pre-exponential factor, E_A the activation energy and k_B the Boltzmann constant. Currently, the most accepted hypothesis is that tensile strain lowers the activation energy for the oxygen-ion hopping by increasing the migration volume[7,8,27]. Consequently, larger ionic conductivities could be achieved at lower temperatures, with important consequences for technological applications.

Here, as a case study for the first MOSS plus RHEED application during PLD of oxide materials, the fundamental question of how strongly homogeneous strain can affect the ionic conductivity in a single crystal is addressed. This fundamental question is of interest as controversial results have been reported on the magnitude of the effect of strain on the ionic conductivity[8,27,28]. The increase in conductivity ascribed to strain effects spreads over several orders of magnitude showing the need to evaluate the effect using samples where any potential side effect is minimized and where the strain/stress state is unambiguously identified.

To achieve this goal, thin films of high crystalline quality—as close to a single crystal as experimentally possible—but with different strain values are employed as model systems. The MOSS is particularly suited for this investigation as the stress state can be probed in-plane, precisely in the direction of the ion migration during conductivity measurements, and *in situ* in a temperature range similar to the operating temperature of SDC for technological applications. This study confirms that tensile strain indeed leads to a decrease in activation energy showing that the effect is relatively small, roughly a factor of 2 at around 350 °C. But more importantly, we show here the potentials of MOSS diagnostic as an invaluable tool for strain engineering in Materials Science.

Results

Strain generation and evolution on different substrates. To study the strain generation and evolution in SDC films, MgO, $NdGaO_3$ (NGO) and $LaAlO_3$ (LAO) single-crystal substrates were used; Table 1 reports their crystallographic properties, the expected epitaxial relations and the in-plane lattice misfit with 15 at.% SDC (cubic fluorite structure with a lattice parameter of 5.43 Å). NGO and LAO provide a lattice mismatch in a suitable range to expect epitaxial films of SDC in tensile and compressive in-plane strain, respectively. On the contrary, the large lattice mismatch between SDC and MgO may prevent any crystallographic matching leading to a fully relaxed growth from the start. Figure 2 compares the MOSS *in situ* stress characterizations with the X-ray diffraction *ex situ* strain measurements of (001)-oriented SDC films on the different substrates. The MOSS measurements are reported as stress-thickness product (directly proportional to the curvature, from the Stoney equation (1)) in function of thickness, a plot commonly used in the literature for wafer curvature measurements[19–25].

As expected, during the growth on MgO, the stress-thickness product, that is, the substrate curvature, remains constant indicating that the film grows without exerting any force on the substrate, that is, stress free. Correspondingly, the X-ray diffraction analysis shows a fully relaxed crystalline structure. The large lattice misfit is accommodated by a high density of interfacial misfit dislocations that fully release the stress as revealed by transmission electron microscopy, discussed below. For the SDC film grown on NGO, the positive curvature indicates that the film exerts a compressive force on the substrate surface and bends it: the film is under in-plane tensile stress, which originates at the very beginning of the growth. The constant slope of the stress-thickness product (Fig. 2a) shows that the stress is constant to a film thickness of 33 nm. An average stress of ~3.3 GPa can be calculated. The tensile in-plane stress is in agreement with the compressive out-of-plane strain of −0.52% measured *ex situ* by XRD. The XRD analysis of the 33-nm-thick SDC film on LAO reveals an average out-of-plane tensile strain of 0.37%. Comparing this value with the theoretical lattice mismatch of 0.49% between the two materials, we conclude that ~25% of the theoretical lattice mismatch is lost, most probably through interfacial misfit dislocations. Accordingly, the *in situ* MOSS measurement (Fig. 2a) shows in this case a constant compressive in-plane stress. In accordance to the slightly smaller strain value measured out-of-plane on LAO compared with NGO, the MOSS shows a smaller value (~3 GPa) of average residual and effective stress for the same total film thickness.

When a strained epitaxial film grows, elastic energy is accumulated until it becomes energetically favourable for dislocations lines to move through the crystal, relieving the strain[9–11]. Experimentally, mismatch values up to around 1% can be expected to be fully accommodated[11]. With the materials and growth condition selected here for SDC, 0.4–0.5% strain was induced starting from a theoretical lattice mismatch of ~0.5–0.6%. One possible explanation for the stress relaxation could be the growth mechanism. During the SDC growth on all the three different substrates, RHEED did not show a two-dimensional layer-by-layer growth. Instead, RHEED revealed a three-dimensional growth mode, also known as island or layer-plus-island growth mode. Such a growth mechanism along the (001) crystallographic orientation of ceria films has been reported and explained[29]. This growth mode facilitates the nucleation of dislocations[12,13], which assist the strain relaxation.

Tensile strained SDC films for electrical characterization. To allow a reliable in-plane electrical characterization of thin ion conducting films, the growth platform must add a negligible contribution to the total conductance of the sample. Among the substrates used in this study, only MgO fulfils this condition. As an example, at around 600 °C, the resistance of an NGO substrate is similar to that of a 10-nm-thick SDC film with the same area[30]. However, although MgO is an ideal substrate for conductivity measurements, it only allows the growth of fully relaxed SDC films.

In previous studies, $SrTiO_3$ (STO) was used as a thin epitaxial buffer layer on MgO to favour a highly ordered growth of ceria films on an insulating substrate[31,32]. More recently, it was found that an additional thin inter-layer of $BaZrO_3$ (BZO) with a thickness of 5 nm grown between MgO and STO improves the crystalline quality of the STO layer[33]. The same sample design described in ref. 33 was used here. The 5-nm-thin STO seed layer does not significantly affect the conductance of the sample while providing a lattice mismatch of ~1.6% (for the fully relaxed structure) with respect to SDC, which will induce an in-plane tensile strain in epitaxially grown SDC.

Figure 3a shows the XRD analysis of three epitaxial SDC films, 13, 20 and 83 nm-thick, grown on the (001)-oriented

Table 1 | Characteristics of the single crystal substrates used for this study.

Substrate	Crystalline structure		Expected epitaxial relation	Lattice mismatch (%)
$NdGaO_3$ (110)	Orthorhombic perovskite $a = 5.43$ Å $c = 5.50$ Å	In-plane	SDC(1$\bar{1}$0)\|\|NGO(1$\bar{1}$0) SDC(110)||NGO(001)	0.64, 0.39
		Out-of-plane	SDC(001)||NGO(110)	
$LaAlO_3$ (001)	Pseudocubic perovskite $a = 3.82$ Å	In-plane	SDC(110)||LAO(100) SDC(1$\bar{1}$0)||LAO(010)	−0.49, −0.49
		Out-of-plane	SDC(001)||LAO(001)	
MgO (001)	face-centered cubic $a = 4.21$ Å		None	≈ −29

Expected epitaxial relations and lattice mismatch with 15 at.% Sm-doped CeO_2 are displayed. The lattice mismatch, given in percent, is calculated as $(a_{substrate} - a_{SDC})/a_{substrate}$.

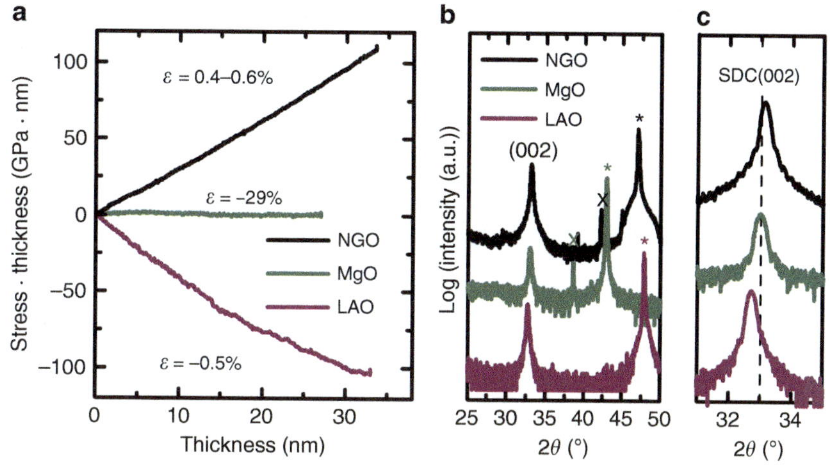

Figure 2 | MOSS and XRD analysis of SDC films grown on NGO and MgO and LAO. (**a**) Stress-thickness product, evaluated using the Stoney equation and the elastic properties of the substrates[57-59], as a function of thickness. A positive (negative) stress value indicates an in-plane tensile (compressive) stress for the film. ε is the lattice mismatch (Table 1). (**b**) $\omega/2\theta$ scans. *indicates the substrate peak and X is an instrumental artefact (K_β reflection of substrate). (**c**) Magnification of the region around the (002) diffraction peak of SDC. The dashed line indicates the angular position for the fully relaxed SDC structure. The peak shifts are in accordance with the sign of the stress-thickness product.

MgO + BZO + STO template platform (hereafter we will refer to this template platform as MgO-BS). In spite of the small thickness of the layer, the $\omega/2\theta$ scans allow a rough estimation of the out-of-plane compressive strain of the STO layer between 0.1 and 0.5%, thus the misfit between the growth platform and SDC is expected to be in the range of 1.7–2.1% in-plane.

The BZO + STO layers were deposited at around 765 °C. The 20- and 83-nm-thick SDC films on MgO-BS were deposited at 650 °C, the same temperature used for the films on MgO, NGO and LAO. To probe the effect of temperature on the developed stress, the 13-nm-thick film was grown at the same temperature used for BZO and STO.

The shift of the diffraction peak of these three films towards larger 2θ angles indicates different out-of-plane compressive strain values (Fig. 3b). To quantify the strain in-plane, reciprocal space maps (RSMs) were recorded. As an example, the RSM acquired along the (204) asymmetric diffraction of the 20-nm-thick SDC film grown on MgO-BS is reported in Fig. 3c. The 20-nm-thick film shows the largest in-plane tensile strain, that is, 0.35%.

As expected, the crystalline structure relaxes with increasing thickness: for the 83-nm film, an in-plane strain of 0.24% is found. The 13-nm-thick film grown at the higher temperature shows the smallest in-plane tensile strain value of ~0.1%. This can be explained by considering that for a film to relax dislocations have to nucleate or existing dislocation lines have to be mobile. The mobility of a dislocation can be described by a periodic energy barrier that can be overcome by accumulating elastic energy or by providing thermal energy[9], which facilitates the relaxation process.

Figure 3d, e shows the results of the *in situ* MOSS plus RHEED measurements performed during the growth of the SDC films on the MgO-BS template platform. Figure 3d reports an example of the wafer curvature measurements showing the capability of MOSS diagnostic to probe *in situ* the in-plane stress of the different layers during the growth of a complex heterostructures.

Owing to the very good lattice matching between MgO and BZO, no significant wafer curvature (stress) can be observed during the growth of the 5-nm-thick BZO layer. On the contrary, the lattice mismatch between BZO and STO (as large as 7%) induces an in-plane tensile stress to the STO film, as revealed by the positive wafer curvature measured by MOSS. It is worth highlighting that for such thin layers the MOSS allows the stress state of the film to be unambiguously identified *in situ* much more effectively than by evaluating *ex situ* the angular position of the broad peaks in the $\omega/2\theta$ scans. Within this experiment, the MOSS diagnostic allows resolving the change of the substrate curvature induced by the deposition of films with thicknesses in the ≥1 nm range.

Finally, the substrate continues to bend in the same direction during the growth of the SDC film on top of STO indicating an in-plane tensile strain of the oxygen ion conductor.

The simultaneous RHEED measurements are shown in Fig. 3e. The RHEED patterns of BZO and STO indicate an almost two-dimensional growth. But, as previously discussed, the RHEED pattern for SDC clearly shows a three-dimensional growth mode from the very beginning of the film nucleation.

Figure 3f shows the *in situ* MOSS measurements of the 20- and 83-nm-thick SDC films grown on MgO-BS, as well as those of the SDC films grown on NGO and MgO for comparison. The data for the 20- and 83-nm-film grown with the same deposition parameters on MgO-BS are in good agreement. At the thickness of 20 nm, the MOSS allows estimating an average in-plane stress of ~1.5 GPa for both films.

The extent of the average tensile stress retained in the SDC layer on MgO-BS was smaller than on NGO (and on LAO in compressive stress). The much larger lattice misfit and the presence of a larger density of crystalline defects in the double buffer layer as compared with a single-crystal substrate can explain the larger stress relaxation.

The stress-thickness curve shows a constant slope up to 65 nm for increasing film thickness until the curve becomes flat between 70 and 75 nm. A constant curvature indicates that the total elastic energy in the film remains constant even though the thickness increases. This can be related to theoretical models for epitaxial thin film relaxation[9-11] according to which above a critical thickness enough elastic energy is accumulated that it becomes favourable for dislocation lines to move. The average stress is then gradually reduced and the total elastic energy remains constant while the film continues to grow. In other words, MOSS measurements allow identifying *in situ* the onset of stress relaxation in the film.

The measured ratio of stress (MOSS) over strain (RSM), equal to $Y/(1-v)$, is the same for both the 20- and 83-nm-thick films and

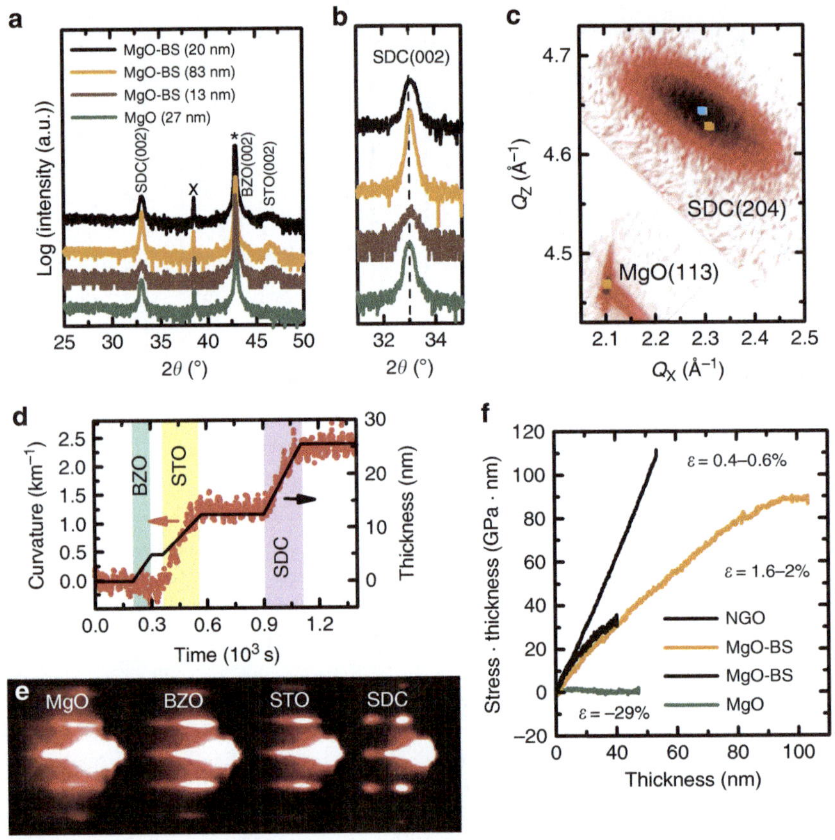

Figure 3 | XRD, MOSS and RHEED analysis of SDC on MgO + BZO + STO. (**a**) $\omega/2\theta$ scans. * and X indicate the substrate peak and an instrumental artefact (K_β reflection of substrate), respectively. (**b**) Magnification of the region around the (002) diffraction peak of SDC. The stress-free film on MgO is included for comparison. (**c**) RSM of the (204) reflection of the 20-nm SDC film. The (113) asymmetric diffraction of the MgO substrate was used for alignment. Orange square indicates the unstrained literature values and blue square the centre of the SDC(204) reflection. (**d**) representative MOSS measurement of wafer curvature during the growth of a complete heterostructure and (**e**) simultaneous RHEED measurement. For the RHEED diagnostic, the electron beam is parallel to the (110) direction of the substrate, thus parallel to (110) of BZO and STO, which show a cube-on-cube growth, but parallel to (100)—equivalently (010)—direction of SDC due to the in-plane 45° rotation of the unit cell of the ionic conductor with respect to the template platform. (**f**) MOSS measurements during the growth of the SDC films with different stress states.

in agreement with the literature[34] showing that the MOSS indeed reflects directly the average stress of the film.

The morphology and microstructure can both influence the evolution of the strain and the activation energy for oxygen ion conduction[35]. The local microstructure of the samples selected for the electrical characterization was investigated using transmission electron microscopy (TEM). Figure 4a, b shows representative examples of SDC films grown on MgO-BS and MgO, respectively.

Figure 4c shows a high-resolution bright-field (BF) STEM image of the cross-section of a SDC film grown on MgO-BS. Selected area electron diffraction and fast Fourier transform analysis reveal very good crystallographic quality showing that the [001] zone axis of the SDC film slightly precesses within only ~1° on the scales of the order of 100 nm. The SDC film grown directly on the MgO (Fig. 4d) demonstrates a cube-on-cube growth in spite of the very large lattice misfit as already reported in the literature[36,37]. Further, the film shows the presence of local isolated domains rotated by 45° around the substrate surface normal. These isolated defects are not present in the film on MgO-BS. The very large lattice mismatch between SDC and MgO results in a lower crystallographic quality of the SDC film, which shows an average grain size of 40 nm and the SDC/MgO interface characterized by an almost continuous line of misfit dislocations that release the stress still preserving the epitaxial relation. The same kind of interface was observed between Gd-doped ceria and MgO (ref. 38). As reported in ref. 33, this difference in morphology does not influence the ionic conduction. These measurements support the XRD and MOSS results that showed a fully relaxed (001) oriented SDC structure for the thin film grown on MgO and the presence of a measurable in-plane tensile strain for the films grown on MgO-BS.

Influence of tensile strain on ionic conductivity. AC impedance spectroscopy was applied for electrical characterization using patterned Pt electrodes on the film surface. Figure 5a shows the grain interior (bulk) ionic conductivity of three different Sm-doped ceria samples from the literature[39,40] compared with the measured conductivity of the MgO-BS template platform, which is more than two orders of magnitudes smaller, that is, negligible. Figure 5b shows the electrical characterization of SDC films with different strain values. Activation energy and conductivity values are in very good agreement with reference data for SDC grain interior[39–41], showing that the grain interior contribution dominates. The complex impedance plane plots (Supplementary Discussion and Supplementary Fig. 1) show a clear polarization of the Pt electrodes at low frequencies, indicating that the dominant charge carriers are indeed oxygen ions, as is well established for SDC under the selected experimental conditions. This leaves no doubt about the nature of the charge carriers.

The activation energy, calculated by fitting the linearized Arrhenius equation (3) to the data, shows a clear trend with respect to the film strain, that is, it decreases with increasing

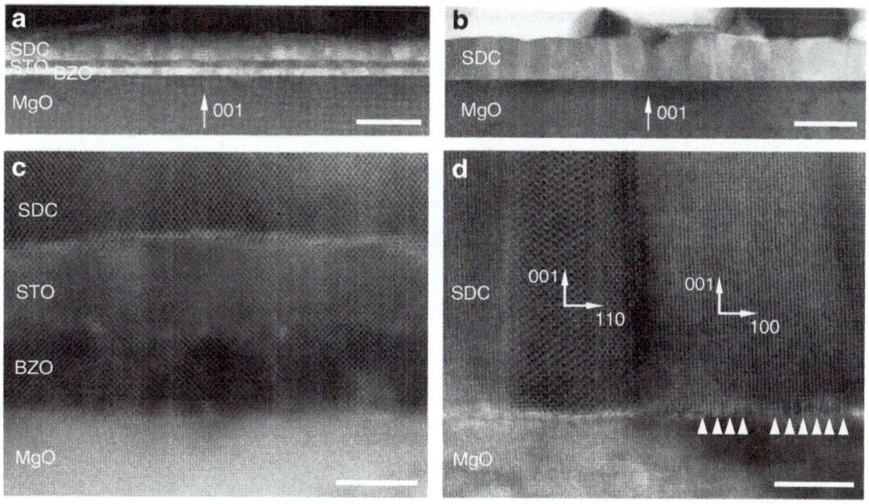

Figure 4 | TEM analysis of SDC on MgO-BS (left) and on MgO (right). Low-magnification cross-sectional HAADF-STEM image of (**a**) ~21-nm-thick SDC film on MgO-BS and (**b**) ~32-nm-thick SDC film on MgO. Scale bars correspond to 50 nm (**a**), 50 nm (**b**), 5 nm (**c**), 5 nm (**d**). Bright-field high-resolution STEM image of the SDC film on MgO-BS (**c**) and on MgO (**d**). One of the local defects (tilted grains) is shown in **d** together with the positions, marked with arrowheads, of some interfacial misfit dislocation at the SDC/MgO interface. The TEM micrographs were acquired along [010] direction of MgO.

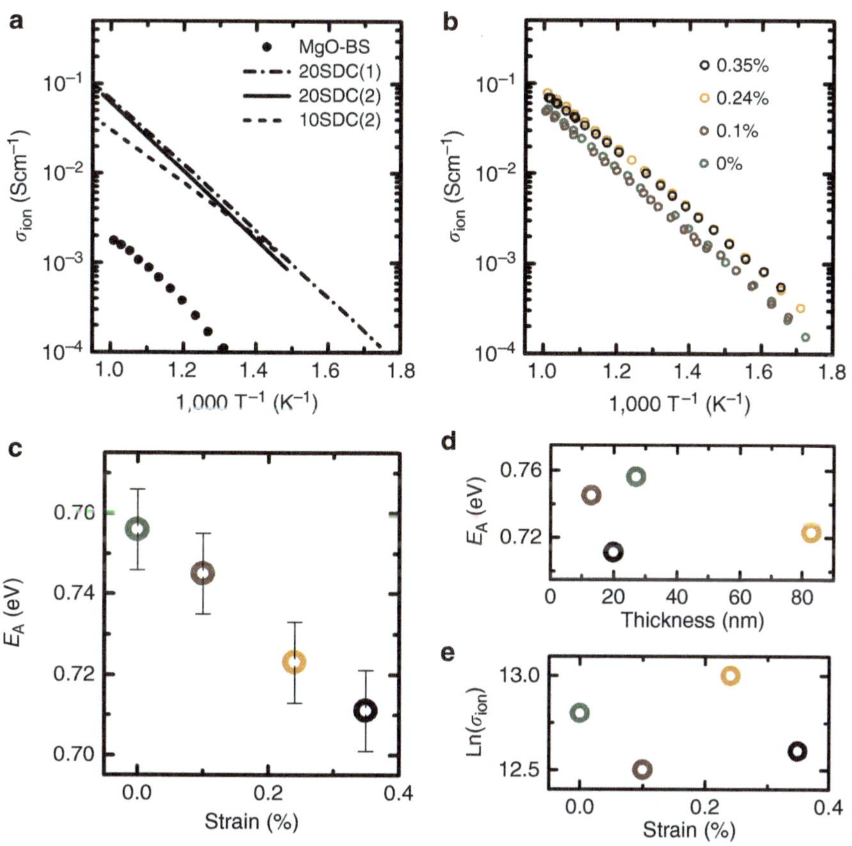

Figure 5 | Impedance spectroscopy measurements of SDC films with different strain. (**a**) Conductivity measurement of the MgO + BZO + STO (MgO-BS) template platform compared with literature references of unstrained Sm-doped ceria bulk: 20 at.% Sm (20SDC(1) (ref. 39)) and 10 and 20 at.% Sm (10SDC(2) and 20SDC(2) (ref. 40)). (**b**) Comparison of conductivity measurements of the SDC films fabricated for this study with different in-plane tensile strain values. (**c**) Activation energy for ion migration as a function of tensile strain. The error bar is determined form the standard deviation of the fit and empirically by comparison of independent measurements. (**d**) Activation energy versus SDC thickness and (**e**) Arrhenius pre-exponential factor of the SDC films with different strain values.

tensile strain (Fig. 5c). Interface effects (for example, dislocations[42]) could also influence the activation energy, resulting in a thickness dependence of the activation energy. This is not the case here (Fig. 5d) suggesting that it is indeed the strain that influences the activation energy. Further, the pre-exponential factor σ_0 of the Arrhenius equation varies little (Fig. 5e), showing that the charge carrier density is not affected by the strain.

The change of conductivity is negligible in the high-temperature range, whereas for example at 350 °C (a temperature relevant for miniaturized SOFCs), the bulk ionic conductivity is a factor of two larger compared with the unperturbed crystalline structure as a result of 0.35% tensile lattice strain, which lowers the activation energy by ~0.05 eV.

The observed influence of the strain values on the activation energy is very similar to that reported in a theoretical study on undoped ceria[43]. From the experimental point of view, literature reports on the effect of strain on conductivity for epitaxial doped ceria films are scarce. In ref. 44, for example, a very large increase in conductivity with increasing tensile strain was reported. Here a surprisingly large value of strain of 2% was estimated for 10 at.% Gd-doped ceria films as thick as 250 nm. However, in this study, larger strain results in higher ionic conductivity, but also in an increased defect density, yielding a much higher activation energy. The same material coupled with Er_2O_3 was used for the fabrication of multi-layered microdots (111)-oriented on sapphire[45]. The reported activation energy of the single doped-ceria layer is in good agreement with the typical values reported for this material, even though the total conductivity is surprisingly smaller than the typical bulk conductivity of Gd-doped ceria[39,41] suggesting that conduction pathways other than through the bulk dominate. However, the authors reported that a change of the compressive in-plane strain of the ceria layers from about 0.1 to 1.1% results in an increase of the activation energy by 0.2 eV. Assuming an almost linear trend of activation energy versus strain for strain values within ± 1%, the reported effect in ref. 45 is in line with our measurements.

Comparing the effect on SDC to a different oxygen ion conductor, that is, yttria-stabilized zirconia (YSZ), a similar increase in conductivity can be extrapolated for this particular strain value (0.35%) from a theoretical study on the influence of strain in YSZ[46]. The result reported here is also in agreement with the ~3.4-fold enhanced conductivity measured for YSZ/Y_2O_3 0.8% strained multilayers with columnar morphology[47]. Few papers report no effect at all, for example, using CeO_2/YSZ multi-layered heterostructures[32], where the typical YSZ bulk conductivity was measured for all samples. In that work, the non-uniform strain along the layers revealed by high-resolution TEM and RSM could not be quantified and the residual and effective strain could be small. Moreover, only the temperature range above 400 °C was investigated, whereas in the present work, impedance measurements were acquired down to a minimum temperature of 280 °C, and it is precisely at lower temperatures where the effect is more evident.

On the contrary, as already pointed out, many other papers report very large strain effect on the ionic conductivity which are not in agreement with this or the above-mentioned studies. Several examples can be found in ref. 8. Finally, the effect of the local strain arising from buckling in free-standing polycrystalline membranes of doped ceria[48], as eventually relevant for devices, is discussed elsewhere, as well as the effect of strain in polycrystalline samples (sintered pellets or film crystallized by post-annealing), where the origin of strain is the specific local micro-morphology[49], which is completely different to the almost single crystalline samples in our study.

Discussion

Our results on the role of strain on the ionic conductivity in Sm-doped ceria shows that, as reported in many literature contributions, lower activation energies indeed result from tensile lattice distortions in the direction of the migration of the charge carriers. Our study was conducted on samples as closely resembling a uniformly strained single crystal as it was experimentally possible showing the characteristic bulk conductivity of the material under investigation.

It should be highlighted that the in-plane geometry of the electrical characterization dictates stringent limitations to the choice of the substrate. As a consequence, a relatively small value of residual and effective strain could be preserved. However, it was shown that significantly larger strain can be achieved. As an example, extrapolating the observed effect of strain on the activation energy, around 350 °C, a 0.5% tensile strained ion conductor, as achieved for films grown on NGO, would show the same conductivity as the unstrained conductor at around 450 °C.

Beside the specific application to oxygen ion conductors, which supports the thesis of a relatively small effect of strain on ionic conductivity, the most important result of our study is that it has been clearly shown that the MOSS is a valuable tool for investigating *in situ* the mechanism of stress generation and evolution in complex multi-layered oxide heterostructures. The stress states arising for different lattice mismatches as well as the relaxation behaviour in the epitaxial oxide films grown by PLD could be clearly monitored and were found to be in very good agreement with standard *ex situ* XRD analysis. Moreover, the stress state of the film can be clearly identified also for very thin layers in the nanometre range, well below the limit of conventional XRD analysis. Finally, the MOSS diagnostic of oxide materials has been coupled for the first time with RHEED during PLD.

Strain-controllable ionic or mixed ionic/electronic conductivity can boost the development of oxide heterostructures as active components for micro-electrochemical devices[45], such as micro-SOFCs[50], electrochemical sensors, resistive switching memories[51], or as active catalytic surfaces. But the potential of the MOSS plus RHEED diagnostics for lattice strain engineering goes far beyond this, providing a tool for tuning material properties for a broad range of applications (multiferroicity, catalysis, nano-device fabrication) in many different disciplines opening unprecedented opportunities in Materials Science.

Methods

Thin film deposition and *in situ* techniques. The thin films were fabricated in a PLD system from Twente Solid State Technology with integrated MOSS and RHEED systems (k-Space Associates, Inc.). The vacuum chamber, equipped with load-lock chamber, has a base pressure of about 10^{-8} mbar. Gas inlet lines allow setting the required background partial pressure during the deposition. A radiant resistive heater sets the desired deposition temperature. The substrate temperature was monitored with a pyrometer. The target carousel can hold up to five different targets. A custom-made substrate holder allows the simultaneous alignment of the substrate surface with the ablation plume, the RHEED and the MOSS systems.

$10 \times 10 \times 0.5\ mm^3$ substrates from CrysTec GmbH are used. For MgO, a special substrate preparation is required where the substrates are cut in small wafers after polishing in order to obtain negligible surface curvature. This procedure results in a not atomically flat surface, as revealed by RHEED. By annealing the substrates in O_2 at 1,000 °C for 12 h, two-dimensional surface reconstruction was obtained. The substrate holder supports the substrate only in the corners without mechanical constrain to allow it to bend freely. As the MgO substrates have a transmission wavelength range of 0.2–8 µm, to ensure an accurate temperature reading and an efficient substrate heating, the unpolished surface of the substrates is coated with Pt (sputtered with 40 W at 3×10^{-2} mbar for 4 min, ~400 nm) used as heat absorber. Such a thin layer does not affect the overall elastic properties of the substrate; neither has it influenced the MOSS measurements of the relative changes in the substrate curvature.

Ceramic sintered pellets of BZO, STO and 15 at.% SDC, prepared in our laboratories, were ablated using a 248-nm KrF excimer laser (Coherent Lambda Physics GmbH) with pulse lenght of 25 ns. A spot size of the laser on the target of $1.2\ mm^2$, a fluence of $1.2\ J\ cm^{-2}$, a frequency of 4 Hz and a target to substrate distance of 5 cm were used. The oxygen back ground pressure was set to 5×10^{-2} mbar. BZO and STO were deposited at ~740 °C and SDC, unless otherwise indicated, at 650 °C. The deposition rates have been accurately calibrated by X-ray reflectometry to be 0.14 Å $pulse^{-1}$ for BZO, 0.09 Å $pulse^{-1}$ for STO and 0.11 Å $pulse^{-1}$ for SDC. The samples were well reproducible, for example, concerning crystallinity, stress state, conductive properties and so on.

For the *in situ* wafer curvature measurement with the MOSS, a 3 × 3 array of laser beams is used. Typically, the measured wafer curvature is converted into the stress-thickness product via Stoney's equation and plotted as a function of thickness (see Supplementary Methods for more details; Supplementary Fig. 2). The main advantages of MOSS over similar techniques, for example, using a cantilever as substrate, are that any substrate can be used, no position-sensitive detector is needed, and the noise is reduced through multiple beams.

***Ex situ* characterization.** XRD characterizations were performed with a Siemens D500 diffractometer. The RSMs were recorded for the MgO(113) and the SDC(204) diffraction peak separately with the integration time for the film peak being 20 times higher. The RSMs consist of rocking curves, recorded for a range of 2θ values. The centre of the diffraction is determined by fitting individual rocking curves in the software Diffract.EVA.

A difference in the thermal expansion coefficients could lead to different *in* and *ex situ* strain values. However, the linear expansion coefficients for the materials used in this work are all around $10^{-5}\,K^{-1}$ (NGO (ref. 52), MgO (ref. 53), LAO (ref. 54), SDC (ref. 55), STO (ref. 54), BZO (ref. 56)) and the expansion mismatch is negligible (for example, 0.003% for SDC grown at 700 °C on MgO).

Cross-sectional TEM specimens were prepared by mechanical polishing and low-voltage Ar^{+} ion milling for the final thinning. TEM and STEM investigations were performed on a Titan 80-300 (FEI) Environmental Transmission Electron Microscope equipped with an imaging-side aberration corrector. The experiments were conducted at an acceleration voltage of 300 kV. Atomic-resolution Z contrast images were obtained by HAADF and BF STEM imaging. The inner and outer collection angles of the detector were 70 and 200 mrad, respectively.

The impedance spectroscopy measurements were carried out under a flow of O_2 in a tube furnace. Rectangular parallel Pt electrodes were used, defined by stainless steel masks except for the characterization of the growth platform, where interdigitated electrodes were patterned by ultraviolet photolithography (Supplementary Methods and Supplementary Fig. 3). The electrodes were deposited by magnetron sputtering at room temperature. Pt was sputtered with 40 W at 3×10^{-2} mbar for 2 min (100–200 nm) on a Ti sticking layer sputtered with 20 W at 7×10^{-2} mbar for 1.5 min ($\sim$5 nm). The target to substrate distance was 4 cm. The electrodes were wired to the read-out electronics using Ag paste and Au wires. The impedance was measured using a Solartron 1260 impedance/gain-phase analyser with a bias voltage of 1 V in the frequency range between 1 Hz and 1 MHz, varying the temperature between 300 and 720 °C. Data analysis was performed with EC-Lab (V10.31), where the data were fit to the response of a RC parallel circuit.

References

1. Healy, N. *et al.* Extreme electronic bandgap modification in laser-crystallized silicon optical fibres. *Nat. Mater.* **13,** 1122–1127 (2014).
2. Ramesh, R. & Spaldin, N. A. Multiferroics: progress and prospects in thin films. *Nat. Mater.* **6,** 21–29 (2007).
3. Strasser, P. *et al.* Lattice-strain control of the activity in dealloyed core-shell fuel cell catalysts. *Nat. Chem.* **2,** 454–460 (2010).
4. Li, S. *et al.* Strain-controlled thermal conductivity in ferroic twinned films. *Sci. Rep.* **4,** 6375 (2014).
5. Llordés, A. *et al.* Nanoscale strain-induced pair suppression as a vortex-pinning mechanism in high temperature superconductors. *Nat. Mater.* **11,** 329–336 (2012).
6. Giri, G. *et al.* Tuning charge transport in solution-sheared organic semiconductors using lattice strain. *Nature* **480,** 504–508 (2011).
7. Kilner, J. A. Ionic conductors: feel the strain. *Nat. Mater.* **7,** 838–839 (2008).
8. Yildiz, B. "Stretching" the energy landscape of oxides - effects on electrocatalysis and diffusion. *MRS Bull.* **39,** 147–156 (2014).
9. Hanbücken, M. *Stress And Strain In Epitaxy: Theoretical Concepts, Measurements and Applications* (Elsevier, 2001).
10. Suresh, S. & Freund, L. B. *Thin Film Materials: Stress, Defect Formation And Surface Evolution* (Cambridge Univ., 2006).
11. Ayers, J. E. *Heteroepitaxy Of Semiconductors: Theory, Growth, and Characterization* (CRC Press Taylor & Francis Group, 2007).
12. Jesson, D. E., Pennycook, S. J., Baribeau, J.-M. & Houghton, D. C. Surface stress, morphological development, and dislocation nucleation during strained-layer epitaxy. *MRS Online Proc. Library* **317,** 31–37 (1993).
13. Yang, W. H. & Srolovitz, D. J. Cracklike surface instabilities in stressed solids. *Phys. Rev. Lett.* **71,** 1593–1596 (1993).
14. Zhu, J., Li, Y. R., Zhang, Y., Liu, X. Z. & Tao, B. W. Effects of compressive and tensile stress on the growth mode of epitaxial oxide films. *Ceram. Int.* **34,** 967–970 (2008).
15. Bauer, S. *et al.* The power of in situ pulsed laser deposition synchrotron characterization for the detection of domain formation during growth of $Ba_{0.5}Sr_{0.5}TiO_3$ on MgO. *J. Synchrotron Radiat.* **21,** 386–394 (2014).
16. Brown, A. S. & Losurdo, M. in *Handbook of Crystal Growth* 2nd Edn (ed. Thomas, F. Kuech) 1169–1209 (Elsevier, 2015).
17. Premper, J., Sander, D. & Kirschner, J. *In situ* stress measurements during pulsed laser deposition of $BaTiO_3$ and $SrTiO_3$ atomic layers on Pt(0 0 1). *Appl. Surf. Sci.* **335,** 44–49 (2015).
18. Jacobs, R. N. *et al.* Dynamic curvature and stress studies for MBE CdTe on Si and GaAs substrates. *J. Electron. Mater.* **44,** 3076–3081 (2015).
19. Chason, E. *et al.* Growth of patterned island arrays to identify origins of thin film stress. *J. Appl. Phys.* **115,** 123519 (2014).
20. Scharf, T., Faupel, J., Sturm, K. & Krebs, H.-U. Intrinsic stress evolution in laser deposited thin films. *J. Appl. Phys.* **94,** 4273–4278 (2003).
21. Floro, J. A., Chason, E., Cammarata, R. C. & Srolovitz, D. J. Physical origins of intrinsic stresses in Volmer-Weber thin films. *MRS Bull.* **27,** 19–25 (2002).
22. Abadias, G., Fillon, A., Colin, J. J., Michel, A. & Jaouen, C. Real-time stress evolution during early growth stages of sputter-deposited metal films: influence of adatom mobility. *Vacuum* **100,** 36–40 (2014).
23. Chason, E., Shin, J. W., Hearne, S. J. & Freund, L. B. Kinetic model for dependence of thin film stress on growth rate, temperature, and microstructure. *J. Appl. Phys.* **111,** 083520 (2012).
24. Floro, J. A. & Chason, E. Measuring Ge segregation by real-time stress monitoring during $Si_{1-x}Ge_x$ molecular beam epitaxy. *Appl. Phys. Lett.* **69,** 3830–3832 (1996).
25. Michotte, S. & Proost, J. *In situ* measurement of the internal stress evolution during sputter deposition of ZnO:Al. *Sol. Energy Mater. Sol. Cells* **98,** 253–259 (2012).
26. Wachsman, E. D. & Lee, K. T. Lowering the temperature of solid oxide fuel cells. *Science* **334,** 935–939 (2011).
27. Jiang, J. & Hertz, J. L. On the variability of reported ionic conductivity in nanoscale YSZ thin films. *J. Electroceram.* **32,** 37–46 (2014).
28. Korte, C. *et al.* Coherency strain and its effect on ionic conductivity and diffusion in solid electrolytes - an improved model for nanocrystalline thin films and a review of experimental data. *Phys. Chem. Chem. Phys.* **16,** 24575–24591 (2014).
29. Pergolesi, D., Fronzi, M., Fabbri, E., Tebano, A. & Traversa, E. Growth mechanisms of ceria- and zirconia-based epitaxial thin films and hetero-structures grown by pulsed laser deposition. *Mater. Renew. Sustain. Energy* **2,** 1–9 (2012).
30. Petric, A. & Huang, P. Oxygen conductivity of $Nd(SrCa)Ga(Mg)O_{3-\delta}$ perovskites. *Solid State Ionics* **92,** 113–117 (1996).
31. Sanna, S. *et al.* Fabrication and electrochemical properties of epitaxial samarium-doped ceria films on $SrTiO_3$-buffered MgO substrates. *Adv. Funct. Mater.* **19,** 1713–1719 (2009).
32. Pergolesi, D. *et al.* Tensile lattice distortion does not affect oxygen transport in Yttria-stabilized zirconia-CeO_2 heterointerfaces. *ACS Nano* **6,** 10524–10534 (2012).
33. Pergolesi, D. *et al.* Probing the bulk ionic conductivity by thin film hetero-epitaxial engineering. *Sci. Tech. Adv.Mater* **16,** 015001 (2015).
34. Wachtel, E. & Lubomirsky, I. The elastic modulus of pure and doped ceria. *Scripta Mater.* **65,** 112–117 (2011).
35. Göbel, M. C., Gregori, G., Guo, X. & Maier, J. Boundary effects on the electrical conductivity of pure and doped cerium oxide thin films. *Phys. Chem. Chem. Phys.* **12,** 14351–14361 (2010).
36. Pérez Casero, R. *et al.* Epitaxial growth of CeO_2 on MgO by pulsed laser deposition. *Appl.Surf. Sci.* **109-110,** 341–344 (1997).
37. Chen, L. *et al.* Electrical properties of a highly oriented, textured thin film of the ionic conductor Gd:$CeO_{2-\delta}$ on (001) MgO. *Appl. Phys. Lett.* **83,** 4737–4739 (2003).
38. Sanna, S. *et al.* Enhancement of the chemical stability in confined δ-Bi_2O_3. *Nat. Mater.* **14,** 500–504 (2015).
39. Sameshima, S. *et al.* Electrical conductivity and diffusion of oxygen ions in rare-earth-doped ceria. *Nippon Seramikkusu Kyokai Gakujutsu Ronbunshi/J. Ceramic Soc. Jpn* **108,** 1060–1066 (2000).
40. Giannici, F. *et al.* Structure and oxide ion conductivity: local order, defect interactions and grain boundary effects in acceptor-doped ceria. *Chem. Mater.* **26,** 5994–6006 (2014).
41. Balazs, G. B. & Glass, R. S ac impedance studies of rare earth oxide doped ceria. *Solid State Ionics* **76,** 155–162 (1995).
42. Sun, L., Marrocchelli, D. & Yildiz, B. Edge dislocation slows down oxide ion diffusion in doped CeO_2 by segregation of charged defects. *Nat. Commun.* **6,** 1–10 (2015).
43. De Souza, R. A., Ramadan, A. & Horner, S. Modifying the barriers for oxygen-vacancy migration in fluorite-structured CeO_2 electrolytes through strain: a computer simulation study. *Energy Environ. Sci.* **5,** 5445–5453 (2012).
44. Mohan Kant, K., Esposito, V. & Pryds, N. Strain induced ionic conductivity enhancement in epitaxial $Ce_{0.9}Gd_{0.1}O_{2-\delta}$ thin films. *Appl. Phys. Lett.* **100,** 033105 (2012).
45. Schweiger, S., Kubicek, M., Messerschmitt, F., Murer, C. & Rupp, J. L. M. A microdot multilayer oxide device: Let us tune the strain-ionic transport interaction. *ACS Nano* **8,** 5032–5048 (2014).
46. Kushima, A. & Yildiz, B. Oxygen ion diffusivity in strained yttria stabilized zirconia: Where is the fastest strain? *J. Mater. Chem.* **20,** 4809–4819 (2010).
47. Aydin, H., Korte, C., Rohnke, M. & Janek, J. Oxygen tracer diffusion along interfaces of strained Y_2O_3/YSZ multilayers. *PCCP* **15,** 1944–1955 (2013).

48. Shi, Y., Bork, A. H., Schweiger, S. & Rupp, J. L. M. The effect of mechanical twisting on oxygen ionic transport in solid-state energy conversion membranes. *Nat. Mater.* **14,** 721–727 (2015).
49. Rupp, J. L. M. *et al.* Scalable oxygen-ion transport kinetics in metal-oxide films: impact of thermally induced lattice compaction in acceptor doped ceria films. *Adv. Funct. Mater.* **24,** 1562–1574 (2014).
50. Evans, A., Bieberle-Hütter, A., Rupp, J. L. M. & Gauckler, L. J. Review on microfabricated micro-solid oxide fuel cell membranes. *J. Power Sources* **194,** 119–129 (2009).
51. Messerschmitt, F., Kubicek, M., Schweiger, S. & Rupp, J. L. M. Memristor kinetics and diffusion characteristics for mixed anionic-electronic $SrTiO_{3-\delta}$ bits: The memristor-based cottrell analysis connecting material to device performance. *Adv. Funct. Mater.* **24,** 7448–7460 (2014).
52. Sasaura, M., Miyazawa, S. & Mukaida, M. Thermal expansion coefficients of high-T_c superconductor substrate $NdGaO_3$ single crystal. *J. Appl. Phys.* **68,** 3643–3644 (1990).
53. Tsay, Y.-f., Bendow, B. & Mitra, S. S. Theory of the temperature derivative of the refractive index in transparent crystals. *Phys. Rev. B* **8,** 2688–2696 (1973).
54. Char, K., Antognazza, L. & Geballe, T. H. Study of interface resistances in epitaxial $YBa_2Cu_3O_{7-x}$/barrier/$YBa_2Cu_3O_{7-x}$ junctions. *Appl. Phys. Lett.* **63,** 2420–2422 (1993).
55. Hyodo, J., Ida, S., Kilner, J. A. & Ishihara, T. Electronic and oxide ion conductivity in $Pr_2Ni_{0.71}Cu_{0.24}Ga_{0.05}O_4$/$Ce_{0.8}Sm_{0.2}O_2$ laminated film. *Solid State Ionics* **230,** 16–20 (2013).
56. Zhao, Y. & Weidner, D. J. Thermal expansion of $SrZrO_3$ and $BaZrO_3$ perovskites. *Phys. Chem. Minerals* **18,** 294–301 (1991).
57. Guennou, M., Bouvier, P., Garbarino, G. & Kreisel, J. Structural investigation of $LaAlO_3$ up to 63 GPa. *J. Phys. Condens. Matter.* **23,** 395401 (2011).
58. Krivchikov, A. I. *et al.* Structure, sound velocity, and thermal conductivity of the perovskite $NdGaO_3$. *Low Temp. Phys.* **26,** 370–374 (2000).
59. Ahrens, T. J. *Mineral Physics & Crystallography: A Handbook of Physical Constants* (American Geophysical Union, 1995).

Acknowledgements

The research leading to these results has received funding from the European Community's Seventh Framework Programme (FP7/2007–2013) under grant agreement no. 290605 (COFUND: PSI-FELLOW), as well as from the Swiss National Foundation for Science (SNFS) under grant agreement no. 200021_126783. We thank Johann Michler, Madoka Hasegawa and Vipin Chawla of the Laboratory for Mechanics of Materials and Nanostructures at EMPA, Thun (CH) for ultraviolet photolithography of interdigitated micro-electrodes.

Author contributions

A.F. and D.P. set up the experimental facility, fabricated and characterized the films, and wrote the manuscript. V.R. performed the TEM measurements. T.L. and D.P. designed and supervised the experiment. A.W. helped in the discussion and interpretation of results. All authors contributed in the revision of the manuscript.

Additional information

Competing financial interests: The authors declare no competing financial interests.

28

Near-field photocurrent nanoscopy on bare and encapsulated graphene

Achim Woessner[1], Pablo Alonso-González[2,3,*], Mark B. Lundeberg[1,*], Yuanda Gao[4], Jose E. Barrios-Vargas[5], Gabriele Navickaite[1], Qiong Ma[6], Davide Janner[1], Kenji Watanabe[7], Aron W. Cummings[5], Takashi Taniguchi[7], Valerio Pruneri[1,8], Stephan Roche[5,8], Pablo Jarillo-Herrero[6], James Hone[4], Rainer Hillenbrand[9,10] & Frank H.L. Koppens[1,8]

Optoelectronic devices utilizing graphene have demonstrated unique capabilities and performances beyond state-of-the-art technologies. However, requirements in terms of device quality and uniformity are demanding. A major roadblock towards high-performance devices are nanoscale variations of the graphene device properties, impacting their macroscopic behaviour. Here we present and apply non-invasive optoelectronic nanoscopy to measure the optical and electronic properties of graphene devices locally. This is achieved by combining scanning near-field infrared nanoscopy with electrical read-out, allowing infrared photocurrent mapping at length scales of tens of nanometres. Using this technique, we study the impact of edges and grain boundaries on the spatial carrier density profiles and local thermoelectric properties. Moreover, we show that the technique can readily be applied to encapsulated graphene devices. We observe charge build-up near the edges and demonstrate a solution to this issue.

[1] ICFO—Institut de Ciencies Fotoniques, The Barcelona Institute of Science and Technology, 08860 Barcelona, Spain. [2] CIC nanoGUNE, 20018 Donostia-San Sebastian, Spain. [3] Institute of Physics, Chinese Academy of Science, Beijing 100190, China. [4] Department of Mechanical Engineering, Columbia University, New York, New York 10027, USA. [5] Catalan Institute of Nanoscience and Nanotechnology (ICN2), CSIC and The Barcelona Institute of Science and Technology, Campus UAB, 08193 Barcelona, Spain. [6] Department of Physics, Massachusetts Institute of Technology, Cambridge, Massachusetts 02139, USA. [7] National Institute for Materials Science, 1-1 Namiki, Tsukuba 305-0044, Japan. [8] ICREA-Institució Catalana de Recerca i Estudis Avançats, 08010 Barcelona, Spain. [9] CIC nanoGUNE and UPV/EHU, 20018 Donostia-San Sebastian, Spain. [10] IKERBASQUE, Basque Foundation for Science, 48011 Bilbao, Spain. * These authors contributed equally to this work. Correspondence and requests for materials should be addressed to F.H.L.K. (email: frank.koppens@icfo.eu).

As large scale integration and wafer scale device processing capabilities of graphene have become available[1–8], technological implementations of electronic and optoelectronic graphene devices are within reach[9–11]. At the same time, to achieve high device performance, any imperfections at the nanometer or even atomic scale need to be minimized or even eliminated. For example, in large area graphene, grown by chemical vapour deposition (CVD), grain boundaries are the stitching regions between different monocrystalline parts of graphene and act as carrier scatterers, limiting the graphene mobility and uniformity[12,13]. In addition, even perfectly monocrystalline graphene is still highly sensitive to its environment, and on typical substrates charge–density inhomogeneities (charge puddles)[14–19] and additional doping near contacts, defects and edges arise, which reduce the device performance as well. Therefore, it is important to efficiently probe the nanoscale optoelectronic properties of graphene devices and to understand their microscopic physical behaviour.

A major challenge is that many of the available characterization techniques are invasive[20], need specifically designed device structures[13,21,22], image only very small areas[14,15,21,23–25], rely on high doping of the graphene,[26] require unhindered electrical access of the probe to the graphene[14,15,21,23,24] or lack the desired nanometer resolution[27] and are expensive and difficult to implement. For the direct quality control of graphene devices, a method that can image electrical and optical properties of graphene devices, at nanoscale resolution, without any special preparation and without modifying the devices is required.

Here we demonstrate fully non-invasive room-temperature scanning near-field photocurrent nanoscopy[28–38] for the first time applied on graphene with infrared frequencies and use it to study the nanoscale optoelectronic properties of graphene devices that can later be used for real applications. This technique is based on electrical probing of the photoresponse due to strongly localized heating. We apply this technique to study the microscopic physics of grain boundaries and charge–density inhomogeneities. In the case of grain boundaries, we were able to identify the magnitude of their Seebeck coefficient, while for charge–density inhomogeneities, we show how they influence the global charge neutrality point of graphene devices. In addition, we study encapsulated graphene devices[39,40], where the encapsulation would prevent many other scanning probe techniques from accessing local properties of graphene. In these devices, we find a charge build-up near the edges and show that using local metal gates instead of a global backgate effectively suppresses this type of edge doping. In general, this technique operates most effectively with mid-infrared light because it does not lead to photodoping[41] and it is more stable in operation, compared with visible light.

Results

Measurement principle. The measurement principle is sketched in Fig. 1a. The setup is based on a scattering-type scanning near-field optical microscope (s-SNOM)[26,42] augmented with electrical contact to the sample to measure currents *in situ*[28–38]. In contrast to conventional s-SNOM, we do not need to measure the outscattered light but rather directly measure current induced by the near-field as explained in the following. A 10.6-µm mid-infrared laser illuminates a metallized atomic force microscope probe, tapping at its mechanical resonance frequency. Part of the incoming light, polarized parallel to the shaft of the probe, excites a strong electric field at the tip apex due to an antenna effect[43]. The spatial extent of this near-field is on the order of 25 nm, limited only by the tip radius and much smaller than the free space wavelength of the impinging light[43].

The near and far fields impinging on the device induce charge flows in the device (by mechanisms discussed below), and drive currents into an external current amplifier via contacts on the device. We isolate the part of the current that is induced by near fields by demodulating the current at the second harmonic of the tip tapping frequency[43]. This demodulated current is denoted I_{PC} and referred to as near-field photocurrent and is obtained together with near-field optical and topography information. A typical map of I_{PC}, obtained by scanning the tip over a CVD graphene device, is shown in Fig. 1b.

We can assess the spatial resolution of the photocurrent maps by comparing a region near the edge (Fig. 1d) with a topographic image from the same region (Fig. 1c). As can be seen, I_{PC} falls to zero for tip locations away from the graphene on a similar length scale as the topography, demonstrating the successful isolation of near-field contributions. In Fig. 1e, we quantify the resolution by observing the change in I_{PC} as the tip is moved over the edge of graphene. The full-width at half maximum of the photocurrent peak at this location is ~100 nm, matching the rise distance in the topographic signal. This resolution is far below any limits relating to the 10.6 µm free space light wavelength.

Photothermoelectric photocurrent generation mechanism. As to the physical mechanism of the photocurrent, we consider the photothermoelectric effect that has been shown to dominate the photoresponse of graphene[11,44–48]: the light (in this case, the tip-enhanced near-field) locally heats the graphene, and this heat acts via non-uniformities in Seebeck coefficient S to drive charge currents within the device and into the contacts (see Methods section). Therefore, we interpret the variations of I_{PC} in terms of microscopic variations in S. The Seebeck coefficient, which depends on material properties such as carrier density and mobility, is a measure of the electromotive force driven by a temperature difference in a material. A complete description of I_{PC} needs to take into account the carrier cooling length[45,46] and overall sample geometry[49]. The carrier cooling length $l_{cool} = \sqrt{\kappa/g}$, where κ the sheet thermal conductivity in plane and g the interfacial thermal conductivity out of plane to the heat sinking substrate, describes how far heat propagates through the charge carriers, before dissipating to the environment (see Supplementary Fig. 2)[46]. A quantitative model of the thermoelectric photocurrent mechanism can be found in the Methods section.

Grain boundary characterization. We first discuss the application of this infrared near-field photocurrent technique to grain boundaries. They are not visible in the simultaneously acquired topography, and are responsible for some of the line-shaped features in the photocurrent map in Fig. 1b. Some of the other features stem from large scale inhomogeneities of the sample. The region within the green frame is shown with higher resolution in Fig. 1d, exhibiting a strong photocurrent signal that changes sign along a sharp boundary, yet the graphene is topographically flat in the vicinity of this boundary (Fig. 1c). We show now that this type of feature indicates a grain boundary.

Figure 2a shows a line profile of I_{PC} across the boundary feature identified in Fig. 1d. This antisymmetric I_{PC} can be explained by a localized deviation in S at the boundary, that is, a line defect within an otherwise uniform thermoelectric medium (Supplementary Figs 3,4 and Supplementary Note 1). Indeed, grain boundaries behave as localized lines of strongly modified electronic properties, within otherwise uniform graphene[20,21,26,50,51]. We remark that the decay of the photocurrent away from the boundary extends over more than

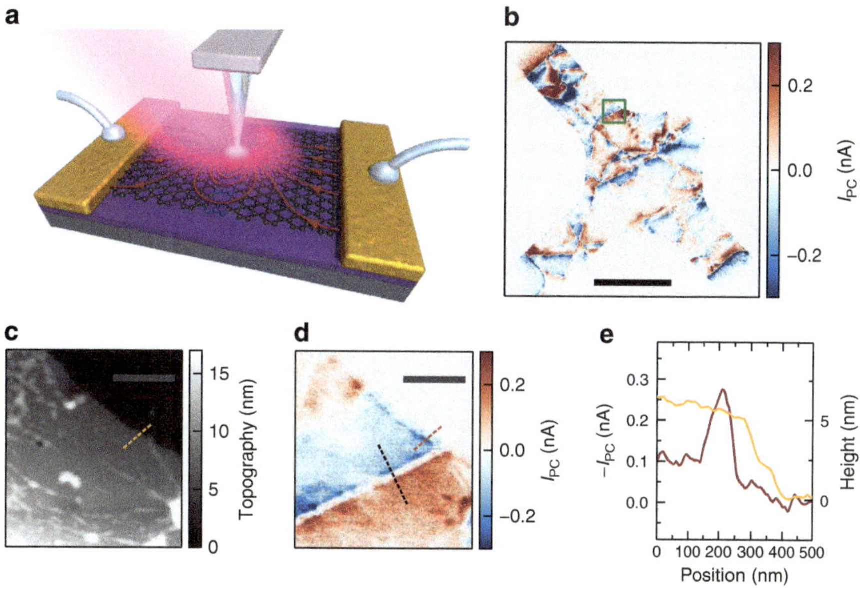

Figure 1 | Near-field photocurrent working principle and photocurrent from grain boundaries. (**a**) Sketch of the scattering-type scanning near-field optical microscope setup. A mid-infrared laser illuminates the atomic force microscope tip, which generates a locally concentrated optical field, which is absorbed by the graphene generating a position dependent photocurrent. The blue region in the graphene lattice represents a grain boundary with a modified Seebeck coefficient. The arrows sketch the photocurrent flow pattern. For each position only the magnitude and direction of the current are measured. The sketch is not to scale. (**b**) I_{PC} map at at backgate voltage $V_{BG} = 0$ V of a single layer CVD graphene device (Supplementary Fig. 1) with three contacts: top left (drain), right (source) and bottom left (ground). Both grain boundaries and wrinkles show characteristic photocurrent patterns. (scale bar, 5 μm) The green box indicates the measurement region in **c**,**d**. (**c**) Topography of etched CVD graphene does not show grain boundary but only wrinkles and other inhomogeneities due to the transfer process. (scale bar, 500 nm) (**d**) I_{PC} at $V_{BG} = 0$ V clearly shows a grain boundary and the expected sign change around it. The black dashed line indicates the measurement positions in Fig. 2a,d. (scale bar, 500 nm) (**e**) Topography (orange) and I_{PC} (red) measured at the orange line in **c** and the red line in **d**, respectively.

100 nm, which is due to a larger hot carrier cooling. We find in this case $l_{cool} = 140$ nm.

To gain more insight in the Seebeck coefficient at the grain boundary, we tune the carrier density by a global gate (Fig. 2d). We observe that the antisymmetric spatial photocurrent profile changes sign as the backgate voltage V_{BG} passes the peak in resistance, that is, the global charge neutrality point V_D. The Seebeck coefficient S_G of graphene itself changes sign at the charge neutrality point[44,45,52,53] (Fig. 2c). Thus, after calibrating the sign to the known sign of the contact photocurrent, our data implies that the Seebeck coefficient of the grain boundary S_{GB} is always smaller in magnitude than S_G, since $I_{PC}(V_{BG}) \propto S_G(V_{BG}) - S_{GB}(V_{BG})$.

Using a polycrystalline graphene model, we compute the resistance due to grain boundaries using a Kubo transport formalism and real space simulations[54]. S_{GB} is the ratio of the first- and zero-order Onsager coefficients (Supplementary Note 2). Indeed, we find that S_{GB} is always smaller in magnitude and has a similar lineshape as S_G in the carrier density range measured (Fig. 2c). Figure 2e shows a simulation of the photocurrent for the calculated Seebeck coefficients, which is in agreement with the measurements (Supplementary Figs 5,6).

Charge puddle characterization. We next examine near-field photocurrent in a typical two-probe exfoliated graphene device (Fig. 3). A strong photocurrent is obtained with the tip near the metal contacts, similar to previous near- and far-field measurements[34,47,55]. In addition, an apparently random pattern of photocurrent is present throughout the device, as in high-resolution far-field measurements[55] but at a much finer scale.

The random photocurrent pattern between the contacts in Fig. 3a indicates random variations in Seebeck coefficient over short length scales (Supplementary Fig. 7). Random variations of the Seebeck coefficient are indeed expected since it depends on carrier density[52], which in turn has fine-scaled inhomogeneities (charge puddles)[14–18]. The photocurrent variations can thus be used to gain insight in the charge puddle distribution. A more detailed view of the photocurrent due to charge puddles in Fig. 3b shows that the length scale that can be resolved is on the order of hundreds of nanometres.

Quantitatively, from the autocorrelation of the photocurrent in comparison with a photothermoelectric model taking into account the size of the charge puddles in Fig. 3c we extract $l_{cool} \sim 200$ nm. The charge puddles are modelled to have a size of ~ 20 nm, in accordance with measurements of graphene on silicon oxide (SiO_2; refs 15–18).

By changing the gate voltage we study the carrier density profile with high spatial resolution (Fig. 4) and highlight the possibility of spatially resolving the charge neutrality point for a large device. I_{PC} from charge puddles is largest around the charge neutrality point and varies with position. This is consistent with the very high sensitivity of the Seebeck coefficient to changes in carrier density, near-zero density (Fig. 4b). The magnitude of photocurrent from charge puddles depends on the difference of Seebeck coefficient between two adjacent charge puddles or in other words the strongest photocurrent from charge puddle appears at the position of highest Seebeck gradient. This allows us to map the local carrier density offset (charge inhomogeneity) throughout the device, as indicated by the extremum of photocurrent in a scan of photocurrent versus gate voltage (Fig. 4c). The photocurrent from adjacent charge puddles with a given charge carrier density offset does not change sign when sweeping through the charge neutrality point. This is because the difference in Seebeck coefficient between these puddles does not change sign.

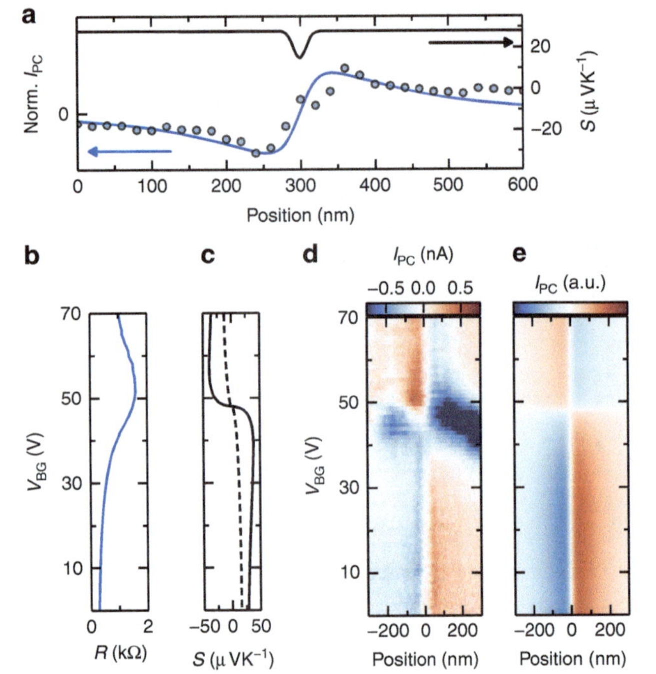

Figure 2 | Photocurrent profile at a grain boundary and its gate voltage dependence. (**a**) Photocurrent profile, measured at the black dashed line in Fig. 1d, perpendicular to the grain boundary at $V_{BG}=0$ V shows good agreement with the photothermoelectric model with $l_{cool}=140$ nm. The modelled spatial Seebeck profile (with FWHM 20 nm) is shown in black. (**b**) Two-probe device resistance as a function of V_{BG}. (**c**) Simulated Seebeck coefficient S_G for pristine graphene (solid line) and S_{GB} for polycrystalline graphene with an average grain size of 25 nm (dashed line; Supplementary Note 2). (**d**) Backgate dependent photocurrent profile perpendicular to the grain boundary shows that the grain boundary changes its sign at the charge neutrality point. (**e**) Simulated backgate dependent photocurrent profile based on the Seebeck profiles in **c** normalized to the simulated photocurrent maximum.

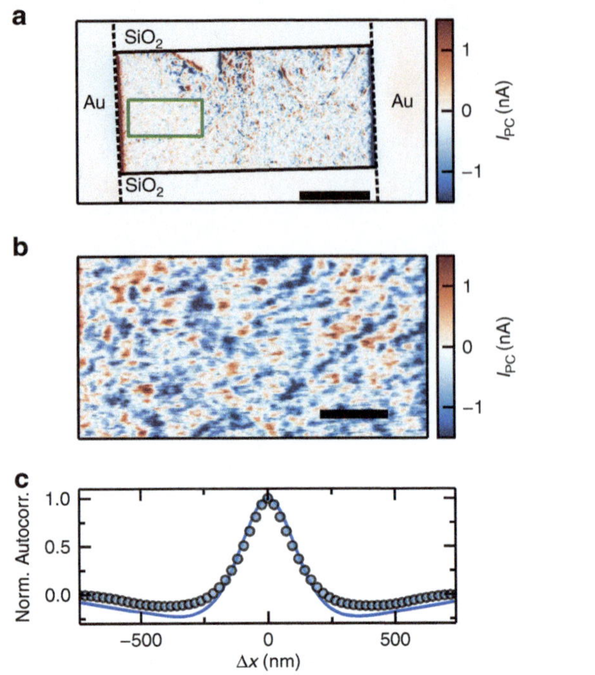

Figure 3 | Photocurrent from charge puddles. (**a**) Near-field photocurrent map of an exfoliated graphene device on 300 nm SiO_2 at $V_{BG}=20$ V. The dashed lines indicate the position of the contacts and solid lines the graphene edges. The green box indicates the measurement region in **b** (scale bar, 5 μm). (**b**) Detailed photocurrent map at the charge neutrality point of the device ($V_{BG}=7$ V) reveals the charge puddles and the high spatial resolution of the technique. (scale bar, 1 μm) (**c**) Autocorrelation of the photocurrent from charge puddles at V_D (data points) compared to photocurrent expected from a random charge puddle distribution and $l_{cool}=200$ nm (blue curve). Autocorrelation is taken along the source drain current path.

We can thus resolve the local charge neutrality point at a given position of the device (green curve, Fig. 4a), which can be different from the global charge neutrality point V_D, the backgate voltage V_{BG} at which the resistance is maximum (blue curve, Fig. 4a). We show that the global charge neutrality point (blue curve, Fig. 4a) is determined by an average of the gate voltages at which the local charge neutrality points appear (red curve, Fig. 4a). Spatially resolved puddle photocurrent can be much narrower (green curve, Fig. 4a) than the average of all possible current paths (red curve, Fig. 4a) This indicates that the graphene locally has less inhomogeneity. Thus the technique gives insight not only in the global but also in the local behaviour of the device.

Characterization of encapsulated devices. Finally, we apply this technique to a graphene device encapsulated between two layers of hexagonal boron nitride (h-BN), using the polymer-free van der Waals assembly technique[39,40] as sketched in Fig. 5e. This device lies on top of an oxidized silicon wafer, used as a backgate. The stack is etched into a triangle and electrically side-contacted by two metal contacts[39]. Recent studies in vacuum and low temperatures[56,57] have shown that the edges affect where current flows in the device, in particular near charge neutrality. In the following we study the build-up of edge doping in ambient conditions and provide a solution to this.

While monitoring the photocurrent of such encapsulated devices, we observe indications of strong carrier density variations near the edges over micrometre scales. These variations are influenced by lighting conditions (Supplementary Fig. 8), gate voltages, and temperature, and evolve over timescales ranging from minutes to weeks. As an example, Fig. 5a–d shows a progression of photocurrent maps, taken after annealing the device at 200 °C for 30 min in ambient conditions to temporarily remove charge density variations near the edges. Initially in Fig. 5a we see very small photocurrents indicating a flat carrier density landscape. After some time (hours), in the dark with only gate voltages smaller than 3 V applied, a small doping gradient between the contacts builds up. This gradient leads to the stronger photocurrent shown in Fig. 5b. The local charge neutrality point, indicated by the maximum of photocurrent, is at the same position close to the edge of the device as further inside the bulk. After keeping the device for 3 h in ambient conditions we can see a change of the local charge neutrality point at the edge of the graphene compared to the bulk in Fig. 5c. The edge is slightly more p-type compared to the bulk. Finally, we apply high gate voltages, of in this case 50 V for ~20 h, to increase the edge doping. A strong p-doping at the edge and an n-doping in the bulk of the graphene is induced in Fig. 5d. This indicates that electric field accelerates the speed and increases this type of edge doping.

We exploit the observed edge doping to create a natural p–n junction along the edge of the device. For this we

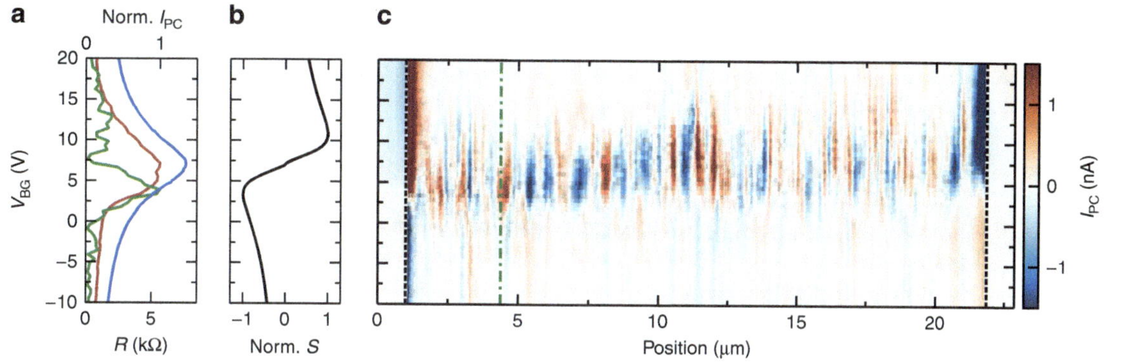

Figure 4 | Dependence of photocurrent profiles on backgate voltage reveals doping inhomogeneities. (**a**) Backgate dependence of the resistance of the device measured simultaneously to the photocurrent in blue. The red curve shows the normalized root mean square of the photocurrent across the device. The green curve shows a single normalized photocurrent backgate trace, corresponding to the green dashed dotted line in **c**. (**b**) Backgate dependent Seebeck coefficient of graphene, calculated from the gate dependent resistance in **a** using the Mott formula[52]. (**c**) Backgate dependence of the photocurrent across the device. Graphene is between the black dashed lines, which indicate the edges of the metal contacts.

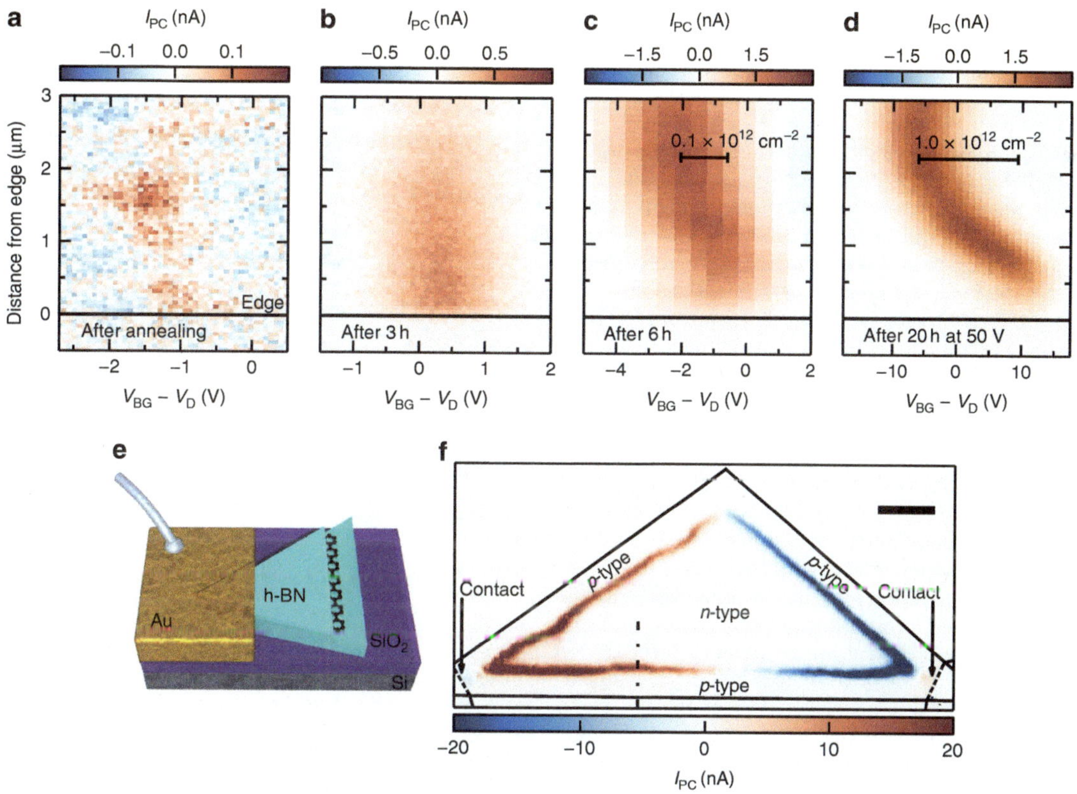

Figure 5 | Near-field photocurrent maps revealing edge doping in encapsulated graphene. (**a**) Spatial photocurrent profile versus backgate voltage V_{BG} (minus voltage of the resistance maximum V_D) near the edge of encapsulated graphene. These data are taken directly after annealing the device. ($V_D = -0.3$ V). (**b**) The same scan on the same device after three hours in air, ($V_D = -0.5$ V) and **c**, after annealing and applying V_{BG} up to 3 V ($V_D = 3$ V). (**d**) The same scan after approximately 20 h in air and after applying V_{BG} up to 50 V ($V_D = 30$ V). (**e**) Sketch of the device, a stack of h-BN(46 nm)/Graphene/h-BN(7 nm) on a Si/SiO_2(300 nm) wafer used as global backgate. In this sketch we only show one of the two contacts used for electrical measurements. (**f**) Photocurrent close to the resistance maximum at $V_{BG} = -28$ V shows a triangular photocurrent pattern, due to edge p-doping. The dashed dotted line indicates where the measurements in **a-d** were taken. (scale bar, 2 μm).

apply a backgate voltage at which the edge of the graphene is p-type and the bulk n-type. This is similar to the situation in Fig. 5d at $V_{BG} = V_D$. We observe photocurrent at the junction in Fig. 5f around the whole device, indicating that the edge doping is uniform around the graphene. The photocurrent decays gradually towards the midline between the electrodes as a result of how the triangular geometry modifies the ability of the contacts to capture photocurrents[49]. The sign change in the middle of the device is because in the current direction the junction changes from a p–n junction to an n–p junction (Supplementary Fig. 9 and Supplementary Note 2). We are able to temporarily reset the edge doping by annealing the device on a hotplate at 200 °C for 30 min as we show in Fig. 5a.

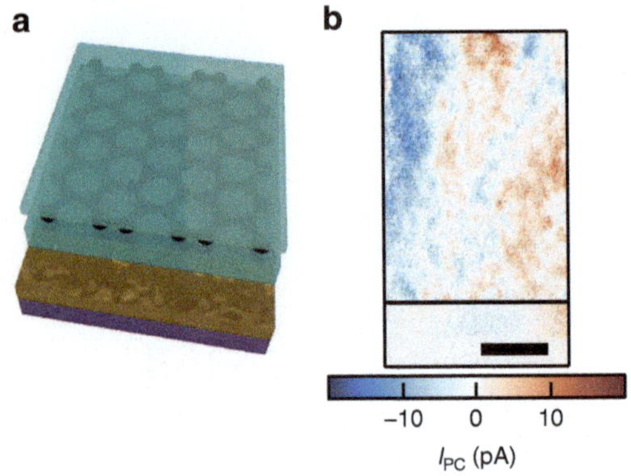

Figure 6 | Edge doping is efficiently suppressed using devices with local metal gates. (**a**) Sketch of the device with local metal gate with the two h-BN layers in light blue and the local metal gate in gold. The layers are from bottom to top SiO_2(300 nm)/AuPd(15 nm)/h-BN(42 nm)/Graphene/h-BN(13 nm). (**b**) Photocurrent from charge puddles in encapsulated graphene on a metal gate close to the charge neutrality point. The electrical contacts are on the left and right outside of this figure. (scale bar, 2 μm).

Local gates prevent charge build-up at the device edges. While we have not been able to precisely identify the origin of the edge doping, we suspect that water molecules are able to penetrate between the boron nitride and the SiO_2 due to the surface of the h-BN being not completely conformal with the underlying SiO_2 substrate. This water molecule penetration then leads to trapped charges is responsible for the observed edge doping. We also present here a technique to completely eliminate the edge doping. We place encapsulated graphene on top of a local conductive gate, such as a 15 nm AuPd alloy sketched in Fig. 6a. As the photocurrent measurement in Fig. 6b shows, we find that edge doping is efficiently suppressed even after extended periods of time at ambient conditions and high gate voltages. Furthermore, such devices efficiently suppress the photodoping effect observed for devices where the h-BN is in contact with SiO_2 (ref. 41; Supplementary Fig. 10 and Supplementary Note 1).

In the device with a local metal gate used to suppress both edge- and photodoping, we find small features due to charge puddles on top of a slowly varying background photocurrent, due to large-scale carrier density inhomogeneities. The size of the features due to charge puddles determined by autocorrelation is ~800 nm. The long length scale of those features is either due to the longer cooling length of the encapsulated graphene compared to the graphene on SiO_2 or due to larger charge puddle size in the encapsulated devices. Further work is required to clearly distinguish these effects.

Discussion

To conclude, we have demonstrated that scanning near-field photocurrent nanoscopy is a versatile technique to characterize the electronic and optoelectronic and even previously inaccessible properties of relevant graphene devices. This technique is highly promising for spatially resolved quality control of regular graphene devices without the need for special device structures and can therefore be readily applied.

Methods

Photocurrent model. Photocurrent I_{PC} in graphene as generated by the photothermoelectric effect and is described as:[45–47]

$$I_{PC} = \frac{1}{RW}\int\int \frac{\partial T}{\partial x} S \, dxdy \quad (1)$$

where R is the total resistance including graphene, contacts and circuitry, W the device width and x the current flow direction. This is valid for rectangular graphene devices and special care needs to be taken for arbitrary shapes, such as in Fig. 5 (ref. 49). For the temperature profile $T(x)$ we consider that the heat spreads in two dimensions with heat sinking to lattice and substrate, producing a $T(x)$ profile described by a modified Bessel function of the second kind, with a finite tip size correction (Supplementary Note 2). A 25-nm finite tip-size correction was used for all simulations.

Measurement details. The s-SNOM used was a NeaSNOM from Neaspec GmbH, equipped with a CO_2 laser operated at 10.6 μm, away from the phonon resonance of SiO_2, which can lead to strong substrate contributions to the photocurrent[48]. The laser power used was ~20 mW. The probes were commercially available metallized atomic force microscopy probes with an apex radius of approximately 25 nm. The tip height was modulated at a frequency of approximately 250 kHz with a 60–80 nm amplitude. A Femto DLPCA-200 current pre-amplifier was used. The probe tip was electrically grounded. Because of the different device geometries and the fact that all the measurements were taken at different times and slightly different device conditions the absolute values of the photocurrents are not comparable.

Device fabrication. The CVD graphene, grown on copper, was transferred onto a self-assembled monolayer[58] on 285 nm of SiO_2 to stabilize the charge neutrality point. The contacts were defined using optical lithography with Ti (5 nm)/Pd (35 nm). The graphene was transferred onto deposited contacts.

The exfoliated graphene device was fabricated on a Si/SiO_2(300 nm) wafer, used as backgate. The Cr(0.8 nm)/Au(80 nm) contacts were defined using electron beam lithography.

The Si/SiO_2(300 nm)/h-BN(46 nm)/Graphene/h-BN(7 nm) and the Si/SiO_2(300 nm)/AuPd(15 nm)/h-BN(42 nm)/Graphene/h-BN(13 nm) stacks, were fabricated using the polymer-free van der Waals assembling technique[39].

References

1. Li, X. *et al.* Large-area synthesis of high-quality and uniform graphene films on copper foils. *Science* **324,** 1312–1314 (2009).
2. Reina, A. *et al.* Large area, few-layer graphene films on arbitrary substrates by chemical vapor deposition. *Nano Lett.* **9,** 30–35 (2009).
3. Bae, S. *et al.* Roll-to-roll production of 30-inch graphene films for transparent electrodes. *Nat. Nanotech.* **5,** 574–578 (2010).
4. Bonaccorso, F. *et al.* Production and processing of graphene and 2d crystals. *Mater. Today* **15,** 564–589 (2012).
5. Ren, W. & Cheng, H.-M. The global growth of graphene. *Nat. Nanotech.* **9,** 726–730 (2014).
6. Lee, J.-H. *et al.* Wafer-scale growth of single-crystal monolayer graphene on reusable hydrogen-terminated germanium. *Science* **344,** 286–289 (2014).
7. Gao, L. *et al.* Face-to-face transfer of wafer-scale graphene films. *Nature* **505,** 190–194 (2014).
8. de Heer, W. A. *et al.* Large area and structured epitaxial graphene produced by confinement controlled sublimation of silicon carbide. *Proc. Natl Acad. Sci. USA* **108,** 16900–16905 (2011).
9. Ferrari, A. C. *et al.* Science and technology roadmap for graphene, related two-dimensional crystals, and hybrid systems. *Nanoscale* **7,** 4598–4810 (2014).
10. Akinwande, D., Petrone, N. & Hone, J. Two-dimensional flexible nanoelectronics. *Nat. Commun.* **5,** 5678 (2014).
11. Koppens, F. H. L. *et al.* Photodetectors based on graphene, other two-dimensional materials and hybrid systems. *Nat. Nanotech.* **9,** 780–793 (2014).
12. Yazyev, O. V. & Louie, S. G. Electronic transport in polycrystalline graphene. *Nat. Mater.* **9,** 806–809 (2010).
13. Yu, Q. *et al.* Control and characterization of individual grains and grain boundaries in graphene grown by chemical vapour deposition. *Nat. Mater.* **10,** 443–449 (2011).
14. Martin, J. *et al.* Observation of electron-hole puddles in graphene using a scanning single-electron transistor. *Nat. Phys.* **4,** 144–148 (2008).
15. Chen, J.-H., Jang, C., Xiao, S., Ishigami, M. & Fuhrer, M. S. Intrinsic and extrinsic performance limits of graphene devices on SiO_2. *Nat. Nanotech.* **3,** 206–209 (2008).
16. Zhang, Y., Brar, V. W., Girit, C., Zettl, A. & Crommie, M. F. Origin of spatial charge inhomogeneity in graphene. *Nat. Phys.* **5,** 722–726 (2009).
17. Decker, R. *et al.* Local electronic properties of graphene on a BN substrate via scanning tunneling microscopy. *Nano Lett.* **11,** 2291–2295 (2011).
18. Xue, J. *et al.* Scanning tunnelling microscopy and spectroscopy of ultra-flat graphene on hexagonal boron nitride. *Nat. Mater.* **10,** 282–285 (2011).
19. Burson, K. M. *et al.* Direct imaging of charged impurity density in common graphene substrates. *Nano Lett.* **13,** 3576–3580 (2013).
20. Duong, D. L. *et al.* Probing graphene grain boundaries with optical microscopy. *Nature* **490,** 235–239 (2012).
21. Huang, P. Y. *et al.* Grains and grain boundaries in single-layer graphene atomic patchwork quilts. *Nature* **469,** 389–392 (2011).

22. Yasaei, P. *et al.* Bimodal phonon scattering in graphene grain boundaries. *Nano Lett.* **15,** 4532–4540 (2015).
23. Deshpande, A., Bao, W., Miao, F., Lau, C. & LeRoy, B. Spatially resolved spectroscopy of monolayer graphene on SiO_2. *Phys. Rev. B* **79,** 205411 (2009).
24. Gibertini, M., Tomadin, A., Guinea, F., Katsnelson, M. I. & Polini, M. Electron-hole puddles in the absence of charged impurities. *Phys. Rev. B* **85,** 201405 (2012).
25. Cho, S. *et al.* Thermoelectric imaging of structural disorder in epitaxial graphene. *Nat. Mater.* **12,** 913–918 (2013).
26. Fei, Z. *et al.* Electronic and plasmonic phenomena at graphene grain boundaries. *Nat. Nanotech.* **8,** 821–825 (2013).
27. Ferrari, A. C. & Basko, D. M. Raman spectroscopy as a versatile tool for studying the properties of graphene. *Nat. Nanotech.* **8,** 235–246 (2013).
28. Hsu, J. W. P., Fitzgerald, E. A., Xie, Y. H. & Silverman, P. J. Near-field scanning optical microscopy imaging of individual threading dislocations on relaxed GexSi1-x films. *Appl. Phys. Lett.* **65,** 344–346 (1994).
29. Buratto, S. K. *et al.* Imaging InGaAsP quantum-well lasers using near-field scanning optical microscopy. *J. Appl. Phys.* **76,** 7720 (1994).
30. Goldberg, B. B., Ünlü, M. S., Herzog, W. D., Ghaemi, H. F. & Towe, E. Near-field optical studies of semiconductor heterostructures and laser diodes. *IEEE J. Sel. Top. Quantum Electron.* **1,** 1073–1081 (1995).
31. Richter, A., Tomm, J. W., Lienau, C. & Luft, J. Optical near-field photocurrent spectroscopy: A new technique for analyzing microscopic aging processes in optoelectronic devices. *Appl. Phys. Lett.* **69,** 3981 (1996).
32. McNeill, C. R. *et al.* Direct photocurrent mapping of organic solar cells using a near-field scanning optical microscope. *Nano Lett.* **4,** 219–223 (2004).
33. Gu, Y. *et al.* Near-field scanning photocurrent microscopy of a nanowire photodetector. *Appl. Phys. Lett.* **87,** 043111 (2005).
34. Mueller, T., Xia, F., Freitag, M., Tsang, J. & Avouris, P. Role of contacts in graphene transistors: a scanning photocurrent study. *Phys. Rev. B* **79,** 245430 (2009).
35. Rauhut, N. *et al.* Antenna-enhanced photocurrent microscopy on single-walled carbon nanotubes at 30 nm resolution. *ACS Nano* **6,** 6416–6421 (2012).
36. Mauser, N. & Hartschuh, A. Tip-enhanced near-field optical microscopy. *Chem. Soc. Rev.* **43,** 1248–1262 (2014).
37. Mauser, N. *et al.* Antenna-enhanced optoelectronic probing of carbon nanotubes. *Nano Lett.* **14,** 3773–3778 (2014).
38. Grover, S., Dubey, S., Mathew, J. P. & Deshmukh, M. M. Limits on the bolometric response of graphene due to flicker noise. *Appl. Phys. Lett.* **106,** 051113 (2015).
39. Wang, L. *et al.* One-dimensional electrical contact to a two-dimensional material. *Science* **342,** 614–617 (2013).
40. Kretinin, A. V. *et al.* Electronic properties of graphene encapsulated with different two-dimensional atomic crystals. *Nano Lett.* **14,** 3270–3276 (2014).
41. Ju, L. *et al.* Photoinduced doping in heterostructures of graphene and boron nitride. *Nat. Nanotech.* **9,** 348–352 (2014).
42. Fei, Z. *et al.* Infrared nanoscopy of dirac plasmons at the graphene-SiO_2 interface. *Nano Lett.* **11,** 4701–4705 (2011).
43. Keilmann, F. & Hillenbrand, R. Near-field microscopy by elastic light scattering from a tip. *Phil. Trans. R Soc. A* **362,** 787–805 (2004).
44. Lemme, M. C. *et al.* Gate-activated photoresponse in a graphene p-n junction. *Nano Lett.* **11,** 4134–4137 (2011).
45. Gabor, N. M. *et al.* Hot carrier-assisted intrinsic photoresponse in graphene. *Science* **334,** 648–652 (2011).
46. Song, J. C. W., Rudner, M. S., Marcus, C. M. & Levitov, L. S. Hot carrier transport and photocurrent response in graphene. *Nano Lett.* **11,** 4688–4692 (2011).
47. Tielrooij, K. J. *et al.* Hot-carrier photocurrent effects at graphene' metal interfaces. *J. Phys. Condens. Matter* **27,** 164207 (2015).
48. Badioli, M. *et al.* Phonon-mediated mid-infrared photoresponse of graphene. *Nano Lett.* **14,** 6374–6381 (2014).
49. Song, J. C. W. & Levitov, L. S. Shockley-Ramo theorem and long-range photocurrent response in gapless materials. *Phys. Rev. B* **90,** 075415 (2014).
50. Tsen, A. W. *et al.* Tailoring electrical transport across grain boundaries in polycrystalline graphene. *Science* **336,** 1143–1146 (2012).
51. Van Tuan, D. *et al.* Scaling properties of charge transport in polycrystalline graphene. *Nano Lett.* **13,** 1730–1735 (2013).
52. Zuev, Y., Chang, W. & Kim, P. Thermoelectric and magnetothermoelectric transport measurements of graphene. *Phys. Rev. Lett.* **102,** 096807 (2009).
53. Wei, P., Bao, W., Pu, Y., Lau, C. N. & Shi, J. Anomalous thermoelectric transport of dirac particles in graphene. *Phys. Rev. Lett.* **102,** 166808 (2009).
54. Cummings, A. W. *et al.* Charge transport in polycrystalline graphene: challenges and opportunities. *Adv. Mater.* **26,** 5079–5094 (2014).
55. Lee, E. J. H., Balasubramanian, K., Weitz, R. T., Burghard, M. & Kern, K. Contact and edge effects in graphene devices. *Nat. Nanotech.* **3,** 486–490 (2008).
56. Ben Shalom, M. *et al.* Quantum oscillations of the critical current and high-field superconducting proximity in ballistic graphene. *Nat. Phys.* doi:10.1038/nphys3592 (2015).
57. Allen, M. T. *et al.* Spatially resolved edge currents and guided-wave electronic states in graphene. *Nat. Phys.* **12,** 128–133 (2016).
58. Chen, S. Y., Ho, P. H., Shiue, R. J., Chen, C. W. & Wang, W. H. Transport/magnetotransport of high-performance graphene transistors on organic molecule-functionalized substrates. *Nano Lett.* **12,** 964–969 (2012).

Acknowledgements

We thank Stijn Goossens, Klaas-Jan Tielrooij and Misha Fogler for many fruitful discussions and Fabien Vialla for the assistance in rendering the graphics in Fig. 1a. Open source software was used (www.matplotlib.org, www.python.org, www.povray.org). F.H.L.K. acknowledges financial support from the Spanish Ministry of Economy and Competitiveness, through the 'Severo Ochoa' Programme for Centres of Excellence in R&D (SEV-2015-0522), support by Fundacio Cellex Barcelona, the ERC Career integration grant (294056, GRANOP), the ERC starting grant (307806, CarbonLight), the Government of Catalonia trough the SGR grant (2014-SGR-1535), the Mineco grants Ramón y Cajal (RYC-2012-12281) and Plan Nacional (FIS2013-47161-P), and project GRASP (FP7-ICT-2013-613024-GRASP). F.H.L.K. and R.H. acknowledge support by the E.C. (European Commission) under Graphene Flagship (contract no. CNECT-ICT-604391). D.J. acknowledges support from the Ramón y Cajal Fellowship Program. Y.G. and J.H. acknowledge support from the US Office of Naval Research N00014-13-1-0662. We acknowledge financial support from the Spanish Ministry of Economy and Competitiveness and 'Fondo Europeo de Desarrollo Regional' through Grant TEC2013-46168-R. Q.M. and P.J.-H. have been supported by AFOSR Grant No. FA9550-11-1-0225 and the Packard Fellowship program. This work made use of the Materials Research Science and Engineering Center Shared Experimental Facilities supported by the National Science Foundation (NSF) (Grant No. DMR-0819762) and of Harvard's Center for Nanoscale Systems, supported by the NSF (Grant No. ECS-0335765). S.R. acknowledges the Spanish Ministry of Economy and Competitiveness for funding (MAT2012-33911), the Secretaria de Universidades e Investigacion del Departamento de Economia y Conocimiento de la Generalidad de Catalunya and the Severo Ochoa Program (MINECO SEV-2013-0295). J.E.B.-V. acknowledges support from SECITI (Mexico, D.F.).

Author contributions

A.W. performed the experiments, analysed the data and wrote the manuscript. P.A.-G. helped with measurements, interpretation and discussion of the results. M.B.L. helped with data analysis, measurements, interpretation, discussion of the results and manuscript writing. Y.G. fabricated the h-BN/graphene/h-BN devices. G.N. and D.J. fabricated the CVD graphene devices. Q.M. fabricated the exfoliated graphene devices. K.W. and T.T. synthesized the h-BN. J.E.B.-V., A.W.C. and S.R. performed the simulations of the Seebeck coefficient at grain boundaries. R.H. and F.H.L.K. supervised the work, discussed the results and co-wrote the manuscript. All authors contributed to the scientific discussion and manuscript revisions.

Additional information

Competing financial interests: R.H. is co-founder of Neaspec GmbH, a company producing scattering-type scanning near-field optical microscope systems such as the ones used in this study. All other authors declare no competing financial interests.

Tailoring the chiral magnetic interaction between two individual atoms

A.A. Khajetoorians[1,2], M. Steinbrecher[1], M. Ternes[3], M. Bouhassoune[4], M. dos Santos Dias[4], S. Lounis[4], J. Wiebe[1] & R. Wiesendanger[1]

Chiral magnets are a promising route towards dense magnetic storage technology due to their inherent nano-scale dimensions and energy efficient properties. Engineering chiral magnets requires atomic-level control of the magnetic exchange interactions, including the Dzyaloshinskii-Moriya interaction, which defines a rotational sense for the magnetization of two coupled magnetic moments. Here we show that the indirect conduction electron-mediated Dzyaloshinskii-Moriya interaction between two individual magnetic atoms on a metallic surface can be manipulated by changing the interatomic distance with the tip of a scanning tunnelling microscope. We quantify this interaction by comparing our measurements to a quantum magnetic model and *ab-initio* calculations yielding a map of the chiral ground states of pairs of atoms depending on the interatomic separation. The map enables tailoring the chirality of the magnetization in dilute atomic-scale magnets.

[1] Department of Physics, Hamburg University, 20355 Hamburg, Germany. [2] Institute for Molecules and Materials, Radboud University, 6525AJ Nijmegen, The Netherlands. [3] Max Planck Institute for Solid State Research, 70569 Stuttgart, Germany. [4] Peter Grünberg Institut and Institute for Advanced Simulation, Forschungszentrum Jülich & JARA, 52425 Jülich, Germany. Correspondence and requests for materials should be addressed to A.A.K. (email: a.khajetoorians@science.ru.nl) or to J.W. (email: jwiebe@physnet.uni-hamburg.de).

Chirality is a peculiar lack of symmetry, which is seeing an increasing importance in various fields of science from the structure of molecules[1] to the magnetic properties of solids[2]. In the field of magnetism, chirality defines a rotational sense for the magnetization, which can rotate clockwise or counterclockwise along an axis of the magnetic object. Such chiral magnets have been recently shown to host many interesting properties that rely on this local rotation, or twisting, of the magnetization such as inducing new quantum information particles based on Majorana fermions[3,4]. So-called skyrmions[5,6], a particular class of chiral magnetization structure, are attractive candidates for information technology as they are inherently nano-sized, and they respond more efficiently to spin-polarized currents[7] as compared with conventional domain walls[2,8,9]. This energy-efficient aspect of chiral magnets, combined with their miniature dimensions, makes them technologically appealing as they can be integrated into current designs for high-density storage[2,10]. And yet, creating chiral magnets with a well-defined size or designed spin texture is a difficult task[11,12] that has not been achieved yet in an atom-by-atom manner.

Central to the understanding of how chiral order emerges at the atomic scale is the Dzyaloshinskii–Moriya interaction (DMI)[13,14], which is defined as $E_{DMI}=\mathbf{D}\cdot(\mathbf{S}_1\times\mathbf{S}_2)$ (Fig. 1). The DMI is a fundamental ingredient responsible for twisting the magnetization of two coupled spins $\mathbf{S}_1$ and $\mathbf{S}_2$, as the overall energy can be reduced between the two spins by canting their relative orientation as dictated by the Dzyaloshinskii–Moriya (DM) vector **D**. Moriya proposed that the origin of this anisotropic magnetic exchange interaction can be derived by considering spin-orbit coupling within a superexchange model[14]. Later, Smith[15] and Fert[16] derived that DMI between $\mathbf{S}_1$ and $\mathbf{S}_2$ can also stem from an indirect exchange mechanism mediated by conduction electrons. Here, similar to the usual Ruderman–Kittel–Kasuya–Yosida (RKKY) interaction[17], conduction electrons locally exchange-couple to the atomic spins becoming spin-polarized and thereby mediate an interaction. However, in addition, the electrons spin-orbit scatter at the non-magnetic host atoms (Fig. 1)[15,16]. This process leads to a long-range DMI between the atomic spins, where not only the amplitude, but also the orientation of the DM vector oscillates as a function of separation. The magnetization of the coupled pair of quantum spins with spin operators $\hat{\mathbf{S}}_1$ and $\hat{\mathbf{S}}_2$ can then be quantified by the following Hamiltonian:

$$\hat{\mathcal{H}}_{\text{imp}} = -J(\hat{\mathbf{S}}_1\cdot\hat{\mathbf{S}}_2) - \mathbf{D}\cdot(\hat{\mathbf{S}}_1\times\hat{\mathbf{S}}_2) + \sum_{i=1}^{2}\left(K_i(\hat{S}_{i,z})^2 - g_i\mu_B\hat{\mathbf{S}}_i\cdot\mathbf{B}\right) \quad (1)$$

It decomposes into an isotropic Heisenberg exchange interaction term (J), which takes into account the usual RKKY-like exchange that favours collinear orientations, a DMI term (**D**), which favours non-collinear spin orientations, a single spin magnetic anisotropy term (K_i), which locks each spin in a given orientation resulting from the interaction with the crystal field, and a Zeeman term, which considers the response of each spin to a magnetic field **B** as determined by the g-factors g_i and the Bohr magneton μ_B. As the orientation of the DM vector in equation (1) dictates a sense of rotation of $\mathbf{S}_2$ compared with $\mathbf{S}_1$, the distance-dependent oscillation of **D** induced by the indirect exchange mechanism inherently contains the possibility to tune the chirality by the separation of the pair. Nevertheless, although the DMI interaction has been studied to a large extent for bulk[5] or thin-film systems[9,18], little is known experimentally about how the indirect conduction electron-mediated DMI emerges between individual atoms, as originally proposed by Smith[15,16]. Central to this understanding are methods that can interrogate exchange interactions[19] with ultimate spatial resolution on surfaces[17].

To this end, we probe the magnetic excitations of coupled pairs of single iron atoms on a Pt(111) surface by low-temperature inelastic scanning tunnelling spectroscopy (ISTS; see Methods section for details). The choice of Pt as a substrate is motivated by a large spin-orbit coupling[20]. It induces a considerable distance dependent, oscillatory DM interaction within the pair, which is of similar strength as the isotropic exchange over a wide range of iron atom distances in this system. We finally show how the chirality of the resulting highly non-collinear magnetization states of the pair can be tuned by tailoring the positions of the two atoms.

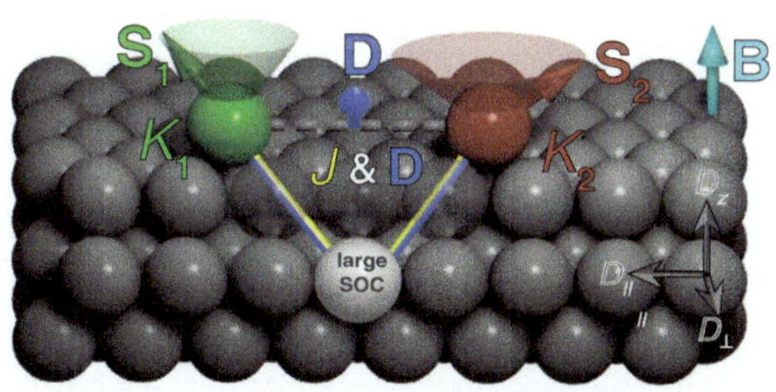

Figure 1 | Cartoon diagram of the indirect conduction electron-mediated DM interaction. The interaction between two magnetic atoms (red/green) is mediated by scattering of conduction electrons at a substrate atom (grey) with strong spin-orbit coupling (SOC)[2,15,16]. The shown scattering process leads to a contribution to the Heisenberg-like exchange (J) and DM vector (**D**), which is oriented perpendicular to the indicated triangle constituted by the two magnetic atoms and the substrate atom. The overall J and **D** are given by the contributions of all substrate atoms resulting in a nonzero $\mathbf{D}=(D_{\parallel}, D_{\perp}, D_z)$ (orientation of the components as indicated) because of broken inversion symmetry at the surface[18]. The orientation of the spins $\mathbf{S}_1$ and $\mathbf{S}_2$ of the coupled pair is determined by the interplay between the single ion magnetic anisotropy of each atom (K_i), J, **D**, and the applied magnetic field (B_z).

Results

Spectroscopy of the uncoupled constituents outside of the pair. To minimize the effect of magnetic anisotropy, which is typically strong for atoms on surfaces[20,21], we use a clean Fe atom and an Fe atom with two attached hydrogen atoms (FeH_2), each with spin $S=5/2$, which exhibit a very weak magnetic anisotropy depending on their adsorption site, face-centered cubic (fcc) or hexagonal close-packed (hcp), on the substrate atomic lattice[22,23]. At zero magnetic field B_z, ISTS on $Fe_{hcp}H_2$ reveals a strong peak at zero bias characteristic for the Kondo effect[23,24], whereas the bare Fe atom shows a symmetric step-like increase in conductance because of spin excitations of the magnetic ground state to the first excited state (Fig. 2a, grey curves). This zero-field splitting originates from the magnetic anisotropy. For Fe_{hcp}, we found $K=0.08$ meV, and for Fe_{fcc}, $K=-0.19$ meV (ref. 22). Owing to the small anisotropy in $Fe_{hcp}H_2$ ($K=0.02$ meV), at zero field all eigenstates are nearly degenerate[23] at our base temperature and the Kondo impurity behaves paramagnetically[25], making it extremely sensitive to the magnetic perturbations produced by J and **D**.

Distance-dependent splitting of the Kondo resonance. By changing the interatomic distance in artificially arranged pairs of atoms with the tip of the scanning tunnelling microscope (STM)[26] (see Methods section and Supplementary Fig. 1 for details), we demonstrate that the magnetic excitations of each spin are modified by the indirect exchange interaction depending on the interaction strength (Fig. 2a,b). We create different pairs, by leaving the $Fe_{hcp}H_2$ stationary and moving the Fe atom to a

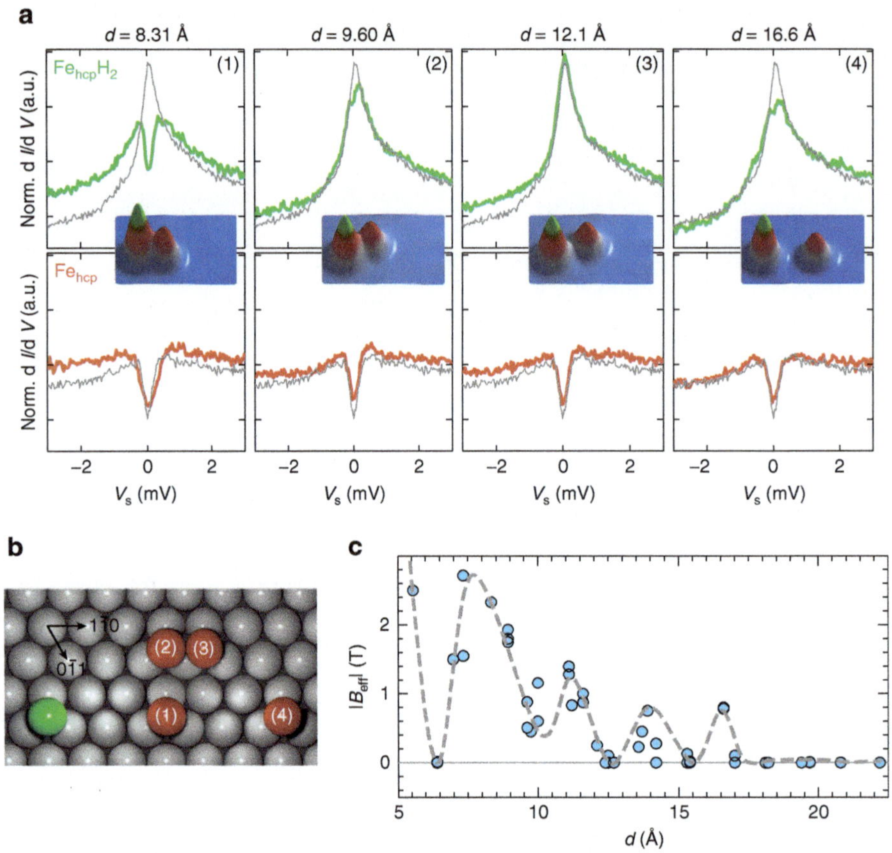

Figure 2 | Detection of the indirect conduction electron-mediated exchange interactions. (**a**) ISTS on $Fe_{hcp}H_2$ (green) and Fe_{hcp} (red) within selected pairs of different configuration (**b**) and separation *d* (topographs shown in insets) in comparison to the spectra measured on the corresponding isolated atoms (grey). The exchange interaction within the pair forces a splitting of the Kondo resonance of $Fe_{hcp}H_2$ and a modification of the magnetic excitation of Fe as compared with the isolated atoms (experimental ISTS parameters: $V_s = 6$ mV, $I_s = 3$ nA, $V_{mod} = 40\,\mu$V, the spectra are normalized (Norm.) by division by a substrate spectrum measured with the same tip). (**b**) Positions of $Fe_{hcp}H_2$ (green) and Fe_{hcp} (red, 1-4) for the pairs of **a** with respect to the substrate atoms (grey). (**c**) The exchange field resulting from magnetic coupling is mimicked as an effective magnetic field ($|B_{eff}|$), which is determined by fitting the splitting of the Kondo resonance of pairs of different distances *d*. The grey dashed line is a guide to the eye and the maximum error bar was determined to be 0.5 T.

chosen distance from the centre of the $Fe_{hcp}H_2$ impurity (Fig. 2b). ISTS on each individual atom of a fabricated pair (Fig. 2a), reveals a splitting of the Kondo resonance on $Fe_{hcp}H_2$ (ref. 27) with a correlated increase in energy of the spin excitation on Fe as compared with the isolated impurities (see grey spectra in the foreground), which indicates the magnetic coupling between the two impurities. To quantify this interaction to first approximation, the splitting of the Kondo resonance at zero magnetic field can be mimicked by the effect of an effective exchange field $|B_{eff}|$ (Fig. 2c; see Methods section for details)[23]. For this purpose, the Kondo spectra are fitted to a Kondo model[25,28], which considers the weak coupling limit within perturbation theory (see Methods section for details). The resulting $|B_{eff}|$ reveals a strong distance-dependent oscillation illustrating the long-range nature of the magnetic coupling reminiscent of the RKKY interaction[17,29]. As the separation of the magnetic atoms in the pairs are too large for a significant contribution of direct or super-exchange, the behaviour in $|B_{eff}|$ strongly suggests that the observed coupling is due to the indirect substrate electron-mediated exchange mechanism.

Determination of Heisenberg and DM exchange. To disentangle the contributions of *J* and **D** to the observed indirect exchange interaction (Fig. 1)[15], we monitor the simultaneous changes in both the Kondo resonance of $Fe_{hcp}H_2$ and the magnetic excitation of Fe as a function of an applied out-of-plane magnetic field (B_z). This is shown in Fig. 3a–d for a particular pair with $d = 8.31$ Å. The corresponding data for pairs with other distances are described in Supplementary Note 1. With increasing B_z the Kondo resonance splits nearly linearly indicating that the impurity easily aligns to the magnetic field direction, with only a small deviation compared with an isolated $Fe_{hcp}H_2$ (ref. 23). However, Fe_{hcp} shows a B_z-dependent evolution, which largely deviates from the isolated impurity[22]. The magnetic excitation shows practically no shift up to $B_z = 6$ T and does shift linearly to higher energies only for larger magnetic fields. For the isolated Fe_{hcp} atom, $B_z = \frac{4K}{g\mu_B} \approx 3.0$ T is required to overcome the easy-plane anisotropy and align the spin fully parallel to the magnetic field direction (that is, into the $S_z = 5/2$ state). In the magnetic field-dependent map of the spin excitation of the isolated Fe_{hcp} atom[22], this is manifested by a minimum in the excitation energy at $B_z = 3.0$ T, which is induced by the level crossing of the $S_z = 5/2$ and $S_z = 3/2$ states, followed by a linear shift in the excitation at higher magnetic fields. The absence of this minimum for the Fe_{hcp} within the pair (Fig. 3d) signifies a strong magnetic coupling, which destroys the level crossing.

To understand how *J* and **D** separately affect the spectra, we simulate the measured ISTS of each atom considering the Hamiltonian equation (1) within the Kondo model (see Methods

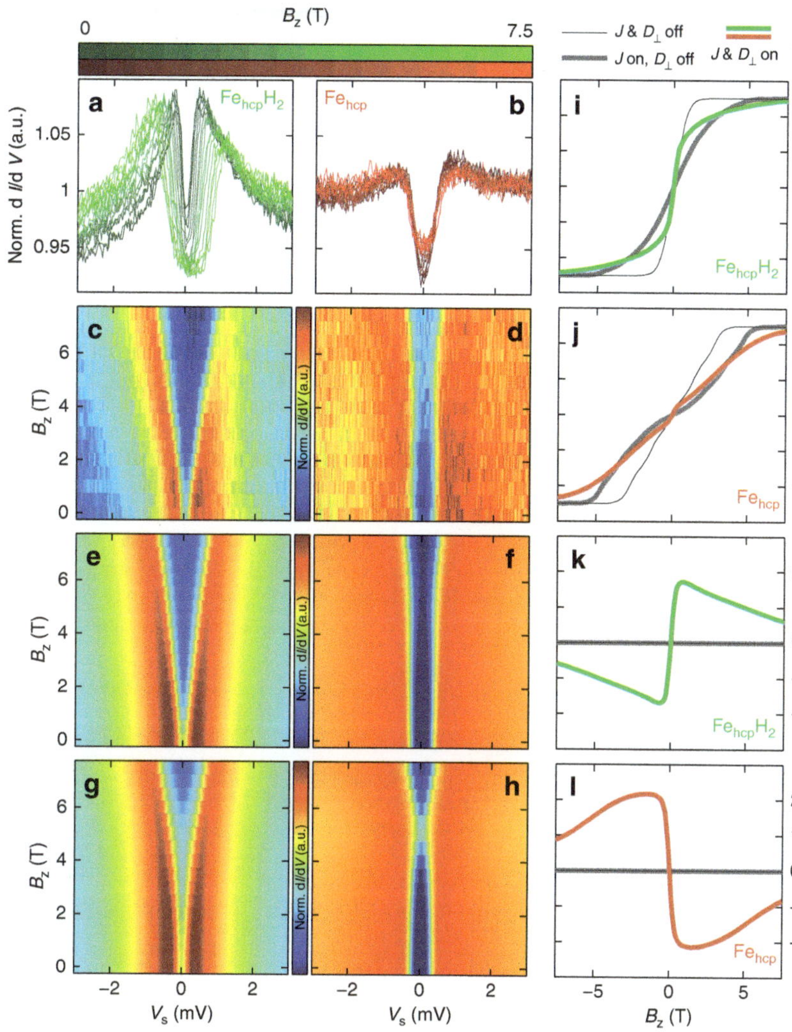

Figure 3 | B-field dependence of spin-excitation and magnetization of a DM-coupled pair. (**a**,**b**) ISTS measured on $Fe_{hcp}H_2$ and Fe_{hcp} within a pair of separation $d = 8.31$ Å as a function of magnetic field as indicated at the top. (**c**,**d**) The same data as a colour intensity plot where the colour indicates the dI/dV signal (experimental ISTS parameters: $V_s = 6$ mV, $I_s = 3$ nA, $V_{mod} = 40$ µV, the spectra are normalized (Norm.) by division by a substrate spectrum measured with the same tip). (**e**-**h**) Simulation of the spectra using a Kondo model that takes into account the exchange interaction within the pair including (**e**,**f**; $|D_\perp| = 0.05$ meV, $J = -0.06$ meV) and excluding ((**g**,**h**) $D_\perp = 0$ meV, $J = -0.06$ meV) the DMI. (**i**-**l**) Simulated expectation values of the out-of-plane (**i**,**j**) and parallel (**k**,**l**) spin components of both atoms in the pair as a function of magnetic field for various settings of J and $|D_\perp|$ as indicated. The parameters within all plots of this figure are $K_{Fe_{hcp}H_2} = 0.02$ meV, $K_{Fe_{hcp}} = 0.08$ meV, $g_{Fe_{hcp}H_2} = g_{Fe_{hcp}} = 2.0$.

section for details)[28] for different values of J and $\mathbf{D}$ and fixed values of K and g, which have been determined by fitting the isolated atoms (see Methods section for details). By considering the simplified superposition of the contributions of all substrate atoms to $\mathbf{D}$ shown in Fig. 1, we expect the out-of-plane component D_z as well as the in-plane component $D_\parallel$, which is parallel to the displacement vector between the two atoms, to be negligible for most of the pairs, whereas only the in-plane component $D_\perp$ pointing perpendicular to the displacement vector survives due to the broken inversion symmetry induced by the surface. This is later confirmed by our first-principles calculations (see below). To reduce the number of parameters, we therefore set $D_z = D_\parallel = 0$ and only consider the absolute value of $D_\perp$. The latter restriction is furthermore justified as equation (1) does not contain any operators besides $\mathbf{D}$ that break the in-plane symmetry. Figure 3e,f exemplarily shows the simulated spectra for the $d = 8.31$ Å pair that best fit the experimental ISTS (Fig. 3c,d) as determined by a variation of both, J and $|D_\perp|$, whereas the same is shown in Fig. 3g,h but with the restriction of zero DMI ($D_\perp = 0$). Obviously, only when considering the combined effect of nonzero $\mathbf{D}$ and J, is it possible to excellently reproduce the trends in the experimental spectra for both $Fe_{hcp}H_2$ and Fe_{hcp}. In particular, the effect of nonzero $\mathbf{D}$ is to avoid the level crossings of the Fe_{hcp} spin states resulting in the experimentally observed absence of the above-mentioned minimum in the excitation energy. Moreover, the almost linear splitting of the $Fe_{hcp}H_2$ Kondo resonance, which behaves very similar to the splitting for the isolated impurity[23], is nicely reproduced. The influence of the DMI on the magnetization is illustrated in Fig. 3i–l by the magnetic field-dependent expectation values of the spin in out-of-plane ($\langle \hat{S}_z \rangle$) and parallel ($\langle \hat{S}_\parallel \rangle$) orientations (see Methods section for details). The most dramatic effect of $\mathbf{D}$ is the appearance of strong parallel components of the magnetization on both atoms in the pair (Fig. 3k,l). The parallel component is positive for $Fe_{hcp}H_2$ and negative for Fe_{hcp} at positive B_z, and vice versa for negative B_z, illustrating the creation of non-collinear, antiferromagnetic magnetization states. Moreover, the difference in the sign of the parallel components defines a rotational sense of the magnetization when going from $Fe_{hcp}H_2$ to Fe_{hcp}, demonstrating the chiral nature of the DMI.

Distance dependency of Dzyaloshinskii–Moriya exchange. It is interesting to investigate how the two indirect conduction electron-mediated exchange components **D** and J vary as a function of distance. It has been shown that the sign and amplitude of J is distance dependent, as determined by the electronic structure of the atom/substrate complex[17]. However, it is not *a priori* clear if the twisting and the chirality induced by **D** oscillates in a similar way[30,31]. To this extent, we vary the separation d between the atoms in the pair and perform a similar measurement and analysis as shown in Fig. 3 in order to extract J and $|D_\perp|$ for all pairs (Supplementary Note 1) within an errorbar whose estimation we explain in Supplementary Note 2. The resulting distance dependence shown in Fig. 4a,b reveals a damped oscillation in J between antiferromagnetic ($J<0$) and ferromagnetic ($J>0$) coupling, which reflects the well-known behaviour of indirect conduction electron-mediated Heisenberg exchange[17]. More importantly, we see that also $|D_\perp|$ shows a strong distance dependence, and long-range oscillations with an apparently smaller wavelength as compared with J, which we discuss below.

This oscillatory behaviour of $|D_\perp|$ further substantiates that the origin for the observed DMI is indirect conduction electron-mediated exchange as proposed by Smith[15]. As we have shown for Co/Pt(111), this indirect exchange interaction between single atoms is intricately related to the electronic structure[17], and first-principles calculations have shown reasonable agreement with the experimental values of J. To understand the distance-dependent evolution of J and **D**, we therefore employed first-principles calculations based on density functional theory as implemented in the full-potential Korringa–Kohn–Rostoker Green function (KKR) method (see Methods section for details)[32], which includes spin-orbit coupling.

The theoretical results are summarized and compared with the experimental values in Fig. 4a–c, where we compare the experimentally extracted values of J and $|D_\perp|$ to calculations for Fe_{hcp}-Fe_{hcp} pairs using a mapping to a classical Heisenberg model (see Methods section for details). The values of J shown in Fig. 4a are obtained after taking the trace of the tensor of exchange interactions for the two adatoms. We note that the variation between the different diagonal terms of the tensor is extremely small, justifying the use of an isotropic J in our model Hamiltonian. Moreover, although in principle also biquadratic exchange interaction can contribute to the Hamiltonian, this

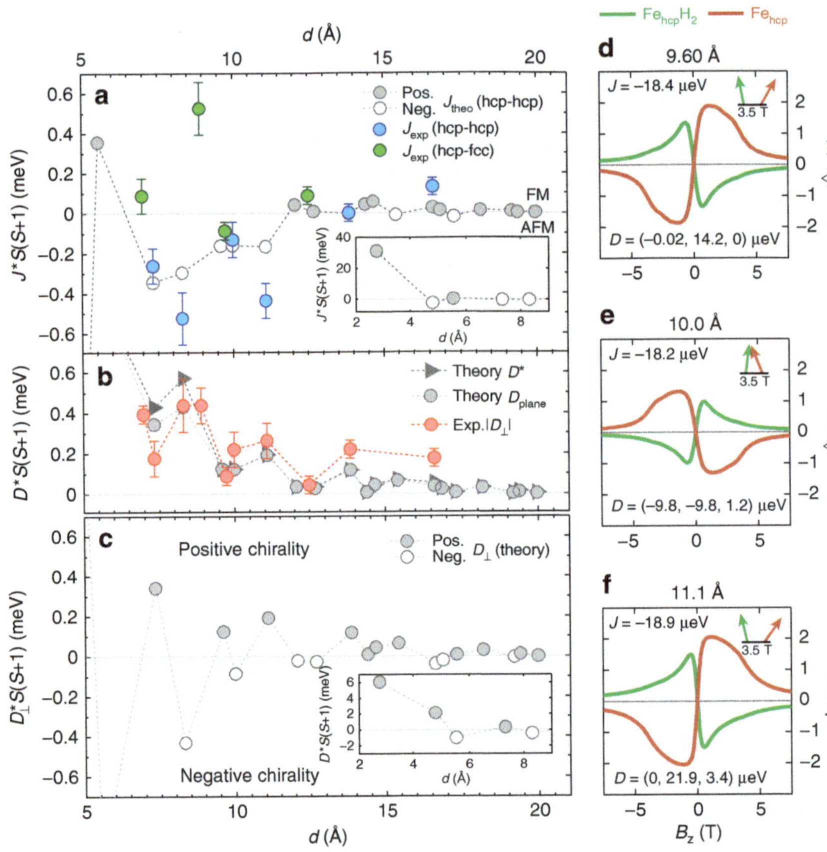

Figure 4 | Distance dependence of indirect conduction electron-mediated DM interaction. **(a-c)** Experimentally determined values of J and $|D_\perp|$ (coloured circles, error bars are estimated by the largest variation of the two parameters, which still results in a reasonable fit to the data) compared with the *ab-initio* calculation of J, $D_\perp$, $D^* = \sqrt{D_\parallel^2 + D_\perp^2 + D_z^2}$ and $D_{plane} = \sqrt{D_\parallel^2 + D_\perp^2}$ using the KKR approach (grey and white circles and triangles). The values are scaled by $S(S+1) = \frac{35}{4}$ for direct comparison to previous publications on related systems[17,29]. The theoretical data are calculated for a pair of two clean Fe atoms and technically restricted to hcp-hcp pairs only. Insets illustrate theoretical data at small separations. The dashed lines are made to guide the eye. **(d-f)** Simulated magnetic field-dependent expectation values of the parallel spin components of both atoms in three selected pairs with negative J and $\mathbf{D} = (D_\parallel, D_\perp, D_z)$ as indicated, and $K_{Fe_{hcp}H_2} = 0.02$ meV, $K_{Fe_{hcp}} = 0.08$ meV, $g_{Fe_{hcp}H_2} = g_{Fe_{hcp}} = 2.0$. The arrows in the insets indicate the orientations of the magnetizations of $Fe_{hcp}H_2$ (green) and Fe_{hcp} (red) defined by the three components of the spin expectation values $\left(\langle\hat{S}_\parallel\rangle, \langle\hat{S}_\perp\rangle, \langle\hat{S}_z\rangle\right)$. Neg.: negative; Pos.: positive.

contribution turns out to be negligible when the distance between the magnetic atoms is large (see Methods section for details). By comparison of the length of **D** $\left(D^* = \sqrt{D_\parallel^2 + D_\perp^2 + D_z^2}\right)$ to the in-plane component $D_{\mathrm{plane}} = \sqrt{D_\parallel^2 + D_\perp^2}$ and to $D_\perp$ (Fig. 4b,c), we see theoretically that the $D_\perp$ component is the dominant term compared with the other components of **D** for most of the pairs. Only at distances $d < 9$ Å, small but significant values of D_z lead to a deviation between D^* and $D_\perp$ (Supplementary Note 3). This justifies the simplified determination of negligible components of **D** via the construction illustrated in Fig. 1 and the restriction to $D_\perp$ used in the simulations of the experimental spectra. Most importantly, the calculated D_{plane} quantitatively reproduces the experimentally determined values of $|D_\perp|$, in particular the apparently smaller wavelength of the oscillation of $|D_\perp|$ as compared with that of J. As the experimental determination is not sensitive to the sign of $D_\perp$, the oscillations are more clearly revealed by the calculated values of $D_\perp$ shown in Fig. 4c. These oscillations indeed show sign changes on a much shorter length scale as the oscillations in J. This is qualitatively expected already from the simple model where the interactions are mediated by a quasi-free electron gas[15]. Here, the long-range form of J is given by $\sin(2k_F d)$, whereas that of **D** is given by $\sin(k_F[R_A + R_B + d])$, where R_A and R_B are the distances of the two atoms A and B of the pair to the substrate atom mediating the DMI (Fig. 1). As $R_A + R_B + d \geq 2d$, the DMI is expected to oscillate on a smaller length scale as the Heisenberg exchange. However, it should be noted that the realistic band structure of the substrate, which yields highly anisotropic energy contours in k-space, results in a more complex behaviour than expected from these simple models[17].

Discussion

The oscillation of the sign of the dominating DM component $D_\perp$ illustrated in Fig. 4c directly implies that the orientation of **D** changes as a function of the separation of the pair, namely between pointing to the left or to the right looking along the displacement vector between the two atoms. As the DMI favours a particular rotational sense of the magnetization in the pair which depends on the orientation of **D** (equation (1)), the oscillations have a direct consequence on the preferred chirality of the magnetization in the pair. This is visualized in Fig. 4d–f, where we show the parallel components of the expectation values of the spins of three antiferromagnetic pairs with positive (d,f) and negative (e) $D_\perp$. At positive B_z, the sign in $D_\perp$ leads to a negative/positive parallel component on $Fe_{hcp}H_2$/Fe_{hcp} for $D_\perp > 0$ and vice versa for $D_\perp < 0$. Taking into account the z-components of the expectation values of the spins, this implies a clockwise or counterclockwise rotation of the magnetization, respectively, when going from $Fe_{hcp}H_2$ to Fe_{hcp}, as indicated by the magnetization vectors in the insets of Fig. 4d–f (see Supplementary Movie 1 for details).

The oscillation in the indirect conduction electron-mediated DM interaction proposed by Smith in 1976 (ref. 15), which we experimentally verify here, consequently enables tailored chirality of the magnetization in pairs of atoms by tuning the interatomic distance. More specifically, by changing the interatomic distance, the strength of the DMI can be of the same magnitude or even larger than that of the isotropic exchange interaction, which drives collinear order. The measured and calculated interaction maps (Fig. 4), combined with small magnetic anisotropy energies[23], enables a class of bottom-up magnets with tunable chirality, for example, of a two-dimensional nature, such as magnetic skyrmions[9], down to one-dimensional systems, such as spin spirals[33].

Methods

Experimental equipment, tip and sample preparation. All experiments were performed with an STM in an ultrahigh vacuum facility at 0.3 K (ref. 34). A magnetic field B_z of up to 12 T can be applied perpendicular to the sample surface. Topography images were obtained in constant-current mode with stabilization current I_s and voltage V_s applied to the sample. The ISTS were obtained by using a lock-in technique to record the differential conductance $dI/dV(V)$ while ramping the bias voltage (V_S) after the feedback loop was switched off, by adding an AC modulation voltage V_{mod} (given in rms) to the bias voltage.

We used a clean tungsten tip and a tungsten tip that was initially coated with nominally 50 monolayers of chromium before inserting it into the STM as described in ref. 34. For the ISTS shown here, both tips were afterwards prepared by voltage-pulsing and dipping into the platinum surface until almost featureless spectra on platinum and close to symmetric spectra on the magnetic atoms were achieved.

The Pt(111) single crystal was cleaned by alternating Ar^+ sputtering (high voltage = 2 kV) and annealing ($\approx$740 °C) cycles of decreasing duration. An oxygen annealing procedure has been performed for 1 h at $\approx$660 °C and an O_2 pressure in the chamber of 2×10^{-6} mbar. The final flash was carried out at $\approx$1000 °C for 1 min and a background pressure of 5.5×10^{-10} mbar.

Finally, Fe was deposited onto the clean Pt(111) surface. To obtain single atoms on the surface, the crystal was cooled to < 10 K during deposition. This resulted in a statistical distribution of Fe atoms adsorbed to the fcc (Fe_{fcc}) or hcp (Fe_{hcp}) hollow sites of the Pt(111) surface, where the adsorption site can be identified by the characteristic ISTS spectra[22]. In addition, we recorded an atomic manipulation image in Supplementary Fig. 1a by recording a constant current image with an adatom partly bound to the tip[35]. This allows us to exactly determine the lattice of the Pt(111) surface and to align the direction of close packed Pt surface atoms parallel to the fast scan direction.

After long-term exposure of the sample to H_2 from the residual gas of the ultrahigh vacuum chamber, some of the Fe atoms form Fe–H complexes with one or two attached H atoms. Using ISTS, the complex with two attached H atoms where the Fe is sitting on the hcp site ($Fe_{hcp}H_2$) can be identified by a characteristic Kondo peak in ISTS spectra[23]. Although clean Fe can be laterally manipulated between the two different hollow adsorption sites with typical manipulation parameters (V_s = 2 mV, I_s = 40–60 nA), it is not possible to manipulate $Fe_{hcp}H_2$ without de-attaching at least one H atom. Therefore, for the creation of the different pairs investigated here, $Fe_{hcp}H_2$ was kept stationary and Fe was moved to different positions. The Pt lattice determined by the atomic manipulation image together with the spectroscopic information from ISTS enables unambiguous determination of the absolute positions of the atoms in the pairs with respect to the Pt lattice underneath. Examples for two pairs with different distances, angles and adsorption sites are given in Supplementary Fig. 1b,c.

Model for simulations of conductance spectra and magnetization curves. To extract the values for the Heisenberg coupling J and the DMI **D** from our experiments, we used a third-order perturbation theory model[25,28] to fit and simulate our spectroscopic data. The Hamiltonian for a number n of impurities in this model consists of a Zeeman and an anisotropy term and looks like this:

$$\hat{\mathcal{H}}_{\mathrm{imp}} = \hat{\mathcal{H}}_{\mathrm{Zeeman}} + \hat{\mathcal{H}}_{\mathrm{ani}}$$

$$\hat{\mathcal{H}}_{\mathrm{imp}} = -\sum_{i=1}^{n} g_i \mu_B \mathbf{B} \cdot \hat{\mathbf{S}}_i + \sum_{i=1}^{n} \left(K_i \left(\hat{S}_{i,z}\right)^2\right),$$

with g_i as the gyromagnetic factors, μ_B the Bohr magneton, **B** the applied magnetic field, $\hat{\mathbf{S}}_i$ the vector of spin matrices $\left(\hat{S}_{i,x}, \hat{S}_{i,y}, \hat{S}_{i,z}\right)$ of the impurity and K_i the uniaxial anisotropy constant.

An additional coupling Hamiltonian considers the isotropic Heisenberg and the anisotropic DMI between all pairs of atoms. The interaction Hamiltonian then looks like the following:

$$\hat{\mathcal{H}}_{\mathrm{int}} = \hat{\mathcal{H}}_{\mathrm{H}} + \hat{\mathcal{H}}_{\mathrm{DM}}$$

$$\hat{\mathcal{H}}_{\mathrm{int}} = -\sum_{i>j=1}^{n} J_{i,j}\left(\hat{\mathbf{S}}_i \cdot \hat{\mathbf{S}}_j\right) - \sum_{i>j=1}^{n} \mathbf{D}_{ij}\left(\hat{\mathbf{S}}_i \times \hat{\mathbf{S}}_j\right).$$

We calculate conductance spectra using a perturbative model where we take into account the scattering matrices up to the third-order using the following Kondo-like interaction matrix between the localized spin system and the electrons, which tunnel between tip and sample:

$$M_{if}^{t\to s} = \left\langle \varphi_f^s, \psi_f \middle| \frac{1}{2}\hat{\boldsymbol{\sigma}} \cdot \hat{\mathbf{S}}_i + U \middle| \varphi_i^t, \psi_i \right\rangle.$$

Here, $\psi_{i,f}$ represent the initial and final wave functions of the probed spin, $\varphi_{i,f}^{s,t}$ are the electron states in sample or tip, respectively, $\hat{\boldsymbol{\sigma}}$ are the standard Pauli spin matrices and U the Coulomb scattering term, which was negligibly small for most calculations. In addition, spin–spin interactions of electrons originating and ending in the substrate were accounted for any spin, not only the one the tip is placed above. This matrix transition element results then in

$$M_{if}^{s\to s}=\sum_i (\Im\rho_S)^{(i)}\left\langle \varphi_f^s,\psi_f\left|\frac{1}{2}\hat{\sigma}\cdot\hat{\mathbf{S}}_i\right|\varphi_i^s,\psi_i\right\rangle.$$

Here, $\Im\rho_S$ is the dimensionless coupling strength between the sample electrons and the localized spin with $\Im$ as the coupling energy and ρ_S the density of states in the sample close to the Fermi energy.

For all simulations we used an electron bandwidth $\omega_0=20\,\mathrm{mV}$ and an effective temperature $T_{eff}=0.6\,\mathrm{K}$, which best describes the effective linewidth of the excitations[22]. The anisotropies as well as the gyromagnetic factors of the three unperturbed impurities (Fe_{hcp}, Fe_{fcc}, $Fe_{hcp}H_2$) are known from earlier work[22,23] and additionally have been extracted here by fitting spectroscopic data on individual impurities to the above model. The resulting parameters of the gyromagnetic factors g and uniaxial anisotropies are $g_{Fe_{hcp}H_2}=2.0$, $K_{Fe_{hcp}H_2}=0.02\,\mathrm{meV}$, $g_{Fe_{hcp}}=2.0$, $K_{Fe_{hcp}}=0.08\,\mathrm{meV}$, $g_{Fe_{fcc}}=2.4$, $K_{Fe_{fcc}}=-0.19\,\mathrm{meV}$. The coupling strengths $\Im\rho_S$, which also resulted from these fits, were $\Im\rho_S=-0.12$ for $Fe_{hcp}H_2$, $\Im\rho_S=-0.02$ for Fe_{hcp} and $\Im\rho_S=-0.002$ for Fe_{fcc}. Thus, we are left with the coupling constants J and **D**, which we determine by calculating the conductance spectra for numerous settings and find the best match to the experimental data. To account for small drifts in the junction during measurements, we additionally allowed a small background slope and offset in dI/dV.

For the determination of the effective exchange field $|B_{eff}|$, which mimics the splitting of the Kondo resonance, we fit the conductance spectra from the above model for a single magnetic atom, that is, without the interaction terms, to the data of the coupled Kondo atom, by adjusting **B** keeping $g=2$.

The same model has been used to calculate the spin-expectation values in out-of-plane ($\langle\hat{S}_z\rangle$), parallel ($\langle\hat{S}_\parallel\rangle$) and perpendicular ($\langle\hat{S}_\perp\rangle$) orientations (see assignment of directions in Fig. 1) as a function of **B**. These expectation values are proportional to the components of the magnetization of the impurities. Please note that, for the Kondo impurity $Fe_{hcp}H_2$, this is strictly only fully accurate as long as **B** is large enough to quench the Kondo screening[23].

First-principles calculations. The electronic and magnetic properties of the investigated dimers were obtained theoretically utilizing the full-potential KKR method[32,36] that is based on density functional theory. Spin-orbit coupling was included self-consistently as it is an essential ingredient in the physics of the DMI.

In this type of investigation, the Pt(111) surface is first computed and then an embedding scheme based on the Green functions is employed to deposit the Fe adatoms on the substrate. The surface is simulated with a slab of 22 Pt layers augmented by two vacuum regions using the experimental lattice parameter of Pt (3.92 Å). To describe the adatoms, a real space cluster surrounding the adatoms is constructed. Pt is a 5d transition element that is initially non-magnetic but thanks to its high magnetic polarizability, it easily can develop a magnetic moment in the vicinity of Fe (ref. 22). This is also the case for the dimers. Thus, around each adatom, there is manifestation of a large induced spin-polarization cloud. Our previous calculations showed that, in order to reasonably describe the magnetic properties of an Fe adatom, the polarization of ~ 60 Pt atoms neighbouring the impurity need to be taken into account[22]. Thus, for the dimers, twice this number of Pt atoms has to be included in our simulations. For the distances between the adatoms considered experimentally, the total magnetic moment of the dimers and substrate, including the spin and orbital components in the ferromagnetic state, reaches a large value of the order of $\sim 10\mu_B$, 95% of which originates from the spin component. As soon as a hydrogen atom is attached to one of the adatoms, the total magnetic moment decreases by $\sim 0.8\mu_B$. These results motivated our choice of attributing a spin of 5/2 for each Fe adatom in our model simulations.

Once the ferromagnetic state is obtained for all the considered dimers, a mapping procedure to a classical Heisenberg model is performed in order to extract the tensor of magnetic exchange interactions $\mathbf{J}^T$ (refs 37,38). In this case, the Heisenberg Hamiltonian without the magnetic anisotropy energy and Zeeman term can be written as

$$\mathcal{H}_{mag}=-\hat{\mathbf{S}}_1\cdot\mathbf{J}^T\cdot\hat{\mathbf{S}}_2$$

The nine elements of the tensor $\mathbf{J}^T$ can be extracted from *ab-initio* using infinitesimal rotations of the magnetic moments[37,38] as:

$$J_{\alpha\beta}^T=\frac{1}{\pi S(S+1)}\mathrm{Im}\int_{-\infty}^{E_F} dE\,\mathrm{Tr}\,\delta V_1^\alpha G_{12}\delta V_2^\alpha G_{21}$$

where $\alpha=(x,y,z)$ and G_{12} being the Green function connecting the atoms 1 and 2. δV^α represents the change of the adatom's potential upon infinitesimal rotation along the direction α. The prefactor $S(S+1)$ takes care of the normalization followed in the model-based calculations presented in the main text. The trace of this tensor divided by 3 gives rise to the isotropic Heisenberg exchange interactions, whereas the antisymmetric part leads to the DMIs[37,38]. As can be seen from the previous equation, the presence of the Green function explains the oscillatory behaviour of J and **D** with respect to the mutual distance between the adatoms. Basically, the elements of the tensor $\mathbf{J}^T$ are described by electrons propagating from atom 1 to atom 2 and back while experiencing scattering at the magnetic part of both potentials.

Although, in principle, the whole tensor can be utilized to build a generalized Heisenberg Hamiltonian, we analysed its properties in order to reduce the quantities to be fitted experimentally to the most relevant ones and ended up with equation (1). Higher-order terms can be in principle added to the Hamiltonian in equation (1), for instance the so-called biquadratic term which is proportional to $(\hat{\mathbf{S}}_1\cdot\hat{\mathbf{S}}_2)^2$. However, and as verified in our simulations, such a term is dramatically small for the distances considered experimentally. This is expected as the biquadratic contribution involves terms containing four Green functions that would describe the double propagation of electrons and their scattering at both adatoms when compared with the regular bilinear term J. Thus, the distance-dependent decay of the biquadratic interactions is expected to be stronger than the one characterizing J.

References

1. Amabilino, D. B. Chiral nanoscale systems: preparation, structure, properties and function. *Chem. Soc. Rev.* **38,** 669–670 (2009).
2. Fert, A., Cros, V. & Sampaio, J. Skyrmions on the track. *Nat. Nanotechnol.* **8,** 152–156 (2013).
3. Beenakker, C. W. J. Search for Majorana Fermions in superconductors. *Annu. Rev. Conden. Ma P* **4,** 113–136 (2013).
4. Mourik, V. *et al.* Signatures of Majorana fermions in hybrid superconductor-semiconductor nanowire devices. *Science* **336,** 1003–1007 (2012).
5. Roessler, U. K., Bogdanov, A. N. & Pfleiderer, C. Spontaneous skyrmion ground states in magnetic metals. *Nature* **442,** 797–801 (2006).
6. Bogdanov, A. N. & Yablonskii, D. A. Thermodynamically stable 'vortices' in magnetically ordered crystals. The mixed state of magnets. *Sov. Phys. JETP* **68,** 101–103 (1989).
7. Brataas, A., Kent, A. D. & Ohno, H. Current-induced torques in magnetic materials. *Nat. Mater.* **11,** 372–381 (2012).
8. Jonietz, F. *et al.* Spin transfer torques in MnSi at ultralow current densities. *Science* **330,** 1648–1651 (2010).
9. Romming, N. *et al.* Writing and deleting single magnetic skyrmions. *Science* **341,** 636–639 (2013).
10. Parkin, S. S., Hayashi, M. & Thomas, L. Magnetic domain-wall racetrack memory. *Science* **320,** 190–194 (2008).
11. Hrabec, A. *et al.* Measuring and tailoring the Dzyaloshinskii-Moriya interaction in perpendicularly magnetized thin films. *Phys. Rev. B* **90,** 020402 (2014).
12. Chen, G. *et al.* Tailoring the chirality of magnetic domain walls by interface engineering. *Nat. Commun.* **4,** 2671 (2013).
13. Dzyaloshinskii, I. E. Thermodynamic theory of weak ferromagnetism in antiferromagnetic substances. *Sov. Phys. JETP* **5,** 1259–1262 (1957).
14. Moriya, T. Anisotropic superexchange interaction and weak ferromagnetism. *Phys. Rev.* **120,** 91–98 (1960).
15. Smith, D. A. New mechanisms for magnetic anisotropy in localised S-state moment materials. *J. Magn. Magn. Mater.* **1,** 214–225 (1976).
16. Fert, A. & Levy, P. M. Role of anisotropic exchange interactions in determining the properties of spin-glasses. *Phys. Rev. Lett.* **44,** 1538–1541 (1980).
17. Zhou, L. *et al.* Strength and directionality of surface Ruderman–Kittel–Kasuya–Yosida interaction mapped on the atomic scale. *Nat. Phys.* **6,** 187–191 (2010).
18. Bode, M. *et al.* Chiral magnetic order at surfaces driven by inversion asymmetry. *Nature* **447,** 190–193 (2007).
19. Dmitrienko, V. E. *et al.* Measuring the Dzyaloshinskii-Moriya interaction in a weak ferromagnet. *Nat. Phys.* **10,** 202–206 (2014).
20. Gambardella, P. *et al.* Giant magnetic anisotropy of single cobalt atoms and nanoparticles. *Science* **300,** 1130–1133 (2003).
21. Rau, I. G. *et al.* Reaching the magnetic anisotropy limit of a 3d metal atom. *Science* **344,** 988–992 (2014).
22. Khajetoorians, A. A. *et al.* Spin excitations of individual Fe Atoms on Pt(111): Impact of the site-dependent giant substrate polarization. *Phys. Rev. Lett.* **111,** 157204 (2013).
23. Khajetoorians, A. A. *et al.* Tuning emergent magnetism in a Hund's impurity. *Nat. Nanotechnol.* **10,** 958–964 (2015).
24. Otte, A. F. *et al.* The role of magnetic anisotropy in the Kondo effect. *Nat. Phys.* **4,** 847–850 (2008).
25. Zhang, Y.-H. *et al.* Temperature and magnetic field dependence of a Kondo system in the weak coupling regime. *Nat. Commun.* **4,** 2110 (2013).
26. Eigler, D. M. & Schweizer, E. K. Positiong single atoms with a scanning tunneling microscope. *Nature* **344,** 524 (1990).
27. Otte, A. *et al.* Spin excitations of a Kondo-Screened atom coupled to a second magnetic atom. *Phys. Rev. Lett.* **103,** 107203 (2009).
28. Ternes, M. Spin excitations and correlations in scanning tunneling spectroscopy. *N. J. Phys.* **17,** 063016 (2015).
29. Khajetoorians, A. A. *et al.* Atom-by-atom engineering and magnetometry of tailored nanomagnets. *Nat. Phys.* **8,** 497–503 (2012).
30. Imamura, H., Bruno, P. & Utsumi, Y. Twisted exchange interaction between localized spins embedded in a one- or two-dimensional electron gas with Rashba spin-orbit coupling. *Phys. Rev. B* **69,** 121303 (2004).
31. Lounis, S., Bringer, A. & Blügel, S. Magnetic adatom induced Skyrmion-like spin texture in surface electron waves. *Phys. Rev. Lett.* **108,** 207202 (2012).

32. Bauer, D. *Development of a relativistic full-potential first-principles multiple scattering Green function method applied to complex magnetic textures of nano structures at surfaces*, PhD thesis, Rheinisch-Westfälische Technische Hochschule (RWTH) Aachen (2013).
33. Menzel, M. *et al.* Information transfer by vector spin chirality in finite magnetic chains. *Phys. Rev. Lett.* **108,** 197204 (2012).
34. Wiebe, J. *et al.* A 300 mK ultra-high vacuum scanning tunneling microscope for spin-resolved spectroscopy at high energy resolution. *Rev. Sci. Instruments* **75,** 4871–4879 (2004).
35. Stroscio, J. A. & Celotta, R. J. Controlling the dynamics of a single atom in lateral atom manipulation. *Science* **306,** 242–247 (2004).
36. Papanikolaou, N., Zeller, R. & Dederichs, P. H. Conceptual improvements of the KKR method. *J. Phys.* **14,** 2799 (2002).
37. Ebert, H. & Mankovsky, S. Anisotropic exchange coupling in diluted magnetic semiconductors: *Ab initio* spin-density functional theory. *Phys. Rev. B* **79,** 045209 (2009).
38. Udvardi, L., Szunyogh, L., Palotás, K. & Weinberger, P. First-prinsciples relativistic study of spin waves in thin magnetic films. *Phys. Rev. B* **68,** 104436 (2003).

Acknowledgements

A.A.K., M.S., J.W. and R.W. acknowledge funding from the SFB668 and the GrK1286 of the DFG. A.A.K. acknowledges funding from the Emmy Noether Program from the DFG, and the Dutch funding agencies FOM, which is part of NWO, and NWO (VIDI program). M.T. acknowledges support from the SFB 767. M.B., M.S.D. and S.L. acknowledge discussions and support from S. Blügel and funding from the HGF-YIG Programme FunSiLab (VH-NG-717) and the DFG project (LO 1659 5-1).

Author contributions

A.A.K., J.W. and M.S. designed the experiment. A.A.K. and M.S. carried out the measurements and M.S., M.T. and A.A.K. did the analysis of the experimental data. M.T. provided the model and software for the data analysis. M.B., M.S.D. and S.L. performed the DFT calculations. A.A.K., J.W., M.S., M.T. and S.L. wrote the manuscript, to which all authors contributed via discussions and criticism. M.S. prepared the figures. All authors contributed to the discussion and interpretation of the results.

Additional information

Competing financial interests: The authors declare no competing financial interests.

Permissions

All chapters in this book were first published in NC, by Nature Publishing Group; hereby published with permission under the Creative Commons Attribution License or equivalent. Every chapter published in this book has been scrutinized by our experts. Their significance has been extensively debated. The topics covered herein carry significant findings which will fuel the growth of the discipline. They may even be implemented as practical applications or may be referred to as a beginning point for another development.

The contributors of this book come from diverse backgrounds, making this book a truly international effort. This book will bring forth new frontiers with its revolutionizing research information and detailed analysis of the nascent developments around the world.

We would like to thank all the contributing authors for lending their expertise to make the book truly unique. They have played a crucial role in the development of this book. Without their invaluable contributions this book wouldn't have been possible. They have made vital efforts to compile up to date information on the varied aspects of this subject to make this book a valuable addition to the collection of many professionals and students.

This book was conceptualized with the vision of imparting up-to-date information and advanced data in this field. To ensure the same, a matchless editorial board was set up. Every individual on the board went through rigorous rounds of assessment to prove their worth. After which they invested a large part of their time researching and compiling the most relevant data for our readers.

The editorial board has been involved in producing this book since its inception. They have spent rigorous hours researching and exploring the diverse topics which have resulted in the successful publishing of this book. They have passed on their knowledge of decades through this book. To expedite this challenging task, the publisher supported the team at every step. A small team of assistant editors was also appointed to further simplify the editing procedure and attain best results for the readers.

Apart from the editorial board, the designing team has also invested a significant amount of their time in understanding the subject and creating the most relevant covers. They scrutinized every image to scout for the most suitable representation of the subject and create an appropriate cover for the book.

The publishing team has been an ardent support to the editorial, designing and production team. Their endless efforts to recruit the best for this project, has resulted in the accomplishment of this book. They are a veteran in the field of academics and their pool of knowledge is as vast as their experience in printing. Their expertise and guidance has proved useful at every step. Their uncompromising quality standards have made this book an exceptional effort. Their encouragement from time to time has been an inspiration for everyone.

The publisher and the editorial board hope that this book will prove to be a valuable piece of knowledge for researchers, students, practitioners and scholars across the globe.

List of Contributors

L.T. Hall and D.A. Simpson
School of Physics, University of Melbourne, Parkville, Victoria 3010, Australia

P. Kehayias, A. Jarmola and D. Budker
Department of Physics, University of California, Berkeley, California 94720, USA

L.C.L. Hollenberg
School of Physics, University of Melbourne, Parkville, Victoria 3010, Australia
Centre for Quantum Computation and Communication Technology, School of Physics, University of Melbourne, Parkville, Victoria 3010, Australia

A. Stacey
Centre for Quantum Computation and Communication Technology, School of Physics, University of Melbourne, Parkville, Victoria 3010, Australia

D.I. Bradley, R.E. George, M. Sarsby, R.P. Haley and J.R. Prance
Department of Physics, Lancaster University, Bailrigg, Lancaster LA1 4YB, UK

Yu. A. Pashkin
Department of Physics, Lancaster University, Bailrigg, Lancaster LA1 4YB, UK

J. Penttilä
Aivon Oy, Valimotie 13A, 00380 Helsinki, Finland

M. Prunnila, H. Heikkinen and D. Gunnarsson
VTT Technical Research Centre of Finland, P.O. Box 1000, 02044 VTT Espoo, Finland

L. Roschier
Lebedev Physical Institute, Moscow 119991, Russia

Kai-Uwe Demasius
IBM Almaden Research Center, San Jose, California 95120, USA
Dresden University of Technology, 01062 Dresden, Germany

Timothy Phung, Weifeng Zhang, Brian P. Hughes, See-Hun Yang, Andrew Kellock and Wei Han and Aakash Pushp
IBM Almaden Research Center, San Jose, California 95120, USA

Stuart S.P. Parkin
IBM Almaden Research Center, San Jose, California 95120, USA
Max Planck Institute for Microstructure Physics, 06120 Halle (Saale), Germany

Amirhossein Khalajhedayati
Department of Chemical Engineering and Materials Science, University of California, Irvine, California 92697, USA

Zhiliang Pan
Department of Mechanical and Aerospace Engineering, University of California, 4200 Engineering Gateway, Irvine, California 92697, USA

Timothy J. Rupert
Department of Chemical Engineering and Materials Science, University of California, Irvine, California 92697, USA
Department of Mechanical and Aerospace Engineering, University of California, 4200 Engineering Gateway, Irvine, California 92697, USA

Wenbin Li
Research Laboratory of Electronics, Massachusetts Institute of Technology, Cambridge, Massachusetts 02139, USA

Ju Li
Department of Nuclear Science and Engineering and Department of Materials Science and Engineering, Massachusetts Institute of Technology, Cambridge, Massachusetts 02139, USA

Erwan Paineau, Mohamed-Salah Amara, Natalia V. Matskova, Ivan Dozov, Stéphan Rouzière, Pascale Launois and Patrick Davidson
Laboratoire de Physique des Solides, CNRS, Univ. Paris-Sud, Université Paris-Saclay, Orsay 91405, France

Marie-Eve M. Krapf and Antoine Thill
LIONS, NIMBE, CEA, CNRS, Université Paris-Saclay, CEA-Saclay, Gif sur Yvette 91191, France

J. Sánchez-Barriga, A. Varykhalov, E. Schierle, E. Weschke, A.A. Ü nal, S. Valencia, E. Golias and O. Rader
Helmholtz-Zentrum Berlin für Materialien und Energie, Elektronenspeicherring BESSY II, Albert-Einstein-Straβe 15, 12489 Berlin, Germany

G. Springholz, H. Steiner, R. Kirchschlager, G. Bauer and A. Ney
Institut für Halbleiter und Festkörperphysik, Johannes Kepler Universität, Altenbergerstr. 69, 4040 Linz, Austria

O. Caha
Department of Condensed Matter Physics, CEITEC, Masaryk University, Kotlarska 2, 61137 Brno, Czech Republic

M. Dunst, J. Braun and H. Ebert
Department Chemie, Ludwig-Maximilians-Universität München, Butenandtstr. 5-13, 81377 München, Germany

J. Minár
Department Chemie, Ludwig-Maximilians-Universität München, Butenandtstr. 5-13, 81377 München, Germany

New Technologies Research Centre, University of West Bohemia, Univerzitni 8, 306 14 Pilsen, Czech Republic

L.V. Yashina
Department of Chemistry, Moscow State University, Leninskie Gory 1/3, 119991 Moscow, Russia

V. Holý
Department of Condensed Matter Physics, Charles University, Ke Karlovu 5, 12116 Prague, Czech Republic

L.J. Zhu
State Key Laboratory of Superlattices and Microstructures, Institute of Semiconductors, Chinese Academy of Sciences, PO Box 912, Beijing 100083, China

Institut für Physik, Martin-Luther-Universität Halle-Wittenberg, von-Danckelmann-Platz 3, Halle 06120, Germany

S.H. Nie and J.H. Zhao
State Key Laboratory of Superlattices and Microstructures, Institute of Semiconductors, Chinese Academy of Sciences, PO Box 912, Beijing 100083, China

P. Xiong and P. Schlottmann
Department of Physics, Florida State University, Tallahassee, Florida 32306, USA

Zengyi Du, Xiong Yang, Hai Lin, Delong Fang, Guan Du, Jie Xing, Huan Yang, Xiyu Zhu and Hai-Hu Wen
Center for Superconducting Physics and Materials, National Laboratory of Solid State Microstructures and Department of Physics, Collaborative Innovation Center for Advanced Microstructures, Nanjing University, Nanjing 210093, China

D. Bossini, Y. Hashimoto, A. Secchi, Th. Rasing and A.V. Kimel
Institute for Molecules and Materials, Spectroscopy of Solids and Interfaces, Radboud University, Heyendaalseweg 135, 6525 AJ Nijmegen, The Netherlands

Institute for Photon Science and Technology Graduate School of Science, The University of Tokyo, 7-3-1 Hongo, Bunkyo-ku, Tokyo 113-0033, Japan (D.B.); Spin Quantum Rectification Project, ERATO, Japan Science and Technology Agency, Sendai 980-8577, Japan (Y.H.)

S. Dal Conte and G. Cerullo
Dipartimento di Fisica, Politecnico di Milano, Piazza Leonardo da Vinci 32, I-20133 Milano, Italy
Istituto di Fotonica e Nanotecnologie, Consiglio Nazionale delle Ricerche, Piazza Leonardo da Vinci 32, I-20133 Milano, Italy

R.V. Pisarev
Ioffe Physical-Technical Institute, Ferroics Physics Laboratory, Russian Academy of Sciences, St Petersburg 194021, Russia

Yanfei Wu, Geoffrey A. Rojas and C. Daniel Frisbie
Department of Chemical Engineering and Materials Science, University of Minnesota, 421 Washington Ave SE, Minneapolis, 55455 Minnesota, USA

Annabel R. Chew and Alberto Salleo
Department of Materials Science and Engineering, Stanford University, 476 Lomita Mall, Stanford, 94305 California, USA

Gjergji Sini
Laboratoire de Physico-chimie des Polymères et des Interfaces, Université de Cergy-Pontoise, 5 Mail Gay Lussac, Neuville sur Oise, Cergy-Pontoise Cedex 95031, France

Physical Science and Engineering Division, Solar & Photovoltaics Engineering Research Center, King Abdullah University of Science and Technology (KAUST), Thuwal 23955-6900, Saudi Arabia

Greg Haugstad
Characterization Facility, University of Minnesota, 100 Union St SE, Minneapolis, 55455 Minnesota, USA

Alex Belianinov and Sergei V. Kalinin
Center for Nanophase Materials Sciences, Oak Ridge National Laboratory, 1 Bethel Valley Rd., Oak Ridge, 37831 Tennessee, USA

Hong Li
School of Chemistry and Biochemistry & Center for Organic Photonics and Electronics, Georgia Institute of Technology, Atlanta, 30332 Georgia, USA

Chad Risko
Department of Chemistry & Center for Applied Energy Research, University of Kentucky, Lexington, 40506 Kentucky, USA

Jean-Luc Brédas
Physical Science and Engineering Division, Solar & Photovoltaics Engineering Research Center, King Abdullah University of Science and Technology (KAUST), Thuwal 23955-6900, Saudi Arabia

Lin Zhao, Aiji Liang, Dongna Yuan, Yong Hu, Defa Liu, Jianwei Huang, Shaolong He, Bing Shen, Yu Xu, Xu Liu, Li Yu, Guodong Liu, Huaxue Zhou, Yulong Huang, Xiaoli Dong and Fang Zhou
National Lab for Superconductivity, Beijing National Laboratory for Condensed Matter Physics, Institute of Physics, Chinese Academy of Sciences, Beijing 100190, China

Kai Liu and Zhongyi Lu
Department of Physics and Beijing Key Laboratory of Opto-electronic Functional Materials and Micro-nano Devices, Renmin University of China, Beijing 100872, China

Zhongxian Zhao and X. J. Zhou
National Lab for Superconductivity, Beijing National Laboratory for Condensed Matter Physics, Institute of Physics, Chinese Academy of Sciences, Beijing 100190, China
Collaborative Innovation Center of Quantum Matter, Beijing 100871, China

Chuangtian Chen4 and Zuyan Xu
Technical Institute of Physics and Chemistry, Chinese Academy of Sciences, Beijing 100190, China

Ju-Hyung Kim
Surface and Interface Science Laboratory, RIKEN, 2-1 Hirosawa, Wako, Saitama 351-0198, Japan

Department of Chemical Engineering, Pukyong National University, 365 Sinseon-ro, Nam-gu, Busan 608-739, Republic of Korea
Center for Organic Photonics and Electronics Research (OPERA), Kyushu University, 744 Motooka, Nishi, Fukuoka 819-0395, Japan

Department of Advanced Materials Science, The University of Tokyo, 5-1-5 Kashiwanoha, Surface and Interface Science Laboratory, RIKEN, 2-1 Hirosawa, Wako, Saitama 351-0198, Japan Kashiwa, Chiba 277-8561, Japan

Jean-Charles Ribierre, Yu Seok Yang and Chihaya Adachi
Center for Organic Photonics and Electronics Research (OPERA), Kyushu University, 744 Motooka, Nishi, Fukuoka 819-0395, Japan

Maki Kawai
Department of Advanced Materials Science, The University of Tokyo, 5-1-5 Kashiwanoha, Surface and Interface Science Laboratory, RIKEN, 2-1 Hirosawa, Wako, Saitama 351-0198, Japan Kashiwa, Chiba 277-8561, Japan

Jaehoon Jung
Surface and Interface Science Laboratory, RIKEN, 2-1 Hirosawa, Wako, Saitama 351-0198, Japan

Department of Chemistry, University of Ulsan, 93 Daehak-ro, Nam-gu, Ulsan 44610, Republic of Korea

Takanori Fukushima
Chemical Resources Laboratory, Tokyo Institute of Technology, 4259 Nagatsuta, Midori-ku, Yokohama 226-8503, Japan

Yousoo Kim
Surface and Interface Science Laboratory, RIKEN, 2-1 Hirosawa, Wako, Saitama 351-0198, Japan

A. Vattré
CEA, DAM, DIF, F-91297 Arpajon, France.

T. Jourdan and M.-C. Marinica
CEA, DEN, Service de Recherches de Métallurgie Physique, Université Paris-Saclay, F-91191 Gif-sur-Yvette, France.

H. Ding
Department of Materials Science and Engineering, Massachusetts Institute of Technology, Cambridge, Massachusetts 02139, USA

M.J. Demkowicz
Department of Materials Science and Engineering, Massachusetts Institute of Technology, Cambridge, Massachusetts 02139, USA
Materials Science and Engineering, Texas A&M University, College Station, Texas 77843, USA

Pranab Kumar Das
Istituto Officina dei Materiali (IOM)-CNR, Laboratorio TASC, in Area Science Park, S.S.14, Km 163.5, I-34149 Trieste, Italy
International Centre for Theoretical Physics (ICTP), Strada Costiera 11, I-34100 Trieste, Italy

D. Di Sante
Consiglio Nazionale delle Ricerche—CNR-SPIN, I-67100 L'Aquila, Italy
Department of Physical and Chemical Sciences, University of L'Aquila, Via Vetoio, I-67100 L'Aquila, Italy

I. Vobornik, J. Fujii, G. Panaccione and R. Ciancio
Istituto Officina dei Materiali (IOM)-CNR, Laboratorio TASC, in Area Science Park, S.S.14, Km 163.5, I-34149 Trieste, Italy

T. Okuda
Hiroshima Synchrotron Radiation Center (HSRC), Hiroshima University, 2-313 Kagamiyama, Higashi-Hiroshima 739-0046, Japan

E. Bruyer and S. Picozzi
Consiglio Nazionale delle Ricerche—CNR-SPIN, I-67100 L'Aquila, Italy

A. Gyenis, B.E. Feldman and A. Yadzani
Joseph Henry Laboratories and Department of Physics, Princeton University, Princeton, New Jersey 08544, USA

J. Tao
Department of Condensed Matter Physics and Materials Science, Brookhaven National Laboratory, Upton, New York 11973, USA.

G. Rossi
Istituto Officina dei Materiali (IOM)-CNR, Laboratorio TASC, in Area Science Park, S.S.14, Km 163.5, I-34149 Trieste, Italy

Dipartimento di Fisica, Università di Milano, Via Celoria 16, I-20133 Milano, Italy

M.N. Ali and R.J. Cava
Department of Chemistry, Princeton University, Princeton, New Jersey 08544, USA

Andreas V. Stier and Scott A. Crooker
National High Magnetic Field Laboratory, Los Alamos National Laboratory, Los Alamos, New Mexico 87545, USA

Kathleen M. McCreary and Berend T. Jonker
Materials Science and Technology Division, Naval Research Laboratory, Washington, Washington DC 20375, USA

Junichiro Kono
Department of Electrical and Computer Engineering, Rice University, Houston, Texas 77005, USA
Department of Physics and Astronomy, Rice University, Houston, Texas 77005, USA Department of Materials Science and NanoEngineering, Rice University, Houston, Texas 77005, USA

Bohdan Senyuk and Owen Puls
Department of Physics and Soft Materials Research Center, University of Colorado, Boulder, Colorado 80309, USA

Oleh M. Tovkach
Bogolyubov Institute for Theoretical Physics, NAS of Ukraine, Kyiv 03680, Ukraine Department of Mathematics, University of Akron, Akron, Ohio 44325, USA

Stanislav B. Chernyshuk
Institute of Physics, NAS of Ukraine, Kyiv 03650, Ukraine

Ivan I. Smalyukh
Department of Physics and Soft Materials Research Center, University of Colorado, Boulder, Colorado 80309, USA
Department of Electrical, Computer, and Energy Engineering, Materials Science and Engineering Program, University of Colorado, Boulder, Colorado 80309, USA
Renewable and Sustainable Energy Institute, National Renewable Energy Laboratory and University of Colorado, Boulder, Colorado 80309, USA

Takuya Higuchi
Department of Physics, The University of Tokyo 7-3-1 Hongo, Bunkyo-ku, Tokyo 113-0033, Japan

Makoto Kuwata-Gonokami
Department of Physics, The University of Tokyo 7-3-1 Hongo, Bunkyo-ku, Tokyo 113-0033, Japan
Photon Science Center, The University of Tokyo 7-3-1 Hongo, Bunkyo-ku, Tokyo 113-0086, Japan

Colin Ophus, Jim Ciston and Peter Ercius
National Center for Electron Microscopy, Molecular Foundry, Lawrence Berkeley National Laboratory, 1 Cyclotron Road, Berkeley, California 94720, USA

Jordan Pierce, Tyler R. Harvey, Jordan Chess and Benjamin J. McMorran
Department of Physics, University of Oregon, 1585 E 13th Avenue, Eugene, Oregon 97403, USA

Cory Czarnik
Gatan Inc., 5794 W Las Positas Boulevard, Pleasanton,California 94588, USA

Harald H. Rose
Department of Physics, Center for Electron Microscopy, Ulm University, Albert-Einstein-Allee 11, 89069 Ulm, Germany

Colin Ophus, Jim Ciston and Peter Ercius
National Center for Electron Microscopy, Molecular Foundry, Lawrence Berkeley National Laboratory, 1 Cyclotron Road, Berkeley, California 94720, USA

Jordan Pierce, Tyler R. Harvey, Jordan Chess and Benjamin J. McMorran
Department of Physics, University of Oregon, 1585 E 13th Avenue, Eugene, Oregon 97403, USA

Cory Czarnik
Gatan Inc., 5794 W Las Positas Boulevard, Pleasanton, California 94588, USA

Harald H. Rose
Department of Physics, Center for Electron Microscopy, Ulm University, Albert-Einstein-Allee 11, 89069 Ulm, Germany

K.P. Krishnaraj and Prabhu R. Nott
Department of Chemical Engineering, Indian Institute of Science, Bangalore 560012, India

D. Sando
Unité Mixte de Physique, CNRS, Thales, Univ. Paris-Sud, Université Paris-Saclay, 91767 Palaiseau, France
School of Materials Science and Engineering, University of New South Wales, Sydney 2052, Australia

Yurong Yang and L. Bellaiche
Department of Physics and Institute for Nanoscience and Engineering, University of Arkansas, Fayetteville, Arkansas 72701, USA

E. Bousquet and Ph. Ghosez
Theoretical Materials Physics, Université de Liège, B-5, B-4000 Sart-Tilman, Belgium

C. Carrétéro, V. Garcia, S. Fusil , A. Barthélémy and M. Bibes
Unité Mixte de Physique, CNRS, Thales, Univ. Paris-Sud, Université Paris-Saclay, 91767 Palaiseau, France

D. Dolfi
Thales Research and Technology France, 1 Avenue Augustin Fresnel, 91767 Palaiseau, France

P.Y. Portnichenko, Y.A. Onykiienko, A.S. Cameron, J.A. Lim, D.S. Inosov and M.A. Surmach
Institut für Festkörperphysik, TU Dresden, Helmholtzstraße 10, D-01069 Dresden, Germany

J. Romhányi
Max Planck Institute for Solid State Research, Heisenbergstraße 1, D-70569 Stuttgart, Germany

A. Henschel, M. Schmidt and H. Rosner
Max Planck Institute for Chemical Physics of Solids, Nöthnitzer Straße 40, D-01187 Dresden, Germany

J.T. Park
Heinz Maier-Leibnitz Zentrum (MLZ), TU München, Lichtenbergstraße 1, D-85747 Garching, Germany

A. Schneidewind
Jülich Centre for Neutron Science (JCNS), Forschungszentrum Jülich GmbH, Outstation at Heinz Maier-Leibnitz Zentrum (MLZ), Lichtenbergstraße 1, D-85747 Garching, Germany

D.L. Abernathy
Quantum Condensed Matter Division, Oak Ridge National Laboratory (ORNL), Oak Ridge, Tennessee 37831, USA

Jeroen van den Brink
Leibniz Institute for Solid State and Materials Research, IFW Dresden, Helmholtzstraße 20, D-01069 Dresden, Germany

J.A. Labastide
Department of Chemistry, University of Massachusetts, Amherst Massachusetts 01003, USA

Department of Physics, University of Massachusetts, Amherst Massachusetts 01003, USA (J.A.L.); Department of Chemistry and Chemical Biology, Harvard University 12 Oxford Street, Cambridge, Massachusetts 02138, USA (N.S.C.)

H.B. Thompson and S.R. Marques
Department of Chemistry, University of Massachusetts, Amherst Massachusetts 01003, USA

N.S. Colella
Department of Polymer Science and Engineering, University of Massachusetts, Amherst Massachusetts 01003, USA

A.L. Briseno
Department of Polymer Science and Engineering, University of Massachusetts, Amherst Massachusetts 01003, USA

M.D. Barnes
Department of Chemistry, University of Massachusetts, Amherst Massachusetts 01003, USA
Department of Physics, University of Massachusetts, Amherst Massachusetts 01003, USA

Li-Xian Wang and Cai-Zhen Li
State Key Laboratory for Mesoscopic Physics, Department of Physics, Peking University, Beijing 100871, China

Da-Peng Yu
State Key Laboratory for Mesoscopic Physics, Department of Physics, Peking University, Beijing 100871, China
Collaborative Innovation Center of Quantum Matter, Beijing, China
Institute of Physics and Electronic Information, Yunnan Normal University, Kunming 650500, China

Zhi-Min Liao
State Key Laboratory for Mesoscopic Physics, Department of Physics, Peking University, Beijing 100871, China
Collaborative Innovation Center of Quantum Matter, Beijing, China

M.A. Khamehchi, M.E. Mossman and P. Engels
Department of Physics and Astronomy,Washington State University, Pullman,Washington 99164, USA

Chunlei Qu and Chuanwei Zhang
Department of Physics, The University of Texas at Dallas, Richardson, Texas 75080, USA

Aline Fluri, Daniele Pergolesi and Alexander Wokaun
Department for Energy and Environment, Paul Scherrer Institut, 5232 Villigen-PSI, Switzerland

Vladimir Roddatis
Institut für Materialphysik, Universität Göttingen, Friedrich-Hund-Platz 1, Göttingen 37077, Germany

Thomas Lippert
Department for Energy and Environment, Paul Scherrer Institut, 5232 Villigen-PSI, Switzerland

Department of Chemistry and Applied Biosciences, Laboratory of Inorganic Chemistry, Vladimir-Prelog-Weg 1-5/10, ETH Zürich, Zürich CH-8093, Switzerland

Achim Woessner, Mark B. Lundeberg, Gabriele Navickaite and Davide Janner
ICFO—Institut de Ciencies Fotoniques, The Barcelona Institute of Science and Technology, 08860 Barcelona, Spain

Pablo Alonso-González
CIC nanoGUNE, 20018 Donostia-San Sebastian, Spain
Institute of Physics, Chinese Academy of Science, Beijing 100190, China

Yuanda Gao and James Hone
Department of Mechanical Engineering, Columbia University, New York, New York 10027, USA

Jose E. Barrios-Vargas and Aron W. Cummings
Catalan Institute of Nanoscience and Nanotechnology (ICN2), CSIC and The Barcelona Institute of Science and Technology, Campus UAB, 08193 Barcelona, Spain

Pablo Jarillo-Herrero and Qiong Ma
Department of Physics, Massachusetts Institute of Technology, Cambridge, Massachusetts 02139, USA

Kenji Watanabe and Takashi Taniguchi
National Institute for Materials Science, 1-1 Namiki, Tsukuba 305-0044, Japan

Valerio Pruneri and Frank H.L. Koppens
ICFO—Institut de Ciencies Fotoniques, The Barcelona Institute of Science and Technology, 08860 Barcelona, Spain
ICREA-Instituci6 Catalana de Recerca i Estudis Avanc¸ats, 08010 Barcelona, Spain

Stephan Roche
Catalan Institute of Nanoscience and Nanotechnology (ICN2), CSIC and The Barcelona Institute of Science and Technology, Campus UAB, 08193 Barcelona, Spain
ICREA-Instituci6 Catalana de Recerca i Estudis Avanc¸ats, 08010 Barcelona, Spain

Rainer Hillenbrand
CIC nanoGUNE and UPV/EHU, 20018 Donostia-San Sebastian, Spain
IKERBASQUE, Basque Foundation for Science, 48011 Bilbao, Spain

A.A. Khajetoorians
Department of Physics, Hamburg University, 20355 Hamburg, Germany
Institute for Molecules and Materials, Radboud University, 6525AJ Nijmegen, The Netherlands

M. Steinbrecher, J. Wiebe and R. Wiesendanger
Department of Physics, Hamburg University, 20355 Hamburg, Germany

M. Ternes
Max Planck Institute for Solid State Research, 70569 Stuttgart, Germany.

M. Bouhassoune, M. dos Santos Dias and S. Lounis
Peter Grünberg Institut and Institute for Advanced Simulation, Forschungszentrum Jülich & JARA, 52425 Jülich, Germany

Index